Photonic Crystals

Photonic Crystals
Theory, Applications, and Fabrication

DENNIS W. PRATHER
SHOUYUAN SHI
AHMED SHARKAWY
JANUSZ MURAKOWSKI
GARRETT J. SCHNEIDER

WILEY
A JOHN WILEY & SONS, INC., PUBLICATION

Published by John Wiley & Sons, Inc., Hoboken, New Jersey

Published simultaneously in Canada

For general information on our other products and services or for technical support, please contact our Customer Care Department within the United States at (800) 762-2974, outside the United States at (317) 572-3993 or fax (317) 572-4002.

Wiley also publishes its books in variety of electronic formats. Some content that appears in print may not be available in electronic format. For more information about Wiley products, visit our web site at www.wiley.com.

ISBN: 978-0-470-27803-1

Library of Congress Cataloging-in-Publication Data is available.

Printed in the United States of America

10 9 8 7 6 5 4 3 2 1

Contents

Preface

Our goal in writing this book was to create a comprehensive guide to the field of photonic crystals from an engineering perspective, which enables its various aspects, e.g., design/simulation and fabrication, to be reduced to practice. In order to achieve this goal, we aimed at providing material essential to understand and appreciate the exciting and rapidly expanding body of research that has been produced since the popular emergence of the field nearly two decades ago. To this end, we hope that we have succeeded in producing a book that is simultaneously the most comprehensive catalog available of relevant research, as well as a complete and broadly accessible presentation of the underlying physics and mathematics required to understand and implement it. In short, we want this book to make photonic crystals accessible to interested students and researchers with minimal prerequisite instruction beyond the undergraduate-level physics, multivariable vector calculus, and complex analysis.

The process of writing this book has taken place over a period of several years, and during that interval, the pace at which new results were being published was often too rapid to keep up with it. At this point, the foundations of the field are mature and well understood; for this reason, we have restricted the scope of the book to a finite range of topics and are confident that the book represents a comprehensive treatment of the more broad field of photonic crystals. In so doing, we have attempted to include what we consider to be the more important aspect of the recent developments of the field, while generally conceding that we had to draw the line somewhere. Therefore, in deciding what to/not include, we have regrettably, but necessarily, left out various aspects of very good work; however, we hope that our extensive lists of references will provide the interested reader with a jump start toward finding just about any aspect of the field in the present-day literature.

The contents of the book are structured as follows: Chapter 1 is a historical overview of the field, briefly charting the course of its evolution and introducing the most important parts of the conceptual framework that has supported its development, particularly the notion of photonic crystals as "semiconductors for light." The relevant analogies between electronics in crystalline semiconductors and photons in periodically structured dielectric materials are discussed, and a roughly chronological survey of some of the more salient publications is presented, which aims to track the progress of the field according to milestone papers in all areas of photonic crystals research: theory, device design, simulation, and fabrication techniques. Chapter 2 contains a thorough tutorial on the foundational concepts that are required to engage the literature in the field. This chapter was written because the field of photonic crystals represents something of a hybrid subject, which has incorporated useful conceptual elements from many areas, including semiconductor and solid-state physics, classical electromagnetic theory, and classical optics. We are not aware of any single volume that presents it in an all-inclusive and consistent framework. The material on Maxwell's equations, waveguides, and optical dispersion is typically found in optics and electromagnetism texts, whereas the contents of the sections dealing with crystal structures, lattices, reciprocal space, and band theory are typically found in solid-state physics and electrical-engineering texts, which invariably present these subjects in the context of electron waves in atomic lattices. Hence, Chapter 2 was included to provide the reader with a consistent and thorough introduction to all the necessary concepts, presented in a consistent framework that is anchored to the goal of understanding the optical properties of photonic crystals. Chapter 3 provides a detailed description of the numerical methods that are most useful, indeed essential, to analyzing photonic crystals as well as to understanding and engineering their properties. After an overview of numerical/computational techniques for electromagnetic modal analysis and device simulation, the chapter focuses on the two predominant techniques, providing a thorough and rigorous exposition of their principles of operation and implementation. The adaptation of these two methods to the gamut of photonic crystal structures, covering one-, two-, and three-dimensional structures, plus quasi-two-dimensional slab photonic crystals and defects for cavities and waveguides, is covered in detail. An extensive bibliography of papers that can be referred to for even more detail is provided, with citations throughout the chapter. Chapters 4 and 5 describe the unique, and uniquely useful, phenomena that photonic crystals produce by virtue of the nature of their interactions with light. Chapter 4 presents fundamental devices and applications based on photonic band gaps, whereas Chapter 5 is focused on the properties of propagating modes within photonic crystals, i.e., the photonic dispersion relation, and the applications that are achievable by engineering photonic dispersion. Chapter 6 is a comprehensive review of the wide array of highly advanced and imaginative methods that researchers have developed toward the goal of fabricating photonic crystals, which has proved particularly challenging in the optical and telecommunications regime.

As is typically the case for sustained research endeavors, many students, colleagues, and sponsors provided key aspects and resources that enabled the pursuit and compilation of knowledge in this book. To provide a full and complete list would be prohibitive, but to name a few, we would like to acknowledge the entire nano-photonics research group (past and present) at the University of Delaware; the software and design teams at EM Photonics, Inc.; Drs. Keith Krapels and Steve Russell at the Office of Naval Research; the Microsystems Technology Office at DARPA; and last, but certainly not least, Dr. Gernot Pomrenke at the Air Force Office of Scientific Research.

Dennis W. Prather
Shouyuan Shi
Ahmed Sharkawy
Janusz Murakowski
Garrett J. Schneider

Newark, Delaware

1

Introduction

Throughout the last several decades, significant activity has occurred in the development of photonic devices that can confine, control, and route light on a scale commensurate with modern electronic devices, namely the nanometer scale. A key motivation for this activity is the promise of photonic devices, and the corresponding circuits comprised thereof, having a density approaching that of modern electronic circuits. However, for this promise to be realized, such devices must possess the ability to confine light on a subwavelength scale, and they must be fabricated in a material system that is compatible with the microelectronics manufacturing infrastructure. Although the latter requirement is readily satisfied through a proper choice of materials, i.e., silicon, the former was historically more elusive. The reason for this originates from the nature of confinement in conventional materials: Reflective, i.e., conducting, materials are lossy at optical wavelengths, whereas refractive, i.e., dielectric, materials do not confine optical modes on a sufficiently small scale. In the search for a solution, researchers have turned to the field of photonic-crystal (PhC) devices and their associated photonic-bandgap devices, which offer both low loss and high confinement, and they can be readily fabricated in silicon.

To a large extent, the field of PhCs can be thought of as having it origins in the conventional one-dimensional (1D) thin-film stack, wherein layers of two or more different optical materials are arranged in a periodic fashion. In such devices, it is well known that certain frequencies are transmitted while others are not. As the index contrast between the alternating layers is increased, the selectivity of the transmitted and reflected wavelengths is also increased. When generalized to two and three dimensions, this selectivity is what gives rise to high mode confinement.

Although 1D thin-film stacks have been known for over a century, their generalization to higher dimensions was not proposed until the 1970s by Bykov [1–4] as a possible means of inhibiting spontaneous emission. In essence, Bykov proposed the use of a periodic structure that served to inhibit certain electromagnetic frequencies, thereby disallowing spontaneous emission. This type of structure has recently come to be called a "photonic crystal," and it can inhibit the propagation of light over a certain band of wavelengths, while allowing other bands to propagate. Such behavior may give rise to a "photonic bandgap" (PBG), which is analogous to the electronic bandgap

Photonic Crystals: Theory, Applications, and Fabrication. By D. W. Prather, S. Shi, A. Sharkawy, J. Murakowski, and G. J. Schneider

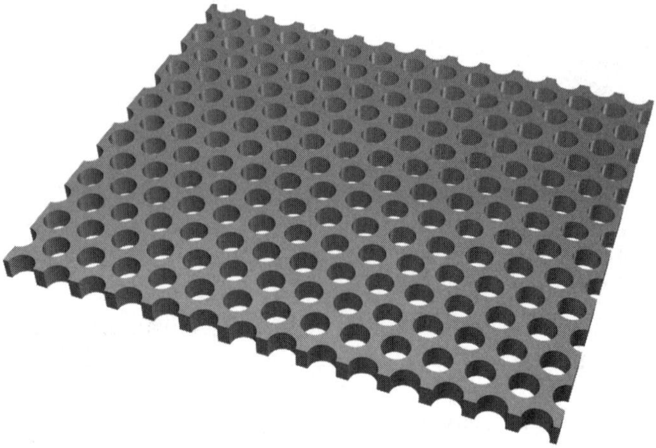

Figure 1.1 Periodic structure of air holes in a high-index finite-height substrate. Such a structure can prohibit propagation for a certain band of frequencies called a photonic bandgap.

in semiconductor materials (see Section 1.2). An example of a simple, easily fabricated periodic structure that may exhibit the property of a bandgap is shown in Fig. 1.1.

These devices ultimately became known as PBG structures, as they suppress band(s) of frequencies from existing within them, the so-called photonic bandgap(s). In a similar vein, Yablonovitch [5] proposed a structure in which an electronic and photonic gap are overlapped in energy; such a structure would be incapable of light emission by recombination, which thereby makes it possible to enhance the performance of lasers, heterojunctions, bipolar transistors, and solar cells. Subsequent to this work, John [6] proposed using such structures for the localization of light in strongly scattering dielectric structures. In each of these cases, the underlying idea was to tailor the properties of photons (i.e., electromagnetic modes) in a material in a manner analogous to that of electrons in atomic crystals. That is to say, in the electronic case, the wave functions of electrons interact with the periodic potential of the atomic lattice, and for a certain range of energies (analogous to frequencies for photons), electronic states do not exist, thus giving rise to an electronic bandgap. For PhCs, the analog of the electronic potential in an atomic crystal is the periodic variation of the dielectric constants of the various constituent materials of the PhC structure. Likewise, because of the periodic interaction, photonic bandgaps appear, in which certain modes, or frequencies, are forbidden. Futhermore, in such a structure, one can introduce one or more line or point defects, which locally disrupt the periodic lattice, wherein a localized mode may exist with a frequency inside the bandgap; such modes are confined to the region of the defect by virtue of being suppressed within the surrounding lattice. For this reason, these devices offer extreme mode confinement as well as the ability to control and route light compactly and efficiently.

With these advances, the field of PhCs, and in particular their realization in silicon-based materials, has been an active field of research over the last several decades. During this time, great success has been achieved in identifying suitable periodic structures, dielectric materials, and both theoretical as well as experimental demonstration and characterization of one-, two-, and three-dimensional (3D) PhC and bandgap structures. Therefore, it is the goal of this book to provide an introduction to these advances and to lay the underlying foundation for the design tools and fabrication processes necessary to promote their continued development.

1.1 HISTORICAL OVERVIEW

To a large extent, the more recent interest in PhCs can be attributed to the early papers of Yablonovitch [5] and John [6]. Both of these papers were, more or less, theoretical assertions; following their publication, the quest was on to demonstrate an actual bandgap device experimentally. To this end, Yablonovitch and Gmitter [7] fabricated a face-centered-cubic (FCC) lattice on a large scale (microwave frequencies) and measured the transmission spectra at various angles, in the hope of observing a complete photonic bandgap. Unfortunately, they did not find one. For this reason, theorists were motivated to investigate the problem systematically [8–12]. Accordingly, they first attempted to solve Maxwell's equations in 3D using the scalar-wave approximation [13], which predicted gaps for FCC the structure, contrary to experimental results. Subsequent solutions incorporating the vector nature of the electromagnetic field were developed, based on the plane-wave expansion method [9, 10, 14, 15]. Application of these tools showed that the FCC structure did not have a complete bandgap [9, 10]. Subsequently, Soukoulis proposed the use of a diamond lattice to reduce the degeneracy of the FCC lattice, and a complete bandgap was found for spheres with a dielectric constant of 3.6. Yablonovitch et al. then proceeded to fabricate the diamond structure by drilling cylindrical holes into the dielectric material [16]. Transmission measurements for this new structure showed a complete bandgap, in agreement with theory.

After this initial success, the field of photonic bandgap devices grew at an exponential rate. Additional structures were proposed to exhibit bandgaps, such as the so-called "woodpile" structure shown in Fig. 1.2, which also has diamond symmetry. The appealing aspect of this structure was that it could be made in a layer-by-layer manner by stacking alumina rods in a criss-cross fashion. To validate earlier theoretical calculations performed on the structure, different transmission measurements were collected and were found to agree with theoretical results. Furthermore, this device was scaled down by stacking perforated silicon wafers in a square pattern, which extended the frequency of the gap into the terahertz regime [17–20]. To obtain a bandgap in the infrared regime [21], the woodpile structure was fabricated using silicon processing techniques. When further scaling was attempted, down to micrometer and submicrometer regime, alignment between subsequent layers became increasingly more challenging and resulted in a marked reduction in sample yield.

For micron-scale photonic-bandgap structures, some researchers turned to the colloidal self-organization of monodisperese spheres [22], knowing that natural opals exhibit the interesting property wherein color of reflected light changed when viewed at different angles under white-light illumination. Along these lines, optical

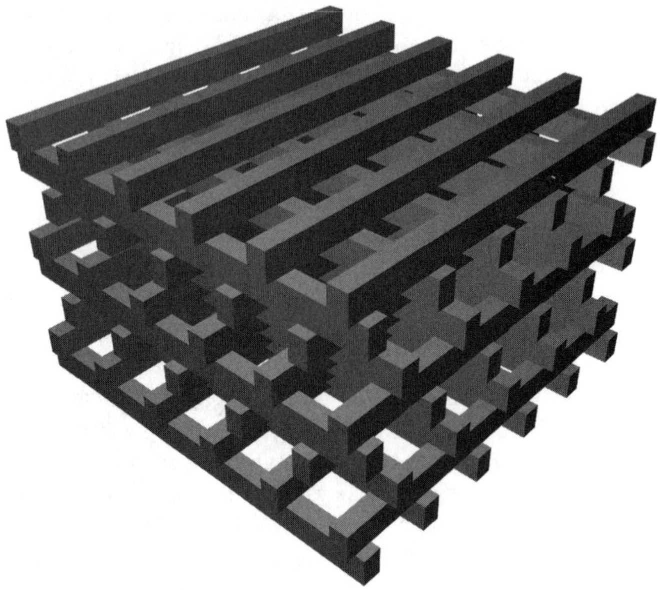

Figure 1.2 A three-dimensional photonic-crystal structure (woodpile).

properties such as transmission, reflection, diffraction, and dispersion were measured in artificial opal structures made from monodisperse polystyrene colloids [23–32]. Transmission data showed dips that corresponded to directional stop bands on a wavelength scale comparable with the spacing of the spheres.

However, a crucial drawback of full three-dimensional and self-organizing photonic-bandgap structures is that, once fabricated, the bandgap behavior cannot be easily altered or engineered for a particular function or application. In addition, the controlled introduction of defects (doping) is hard to achieve, which in the end has significantly limited their use.

To overcome this limitation, Joannopoulos et al. [33] introduced the planar PhC waveguide to guide optical light in air along narrow channels and around very tight bends with minimal loss [34–39]. This structure uses two-dimensional periodicity that consists of infinitely long silicon pillars in an air background [40]. By introducing a point defect, Villeneuve et al. [41] were able to design a high-Q nanocavity in the same structure, which supported a single localized mode. The cavity was made by introducing a point defect locally into a lattice cell surrounded by a sufficient number of layers to achieve high Q value. Villeneuve et al. also fabricated a one-dimensional optical filter from an air-bridge nanocavity that consisted of a channel waveguide and a one-dimensional PhC suspended in air [42, 43]. Channel-drop filters were later studied by Fan and colleagues [44, 45], who used two waveguides and an optical resonator system of two single-mode cavities, and another structure with a single cavity that supports doubly degenerate hexapole states, to achieve wavelength filtering [46, 47]. Fan et al. [48] later analyzed the effect of waveguide branching in PhCs using two resonant cavities with equal decay rates in the branched waveguides. Povinelli et al. [49] presented a study in which they constructed quasi-two-dimensional (2D) defect modes in a three-dimensional (3D) PhC structure [49]. Such modes had a striking similarity to those found in 2D PhCs in terms of polarization, field profile, and projected band structures.

Following this work, Johnson studied the effect of finite height on two-dimensional photonic-bandgap structures in terms of bandgap size and confinement in photonic-crystal slabs [50]. He found that, when going from infinite- to finite-height photonic-bandgap structures, the size of the bandgap would shrink about 40% because of radiation losses imposed by the finite height of the structure. Johnson et al. [51] also studied the effect of cross talk in waveguide intersections. He used resonant tunneling to achieve nearly 100% throughput and 0% cross talk between crossing perpendicular PhC waveguides.

It is also important to point out key contributions to materials systems, such as III–V materials. In this area, Scherer and colleagues [52–55] demonstrated photonic-crystal structures in perforated slabs of GaAsP. They used electron-beam lithography followed by a sequence of dry etching steps to define PhC structures with 350-nm spacings in GaAsP and GaAs epilayers. Similar structures were later fabricated also in InGaAsP, which led to the fabrication of a two-dimensional photonic-bandgap defect-mode laser designed for a peak emission wavelength of 1.55 micrometers at room temperature [56–61]. Lee et al. [62, 63] measured the spontaneous emission power spectrum of the fabricated structure. Theoretical studies of the fabricated structures were presented by Vučković and colleagues [64, 65].

Following this work, Yoshie et al. [66] later characterized different modes within two-dimensional PhC nanocavities with self-organized InAs quantum dots as an active material. They observed highly localized donor-mode resonances with 3–5-nm line widths. Additionally, Lončar et al. [67, 68] introduced line defects to PhC slab structures to resemble an optical wire or waveguide. Meanwhile, Adibi et al. [69, 70] explored different techniques to control the positions of guided modes as well as the number of guided modes within a waveguide, eventually leading to the design of single-mode PhC waveguides in a perforated slab of air holes. Photonic-crystal waveguides were also presented using coupled cavities by Bayindir et al. [71–74] and Özbay et al. [75]. They demonstrated theoretically and experimentally that photons can propagate through strongly localized defect cavities because of coupling between adjacent cavity modes based on the tight-binding formalism also used in solid-state physics [76]. They observed a high transmission of electromagnetic waves through the cavities even when they were placed along an arbitrarily shaped path.

Strong directionally dispersive optical properties in PhCs were studied and explored for so-called "superprism" applications by Kosaka et al. [77–79], who demonstrated extraordinary angle sensitivity of light propagation. In this work, a transmitted beam was swung from -70° to $+70^\circ$ with a slight change in the incident angle within $\pm 7^\circ$. This phenomenon implies the presence of a negative effective refractive [80] index, and it was used for applications such as beam steering, spot size conversion, and self-collimation [81].

Two-dimensional photonic crystals can be realized by using either a periodic array of dielectric rods of any shape and/or geometry or a perforated dielectric slab of air holes. Such structures can be further optimized to achieve either a wider or a narrower bandgap based on the desired application. Two-dimensional photonic crystals impose

Figure 1.3 Scanning electron microscopy (SEM) picture of a periodic array of air holes in a suspended Si membrane (slab photonic crystal).

periodicity in two dimensions, whereas the third dimension either is practically infinitely long (photonic-crystal fibers) or has a finite height (photonic-crystal slabs). Three-dimensional photonic crystals impose periodicity in all three dimensions. An example of a three-dimensional photonic crystal is the woodpile structure in Fig. 1.2.

Because they are easier to analyze and fabricate, two-dimensional photonic crystals have attracted the attention of many researchers and engineers. An example of a fabricated two-dimensional photonic crystal is shown in Fig. 1.3. Planar photonic-crystal devices, such as splitters [48, 71, 82–84], high-Q microcavities [35, 41, 56, 58, 85–87], and channel drop/add filters [44, 45, 88, 89], have been investigated both theoretically [50, 69, 90] and experimentally [52, 55, 59, 91–94].

1.2 ANALOGY BETWEEN PHOTONIC AND SEMICONDUCTOR CRYSTALS

Understanding the relationship between electronic band structure and photonic band structure requires an examination of the basic equations that govern the motion of electrons and photons, which are called the Schrödinger equation and Maxwell's equations, respectively. When applied to the problems of electrons in crystalline solids and photons in periodic dielectric media, respectively, analogies between the two governing relations become apparent. To make the most of these analogies, it is important to examine both the similarities and the differences.

In a semiconductor crystal, electron behavior is governed by quantum mechanics [95]. The electron is described using the Schrödinger equation

$$\underbrace{\left[\frac{-h^2}{8\pi^2 m^*}\nabla^2 + V(\mathbf{r})\right]}_{\text{Hamiltonian operator}} \varphi(\mathbf{r}) = E\varphi(\mathbf{r}), \tag{1.1}$$

where h is Planck's constant, m^* is the effective mass [76] of the electron, $V(\mathbf{r})$ is the potential function, $\varphi(\mathbf{r})$ is the wave function of the electron, and E is the total energy. The wave function $\varphi(\mathbf{r})$ in quantum physics can be interpreted as a *probability amplitude* function, i.e., the probability of a measurement finding the electron within a volume element dV, located at a position denoted by $\mathbf{r}$, is given by $|\varphi(\mathbf{r})|^2 dV$. For a crystal lattice, in the absence of defects, the potential is periodic; i.e., $V(\mathbf{r}) = V(\mathbf{r} + \mathbf{R})$, where $\mathbf{R}$ is a vector that represents the displacement between any two lattice sites. In this case, solutions $\varphi(\mathbf{r})$ take the form of propagating waves called *Bloch waves*: plane waves modulated by an amplitude function with the same periodicity as the crystal lattice (see Chapter 2).

The electromagnetic fields are governed by classic electromagnetic theory [96]. The fields are described using Maxwell's equations: four coupled equations that relate the electric and magnetic fields to one another and to any sources of charge or current present. In the absence of magnetic media, two of these four equations can be combined to yield

$$\underbrace{\left[\nabla \times \frac{1}{\varepsilon(\mathbf{r})} \nabla \times\right]}_{\text{Maxwell operator}} \mathbf{H}(\mathbf{r}) = (\omega/c)^2 \mathbf{H}(\mathbf{r}), \tag{1.2}$$

where $\mathbf{H}(\mathbf{r})$ is the magnetic field, ω is the angular frequency, c is the speed of light in vacuum, and $\varepsilon(\mathbf{r})$ is the dielectric permittivity. In a photonic crystal, the optical properties of the medium are periodic, i.e., $\varepsilon(\mathbf{r}) = \varepsilon(\mathbf{r} + \mathbf{R})$, and solutions are also Bloch waves.

Equations (1.1) and (1.2) are eigenvalue equations with similar forms. Inspection of the two equations reveals certain analogous features. Each equation describes a wavelike function in space: $\varphi(\mathbf{r})$ and $\mathbf{H}(\mathbf{r})$, respectively. Each wave exists in a medium characterized by a periodic material function: the periodic potential V and the periodic permittivity ε, respectively. These periodic functions each appear within second-order space-differential operators, the Hamiltonian and Maxwell operators, respectively, where both of these operators are Hermitian. The corresponding real eigenvalues are the electron energy E and the squared frequency of the electromagnetic wave ω^2, respectively.

In both the electronic and photonic cases, because of the coherent superposition of the partial scattering of the wave at each lattice site, certain bands exist in the eigenvalue spectrum in which propagation is forbidden, which are called bandgaps. Likewise, in both cases, defects can occur in the periodic medium, which can support trapped or "localized" states with eigenvalues that would be forbidden in the defect-free lattice. As the defect density increases, the probability for the trapped wave to tunnel to a nearby defect site also increases; this permits the propagation of states with eigenvalues inside the bandgap of the bulk crystal. This effect, which is well known in semiconductor applications, is called doping, and it is vital to the usefulness of nearly all semiconductor devices like diodes, transistors, light-emitting diodes (LEDs), and so on. Not surprisingly, defects are also essential to the usefulness of many of the proposed applications of photonic crystals (see Chapter 4).

These analogies between the Schrödinger and Maxwell eigenvalue equations have allowed researchers to apply to PhCs many useful conceptual tools of solid state physics, such as Bloch waves, reciprocal space, Brillouin zones, band structure, conduction and valence bands, and dispersion diagrams. Some of these have already been mentioned briefly in this discussion; a more comprehensive discussion of all these tools can be found in Chapter 2.

Having presented the useful similarities between semiconductor and photonic crystals, completeness demands that we consider the important differences as well. First is the fundamentally different physical nature of the wave: The magnetic field is a vector quantity, whereas the electronic wave function can usually be treated as a scalar (i.e., spin can be ignored). Equation (1.2) contains three scalar equations, one for each component of the vector field. Scalar-wave approximations can yield valuable insights into photonic crystals, but for quantitative comparisons with experiments, a rigorous full vector-wave calculation is required.

Second, no characteristic length scale is inherent in Maxwell's equations. This contrasts with the Schrödinger equation, in which fundamental physical constants, such as Planck's constant and the electron mass, conspire to create fundamental scales of length like the Bohr radius. Scaling the dimensions of a particular electronic problem by a given factor can profoundly change the outcome. The same operation applied to a photonic problem will change only the frequencies of the solution by the scale factor; the solution is otherwise unchanged. This important distinction is of great importance because unlike the atoms that comprise crystalline solids such as semiconductors, the periodically arranged bits of material that constitute the "photonic atoms" of a PhC are generally not naturally occurring and must be designed and fabricated. This process has proved challenging at optical length scales (see Chapter 6). The scale invariance of the Maxwell eigenvalue equation allows proposed structures to be tested at more accessible microwave scales, with confidence that desired properties will be retained if an identical structure is fabricated on a smaller scale (provided that the optical material properties are retained as the frequencies are scaled).

Third, an important distinction exists between the two respective eigenvalues. Because the photonic eigenvalue is proportional to the squared frequency, and $E_{\text{photon}} = h\omega/2\pi$, there are no negative-energy photonic solutions; i.e., no truly bound states are allowed. Nevertheless, it is possible to realize microcavities in PhCs with very high Q values, with correspondingly long lifetimes of confined optical modes.

Finally, in the limit of linear response, the solutions of the photonic eigenvalue problem are uncoupled. In other words, electromagnetic waves obey the principle of superposition and do not interact. (Exceptions to this principle fall in the realm of nonlinear optics.) This behavior is in contrast to the often strong interactions between electrons. For this reason, it has been noted by Yablonovitch [97] that band theory, which assumes that the wave functions of multiple particles are uncorrelated, actually makes more sense for photons than for electrons.

1.3 ANALYZING PHOTONIC-BANDGAP STRUCTURES

Only a precious few geometries can be analyzed using exact analytical methods. For most photonic crystals with practical importance, numerical simulations are required to gain insight into the behavior of light that interacts with them. One method often employed in this case is the plane-wave expansion method (PWEM, described in detail in Chapter 3), which makes use of the fact that normal modes in periodic structures can be expressed as a superposition of a set of plane waves. This way, Eq. (1.2) is converted to a *matrix* eigenvalue problem, which is solved using standard numerical techniques to obtain the dispersion relations inside a photonic crystal [9, 10, 15, 98]. The solutions are eigenfrequencies, which are plotted as a function of wave vectors tracing the edges of an irreducible Brillouin zone, to form a dispersion diagram; examples are shown in Fig. 1.4. The dispersion diagram is a graphical representation of the frequencies that correspond to waves propagating within a photonic-crystal lattice with various wave vectors. Although a two-dimensional dispersion diagram (depicting only those wave vectors on the irreducible Brillouin-zone edge) is sufficient to determine whether a bandgap exists for a certain photonic crystal, it is not sufficient to indicate the presence of other interesting and useful dispersive behaviors, such as negative refractive-index phenomena or the superprism effect (see Chapter 5). In these cases, a three-dimensional plot of the dispersion surface, depicting eigenfrequencies as a function of all wave vectors within the full Brillouin zone, provides a more detailed view of a photonic crystal's supported modes at frequencies both inside and outside any bandgap. An example of such a diagram is shown in Fig. 1.5, where a dispersion surface that represents just one of the higher bands is displayed.

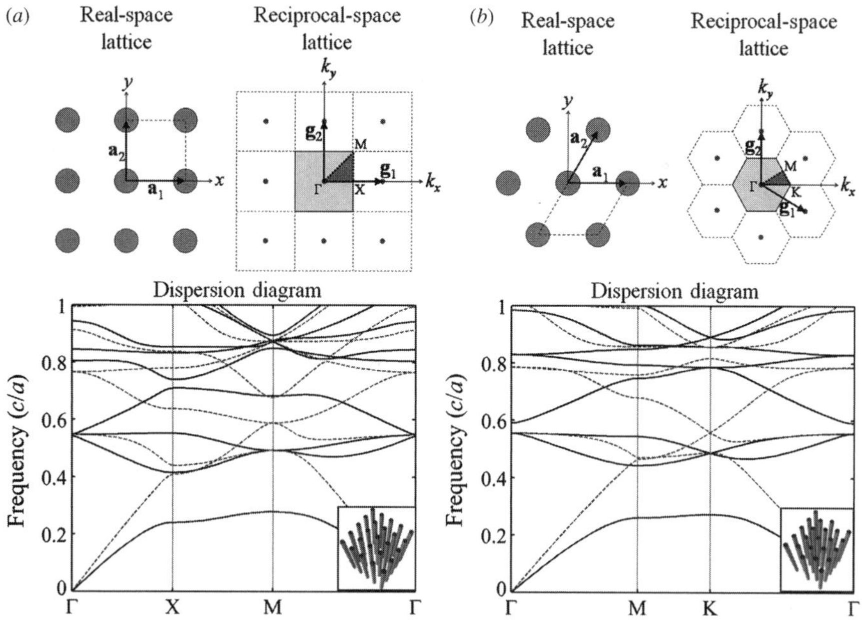

Figure 1.4 PWEM used to analyze different two-dimensional PhC structures with (a) square or (b) triangular lattices. Eigenfrequencies are determined for Bloch wave vectors around the edge of an irreducible Brillouin zone, which is indicated by a dark shaded triangle for each lattice. The result is a dispersion diagram showing the allowed frequencies for different wave vectors within the lattice. The two separable polarization modes, known as transverse electric (TE) and transverse magnetic (TM), are indicated by dashed and solid curves, respectively.

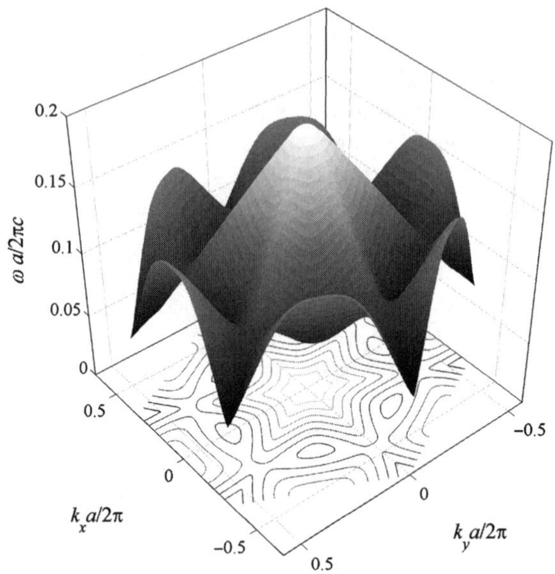

Figure 1.5 Dispersion surface. The horizontal plane spans both Bloch wave vector components k_x and k_y. The vertical axis gives the normalized frequency $\omega a/2\pi c = a/\lambda$.

Although the PWEM allows for analyzing the dispersion properties of a PhC structure, it is still limited because transmission spectra, field distributions, and backreflections from finite-size structures cannot be directly determined. Therefore, an alternative approach has been adopted, which is based on a numerical solution of Maxwell's equations in real space using the finite-difference time-domain (FDTD) method (also described in detail in Chapter 3). The FDTD method has been used to

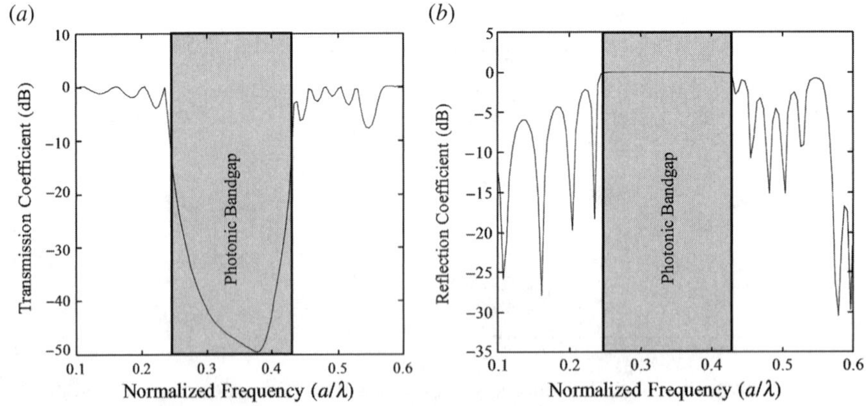

Figure 1.6 Transmission (a) and reflection (b) for six layers of a 2D photonic-crystal structure composed of a square lattice of silicon ($\varepsilon_r = 11.56$) rods in air ($\varepsilon_r = 1$), with $r/a = 0.2$, obtained using FDTD. Low transmission and near-perfect reflectivity is observed within the bandgap.

analyze multichannel drop/add filters [88] and calculate the transmission through sharp waveguide bends [36], as well as in the study of a waveguiding mechanism through localized coupled cavities in three-dimensional photonic crystals [72, 99].

To illustrate the use of FDTD on finite-size PhC structures, consider a 2D PhC structure consisting of a square lattice of silicon rods in air background. The transmission coefficient as a function of frequency for six layers of such a photonic crystal is shown in Fig. 1.6a, and the reflection coefficient is shown in Fig. 1.6b. The photonic bandgap can be observed in these figures as a frequency range of low transmission and high reflectivity, because the electromagnetic wave penetrates into the PhC only to a limited depth before being attenuated by the destructive interference of the forward-scattered waves that emerge from the periodic lattice of scattering objects. The bandgap has a center frequency at $a/\lambda_{\text{center}} = 0.339$ and a bandgap size $\Delta f/f_{\text{center}}$ equal to 55.4%. This structure can in principle operate for any range of frequencies, from microwave to optical; all results are expressed in units normalized to the operating wavelength. For example, consider the near-infrared frequency regime widely used in telecommunications. If we wish to operate at $\lambda_{\text{center}} = 1550$ nm, then the structure will have a lattice constant equal to $a = 525.5$ nm, and the silicon rods will have a radius of $r = 105$ nm.

In the remainder of this book, following a brief introduction to the interaction of electromagnetic field with periodic dielectric structures in Chapter 2, the numerical methods employed in analyzing them are presented in Chapter 3. Chapters 4 and 5 discuss the applications of photonic crystals taking advantage of the photonic bandgap and of the unique dispersive properties, respectively. A variety of methods developed for the fabrication of 2D and 3D photonic crystals are described in Chapter 6.

REFERENCES

1. V. Bykov, "Spontaneous Emission in a Periodic Structure," *Zhurnal Experimentalnoi i Teoreticheskoi Fiziki*, vol. 62, 1972.
2. V. Bykov, "Sponteneous Emission in a Periodic Structure," *Soviet Physics JETP*, vol. 35, pp. 269–273, 1972.
3. V. Bykov, "Spontaneous Emission from a Medium with a Band Spectrum," *Sov. J. Quant. Electron.*, vol. 4, pp. 861–871, 1975.
4. V. Bykov, "Free Oscillations of Elastic Dipole Located in Periodic Structure," *Zhurnal Tekhnicheskoi Fiziki*, vol. 48, 1978.
5. E. Yablonovitch, "Inhibited Spontaneous Emission in Solid-State Physics and Electronics," *Phys. Rev. Lett.*, vol. 58, pp. 2059–2062, 1987.
6. S. John, "Strong Localization of Photons in Certain Disordered Dielectric Superlattices," *Phys. Rev. Lett.*, vol. 58, pp. 2486–2489, 1987.
7. E. Yablonovitch and T. J. Gmitter, "Photonic Band Structure: The Face-Cenetered-Cubic Case," *Phys. Rev. Lett.*, vol. 63, pp. 1950–1953, 1989.
8. J. D. Joannopoulos, P. R. Villeneuve, and S. H. Fan, "Photonic Crystals," *Sol. Stat. Comm.*, vol. 102, pp. 165–173, 1997.

9. Z. Zhang and S. Satpathy, "Electromagnetic Wave Propagation in Periodic Structures: Bloch Wave Solution of Maxwell's Equations," *Phys. Rev. Lett.*, vol. 65, pp. 2650–2653, 1990.
10. K. M. Leung and Y. F. Liu, "Photon Band Structures: The Plane-Wave Method," *Phys. Rev. B*, vol. 41, pp. 10188–10190, 1990.
11. S. John and R. Rangarajan, "Optimal Structures for Classical Wave Localization—an Alternative to the Ioffe-Regel Criterion," *Phys. Rev. B*, vol. 38, pp. 10101–10104, 1988.
12. E. N. Economou and A. Zdetsis "Classical Wave Propagation in Periodic Structures," *Phys. Rev. B*, vol. 40, pp. 1334–1337, 1989.
13. S. Datta, C. Chan, K. Ho, and C. M. Soukoulis, "Photonic Bandgaps in Periodic Dielectric Structures—The Scalar Wave Approximation," *Phys. Rev. B*, vol. 46, pp. 10650–10656, 1992.
14. M. Plihal and A. A. Maradudin, "Photonic Band Structure of Two-Dimensional Systems: The Triangular Lattice," *Phys. Rev. B*, vol. 44, pp. 8565–8571, 1991.
15. L. Liu and J. T. Liu, "Photonic Band Structure in the Nearly Plane Wave Approximation," *Eur. Phys. J. B*, vol. 9, pp. 381–388, 1999.
16. E. Yablonovitch, T. J. Gmitter, and K. M. Leung, "Photonic Band-Structure—the Face-Centered-Cubic Case Employing Nonspherical Atoms," *Phys. Rev. Lett.*, vol. 67, pp. 2295–2298, 1991.
17. E. Özbay, A. Abeyta, G. Tuttle, M. C. Tringides, R. Biswas, M. M. Sigalas, and C. M. Soukoulis, "Measuerment of a Three-Dimensional Photonic Band Gap in a Crystal Structure Made of Dielectric Rods," *Phys. Rev. B*, vol. 50, pp. 1945–1948, 1994.
18. E. Özbay, E. Michel, G. Tuttle, R. Biswas, M. M. Sigalas, and K. M. Ho, "Micromachined Millimeter-Wave Photonic Band-Gap Crystals," *Appl. Phys. Lett.*, vol. 64, pp. 2059–2061, 1994.
19. E. Özbay, E. Michel, G. Tuttle, R. Biswas, K. M. Ho, J. Bostak, and M. Bloon, "Double-Etch Geometry for Millimeter-Wave Photonic Band-Gap Crystals," *Appl. Phys. Lett.*, vol. 65, pp. 1617–1619, 1994.
20. E. Özbay, E. Michel, G. Tuttle, R. Biswas, K. M. Ho, J. Bostak, and M. Bloom, "Terahertz Spectroscopy of Three-Dimensional Photonic Band-Gap Crystal," *Opt. Lett.*, vol. 19, pp. 1155, 1994.
21. S.-Y. Lin and J. G. Fleming, "A Three-Dimensional Optical Photonic Crystal," *J. Lightwave Technol.*, vol. 17, pp. 1944–1947, 1999.
22. S. John and K. Busch, "Photonic Band Gap Formation in Certain Self-Organizing Systems," *Phys. Rev. E*, vol. 58, pp. 3896–3908, 1998.
23. İ. İ. Tarhan and G. Watson, "Analytical Expression for the Optimized Stop Bands of Fcc Photonic Crystals in the Scalar-Wave Approximation," *Phys. Rev. B*, vol. 54, pp. 7593–7597, 1996.
24. İ. İ. Tarhan, M. P. Zinkin, and G. Watson, "An Interferometric Technique for Measurement of Photonic Band Structure in Colloidal Crystals," *Opt. Lett.*, vol. 20, pp. 1571–1574, 1995.
25. İ. İ. Tarhan and G. Watson, "Photonic Band Structure of Fcc Colloidal Crystals," *Phys. Rev. Lett.*, vol. 76, pp. 315–318, 1996.
26. İ. İ. Tarhan and G. Watson, "Optical Measurements of Photonic Band Structure in Fcc Colloidal Crystals," C. M. Soukoulis, Ed. in: *Photonic Band Gap Materials*, Norwell, MA, Kluwer Academic, 1996, pp. 93–106.

27. R. D. Pradhan, İ. İ. Tarhan, and G. Watson, "Impurity Modes in the Optical Stop Bands of Doped Colloidal Crystals," *Phys. Rev. B*, vol. 54, pp. 13721–13724, 1996.
28. R. D. Pradhan, J. A. Bloodgood, and G. Watson, "Photonic Band Structure of BCC Colloidal Crystals," *Phys. Rev. B*, vol. 55, pp. 9503–9506, 1997.
29. B. T. Rosner, G. J. Schneider, and G. Watson, "Interferometric Investigation of Photonic Band Structure Effects in Pure and Doped Colloidal Crystals," *J. Opt. Soc. Am. B*, vol. 15, pp. 2654–2658, 1998.
30. Y. Xia, B. Gates, and S. H. Park, "Fabrication of Three-Dimensional Photonic Crystals for Use in the Spectral Region from Ultraviolet to Near-Infrared," *J. Lightwave Technol.*, vol. 17, pp. 1956–1962, 1999.
31. P. Jiang, G. N. Ostojic, R. Narat, D. M. Mittleman, and V. L. Colvin, "The Fabrication and Bandgap Engineering of Photonic Multilayers," *Adv. Mater.*, vol. 13, pp. 389–393, 2001.
32. Y. Xia, B. Gates, and Z.-Y. Li, "Self-Assembly Approaches to Three-Dimensional Photonic Crystals," *Adv. Mater.*, vol. 13, pp. 409–413, 2001.
33. J. D. Joannopoulos, P. R. Villeneuve, and S. Fan, "Photonic Crystals: Putting a New Twist on Light," *Nature*, vol. 386, pp. 143, 1997.
34. E. Chow, S. Y. Lin, J. R. Wendt, S. G. Johnson, and J. D. Joannopoulos, "Quantitative Analysis of Bending Efficiency in Photonic-Crystal Waveguide Bends at 1.55 μm Wavelengths," *Opt. Lett.*, vol. 26, pp. 286–288, 2001.
35. R. D. Meade, A. Devenyi, J. D. Joannopoulos, O. L. Alerhand, D. A. Smith, and K. Kash, "Novel Applications of Photonic Band Gap Materials: Low-Loss Bends and High Q Cavities," *J. Appl. Phys.*, vol. 75, pp. 4753–4755, 1994.
36. A. Mekis, J. C. Chen, I. Kurland, S. Fan, P. R. Villeneuve, and J. D. Joannopoulos, "High Transmission Through Sharp Bends in Photonic Crystal Waveguides," *Phys. Rev. Lett.*, vol. 77, pp. 3787–3790, 1996.
37. S. Fan, J. N. Winn, A. Devenyi, J. C. Chen, R. D. Meade, and J. D. Joannopoulos, "Guided and Defect Modes in Periodic Dielectric Waveguides," *J. Opt. Soc. Am. B*, vol. 12, p. 1267, 1995.
38. A. Mekis, J. Fan, and J. D. Joannopoulos, "Bound States in Photonic Crystal Waveguides and Waveguide Bends," *Phys. Rev. B*, vol. 58, pp. 4809–4917, 1998.
39. S. Y. Lin, E. Chow, V. Hietch, P. R. Villeneuve, and J. D. Joannopoulos, "Experimental Demonstration of Guiding and Bending of Light in a Photonic Crystal," *Science*, vol. 282, p. 274, 1998.
40. J. N. Winn, R. D. Meade, and J. D. Joannopoulos, "Two-Dimensional Photonic Band-Gap Materials," *J. Mod. Opt.*, vol. 41, pp. 257–273, 1994.
41. P. R. Villeneuve, S. Fan, and J. D. Joannopoulos, "Microcavities in Photonic Crystals: Mode Symmetry, Tunability, and Coupling Efficiency," *Phys. Rev. B*, vol. 54, pp. 7837–7842, 1996.
42. P. R. Villeneuve, S. Fan, and J. D. Joannopoulos, "Air-Bridge Microcavities," *Appl. Phys. Lett.*, vol. 67, pp. 167–169, 1995.
43. J. C. Chen, H. A. Haus, J. Fan, P. R. Villeneuve, and J. D. Joannopoulos, "Optical Filters from Photonic Band Gap Air Bridges," *J. Lightwave Technol.*, vol. 14, p. 2575, 1996.
44. H. A. Haus, S. Fan, P. R. Villeneuve, and J. D. Joannopoulos, "Channel Drop Filters in Photonic Crystals," *Opt. Express*, vol. 3, pp. 4–11, 1998.

45. R. Fan, R. Villeneuve, J. D. Joannopoulos, and H. A. Haus, "Channel Drop Tunneling Through Localized States," *Phys. Rev. Lett.*, vol. 80, pp. 960–963, 1998.
46. E. Centeno, B. Guizal, and D. Felbacq, "Multiplexing and Demultiplexing with Photonic Crystals," *J. Opt. A*, vol. 1, pp. L10–L13, 1999.
47. C. Zhang, F. Qiao, J. Wan, and J. Zi, "Enlargement of Nontransmission Frequency Range in Photonic Crystals by Using Multiple Heterostructures," *J. Appl. Phys.*, vol. 87, pp. 3174–3176, 2000.
48. S. Fan, S. G. Johnson, and J. D. Joannopoulos, "Waveguide Branches in Photonic Crystals," *J. Opt. Soc. Am. B*, vol. 18, pp. 162–165, 2001.
49. M. L. Povinelli, S. G. Johnson, J. Fan, and J. D. Joannopoulos, "Emulation of Two-Dimensional Photonic Crystal Defect Modes in a Photonic Crystal with a Three-Dimensional Photonic Band Gap," *Phys. Rev. B*, vol. 64, pp. 753131–753138, 2001.
50. S. G. Johnson, S. Fan, P. R. Villeneuve, and J. D. Joannopoulos, "Guided Modes in Photonic Crystal Slabs," *Phys. Rev. B*, vol. 60, pp. 5751–5758, 1999.
51. S. G. Johnson, C. Manolatou, S. H. Fan, P. R. Villeneuve, J. D. Joannopoulos, and H. A. Haus, "Elimination of Cross Talk in Waveguide Intersections," *Opt. Lett.*, vol. 23, pp. 1855–1857, 1998.
52. C. C. Cheng and A. Scherer, "Fabrication of Photonic Band-Gap Crystals," *J. Vac. Sci. Technol B.*, vol. 13, pp. 2696–2700, 1995.
53. C. C. Cheng, A. Scherer, V. Arbet-Engels, and E. Yablonovitch, "Lithographic Band Gap Tuning in Photonic Band Gap Crystals," *J. Vac. Sci. Technol. B*, vol. 14, pp. 4110–4114, 1996.
54. J. O'Brien, O. Painter, R. Lee, C. C. Cheng, A. Yariv, and A. Scherer, "Lasers Incorporating 2D Photonic Bandgap Mirrors," *Electron. Lett.*, vol. 32, pp. 2243–2244, 1996.
55. C. C. Cheng, A. Scherer, R.-C. Tyan, Y. Fainman, G. Witzgall, and E. Yablonovitch, "New Fabrication Techniques for High Quality Photonic Crystals," *J. Vac. Sci. Technol. B*, vol. 15, pp. 2764–2767, 1997.
56. B. D'Urso, O. Painter, J. O'Brien, T. Tombrello, A. Yariv, and A. Scherer, "Modal Reflectivity in Finite-Depth Two-Dimensional Photonic-Crystal Microcavities," *J. Opt. Soc. Am. B*, vol. 15, pp. 1155–1159, 1998.
57. T. F. Krauss, O. Painter, A. Scherer, J. S. Roberts, and R. M. De La Rue, "Photonic Microstructures as Laser Mirrors," *Opt. Eng.*, vol. 37, pp. 1143–1148, 1998.
58. A. Scherer, O. Painter, B. D'Urso, R. Lee, and A. Yariv, "InGaAsP Photonic Band Gap Crystal Membrane Microresonators," *J. Vac. Sci. Tech. B*, vol. 16, pp. 3906–3910, 1998.
59. O. Painter, J. Vučković, and A. Scherer, "Defect Modes of a Two-Dimensional Photonic Crystal in an Optically Thick Dielectric Slab," *J. Opt. Soc. Am. B*, vol. 16, pp. 275–285, 1999.
60. O. Painter, A. Husain, A. Scherer, J. O'Brien, I. Kim, and P. D. Dapkus, "Room Temperature Photonic Crystal Defect Lasers at Near-Infrared Wavelengths in InGaAsP," *J. Lightwave Technol.*, vol. 17, pp. 2082–2088, 1999.
61. O. Painter, R. K. Lee, A. Scherer, A. Yariv, J. O'Brien, P. D. Dapkus, and I. Kim, "Two-Dimensional Photonic Band-Gap Defect Mode Laser," *Science*, vol. 284, pp. 1819–1821, 1999.
62. R. K. Lee, O. J. Painter, B. Kitzke, A. Scherer, and A. Yariv, "Photonic Bandgap Disk Laser," *Electron. Lett.*, vol. 35, pp. 569–570, 1999.

63. R. K. Lee, O. J. Painter, B. D'Urso, A. Scherer, and A. Yariv, "Measurement of Spontaneous Emission from a Two-Dimensional Photonic Band Gap Defined Microcavity at Near-Infrared Wavelengths," *Appl. Phys. Lett.*, vol. 74, pp. 1522–1524, 1999.

64. Y. Xu, J. Vučković, R. K. Lee, O. J. Painter, A. Scherer, and A. Yariv, "Finite-Difference Time-Domain Calculation of Spotaneous Emission Lifetime in a Microcavity," *J. Opt. Soc. Am. B*, vol. 16, pp. 465–474, 1999.

65. J. Vučković, O. Painter, Y. Xu, A. Yariv, and A. Scherer, "Finite-Difference Time-Domain Calculation of the Spontaneous Emission Coupling Factor in Optical Microcavities," *IEEE J. Quantum Electron.*, vol. 35, pp. 1168–1175, 1999.

66. T. Yoshie, A. Scherer, H. Chen, D. Huffaker, and D. Deppe, "Optical Characterization of Two-Dimensional Photonic Crystal Cavities with Indium Arsenide Quantum Dot Emitters," *Appl. Phys. Lett.*, vol. 79, pp. 114–116, 2001.

67. M. Lončar, D. Nedeljkovic, T. Doll, J. Vučković, A. Scherer, and T. P. Pearsall, "Waveguiding in Planar Photonic Crystals," *Appl. Phys. Lett.*, vol. 77, pp. 1937–1939, 2000.

68. M. Lončar, T. Doll, J. Vučković, and A. Scherer, "Design and Fabrication of Silicon Photonic Crystal Optical Waveguides," *J. Lightwave Technol.*, vol. 18, pp. 1402–1411, 2000.

69. A. Adibi, Y. Xu, R. K. Lee, A. Yariv, and A. Scherer, "Properties of the Slab Modes in Photonic Crystal Optical Waveguides," *J. Lightwave Technol.*, vol. 18, pp. 1554–1564, 2000.

70. A. Adibi, R. K. Lee, Y. Xu, A. Yariv, and A. Scherer, "Design of Photonic Crystal Optical Waveguides with Single Mode Propagation in the Photonic Bandgap," *Electron. Lett.*, vol. 36, pp. 1376–1378, 2000.

71. M. Bayindir, B. Temelkuran, and E. Özbay, "Photonic Crystal Based Beam Splitters," *Appl. Phys. Lett.*, vol. 77, pp. 3902–3904, 2000.

72. M. Bayindir, B. Temelkuran, and E. Özbay, "Propagation of Photons by Hopping: A Waveguiding Mechanism Through Localized Coupled Cavities in Three-Dimensional Photonic Crystals," *Phys. Rev. B*, vol. 61, pp. R11855–R11858, 2000.

73. M. Bayindir, B. Temelkuran, and E. Özbay, "Tight-Binding Description of the Coupled Defect Modes in Three-Dimensional Photonic Crystals," *Phys. Rev. Lett.*, vol. 84, pp. 2140–2143, 2000.

74. M. Bayindir and E. Özbay, "Heavy Photons at Coupled-Cavity Waveguide Band Edges in a Three-Dimensional Photonic Crystal," *Phys. Rev. B*, vol. 62, pp. R2247–R2250, 2000.

75. E. Özbay, M. Bayindir, I. Bulu, and E. Cubukcu, "Investigation of Localized Coupled-Cavity Modes in Two-Dimensional Photonic Bandgap Structures," *IEEE J. Quantum Electron.*, vol. 38, pp. 837–843, 2002.

76. N. W. Ashcroft and N. D. Mermin, *Solid State Physics*, Philadelphia: Saunders College, 1976.

77. H. Kosaka, T. Kawashima, A. Tomita, M. Notomi, T. Tamamura, T. Sato, and S. Kawakami, "Superprism Phenomena in Photonic Crystals," *Phys. Rev. B*, vol. 58, pp. R10096–R10099, 1998.

78. H. Kosaka, T. Kawashima, A. Tomita, M. Notomi, T. Tamamura, T. Sato, and S. Kawakami, "Photonic Crystals for Micro Lightwave Circuits Using Wavelength-Dependent Angular Beam Steering," *Appl. Phys. Lett.*, vol. 74, pp. 1370–1372, 1999.

79. H. Kosaka, T. Kawashima, A. Tomita, M. Notomi, T. Tamamura, T. Sato, and S. Kawakami, "Superprism Phenomena in Photonic Crystals: Toward Microscale Lightwave Circuits," *J. Lightwave Technol.*, vol. 17, pp. 2032–2034, 1999.

80. R. A. Shelby, D. R. Smith, and S. Schultz, "Experimental Verification of a Negative Index of Refraction," *Science*, vol. 292, pp. 77–79, 2001.
81. H. Kosaka, T. Kawashima, A. Tomita, M. Notomi, T. Tamamura, T. Sato, and S. Kawakami, "Self-Collimating Phenomena in Photonic Crystals," *Appl. Phys. Lett.*, vol. 74, pp. 1212–1214, 1999.
82. G. Parker and M. Charlton, "Photonic Crystals," *Physics World*, vol. 13, pp. 29–34, 2000.
83. T. Sondergaard and K. H. Dridi, "Energy Flow in Photonic Crystal Waveguides," *Phys. Rev. B*, vol. 61, pp. 15688–15696, 2000.
84. R. W. Ziolkowski, "FDTD Analysis of PBG Waveguides, Power Splitters and Switches," *Opt. Quantum Electron.*, vol. 31, pp. 843–855, 1999.
85. G. Rarity and C. Weisbuch, Eds. *Microcavities and Photonic Bandgaps: Physics and Applications*. Elounda, Crete: Kluwer Academic Publishers, 1995.
86. H.-B. Sun, V. Mizeikis, Y. Xu, S. Juodkazis, J.-Y. Ye, S. Matsuo, and H. Misawa, "Microcavities in Polymeric Photonic Crystals," *Appl. Phys. Lett.*, vol. 79, pp. 1–3, 2001.
87. S. G. Johnson, F. Shanhui, A. Mekis, and J. D. Joannopoulos, "Multipole-Cancelation Mechanism for High-Q Cavities in the Absence of a Complete Photonic Band Gap," *Appl. Phys. Lett.*, vol. 78, pp. 3388–3390, 2001.
88. A. Sharkawy, S. Shi, and D. W. Prather, "Multichannel Wavelength Division Multiplexing Using Photonic Crystals," *Appl. Opt.*, vol. 40, pp. 2247–2252, 2001.
89. R. Stoffer, H. J. W. M. Hoekstra, R. M. D. Ridder, E. V. Groesen, and F. P. H. V. Beckum, "Numerical Studies of 2D Photonic Crystals: Waveguides, Coupling Between Waveguides and Filters," *Opt. Quantum Electron.*, vol. 32, pp. 947–961, 2000.
90. S. G. Johnson, P. R. Villeneuve, S. Fan, and J. D. Joannopoulos, "Linear Waveguides in Photonic-Crystal Slabs," *Phys. Rev. B*, vol. 62, pp. 8212–8222, 2000.
91. M. Charlton, M. E. Zoorob, G. Parker, M. C. Netti, J. J. Baumberg, J. A. Cox, and H. Kemhadjian, "Experimental Investigation of Photonic Crystal Waveguide Devices and Line-Defect Waveguide Bends," *Mater. Sci. Eng.*, vol. B74, pp. 17–24, 2000.
92. M. Gander, R. McBride, J. Jones, D. Mogilevtsev, T. Birks, J. Knight, and P. Russell, "Experimental Measurement of Group Velocity Dispersion in Photonic Crystal Fibre," *Electron. Lett.*, vol. 35, pp. 63–64, 1999.
93. C. J. M. Smith, R. M. D. L. Rue, M. Rattier, S. Olivier, H. Benisty, C. Weisbuch, P. R. Krauss, R. Houdre, and U. Oesterle, "Coupled Guide and Cavity in a Two-Dimensional Photonic Crystal," *Appl. Phys. Lett.*, vol. 78, pp. 1487–1489, 2001.
94. H. Hirayama, T. Hamano, and Y. Aoyagi, "Novel Surface Emitting Laser Diode Using Photonic Band-Gap Crystal Cavity," *Appl. Phys. Lett.*, vol. 69, pp. 791–793, 1996.
95. R. L. Liboff, Introductory Quantum Mechanics, 4th ed. San Francisco: Addison Wesley, 2003.
96. J. D. Jackson, Classical Electrodynamics, 3rd ed. New York: Wiley, 1999.
97. E. Yablonovitch, "Photonic Band-Gap Structures," *J. Opt. Soc. Am. B*, vol. 10, pp. 283–295, 1993.
98. K. M. Ho, C. T. Chan, and C. M. Soukoulis, "Existence of a Photonic Gap in Periodic Dielectric Structures," *Phys. Rev. Lett.*, vol. 65, pp. 3152–3155, 1990.
99. A. R. McGurn, "Photonic Crystal Circuits: A Theory for Two- and Three Dimensional Networks," *Phys. Rev. B*, vol. 61, pp. 13235–13249, 2000.

2

Preliminary Concepts of Electromagnetic Waves and Periodic Media

This chapter introduces several basic concepts used in subsequent chapters where the propagation of electromagnetic waves in photonic crystals is analyzed. The discussion is limited to topics directly applicable to photonic crystals—for a general introduction to electromagnetism, the reader is referred to one of the standard textbooks, such as those written by Jackson [1] or Balanis [2]. Thus, Section 2.1 shows how a wave equation is a consequence of Maxwell's equations. Analysis of the solutions to the wave equation in unbounded homogeneous media leads to the concepts of k-space, light cone, equifrequency surface, as well as phase and group velocity. It is also shown how the interaction of the wave with material boundaries allows for guiding of electromagnetic energy in confined geometries. Section 2.2 formalizes the framework for analyzing periodic media, using the real-space lattice and unit cell, and it introduces the concept of the reciprocal lattice and Brillouin zone. In Section 2.3, the interaction of waves with periodic media is introduced, which leads to Bloch's theorem and to the concepts of dispersion diagrams, dispersion surfaces, and bandgaps.

2.1 ELECTROMAGNETIC WAVES

A time-varying electric field creates a magnetic field. Similarly, a time-varying magnetic field produces an electric field. This mutual interdependence between time-varying fields enables the existence of electromagnetic waves that propagate throughout space and within materials.

The behavior of electric and magnetic fields, and indeed all classical electromagnetic phenomena, are governed by a compact and elegant set of fundamental rules known as Maxwell's equations. These coupled partial differential equations, which were put forth as the complete classical theory of electromagnetism by James Clerk Maxwell, are discussed below.

Photonic Crystals: Theory, Applications, and Fabrication. By D. W. Prather, S. Shi, A. Sharkawy, J. Murakowski, and G. J. Schneider

2.1.1 Maxwell's Equations in Linear, Homogeneous Media

Maxwell's equations are based on experimentally established facts named after their discoverers: Coulomb's law, Ampère's law, and Faraday's law. The physical quantities that appear in Maxwell's equations are the electric field **E**, the magnetic field **B**, the electric displacement field **D**, the magnetic field intensity **H**, the electric current density **J**, and the charge density ρ. Both integral and differential forms of these equations are listed below.

1. *Faraday's law* is based on the experimental fact that the time-varying magnetic flux induces electromotive force:

$$\oint_C \mathbf{E} \cdot d\mathbf{l} = -\int_S \frac{\partial \mathbf{B}}{\partial t} \cdot d\mathbf{s}, \qquad \nabla \times \mathbf{E} = -\frac{\partial \mathbf{B}}{\partial t}, \tag{2.1}$$

   where S is an arbitrary two-dimensional (2D) surface in three-dimensional (3D) space and C is its boundary.

2. *Gauss' law* is a mathematical expression that can be derived from *Coulomb's law*, the observation that the interaction strength between electric charges is inversely proportional to the square of the distance between them

$$\oint_S \mathbf{D} \cdot d\mathbf{s} = \int_V \rho\, dv, \qquad \nabla \cdot \mathbf{D} = \rho, \tag{2.2}$$

   where V is a volume enclosed by surface S.

3. Maxwell's third equation is a generalization of *Ampère's law*, which states that the line integral of the magnetic field along any enclosed contour equals the total electric current enclosed by that contour

$$\oint_C \mathbf{H} \cdot d\mathbf{l} = \int_S \mathbf{J} \cdot d\mathbf{s} + \int_S \frac{\partial \mathbf{D}}{\partial t} \cdot d\mathbf{s}, \qquad \nabla \times \mathbf{H} = \mathbf{J} + \frac{\partial \mathbf{D}}{\partial t}, \tag{2.3}$$

   where S is a surface with boundary C.

4. Maxwell's fourth equation, sometimes called *Gauss' law for magnetism*, codifies the absence of magnetic charges (magnetic monopoles)

$$\oint_S \mathbf{B} \cdot d\mathbf{s} = 0, \qquad \nabla \cdot \mathbf{B} = 0. \tag{2.4}$$

In linear, isotropic, and homogeneous media, the following relations hold between the fields:

$$\mathbf{D} = \varepsilon_r \varepsilon_0 \mathbf{E}, \qquad \mathbf{B} = \mu_r \mu_0 \mathbf{H}, \tag{2.5}$$

where ε_0 and μ_0 are the permittivity and the permeability of the free space, whereas ε_r and μ_r are the relative permittivity and permeability of the material.

If the analyzed region of interest contains objects made of different materials, then the constitutive equations (2.5) hold for each material separately and the fields are continuous within the objects. However, at material boundaries, the fields may undergo discontinuities. The exact form of such field discontinuities depends on the material properties, such as permittivity and permeability, and can be obtained by applying the integral form of Eqs. (2.1) through (2.4) to small spatial regions that encompass the material interface.

For example, consider a boundary between materials 1 and 2 characterized by permittivities ε_1, ε_2, and permeabilities μ_1, μ_2, respectively, as depicted in Fig. 2.1. Application of Faraday's law, Eq. (2.1), to the loop *abcd* in Fig. 2.1a yields

$$
\begin{aligned}
&E_{2t}\Delta l + E_{2n}\frac{\Delta h}{2} + E_{1n}\frac{\Delta h}{2} + E_{1t}(-\Delta l) + E_{1n}\left(-\frac{\Delta h}{2}\right) + E_{2n}\left(-\frac{\Delta h}{2}\right) \\
&\quad = -\frac{\partial B_\perp}{\partial t}\Delta h \Delta l, \qquad (2.6)
\end{aligned}
$$

where $B_\perp$ is a component of the magnetic field perpendicular to the rectangle *abcd* and other symbols follow the notation of Fig. 2.1a. By taking the limit of vanishing Δh with small but finite Δl, Eq. (2.6) is reduced to

$$
E_{2t}\Delta l + E_{1t}(-\Delta l) = 0, \qquad (2.7)
$$

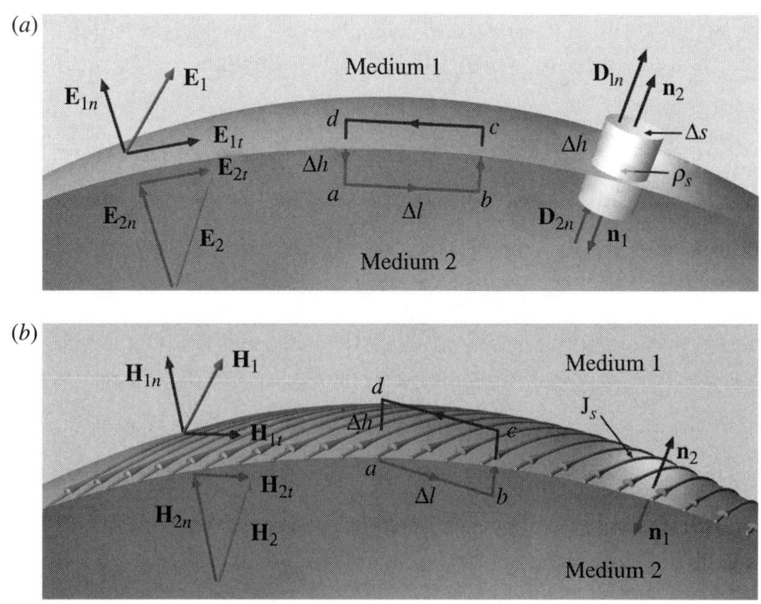

Figure 2.1 Boundary conditions at the interface of two materials characterized by relative permittivities ε_1, ε_2, and permeabilities μ_1, μ_2, respectively.

which yields the continuity of the tangential component of the electric field across a boundary between two media

$$\mathbf{E}_{2t} = \mathbf{E}_{1t}, \tag{2.8}$$

now written in vectorial form because the orientation of the loop *abcd* on the surface can be chosen arbitrarily. Given that

$$\mathbf{D}_{1t} = \varepsilon_1 \mathbf{E}_{1t} \qquad \text{and} \qquad \mathbf{D}_{2t} = \varepsilon_2 \mathbf{E}_{2t}, \tag{2.9}$$

where $\varepsilon_1 = \varepsilon_{r1}\varepsilon_0$, and $\varepsilon_1 = \varepsilon_{r2}\varepsilon_0$, the tangential component of the electric displacement field $\mathbf{D}$ is generally discontinuous at the boundary, and

$$\varepsilon_1 \mathbf{D}_{2t} = \varepsilon_2 \mathbf{D}_{1t}. \tag{2.10}$$

The boundary conditions for the normal components of $\mathbf{E}$ and $\mathbf{D}$ can be determined by applying Gauss' law, Eq. (2.2). When applied to the cylinder shown in Fig. 2.1a, in the limit of vanishing cylinder height $\Delta h \rightarrow 0$ so that the contribution to the total flux by the side surface is negligible, it leads to

$$\mathbf{D}_1 \cdot \mathbf{n}_2 \Delta s + \mathbf{D}_2 \cdot \mathbf{n}_1 \Delta s = \rho_s \Delta s, \tag{2.11}$$

where $\mathbf{n}_1$ and $\mathbf{n}_2$ are the unit vectors normal to the bottom and top surfaces of the cylinder, as shown in Fig. 2.1a, whereas ρ_s is the surface charge density. Because the normal unit vector at the surface of an enclosed volume is conventionally defined in the outward direction, $\mathbf{n}_1 = -\mathbf{n}_2$ and Eq. (2.11) simplifies to

$$\mathbf{n}_2 \cdot (\mathbf{D}_1 - \mathbf{D}_2) = \rho_s. \tag{2.12}$$

With D_{1n} and D_{2n} defined as the normal components of $\mathbf{D}_1$ and $\mathbf{D}_2$, Eq. (2.12) can be rewritten as

$$D_{1n} - D_{2n} = \rho_s. \tag{2.13}$$

This means that the normal component of $\mathbf{D}$ changes abruptly at a charged boundary between two different media, and the amount of change is equal to the surface charge density ρ_s.

By using constituent relations of Eq. (2.9), the boundary conditions for the normal component of the electric field $\mathbf{E}$ can be easily obtained

$$\varepsilon_1 E_{1n} - \varepsilon_2 E_{2n} = \rho_s. \tag{2.14}$$

Similarly, the application of Gauss' law to magnetism, Eq. (2.4), results in

$$B_{1n} = B_{2n}, \tag{2.15}$$

which means that the normal component of the magnetic field **B** is continuous across the boundary between adjacent media. By using relations

$$\mathbf{B}_1 = \mu_1 \mathbf{H}_1, \qquad \mathbf{B}_2 = \mu_2 \mathbf{H}_2, \tag{2.16}$$

where $\mu_1 = \mu_{r1}\mu_0$, and $\mu_2 = \mu_{r2}\mu_0$, the boundary conditions for the normal component of the **H**-field are obtained

$$\mu_1 H_{1n} = \mu_2 H_{2n}. \tag{2.17}$$

Finally, if Ampère's law, Eq. (2.3), is applied to a closed rectangular path with sides Δl and Δh as shown in Fig. 2.1b, and the limit of vanishing Δh is taken, $\Delta h \to 0$, then the following equation ensues:

$$H_{2t}\Delta l + H_{1t}(-\Delta l) = J_s \Delta l \implies H_{2t} - H_{1t} = J_s, \tag{2.18}$$

where H_{1t} and H_{2t} are the tangential components of the magnetic field intensity **H** in the respective media pointing in the direction of the long sides of the rectangle *abcd* in Fig. 2.1b, and J_s is the magnitude of the component of the surface current density normal to the rectangle *abcd*. The last of Eq. (2.18) can be expressed in vectorial form as

$$\mathbf{n}_2 \times (\mathbf{H}_1 - \mathbf{H}_2) = \mathbf{J}_s, \tag{2.19}$$

where $\mathbf{n}_2$ is the normal unit vector pointing away from medium 2. In the absence of surface currents, the last condition simplifies to the continuity of the tangential component of the magnetic field intensity

$$H_{1t} = H_{2t}. \tag{2.20}$$

By using constitutive relations, Eq. (2.16), the boundary condition for the tangential component of the magnetic field **B** is obtained

$$\mu_1 \mathbf{B}_{2t} = \mu_2 \mathbf{B}_{1t}. \tag{2.21}$$

2.1.2 Electromagnetic Waves

Maxwell's equations (2.1) through (2.4) admit solutions in the form of propagating waves. To observe this, we start by eliminating fields **D** and **B** using constitutive equations (2.5), which hold for isotropic homogeneous materials, assuming that the region is devoid of free charges and electric currents. This leads to

$$\nabla \times \mathbf{E} = -\mu \frac{\partial \mathbf{H}}{\partial t}, \tag{2.22}$$

$$\nabla \cdot \mathbf{E} = 0, \tag{2.23}$$

$$\nabla \times \mathbf{H} = \varepsilon \frac{\partial \mathbf{E}}{\partial t}, \tag{2.24}$$

$$\nabla \cdot \mathbf{H} = 0, \tag{2.25}$$

where $\varepsilon = \varepsilon_r \varepsilon_0$ and $\mu = \mu_r \mu_0$, and it is implicitly assumed that the medium properties do not vary in time. Taking the curl of Eq. (2.22) and changing the order of differentiation on the right-hand side

$$\nabla \times (\nabla \times \mathbf{E}) = -\mu \frac{\partial(\nabla \times \mathbf{H})}{\partial t}, \tag{2.26}$$

allows the substitution for $\nabla \times \mathbf{H}$ from Eq. (2.24), which yields

$$\nabla \times (\nabla \times \mathbf{E}) = -\mu\varepsilon \frac{\partial^2 \mathbf{E}}{\partial t^2}, \tag{2.27}$$

The left-hand side of Eq. (2.27) is expanded using vector identity $\nabla \times (\nabla \times \mathbf{E}) = \nabla(\nabla \cdot \mathbf{E}) - \nabla^2 \mathbf{E}$, where the first term vanishes because of the absence of free charges as expressed by Gauss' law in Eq. (2.23). As a result, the wave equation is obtained for the electric field

$$\nabla^2 \mathbf{E} = \frac{1}{v^2} \frac{\partial^2 \mathbf{E}}{\partial t^2}, \tag{2.28}$$

where the speed of wave propagation v is defined by

$$v = \frac{1}{\sqrt{\mu\varepsilon}}. \tag{2.29}$$

The wave equation for the magnetic field intensity $\mathbf{H}$ is obtained in a similar fashion.

Elementary solutions to Eq. (2.28) are plane waves

$$\mathbf{E}(t, \mathbf{r}) = \mathbf{E}_0 e^{j(\omega t - \mathbf{k} \cdot \mathbf{r})} + \text{c.c.}, \tag{2.30}$$

where "c.c." indicates the presence of a complex-conjugate term, which ensures that the (physical) electric field is $\mathbf{E}$ real.[1] In Eq. (2.30), ω is the frequency of the plane wave and $\mathbf{k}$ is its wave vector. Substituting solution (2.30) into wave equation (2.28) yields an algebraic equation that relates the wave vector $\mathbf{k}$ with the angular frequency ω

$$|\mathbf{k}|^2 = \frac{\omega^2}{v^2}, \tag{2.31}$$

where $|\mathbf{k}|^2 = k_x^2 + k_y^2 + k_z^2$. Thus, Eq. (2.30) provides a solution to the wave equation (2.28) if and only if parameters $\mathbf{k}$ and ω satisfy Eq. (2.31). The latter equation defines a cone in the hybrid space of ω, k_x, and k_y, which is commonly referred to as the light cone. Because Eq. (2.31) has no intrinsic scale, and neither does the wave equation (2.30), electromagnetic waves with arbitrary frequencies are allowed, and their

[1] "c.c." usually will be omitted in the text. However, it should be understood that all physically measurable quantities (observables) are real, and that the presence of a complex-conjugate term is implied whenever it is necessary to ensure a real result.

behavior is governed by the same principles. Thus, phenomena observed in one part of the spectrum will be reproduced identically in another part of the spectrum as long as material properties are preserved and the dimensions of objects are scaled accordingly. This provides an invaluable tool in experimental validation and testing of hypotheses as one can build devices at scales that make them easy to manipulate and can rest assured that observations made there will hold for their scaled versions when subjected to commensurately longer or shorter wavelength radiation. Indeed, this freedom has been frequently taken advantage of in the field of photonic crystals where large versions of three-dimensional structures are often tested in the radio-frequency (RF) range before an attempt is made to fabricate them for the optical regime.

2.1.3 Optical Waves

Optical waves are typically identified as electromagnetic waves with frequency that ranges from about 10^{12} Hz to $\sim 10^{17}$ Hz or, in terms of free-space wavelength, from ~ 3 nm (ultraviolet) to 300 μm (infrared). Outside of this range lie microwaves in the long-wave portion of the spectrum and X-rays in the short-wave portion. The boundaries of the optical region are fluid and depend mostly on what techniques are used to generate, manipulate, and detect the waves. They are listed here merely to establish a general reference for the reader and to define, at least loosely, the meaning of the term "optical wave" used throughout this book.

2.1.3.1 Reflection, Refraction, and Diffraction. When an electromagnetic wave encounters a boundary between two different homogenous media, it splits into two waves: a reflected wave that propagates back to the first medium and a transmitted (or refracted) wave that proceeds into the second medium. Consider a plane electromagnetic wave propagating in medium 1 (ε_1, μ_1) obliquely incident on the boundary with medium 2 (ε_2, μ_2). k-vectors of the three waves, incident, reflected, and transmitted, lie in the same plane (perpendicular to the boundary) and form angles θ_i, θ_r, and θ_t with the normal to the boundary. The three angles are related by Snell's law of reflection and refraction

$$\begin{array}{c} \theta_i = \theta_r \\ n_1 \sin\theta_i = n_2 \sin\theta_t \end{array}, \tag{2.32}$$

where n_1 and n_2 are the refractive indices of medium 1 and medium 2, respectively. The refractive index of a particular medium is defined as the ratio of the phase velocity in free space over the phase velocity in the medium

$$n = \frac{c}{v} = \frac{\sqrt{\mu_r \mu_0 \varepsilon_r \varepsilon_0}}{\sqrt{\mu_r \mu_0}} = \sqrt{\mu_r \varepsilon_r}. \tag{2.33}$$

The fraction of energy carried by the reflected and transmitted waves define the reflection and transmission coefficients R and T, which are obtained by applying boundary conditions given in Eqs. (2.8), (2.13), (2.15), and (2.20) to the case of an incident plane wave impinging on a planar material boundary:

$$R_s = \left[\frac{\sin(\theta_t - \theta_i)}{\sin(\theta_t + \theta_i)}\right]^2 = \left[\frac{n_1 \cos\theta_i - n_2 \cos\theta_t}{n_1 \cos\theta_i + n_2 \cos\theta_t}\right]^2 = \left[\frac{n_1 \cos\theta_i - n_2\sqrt{1 - \left(\frac{n_1}{n_2}\sin\theta_i\right)^2}}{n_1 \cos\theta_i + n_2\sqrt{1 - \left(\frac{n_1}{n_2}\sin\theta_i\right)^2}}\right]^2, \tag{2.34}$$

$$R_p = \left[\frac{\tan(\theta_t - \theta_i)}{\tan(\theta_t + \theta_i)}\right]^2 = \left[\frac{n_1 \cos\theta_t - n_2 \cos\theta_i}{n_1 \cos\theta_t + n_2 \cos\theta_i}\right]^2 = \left[\frac{n_1\sqrt{1 - \left(\frac{n_1}{n_2}\sin\theta_i\right)^2} - n_2 \cos\theta_i}{n_1\sqrt{1 - \left(\frac{n_1}{n_2}\sin\theta_i\right)^2} + n_2 \cos\theta_i}\right]^2. \tag{2.35}$$

The subscripts s and p indicate the dependence of the reflection coefficients on the polarization of the incident wave, where s corresponds to the polarization perpendicular to the plane defined by the incident k-vector and the boundary normal, whereas p corresponds to in-plane polarization. Transmission coefficients are given by $T_s = 1 - R_s$ and $T_p = 1 - R_p$.

If $n_1 > n_2$, a range of incidence angles θ_i exists such that

$$\frac{n_1}{n_2}\sin\theta_i > 1. \tag{2.36}$$

In this case, no real angle of transmission θ_t exists to satisfy Snell's law (2.32), and all the electromagnetic energy is reflected back to medium 1. Accordingly, such a situation is described as total internal reflection, or TIR. Because the propagation takes place entirely in one medium (the wave never escapes to the low-index material), TIR can be used to guide electromagnetic waves.

Although in TIR all energy is reflected back to the material of higher index, the term "total" is not meant to imply that the electromagnetic field is completely within medium 1. Indeed, careful analysis shows that an evanescent tail extends into the lower-index material, in which the fields decay exponentially.

The most explicit manifestation of the wave nature of light is diffraction. Waves, in general, tend to "bend" around objects whose dimensions are comparable with the wavelength. Such bending is referred to as diffraction. An intuitive picture of the process is provided by Huygens' principle, which states that every point reached by an advancing wavefront itself becomes a point source of the wave. By adding waves produced by these points, a new wavefront is obtained. Such a picture aids in understanding the spreading of light coming out of a small aperture or a single slit, which cannot be explained using ray optics in which light is assumed to propagate along straight lines between material boundaries.

By periodically distributing small objects in space, with the period comparable with the wavelength, the effect of diffraction can be "amplified" thanks to constructive and destructive interference of waves diffracted by the individual objects. For example, periodic placement of slits produces a diffraction grating, which can be used to separate different wavelengths. This effect can be easily observed with everyday items such as compact discs, which have grooves that carry data densely packed behind a transparent polycarbonate layer. By holding it under a bright white light source, bands of color are produced, which are the result of light diffraction by the periodic arrangement of grooves that redirect different wavelengths (colors) along different directions. In photonic crystals, the concept of periodicity is extended to two and three dimensions to produce even more interesting and useful effects.

2.1.3.2 Dielectric Tensor and Equifrequency Surfaces of Anisotropic Materials. The wave equation (2.28) was derived under the assumption that the material is homogeneous, static, and isotropic, i.e., when ε and μ in the constitutive equations (2.5) are *scalars* independent on the spatial or temporal coordinates. Using scalars in Eq. (2.5) implies that the material response to the external electromagnetic field is independent of the field orientation. In this section, we extend the treatment to anisotropic materials. In particular, we are interested in electromagnetic waves in the optical range of the spectrum propagating in such materials. Because the magnetic response of typical materials at optical frequencies is slow, we assume that the relative permeability is one, i.e., $\mu_r = 1$. In contrast, the permittivity ε is assumed to be a second-rank tensor, which takes the form of a 3×3 matrix, denoted by $\bar{\bar{\varepsilon}}$, once the system of coordinates is chosen. To simplify the analysis, it is assumed that the material is nonlossy, which together with general symmetry considerations implies that ε is real and symmetric. Such a matrix has real eigenvalues and can be diagonalized by properly rotating the system of coordinates. Thus, without the loss of generality, it can be assumed that $\bar{\bar{\varepsilon}}$ is diagonal.

To see how electromagnetic waves propagate in materials characterized by such permittivity, we begin with the differential form of Maxwell's equations (2.1) through (2.4) in the absence of free charges and electric currents

$$\begin{aligned} \nabla \times \mathbf{E} &= -\frac{\partial \mathbf{B}}{\partial t} \\ \nabla \cdot \mathbf{D} &= 0 \\ \nabla \times \mathbf{H} &= \frac{\partial \mathbf{D}}{\partial t} \\ \nabla \cdot \mathbf{B} &= 0 \end{aligned}, \tag{2.37}$$

and use constitutive relations

$$\begin{aligned} \mathbf{B} &= \mu \mathbf{H} \\ \mathbf{D} &= \bar{\bar{\varepsilon}}\,\mathbf{E} \end{aligned}, \tag{2.38}$$

where $\bar{\bar{\varepsilon}}$ is a second-rank tensor as discussed above. By taking the curl of the first of Eq. (2.37), interchanging the temporal and spatial derivatives on the right-hand side, and using the third of Eq. (2.37) with the help of constitutive relations (2.38), we arrive at

$$\nabla \times (\nabla \times \mathbf{E}) = -\mu \bar{\bar{\varepsilon}} \frac{\partial^2 \mathbf{E}}{\partial t^2}. \tag{2.39}$$

Thanks to translational invariance (both in space and in time) of the partial differential equation (2.39), the solutions can be sought in the form of plane waves

$$\mathbf{E} = E_0 \hat{\mathbf{s}} e^{j(\omega t - \mathbf{k} \cdot \mathbf{r})}, \tag{2.40}$$

where E_0 is the field amplitude and $\hat{\mathbf{s}}$ is a unit vector of field polarization. Substituting the expression for the electric field Eq. (2.40) into Eq. (2.39) yields, after simple algebra,

$$\mathbf{k} \times (\mathbf{k} \times \hat{\mathbf{s}}) = -\omega^2 \mu \bar{\bar{\varepsilon}}\, \hat{\mathbf{s}}, \tag{2.41}$$

which is an algebraic vector equation. It can be cast in a matrix form

$$\sum_{m=1}^{3} K_{nm} s_m = 0, \tag{2.42}$$

where matrix K is

$$K = \begin{pmatrix} \mu\omega^2 \varepsilon_{11} - k_2^2 - k_3^2 & k_1 k_2 & k_1 k_3 \\ k_1 k_2 & \mu\omega^2 \varepsilon_{22} - k_1^2 - k_3^2 & k_2 k_3 \\ k_1 k_3 & k_2 k_3 & \mu\omega^2 \varepsilon_{33} - k_1^2 - k_2^2 \end{pmatrix}, \tag{2.43}$$

and we have assumed that $\bar{\bar{\varepsilon}}$ is diagonal, with respective entries ε_{11}, ε_{22}, and ε_{33}. The linear equation (2.42) allows nontrivial solutions (where not all s_m are zero) if and only if the determinant of matrix K vanishes, $\det(K) = 0$, which leads to the following fourth-order equation for the components of the wave vector $\mathbf{k}$:

$$\frac{k_1^2}{\varepsilon_{22}} + \frac{k_1^2}{\varepsilon_{33}} + \frac{k_2^2}{\varepsilon_{11}} + \frac{k_2^2}{\varepsilon_{33}} + \frac{k_3^2}{\varepsilon_{11}} + \frac{k_3^2}{\varepsilon_{22}} - \frac{k_1^2 k^2}{\mu\omega^2 \varepsilon_{22}\varepsilon_{33}} - \frac{k_2^2 k^2}{\mu\omega^2 \varepsilon_{11}\varepsilon_{33}} - \frac{k_3^2 k^2}{\mu\omega^2 \varepsilon_{11}\varepsilon_{22}} = \mu\omega^2, \tag{2.44}$$

where $k^2 = k_1^2 + k_2^2 + k_3^2$. Equation (2.44) defines, for a given set of material properties set by the permittivity tensor $\bar{\bar{\varepsilon}}$, a surface in the k-space at every frequency ω.

If the diagonal elements of the matrix $\bar{\bar{\varepsilon}}$ are identical, $\varepsilon_{11} = \varepsilon_{22} = \varepsilon_{33} = \varepsilon$, then Eq. (2.44) reduces to

$$\frac{(k^2 - \mu\omega^2 \varepsilon)^2}{(\mu\omega^2)^2} = 0, \tag{2.45}$$

and further to

$$k^2 = \omega^2 \mu \varepsilon, \tag{2.46}$$

which defines a sphere with a radius $\sqrt{\mu\varepsilon\omega}$ in the three-dimensional k-space for any fixed value of angular frequency ω. This sphere is the equifrequency surface (EFS) for frequency ω, where the shape is the manifestation of the fact that the material is isotropic for $\varepsilon_{11} = \varepsilon_{22} = \varepsilon_{33} \equiv \varepsilon$. Furthermore, in this case, for each k-vector solving Eq. (2.46), there are two mutually orthogonal vectors $\hat{\mathbf{s}}$ that satisfy Eq. (2.41), which means that the phase velocity of a wave is independent of its polarization.

If exactly two of the three eigenvalues of $\bar{\bar{\varepsilon}}$ are identical, $\varepsilon_{11} = \varepsilon_{22} \equiv \varepsilon \neq \varepsilon_{33}$, then Eq. (2.44) reduces to

$$\left(\frac{k^2}{\mu\omega^2\varepsilon} - 1\right)\left(\frac{k_1^2 + k_2^2}{\varepsilon_{33}} + \frac{k_3^2}{\varepsilon} - 1\right) = 0, \tag{2.47}$$

which defines two surfaces in the three-dimensional k-space: a sphere and an ellipsoid, which have exactly two common points, as shown in Fig. 2.2. Materials characterized by such a permittivity tensor, which leads to the EFS of Fig. 2.2, are known as axial because of the existence of a preferred direction (axis) such that the phase velocity of waves propagating along it is independent of polarization.

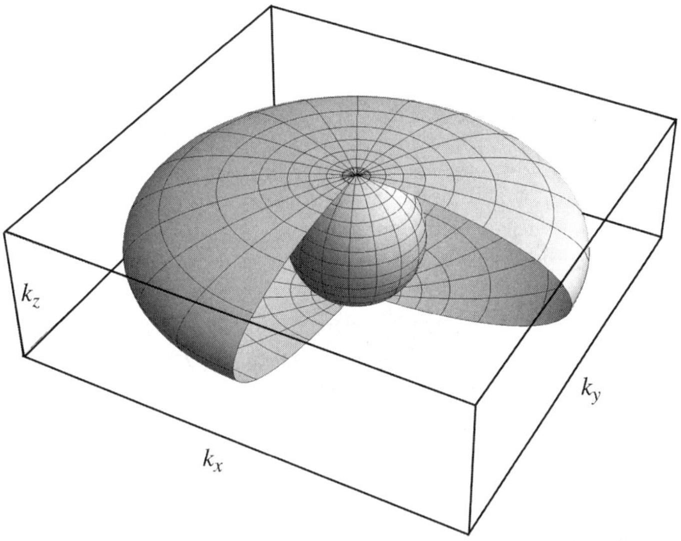

Figure 2.2 Equifrequency surfaces of axial materials. For every frequency and every direction of wave propagation (with the exception of propagation along z-direction), there are two different phase velocities that correspond to two polarizations of the wave. Along the z-direction, plane waves propagate with phase velocity independent of polarization. For illustration, the ratios of the eigenvalue are chosen as $\varepsilon_{11} : \varepsilon_{22} : \varepsilon_{33} = 1 : 1 : 11$. Part of the outer surface is cut out to show the inner surface.

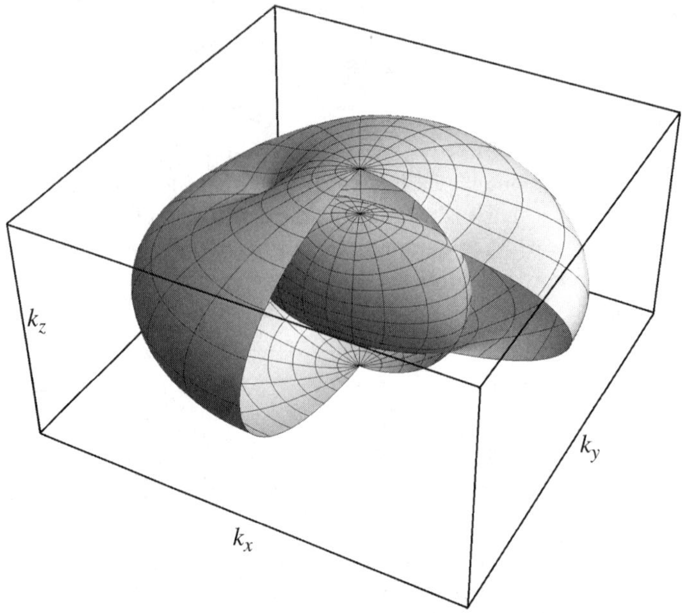

Figure 2.3 Equifrequency surface of a material with the eigenvalues of the dielectric tensor all different. For illustration, the ratios of the eigenvalues are chosen as $\varepsilon_{11} : \varepsilon_{22} : \varepsilon_{33} = 1 : 3 : 11$. Part of the outer surface is cut out to show the inner surface. With the exception of directions defined by the four degeneracy points lying in the k_x-k_z plane ($k_y = 0$), the electromagnetic plane wave can propagate with one of two different phase velocities depending on polarization. For propagation directions that correspond to any of the four degeneracy points, the phase velocity is independent of polarization.

When all three eigenvalues of the permittivity tensor $\bar{\bar{\varepsilon}}$ are different, $\varepsilon_{11} \neq \varepsilon_{22} \neq \varepsilon_{33} \neq \varepsilon_{11}$, the fourth-order surface defined by Eq. (2.44) in the three-dimensional k-space is graphically illustrated in Fig. 2.3. Except for the four degeneracy points in the plane of the lowest and the highest eigenvalues of tensor $\bar{\bar{\varepsilon}}$, every propagation direction admits two waves with mutually orthogonal polarizations and different phase velocities; at the degeneracy points, the phase velocity is independent of polarization.

Although the most general equifrequency surface of ordinary (homogeneous), albeit anisotropic, materials shown in Fig. 2.3 may already appear somewhat complex, photonic crystals admit a considerably wider variety. The study of EFSs in photonic crystals, its engineering, and use in optical devices is presented in Chapter 4.

2.1.4 Guided Waves

When the energy of the propagating wave is confined laterally to a region smaller than would be possible in unbounded space, it is said that guiding of electromagnetic waves takes place; the device or structure that effects such confinement is referred to as a waveguide. In the RF portion of the electromagnetic spectrum, examples of waveguides include a coaxial cable, coplanar waveguide, strip waveguide, and hollow metal tube. In the case of optical wavelengths, metals introduce considerable loss, and therefore, purely dielectric waveguides are preferred—an optical fiber is the most commonly cited example. Dielectric waveguides are possible thanks to the phenomenon of total internal reflection mentioned in Section 2.1.3.1, whereby the electromagnetic wave reflects back to the high-index material as long as the incident angle is greater than the critical angle determined by Snell's law of refraction, Eq. (2.32). Thus, if the high-index region is surrounded by low-index material, and a wave is launched into the high-index region so that the incidence angle with respect to the lateral boundary is greater than the critical angle for a particular material system, then waveguiding will take place.

In this section, the simplest example of a dielectric waveguide is analyzed in which a high-index dielectric slab is surrounded, or clad, by low-index material. Before examining the slab itself, consider again the boundary between high- and low-index materials with the wave launched from the high-index side and the angle of incidence greater than the critical angle, as shown in Fig. 2.4. In the absence of surface charges and surface currents, the boundary conditions can be written in a compact form as

$$\begin{aligned} (\mathbf{E}-\mathbf{E}')\times\hat{\mathbf{z}}&=0\\ (\mathbf{D}-\mathbf{D}')\cdot\hat{\mathbf{z}}&=0\\ (\mathbf{B}-\mathbf{B}')\cdot\hat{\mathbf{z}}&=0\\ (\mathbf{H}-\mathbf{H}')\times\hat{\mathbf{z}}&=0 \end{aligned}, \tag{2.48}$$

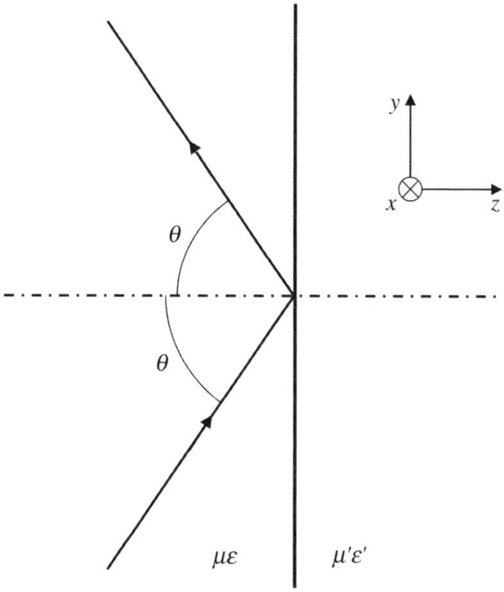

Figure 2.4 Total internal reflection at a dielectric boundary.

where the unprimed and primed symbols refer to fields on the left- and right-hand side of the dielectric boundary, respectively, as in Fig. 2.4.

Assume that the incident field is a plane wave described by the following formulas:

$$\begin{aligned} \mathbf{E}(t,\mathbf{r}) &= \mathbf{E}_0 e^{j(\omega t - \mathbf{k}\cdot\mathbf{r})} \\ \mathbf{D}(t,\mathbf{r}) &= \mathbf{D}_0 e^{j(\omega t - \mathbf{k}\cdot\mathbf{r})} \\ \mathbf{B}(t,\mathbf{r}) &= \mathbf{B}_0 e^{j(\omega t - \mathbf{k}\cdot\mathbf{r})} \\ \mathbf{H}(t,\mathbf{r}) &= \mathbf{H}_0 e^{j(\omega t - \mathbf{k}\cdot\mathbf{r})} \end{aligned}, \tag{2.49}$$

where $\mathbf{E}_0$, $\mathbf{D}_0$, $\mathbf{B}_0$, and $\mathbf{H}_0$ are field amplitudes and, for simplicity, the wave vector $\mathbf{k}$ is assumed to have no component along the x-direction: $\mathbf{k}\cdot\hat{\mathbf{x}} = 0$. Because the fields must have the same functional dependence on time t and on the x and y coordinates at the dielectric boundary ($z = 0$), the reflected field (denoted by double prime) and the evanescent field on the other side of the boundary (denoted by prime) have the form

$$\begin{aligned} \mathbf{E}''(t,\mathbf{r}) &= \mathbf{E}_0'' e^{j(\omega t - \mathbf{k}''\cdot\mathbf{r})} \\ \mathbf{D}''(t,\mathbf{r}) &= \mathbf{D}_0'' e^{j(\omega t - \mathbf{k}''\cdot\mathbf{r})} \\ \mathbf{B}''(t,\mathbf{r}) &= \mathbf{B}_0'' e^{j(\omega t - \mathbf{k}''\cdot\mathbf{r})} \\ \mathbf{H}''(t,\mathbf{r}) &= \mathbf{H}_0'' e^{j(\omega t - \mathbf{k}''\cdot\mathbf{r})} \end{aligned}, \qquad \begin{aligned} \mathbf{E}'(t,\mathbf{r}) &= \mathbf{E}_0' e^{j(\omega t - \mathbf{k}'\cdot\mathbf{r})} \\ \mathbf{D}'(t,\mathbf{r}) &= \mathbf{D}_0' e^{j(\omega t - \mathbf{k}'\cdot\mathbf{r})} \\ \mathbf{B}'(t,\mathbf{r}) &= \mathbf{B}_0' e^{j(\omega t - \mathbf{k}'\cdot\mathbf{r})} \\ \mathbf{H}'(t,\mathbf{r}) &= \mathbf{H}_0' e^{j(\omega t - \mathbf{k}'\cdot\mathbf{r})} \end{aligned}, \tag{2.50}$$

where the components of the k-vectors tangential to the boundary are all identical

$$\hat{\mathbf{z}} \times \mathbf{k} = \hat{\mathbf{z}} \times \mathbf{k}' = \hat{\mathbf{z}} \times \mathbf{k}'', \tag{2.51}$$

and the fields satisfy the following constitutive relations:

$$\begin{aligned} \mathbf{D} &= \varepsilon\mathbf{E} \\ \mathbf{D}' &= \varepsilon'\mathbf{E}' \\ \mathbf{D}'' &= \varepsilon\mathbf{E}'' \end{aligned}, \qquad \begin{aligned} \mathbf{B} &= \mu\mathbf{H} \\ \mathbf{B}' &= \mu'\mathbf{H}' \\ \mathbf{B}'' &= \mu\mathbf{H}'' \end{aligned}. \tag{2.52}$$

With the fields defined in Eqs. (2.49) and (2.50), Maxwell's equations (2.37) for the three waves reduce to

$$\begin{aligned} \mathbf{k} \times \mathbf{E}_0 &= \omega\mathbf{B}_0 \\ \mathbf{k}\cdot\mathbf{D}_0 &= 0 \\ \mathbf{k} \times \mathbf{H}_0 &= -\omega\mathbf{D}_0 \\ \mathbf{k}\cdot\mathbf{B}_0 &= 0 \end{aligned}, \qquad \begin{aligned} \mathbf{k}' \times \mathbf{E}_0' &= \omega\mathbf{B}_0' \\ \mathbf{k}'\cdot\mathbf{D}_0' &= 0 \\ \mathbf{k}' \times \mathbf{H}_0' &= -\omega\mathbf{D}_0' \\ \mathbf{k}'\cdot\mathbf{B}_0' &= 0 \end{aligned}, \qquad \begin{aligned} \mathbf{k}'' \times \mathbf{E}_0'' &= \omega\mathbf{B}_0'' \\ \mathbf{k}''\cdot\mathbf{D}_0'' &= 0 \\ \mathbf{k}'' \times \mathbf{H}_0'' &= -\omega\mathbf{D}_0'' \\ \mathbf{k}''\cdot\mathbf{B}_0'' &= 0 \end{aligned}, \tag{2.53}$$

and the boundary conditions (2.48) can be restated as follows:

$$\begin{aligned}(\mathbf{E}_0+\mathbf{E}_0''-\mathbf{E}_0')\times\hat{\mathbf{z}}&=0\\ [\varepsilon(\mathbf{E}_0+\mathbf{E}_0'')-\varepsilon'\mathbf{E}_0']\cdot\hat{\mathbf{z}}&=0\\ (\mathbf{k}\times\mathbf{E}_0+\mathbf{k}''\times\mathbf{E}_0''-\mathbf{k}'\times\mathbf{E}_0')\cdot\hat{\mathbf{z}}&=0\,,\\ \left[\frac{1}{\mu}(\mathbf{k}\times\mathbf{E}_0+\mathbf{k}''\times\mathbf{E}_0'')-\frac{1}{\mu'}\mathbf{k}'\times\mathbf{E}_0'\right]\times\hat{\mathbf{z}}&=0\end{aligned} \tag{2.54}$$

where constitutive relations (2.52) have been used to eliminate all fields other than **E**.

Equations (2.53) and (2.52) yield the following conditions for the k-vectors:

$$\mathbf{k}^2=\mu\varepsilon\omega^2=n^2\frac{\omega^2}{c^2},\qquad \mathbf{k}'^2=\mu\varepsilon\omega^2=n'^2\frac{\omega^2}{c^2},\qquad \mathbf{k}''^2=\mu\varepsilon\omega^2=n^2\frac{\omega^2}{c^2}, \tag{2.55}$$

where c is the speed of light in vacuum, whereas n and n' are refractive indices of the respective materials. With Eqs. (2.51) and (2.55) and the Pythagorean theorem $|\hat{\mathbf{z}}\times\mathbf{k}'|^2+|\hat{\mathbf{z}}\cdot\mathbf{k}'|^2=\mathbf{k}'^2$, the TIR condition $\sin\theta_i\geq n'/n$ can be understood as such that the incidence angle θ_i requires a purely imaginary component of $\mathbf{k}'$ normal to the dielectric boundary. According to Eq. (2.50), with imaginary k_z', the electromagnetic field in the second dielectric decays exponentially

$$\mathbf{E}'(t,\mathbf{r})=\mathbf{E}_0'e^{j(\omega t-k_x'x-k_y'y)}e^{-\beta z}, \tag{2.56}$$

where $\beta=jk_z'$ is real and positive to ensure that the field is finite as $z\rightarrow\infty$.

Fields propagating in the first material are stitched at the dielectric boundary with the evanescent fields in the second material using boundary conditions (2.54). To this end, it is convenient to consider two cases separately: one with the incident wave polarized in the x-direction, or perpendicularly to the incidence plane defined by the incidence k-vector and the surface normal, and the second case in which the wave is polarized parallel to the incidence plane. In the first case, which is referred to as TE (for "transverse electric"), the first and last of Eqs. (2.54) reduce to

$$\begin{aligned}E_{0x}+E_{0x}''&=E_{0x}'\\ \frac{1}{\mu}(k_zE_{0x}-k_zE_{0x}'')&=\frac{1}{\mu'}k_z'E_{0x}'\end{aligned}\,, \tag{2.57}$$

where the vector identity $(\mathbf{A}\times\mathbf{B})\times\mathbf{C}=(\mathbf{A}\cdot\mathbf{C})\mathbf{B}-(\mathbf{B}\cdot\mathbf{C})\mathbf{A}$ has been used.[2] By substituting E_{0x}' from the first equation into the second one and solving for E'', the following equation for the reflected amplitude is obtained:

$$E_{0x}''=\left(\frac{k_z}{\mu}-\frac{k_z'}{\mu'}\right)\left(\frac{k_z}{\mu}+\frac{k_z'}{\mu'}\right)^{-1}E_{0x}. \tag{2.58}$$

[2]It can be shown that if the incident wave has only the x-component of the electric field, then so do the reflected and transmitted waves.

With $k'_z = -j\beta$, it becomes

$$E''_{0x} = \left(\frac{k_z}{\mu} + j\frac{\beta}{\mu'}\right)\left(\frac{k_z}{\mu} - j\frac{\beta}{\mu'}\right)^{-1} E_{0x}. \tag{2.59}$$

In Eq. (2.59), the field amplitude of the reflected wave is obtained from the amplitude of the incident wave by multiplying it by a ratio of two mutually conjugate complex numbers. Such a ratio is simply a phase factor, which means that in total internal reflection, the wave picks up a phase factor $\exp(2j\phi)$ on reflection from the boundary with a lower index material. The phase angle 2ϕ is given by

$$\tan(\phi) = \frac{\beta/\mu'}{k_z/\mu}, \tag{2.60}$$

which at optical frequencies, where the magnetic response of materials can be neglected, becomes

$$\tan(\phi) = \frac{\beta}{k_z}. \tag{2.61}$$

Thus, the total field in the first medium is

$$\begin{aligned}\mathbf{E}(t, \mathbf{r}) + \mathbf{E}''(t, \mathbf{r}) &= \mathbf{E}_0 e^{j(\omega t - \mathbf{k}\cdot\mathbf{r})} + \mathbf{E}''_0 e^{j(\omega t - \mathbf{k}''\cdot\mathbf{r})} \\ &= \mathbf{E}_0 e^{j(\omega t - \mathbf{k}\cdot\mathbf{r})} + \mathbf{E}_0 e^{2j\phi} e^{j(\omega t - \mathbf{k}''\cdot\mathbf{r})} \\ &= [e^{-jk_z z} e^{-j\phi} + e^{jk_z} e^{j\phi}]\mathbf{E}_0 e^{j\phi} e^{j(\omega t - k_x x - k_y y)} \\ &= 2\mathbf{E}_0 e^{j\phi} \cos(k_z z + \phi) e^{j(\omega t - k_x x - k_y y)}.\end{aligned} \tag{2.62}$$

The overall phase factor $e^{j\phi}$ can be absorbed into the field amplitude $\mathbf{E}_0$. As a result, the total field in the high-index material ($z < 0$) is

$$\mathbf{E}(t, \mathbf{r}) + \mathbf{E}''(t, \mathbf{r}) = 2\mathbf{E}_0 \cos(k_z z + \phi) e^{j(\omega t - k_x x - k_y y)}. \tag{2.63}$$

In the low-index medium ($z > 0$), the field, which was obtained from Eq. (2.56) and the first of boundary conditions (2.57), falls exponentially in the z direction

$$\mathbf{E}'(t, \mathbf{r}) = 2\mathbf{E}_0 \cos(\phi) e^{j(\omega t - k'_x x - k'_y y)} e^{-\beta z}. \tag{2.64}$$

Figure 2.5 shows the field profile along the z-direction. Note that at the dielectric boundary, the field is continuous and smooth (the first derivative along the z direction is continuous). The latter assertion can be verified by evaluating the first derivative of

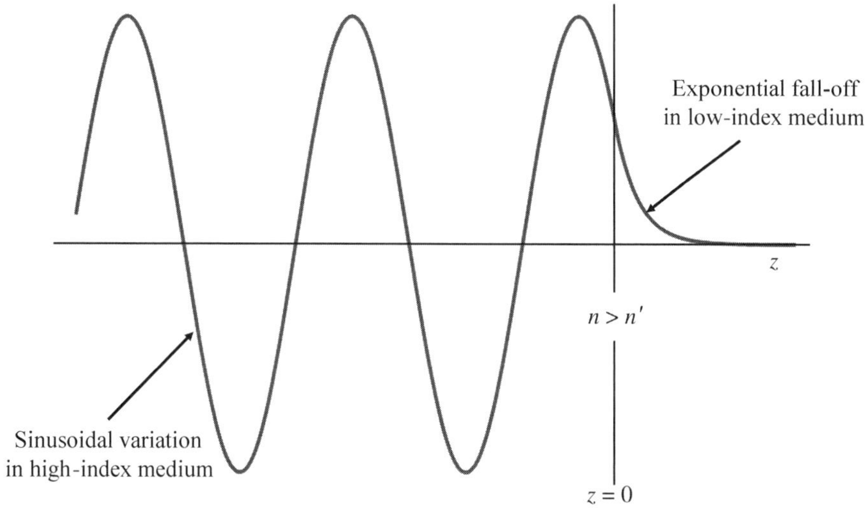

Figure 2.5 Profile of instantaneous electric field amplitude corresponding to the TE wave undergoing total internal reflection at a dielectric boundary.

expressions (2.63) and (2.64) in the z direction at $z = 0$ and by taking the ratio of the numerical components, i.e.,

$$\left.\frac{\partial[\mathbf{E}(t,\mathbf{r})+\mathbf{E}''(t,\mathbf{r})]}{\partial z}\right|_{z=0} = 2\mathbf{E}_0 k_z(-\sin\phi)e^{j(\omega t - k_x x - k_y y)}, \tag{2.65}$$

$$\left.\frac{\partial\mathbf{E}'(t,\mathbf{r})}{\partial z}\right|_{z=0} = 2\mathbf{E}_0\cos(\phi)e^{j(\omega t - k'_x x - k'_y y)}(-\beta), \tag{2.66}$$

$$\frac{2k_z(-\sin\phi)e^{j(\omega t - k_x x - k_y y)}}{2\cos\phi\, e^{j(\omega t - k_x x - k_y y)}(-\beta)} = \frac{k_z}{\beta}\tan\phi = 1, \tag{2.67}$$

where the last equality in Eq. (2.67) was obtained from Eq. (2.61). In other words,

$$\left.\frac{\partial[\mathbf{E}(t,\mathbf{r})+\mathbf{E}''(t,\mathbf{r})]}{\partial z}\right|_{z=0} = \left.\frac{\partial\mathbf{E}(t,\mathbf{r})}{\partial z}\right|_{z=0}, \tag{2.68}$$

which means that the field is smooth at the dielectric boundary.

Now, if the half-space of high-index material is replaced with a finite-thickness slab, and the second dielectric boundary is placed such a distance h from the first one that it falls precisely at the same phase point of the field profile as the first dielectric boundary, then by symmetry, the field in the low-index region on the other side of the slab will follow the same exponential decay as one calculated above. This situation is illustrated in Fig. 2.6. Accordingly, this condition can be stated as follows:

$$k_z h = 2\pi m + 2\phi, \tag{2.69}$$

where m is an integer.

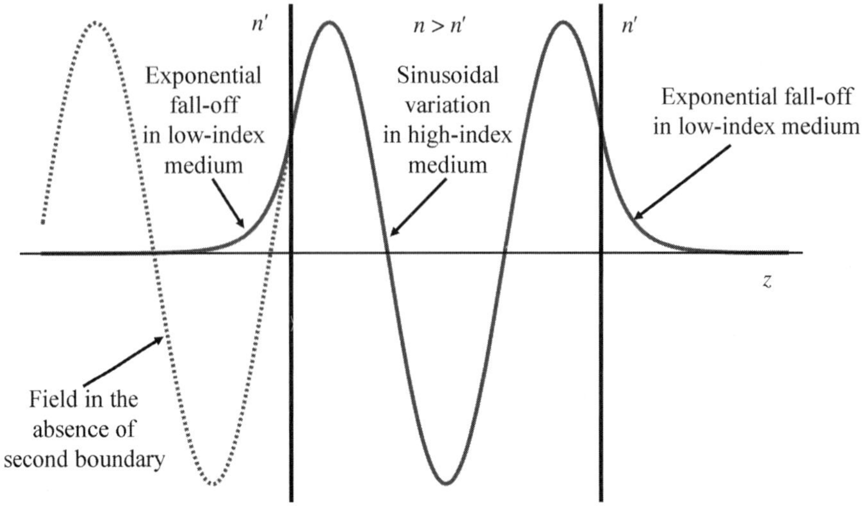

Figure 2.6 Waveguiding in a dielectric slab. High-index material confines the propagating electromagnetic wave laterally by total internal reflection.

Equation (2.69) can be understood as a condition for slab thickness h given k_z, or as a condition for k_z given slab thickness h. In the latter case, it is clear that only a discrete set of k_z for the waves guided in the slab at any given frequency ω is allowed. Because the magnitude of k_z is bounded for waves propagating in the slab, $k_z < k = n(\omega/c)$, only a finite number of propagating modes exist. To proceed further, we first solve Eq. (2.69) for ϕ, and then take the tangent of both sides of the equation

$$\begin{aligned} \phi &= \frac{k_z h}{2} - \pi m \\ \tan\phi &= \tan\left(\frac{k_z h}{2} - \pi m\right) \end{aligned} . \tag{2.70}$$

Because the periodicity of the tangent function is exactly π, i.e., $\tan(x+\pi) = \tan(x)$, the last equation can be simplified to

$$\tan\phi = \tan\left(\frac{k_z h}{2}\right). \tag{2.71}$$

Equation (2.61) can be used to substitute for the left-hand side of Eq. (2.71). By using the boundary condition at the dielectric interface, and Eq. (2.55) where $-j\beta$ is substituted for k_z', β is obtained in the form

$$\beta = \frac{\omega}{c}\sqrt{n^2 - n'^2 - \left(\frac{c}{\omega}k_z\right)^2}. \tag{2.72}$$

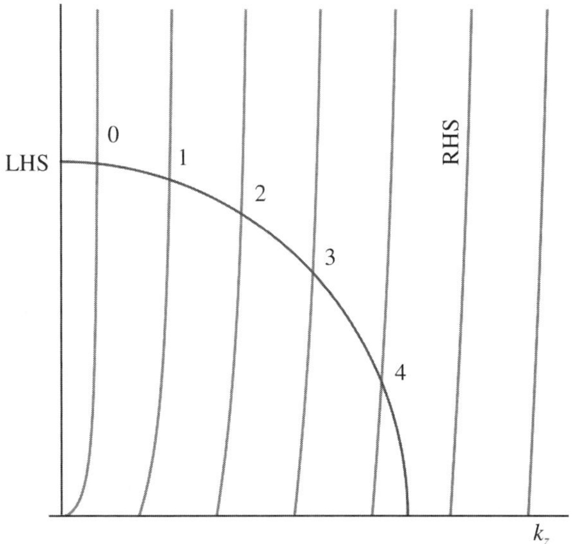

Figure 2.7 Illustration of Eq. (2.73). The circular arc shows the evaluation of the left-hand side (LHS), whereas the nearly vertical curves show the evaluation of the right-hand side (RHS) as a function of k_z. The intersection points indicate possible solutions. In this example, five modes, which are numbered from 0 to 4, are possible for a chosen set of slab parameters n, n', h, and frequency ω.

As a result, the guiding condition (2.71) takes the form

$$\frac{\omega}{c}\sqrt{n^2 - n'^2 - \left(\frac{c}{\omega}k_z\right)^2} = k_z \tan\left(\frac{k_z h}{2}\right), \tag{2.73}$$

which is an equation to be solved for k_z. Because Eq. (2.73) is transcendental, no analytical solution exists, and numerical methods must be employed to obtain allowed values of k_z. Figure 2.7 illustrates Eq. (2.73) by showing the left-hand side and the right-hand side as a function of k_z.

Once k_z is found, the magnitude of the k-vector in the direction of propagation, or the propagation constant k_{slab}, is obtained according to Eq. (2.55),

$$k_{\text{slab}} = \sqrt{\left(n\frac{\omega}{c}\right)^2 - k_z^2}. \tag{2.74}$$

Because k_z is a function of frequency ω, as a result of Eq. (2.73), k_{slab} and ω are related via functional dependence provided by Eq. (2.74). By inverting this relation, frequency ω as a function of k_{slab} is obtained, which is the *dispersion relation* for modes propagating in the slab. As an example, a dispersion curve for the fundamental

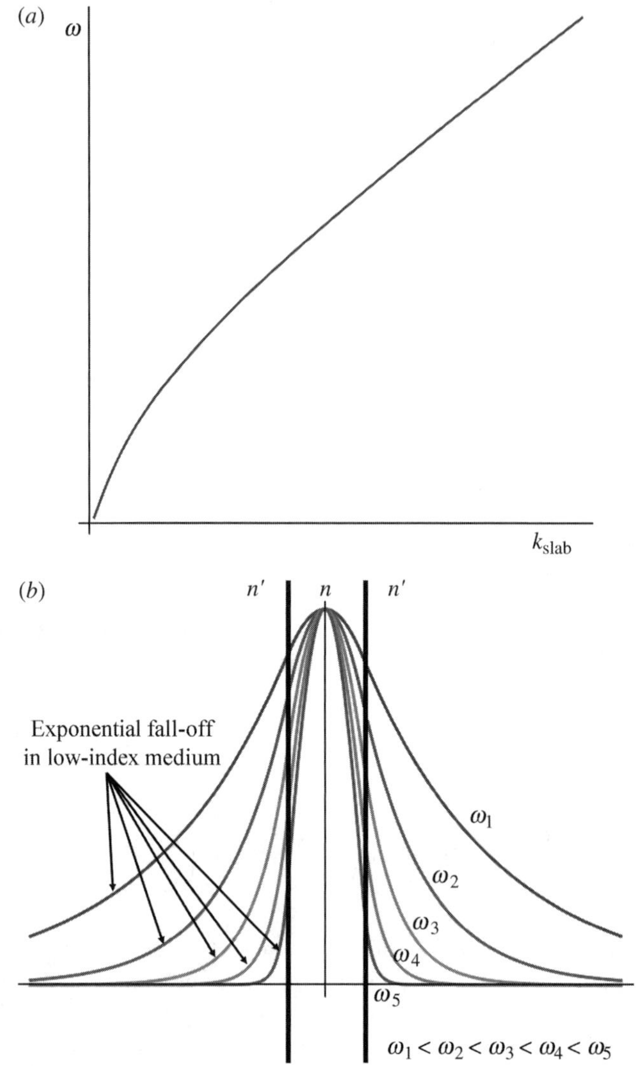

Figure 2.8 (a) Dispersion curve for a fundamental TE mode of a slab waveguide. The initially low effective index [steeper curve at low frequencies—see Eq. (2.75)] is the manifestation of the electromagnetic field extension to the low-index region of the cladding as shown in (b). At sufficiently high frequency, the effective index of the slab approaches that of bulk material because most of the field is concentrated in the high-index material with diminishingly small evanescent tails extending out.

mode of a slab waveguide is shown in Fig. 2.8a, and the mode profiles for different frequencies are shown in Fig. 2.8b. The effective index of a slab

$$n_{\text{eff}} = \frac{k_{\text{slab}}}{\omega/c}, \tag{2.75}$$

is within the range $n' < n_{\text{eff}} < n$.

A similar procedure can be applied to the TM mode, in which the wave is polarized in the plane of incidence to obtain a corresponding dispersion relation.

2.1.5 Group Velocity in Homogeneous Media

Group velocity is the measure of how fast a wave packet traverses the medium. In general, it is different from phase velocity, which is concerned only with the propagation of plane waves. Because we know that phase velocity is well defined, we choose wave packets that deviate from plane waves as little as possible. In principle, it is not possible because plane waves occupy all space and have constant amplitude throughout, whereas for a wave packet, we typically require spatial localization. In practical terms, we choose a wave that is well localized in the spatial frequency space, i.e., the Fourier domain. This means that in the real space, it is going to be delocalized and can therefore be expressed approximately as the carrier wave corresponding to the k-vector around which the wave is localized in the spatial frequency space, multiplied by an envelope function, which is the slow undulation of the carrier wave amplitude. Then, the group velocity is the velocity at which the envelope function propagates through the real space.

Turning into mathematical details, the starting point is the observation that any wave packet can be decomposed into a superposition of plane waves

$$\Psi(\mathbf{r}, t) = \int d\mathbf{k}\tilde{\Psi}(\mathbf{k}) \exp[j(\omega(\mathbf{k})t - \mathbf{k} \cdot \mathbf{r})], \tag{2.76}$$

where $\omega(\mathbf{k})$ represents the dispersion relation of the medium, i.e., the dependence of the frequency on the wave vector. If the spatial-frequency distribution $\tilde{\Psi}(\mathbf{k})$ of the wave packet $\Psi(\mathbf{r}, t)$ is narrowly concentrated around a central wave vector $\mathbf{k}_0$, then the function $\omega(\mathbf{k})$ can be Taylor-expanded around $\mathbf{k}_0$, and the lowest order term accurately describes the behavior of $\omega(\mathbf{k})$

$$\omega(\mathbf{k}) = \omega_0 + \left(\nabla_k \omega|_{\mathbf{k}=\mathbf{k}_0}\right) \cdot (\mathbf{k} - \mathbf{k}_0) + O(\mathbf{k} - \mathbf{k}_0)^2, \tag{2.77}$$

where $\omega_0 = \omega(\mathbf{k}_0)$ and $O(\mathbf{k} - \mathbf{k}_0)^2$ represents higher order terms of $(\mathbf{k} - \mathbf{k}_0)$. With the help of the expansion in Eq. (2.77), the Eq. (2.76) can be rewritten as

$$\Psi(\mathbf{r}, t) = \exp[j(\omega_0 t - \mathbf{k}_0 \cdot \mathbf{r})] \times \int d\mathbf{k}\tilde{\Psi}(\mathbf{k}) \exp\left[j\left(\left(\nabla_k \omega|_{\mathbf{k}=\mathbf{k}_0}\right)t - \mathbf{r}\right) \cdot (\mathbf{k} - \mathbf{k}_0) + jO(\mathbf{k} - \mathbf{k}_0)^2 t\right],$$

and further as

$$\Psi(\mathbf{r}, t) = \exp[j(\omega_0 t - \mathbf{k}_0 \cdot \mathbf{r})] \times \int d\mathbf{k}\tilde{\Psi}(\mathbf{k}) \left[1 + j\left(\left(\nabla_k \omega|_{\mathbf{k}=\mathbf{k}_0}\right)t - \mathbf{r}\right) \cdot (\mathbf{k} - \mathbf{k}_0) + jO(\mathbf{k} - \mathbf{k}_0)^2 t\right], \tag{2.78}$$

where we Taylor-expanded the exponential function under the integral. In Eq. (2.78), we can identify a high-frequency part—the carrier plane wave $\exp[j(\omega_0 t - \mathbf{k}_0 \cdot \mathbf{r})]$ before the integral, and the envelope function

$$A(\mathbf{r}, t) = \int d\mathbf{k}\tilde{\Psi}(\mathbf{k})\Big[1 + j\Big(\Big(\nabla_k \omega|_{\mathbf{k}=\mathbf{k}_0}\Big)t - \mathbf{r}\Big) \cdot (\mathbf{k} - \mathbf{k}_0) + jO(\mathbf{k} - \mathbf{k}_0)^2 t\Big]. \quad (2.79)$$

Group velocity is defined as the velocity of the envelope function of a wave packet. As time progresses, the envelope will shift in real space. This shift can be represented by introducing a time dependence of the spatial argument of the envelope $A(\mathbf{r}, t) = A[\mathbf{r}(t), t]$. As a result, the total time derivative of the envelope is

$$\frac{dA}{dt} = (\nabla_r A)\frac{d\mathbf{r}}{dt} + \frac{\partial A}{\partial t} = \int d\mathbf{k}\tilde{\Psi}(\mathbf{k})\Big[j\Big(\Big(\nabla_k \omega|_{\mathbf{k}=\mathbf{k}_0}\Big) - \dot{\mathbf{r}}\Big) \cdot (\mathbf{k} - \mathbf{k}_0) + jO(\mathbf{k} - \mathbf{k}_0)^2\Big], \quad (2.80)$$

where $\dot{\mathbf{r}} = d\mathbf{r}/dt$.

Movement in space of the envelope function means that its shape remains unchanged and only its position changes. For the envelope to remain unchanged and given the arbitrariness of $\tilde{\Psi}(\mathbf{k})$, it is necessary that the expression in square brackets in Eq. (2.80) vanishes. To the first order in $(\mathbf{k} - \mathbf{k}_0)$, it means that $(\nabla_k \omega|_{\mathbf{k}=\mathbf{k}_0}) = \dot{\mathbf{r}}$ or, in other words, that the group velocity is

$$\mathbf{v}_g(\mathbf{k}_0) = \nabla_k \omega|_{\mathbf{k}=\mathbf{k}_0}. \quad (2.81)$$

2.2 PERIODIC MEDIA

The propagation of waves through a medium with a periodic modulation of material properties on the scale of the wavelength is often profoundly different from that of the homogeneous case. This fact is well known in solid-state physics, in which the periodicity in the electrostatic potential associated with atoms in crystalline solids is responsible for the formation of energy bands and energy gaps of electronic states in metals, semiconductors, and insulators.

Electromagnetic waves that propagate through a periodic dielectric medium may exhibit many interesting and potentially useful phenomena. Examples include diffraction of light from the periodic strain accompanying an acoustic wave and the diffraction of X-rays. These phenomena have been used in many optical devices, such as diffraction gratings, holograms, distributed-feedback (DFB) lasers, distributed-Bragg reflector (DBR) lasers, high-reflectance Bragg mirrors, and acousto-optic modulators. To understand these phenomena and to take advantage of them to realize novel and useful devices, we will require a framework in which to describe the unique and important properties of periodic media concisely. This framework is derived largely from the mature field of solid-state physics and is used extensively throughout this book. Hence, the role of this section is to present the framework for readers who may not be familiar with solid-state physics.

2.2.1 Real-Space Lattices, Lattice Vectors

An ideal crystal can be defined as the regular repetition of a basic building block of objects throughout all space. This building block is commonly referred to as a *unit cell.* Such ideal crystals may occupy one-, two-, or three-dimensional spaces. A collection of points in the space such that a single point is associated with each single unit cell is the called a *space lattice*. In three dimensions, a lattice is defined by three fundamental translation vectors $\mathbf{a}_1$, $\mathbf{a}_2$, and $\mathbf{a}_3$ so that arrangement of objects in an ideal crystal looks identical when observed from any point $\mathbf{r}$ and when it is observed from another point $\mathbf{r}'$ given by

$$\mathbf{r}' = \mathbf{r} + m_1\mathbf{a}_1 + m_2\mathbf{a}_2 + m_3\mathbf{a}_3 = \mathbf{r} + \mathbf{R}, \tag{2.82}$$

where m_1, m_2, and m_3 are any arbitrary integers. The translation vectors $\mathbf{R}$ that relate $\mathbf{r}$ to $\mathbf{r}'$ in this expression are called *lattice vectors*. A major value of the notion of a unit cell obtains from Eq. (2.82), namely that any arbitrary point in the crystal structure can be considered identical to a corresponding point within a chosen unit cell, differing only by a lattice vector. This means that a complete analysis of a crystal may be performed by considering only those points within the single unit cell.

2.2.1.1 Primitive Lattice Vectors and Primitive Cells. It should be noted that for any given crystal, there are many suitable choices of the unit cell and, likewise, many potential choices for the set of fundamental translation vectors. A useful distinction among the choices is provided by the notion of *primitive translation vectors*. Any set of translation vectors $\{\mathbf{a}_1, \mathbf{a}_2, \mathbf{a}_3\}$ is said to be a *primitive* set if any two lattice points $\mathbf{r}$ and $\mathbf{r}'$ can be connected by Eq. (2.82) for a suitable choice of integers $\{m_1, m_2, m_3\}$. On the contrary, if any pair of lattice points exists for which no set of integers can connect them via Eq. (2.82), then the set of translation vectors is not a primitive set. A primitive set of translation vectors defines the volume of the smallest possible unit cell that can be used to describe a crystal. Such a unit cell, which is the so-called *primitive cell*, can be constructed by taking the primitive translation vectors as the edges of a general parallelepiped (or parallelogram in 2D). The primitive-cell volume is thus given by

$$V = |\mathbf{a}_1 \cdot \mathbf{a}_2 \times \mathbf{a}_3|. \tag{2.83}$$

Note, however, that a primitive cell thus specified by the primitive translation vectors is not unique; indeed, many primitive cells can be defined for any crystal, even for the same choice of primitive lattice vectors. Let us consider some simple examples in two dimensions to illustrate.

Figure 2.9a depicts a square lattice in two dimensions. A pair of fundamental translation vectors $\mathbf{a}_1$, $\mathbf{a}_2$ is indicated, along with the corresponding unit cell (shaded region). This lattice is square because $|\mathbf{a}_1| = |\mathbf{a}_2|$, and $\mathbf{a}_1 \perp \mathbf{a}_2$. It is clear

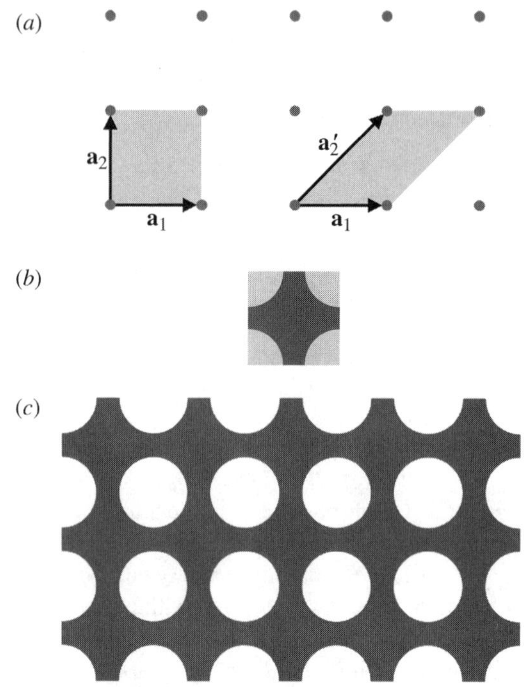

Figure 2.9 (a) A two-dimensional square lattice, with two choices for the primitive translation vectors and corresponding primitive cells (shaded regions). (b) Example of a basis suitable for the square lattice in (a), with shading to indicate the bounds of the primitive cell. (c) Several periods of the crystal structure resulting from the combination of the lattice in (a) with the basis in (b).

by inspection that any lattice point can be related to any other by this choice of fundamental translation vectors, and that these are therefore primitive translation vectors. Another possible choice of primitive translation vectors is also shown by the pair of vectors $\mathbf{a}_1$, $\mathbf{a}'_2$. The corresponding unit cell is in this case a parallelogram. Note that for both choices, the volume of the unit cell (area in 2D) is the same, and that both unit cells fill the entire space when repeated in increments of the respective primitive translation vectors.

The lattice alone is a mere abstraction used to describe certain crystals; to specify a crystal fully, the contents of the unit cell, or *basis*, must also be specified.[3] Figure 2.9b depicts an example of the unit cell with a simple basis, in which the dark shaded region might represent a dielectric material, such as silicon or glass, whereas the remaining area might be vacuum, air, or some other material. Figure 2.9c shows what obtains when this basis is repeated at every lattice point.

[3]Note that the term "basis" is also commonly used to refer to a set of objects, e.g., vectors or functions, that span a particular space. For example, when writing a Fourier expansion, the Fourier components are plane waves, and the expansion is said to be in a "plane-wave basis." Likewise, a set of translation vectors, such as the primitive vectors defining a lattice and spanning a Cartesian space, is often called a "basis." The usage of a single term for two distinct concepts is unfortunate but driven by convention.

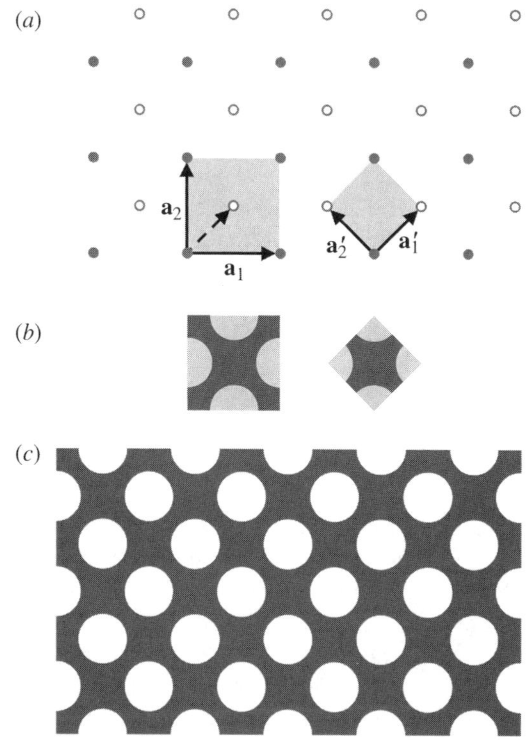

Figure 2.10 (a) A two-dimensional lattice that can be described in two ways: as a square lattice with a two-atom basis (left) or a smaller square lattice, rotated 45° with respect to the page (right). (b) Examples of basis choices for the two descriptions in (a) are superimposed on the respective unit cells (shaded), both of which yield the same final crystal structure, (c).

Next, consider the lattice depicted in Fig. 2.10a. This lattice consists of points indicated by filled circles that replicate the lattice of Fig. 2.9a, along with a second set of points indicated by open circles, each located at the center of a unit cell in the original lattice. This combined lattice can be described by the same fundamental translation vectors $\mathbf{a}_1$ and $\mathbf{a}_2$, which are indicated within the lattice to the left. However, if we consider the open lattice points and filled lattice points to be identical, then these translation vectors are not primitive, as shown by the dashed arrow that connects two lattice points that cannot be related by Eq. (2.82) for any choice of integers m. Within the lattice to the right is a different choice of fundamental translation vectors $\mathbf{a}_1'$ and $\mathbf{a}_2'$ that are primitive. From this choice and the corresponding shaded unit cell, we can see that this lattice (when open and filled points are identical) is also a square lattice, smaller than the original and rotated by 45°. Figure 2.10b depicts an example of basis choices that, along with the lattice, would define a crystal. The left and right depictions correspond, respectively, to the choices of fundamental translation vectors directly above them. In each, the corresponding combination of fundamental translation vectors and basis results in the same crystal, which is depicted in Fig. 2.10c. This example serves as an illustration of the fact that a single crystal can be described in more than one way.

Now let us consider the lattice of Fig. 2.10a in the case when the open points and filled points are not identical. In this case, the translation vectors $\mathbf{a}'_1$ and $\mathbf{a}'_2$ are not fundamental translation vectors at all; the crystal will not appear the same from all points connected by Eq. (2.82) using these translation vectors. This is an example of a crystal with the same lattice as the original case considered in Fig. 2.9, but with a more complicated basis, which is characterized by two nonidentical "atoms" as illustrated in Fig. 2.11. The original vectors $\mathbf{a}_1$ and $\mathbf{a}_2$ are again a primitive set in this case. The two distinct "atoms" that comprise the basis are indicated in Fig. 2.11a. When a crystal is formed according to the lattice of Fig. 2.10a, with these atoms as the basis, the structure depicted in Fig. 2.11b is obtained. The shaded rectangular regions in (b) represent several possible choices for primitive cells in this crystal: one (I) centered on the filled lattice points and another (II) centered on the open lattice points. The third (III) and fourth (IV) are unit cells centered on neither; however, they make more readily apparent the two-atom basis of this crystal. In addition, regions III and IV lack the obvious rotation and mirror symmetries of the other two choices shown. This example illustrates some other factors to consider when describing a crystal structure, namely the ease of identifying the structure and symmetry of the basis.

The various types of crystal structures can be organized and classified according to their symmetry, i.e., operations such as translations, rotations, and reflections about certain points that leave the ideal crystal unchanged. Numerous classification schemes are available, a complete description of them lies within the scope of

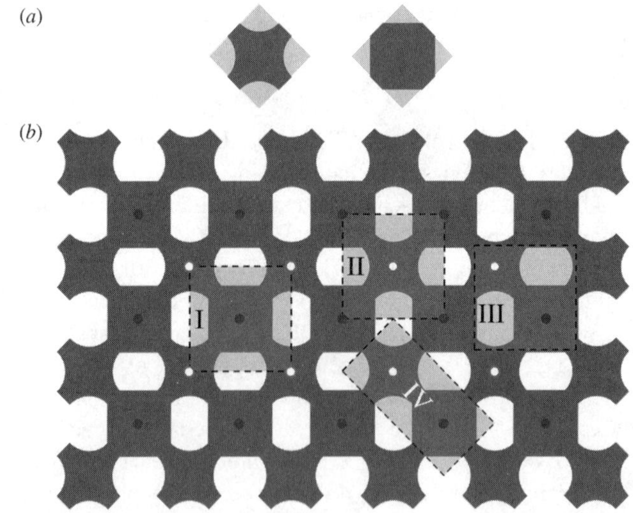

Figure 2.11 (a) Two different "atoms" that can be used to fill the lattice cells depicted in Fig. 2.10, for the case when the open points and filled points are not identical. (b) Resulting crystal structure, along with four different, equally valid choices for the primitive cell of this crystal (I–IV). Note that all four primitive cells fill the entire 2D space when repeated over integer multiples of the same pair of fundamental translation vectors.

crystallography and involves elements of group theory. This discussion is beyond the scope of this book (see Refs. 3 and 4).

There is a systematic method for constructing a primitive cell that is worth examining. The procedure is illustrated in Fig. 2.12 for three different 2D lattices: square ($|\mathbf{a}_1| = |\mathbf{a}_2|$ and $\mathbf{a}_1 \perp \mathbf{a}_2$), triangular ($|\mathbf{a}_1| = |\mathbf{a}_2|$ and $\angle \mathbf{a}_1, \mathbf{a}_2 = 60°$ or $120°$), and general oblique ($|\mathbf{a}_1| \neq |\mathbf{a}_2|$ and $\angle \mathbf{a}_1, \mathbf{a}_2$ arbitrary). The construction proceeds as follows (in 2D): First, a lattice point is chosen as the origin. Then, the origin point is connected to its nearest neighboring lattice points by straight lines (solid lines in Fig. 2.12). Next, perpendicular lines are drawn from the midpoints of these connecting lines, which are extended until they intersect one another (dashed lines in Fig. 2.12). The region enclosed by the intersecting perpendiculars is a primitive cell, which is known as the *Wigner-Seitz cell.* The procedure for constructing the Wigner-Seitz cell of a 3D lattice is the same as for 2D lattices, except that the perpendicular lines must be replaced with planes perpendicular to the nearest-neighbor connecting lines; hence, the Wigner-Seitz cell becomes the volume enclosed by the intersecting planes.

The Wigner-Seitz primitive cell is interesting for several reasons, including: (1) it shares the symmetry of the lattice, (2) it encloses the region containing all points that are closer to the origin lattice point than to any other lattice point, and most importantly,

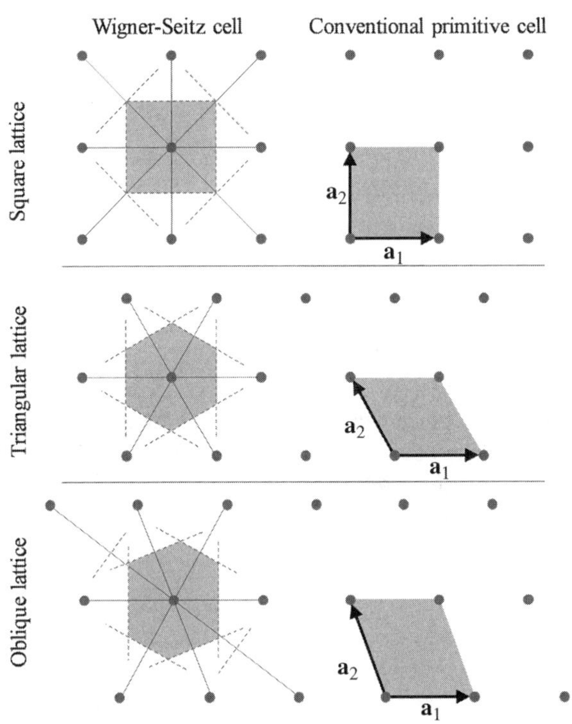

Figure 2.12 Illustration of the procedure for constructing the Wigner-Seitz primitive cells for a selection of 2D lattices. (The procedure is described in the text.) Also shown for each lattice is a conventional primitive cell, as defined by the primitive lattice vectors indicated, for comparison.

(3) as we shall see in Section 2.2.2, the Wigner-Seitz cell in reciprocal space, which is known as the Brillouin zone, has special significance.

2.2.1.2 Common Lattice Types and Crystal Structures in Two and Three Dimensions. In this section, we briefly examine the features of the lattices most commonly found in photonic crystals. We begin with two of the two-dimensional lattices illustrated in Fig. 2.12, namely the square lattice and the triangular (or hexagonal) lattice. Any 2D lattice can be characterized by the relative lengths of, and the angle between, its two primitive translation vectors. The square lattice has $|\mathbf{a}_1| = |\mathbf{a}_2|$ and $\mathbf{a}_1 \perp \mathbf{a}_2$, i.e., $\angle \mathbf{a}_1, \mathbf{a}_2 = (2n - 1)\pi/2$, where n is any integer (odd multiples of 90°). Its Wigner-Seitz cell is a square. The triangular (or hexagonal) lattice has $|\mathbf{a}_1| = |\mathbf{a}_2|$ and $\angle \mathbf{a}_1, \mathbf{a}_2 = (3n \pm 1)\pi/3$. Its Wigner-Seitz cell is a regular hexagon. An important crystal structure derived from the hexagonal lattice is the so-called graphite lattice, which possesses a two-atom basis in a hexagonal lattice. The second atom is shared among three corners of the Wigner-Seitz hexagon; i.e., it is centered within one equilateral triangle defined by three mutual-nearest-neighbor lattice points. Hence, this lattice is composed of one atom at each vertex of an array of regular hexagons.

Three other distinct types of 2D lattice are available for a total of five such distinct lattices in two dimensions. Each distinct lattice, which is termed a *Bravais lattice*, either possesses certain symmetries that make it distinct or, in one case, lacks any symmetry at all. The three remaining 2D Bravais lattices are as follows: rectangular ($|\mathbf{a}_1| \neq |\mathbf{a}_2|$, $\mathbf{a}_1 \perp \mathbf{a}_2$), centered rectangular ($|\mathbf{a}_1| \neq |\mathbf{a}_2|$, $|\mathbf{a}_1 \cdot \mathbf{a}_2| = a_1/2$), and oblique (no symmetry). The translation vectors are assumed to be primitive, and $a_1 > a_2$. These other 2D lattices are not typically of interest in the field of photonic crystals.

In three dimensions, there are 14 Bravais lattices, but we shall consider only the four that are of greatest interest in photonic crystals: simple cubic (SC), body-centered cubic (BCC), face-centered cubic (FCC), and diamond.

The simple cubic structure has an atom located at each corner of a cube. Each atom is shared among eight cubic unit cells, and there are eight corners per cube, hence, one atom per unit cell. This cube is identical in size and orientation to the Wigner-Seitz cell, which has one atom at its center. Hence, it differs from the conventional unit cell by a displacement of half a unit along each cube axis. The dimension of this cube a is called the lattice constant or lattice parameter. The volume of the simple cubic primitive cell is a^3. The conventional choice for the primitive vectors, in terms of a Cartesian system aligned with the unit cell, is simply $\mathbf{a}_1 = a\hat{\mathbf{x}}$, $\mathbf{a}_2 = a\hat{\mathbf{y}}$, and $\mathbf{a}_3 = a\hat{\mathbf{z}}$.

The body-centered-cubic structure adds a second atom to the simple cubic unit cell, which is located at the center of the cube; this defines the "conventional" BCC unit cell, which can be equivalently viewed as a simple cubic lattice with a two-atom basis. But the BCC lattice is a distinct Bravais lattice, and the primitive cell has a volume of one half the volume of the cubic conventional unit cell, i.e., $\frac{1}{2}a^3$. The BCC Wigner-Seitz cell is a truncated regular octahedron. It is composed of six squares lying centered on the faces of the conventional cubic cell, rotated such that their sides are parallel to the face diagonals, along with eight hexagons, two of which are oriented normal to each of the four body diagonals of the cubic cell. The most common conventional choice for the primitive vectors of the BCC lattice, in terms of a Cartesian system aligned to the cubic unit cell, is $\mathbf{a}_1 = \frac{1}{2}a(\hat{\mathbf{x}} + \hat{\mathbf{y}} - \hat{\mathbf{z}})$, $\mathbf{a}_2 = \frac{1}{2}a(-\hat{\mathbf{x}} + \hat{\mathbf{y}} + \hat{\mathbf{z}})$, and $\mathbf{a}_3 = \frac{1}{2}a(\hat{\mathbf{x}} - \hat{\mathbf{y}} + \hat{\mathbf{z}})$.

The face-centered cubic structure possesses an atom centered on each face of the conventional cubic cell, along with the atoms at the corners it inherits from the SC lattice. Hence, it can be viewed as simple cubic with a four-atom basis. But the FCC lattice is also a distinct Bravais lattice, whose primitive cell has a volume of one fourth the volume of the conventional cube, i.e., $\frac{1}{4}a^3$. The FCC Wigner-Seitz cell is a rhombic dodecahedron composed of 12 rhombuses oriented such that each one faces the midpoint of one of the conventional cube's 12 edges, and is located in the plane that normally bisects a line segment that connects the body center of the cube to the respective edge midpoint. The most common conventional choice for the primitive vectors of the FCC lattice, in terms of a cartesian system aligned to the cubic unit cell, is $\mathbf{a}_1 = \frac{1}{2}a(\hat{\mathbf{x}} + \hat{\mathbf{y}})$, $\mathbf{a}_2 = \frac{1}{2}a(\hat{\mathbf{y}} + \hat{\mathbf{z}})$, and $\mathbf{a}_3 = \frac{1}{2}a(\hat{\mathbf{x}} + \hat{\mathbf{z}})$. The FCC lattice is noteworthy because it has the largest packing density of any 3D Bravais lattice for hard-sphere atoms; this property plays a prominent role in the formation of self-assembled photonic crystals, which are often called synthetic opals (see Chapter 6).

The diamond structure consists of an FCC lattice with a two-atom basis, in which the second atom is displaced from the corresponding atom of the plain FCC lattice by a distance of one fourth the cube dimension in each Cartesian direction, i.e., by $\frac{1}{4}a(\hat{\mathbf{x}} + \hat{\mathbf{y}} + \hat{\mathbf{z}})$. Unlike the other lattices discussed, the diamond structure is not a Bravais lattice; a primitive cell cannot be found such that there is only one atom per cell while still filling the entire space when repeated according to the set of lattice vectors $\mathbf{R}$ defined by all combinations of integers in Eq. (2.82). The diamond lattice is important for photonic crystals, as it has been shown to be the optimum lattice for many useful properties of photonic crystals described in later chapters of this book.

Figure 2.13 contains a summary view of the lattices just discussed, depicting each lattice's conventional unit cell using a basis composed of spherical atoms whose radius are 0.2 times the conventional cube dimension. Also depicted are the Wigner-Seitz cells for the BCC and FCC lattices, along with some notable variations that are obtained when the atoms within the conventional cell are not all identical. These include the *zincblende* structure, which is a diamond lattice where the second atom in the basis is not identical to the first. This structure gets its name from ZnS, but it is the crystal structure of several important compound semiconductor materials, such as GaAs.

2.2.1.3 Miller's Index System for Crystals. It is often necessary to specify various directions in a crystal in a frame of reference that is tied to the crystal lattice. The notion of crystallographic planes, which are defined in general by any three noncollinear lattice points, is significant historically owing to the discovery of X-ray diffraction and the theories of coherent scattering by atomic planes put forth to explain it by Bragg [5], Ewald [6], and Laue [7]. The field of photonic crystals has adopted the same set of conventions for discussing crystalline orientations and symmetry directions. By these conventions, crystalline planes and the directions normal

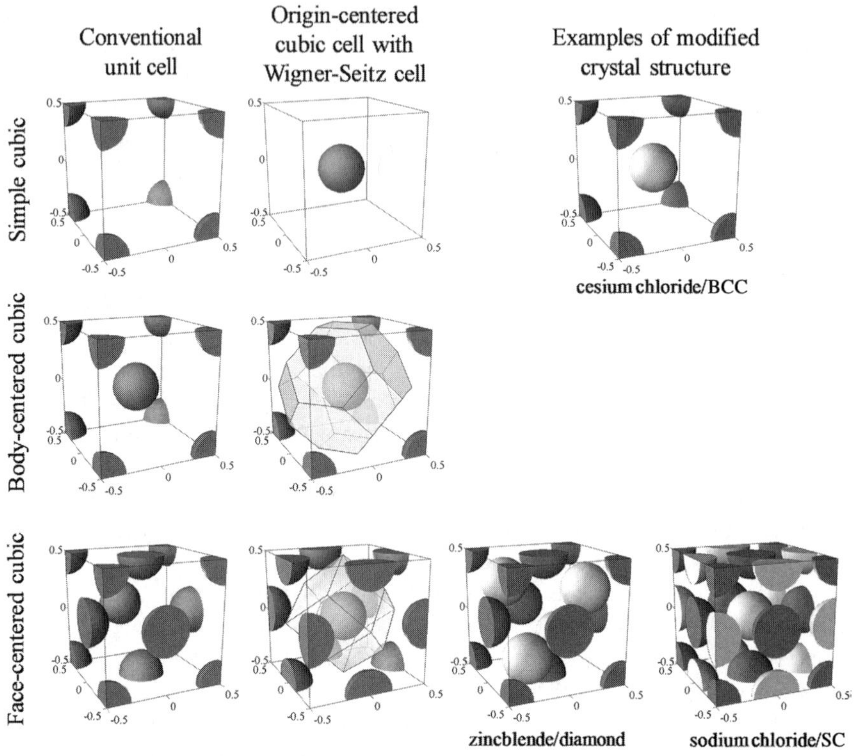

Figure 2.13 Depictions of conventional cubic unit cells for a variety of 3D lattices: simple cubic (SC: top row), body-centered cubic (BCC: middle row), and face-centered cubic (FCC: bottom row). All cells contain spherical atoms of radius 0.2, in units of the cubic cell dimension indicated on the plot axes. The left column contains the conventional unit cell, which contains one, two, and four atoms per cell in the SC, BCC, and FCC case, respectively. The center column shows the alternative cell with its atom at the origin, and the Wigner-Seitz (W-S) primitive cell. Note that the W-S cell for SC is identical to the conventional cell apart from a displacement of the origin. The right column shows some common crystal structures that can be obtained by adding atoms to the conventional SC and FCC unit cells: By adding the lighter atom to the center of the SC cell, we obtain the cesium chloride structure (or simply a BCC cell in the case that lighter and darker atoms are the same). By adding four lighter atoms along the body diagonals of the FCC cell, we obtain either zincblende (if lighter and darker are different) or diamond (if lighter and darker are the same). By adding four lighter atoms to the center and edges of the FCC cell, we obtain the sodium chloride structure (or simply a smaller SC cell if lighter and darker atoms are the same).

to them are specified by sets of numbers called the *Miller indices*. The Miller indices are determined for a given plane as follows:

1. Determine the intercepts b_1, b_2, and b_3 of the plane with the three axes defined by the fundamental translation vectors, $\mathbf{a}_1$, $\mathbf{a}_2$, and $\mathbf{a}_3$, respectively.
2. Take the reciprocals of these intercepts and multiply by a common integer (the least common multiple of b_1, b_2, b_3), so as to obtain the three smallest integers with the same ratio as the reciprocal intercepts.

The resulting three integers, which are often designated by h, k, and l, are the Miller indices of the plane. If for one or more axes no intercept exists, then the plane is parallel to those axes, and the intercept may be taken to be infinity, which yields an index of zero. By convention, the indices are enclosed in parentheses to indicate a plane (hkl) and in square brackets to indicate the direction normal to the plane $[hkl]$. Furthermore, if an intercept is on the negative side of the origin, that index is denoted by a bar above the index: $(\bar{h}kl)$. Finally, families of planes with the same indices in permuted order are collectively denoted by curly braces; i.e., $\{hkl\}$ refers to the set of planes (hkl), (hlk), (khl), (klh), (lhk), and (lkh), along with all combinations that involve negative indices. Likewise, angled brackets are used to denote families of directions, $\langle hkl \rangle$. For example, the planes parallel to the $\mathbf{a}_2$-$\mathbf{a}_3$ plane are indexed as (100). As another example, if the fundamental translation vectors correspond to a cubic unit cell, as is typical for the lattices presented in Section 2.2.1.2, then we can denote the set of face diagonal directions as $\langle 110 \rangle$ and the family of planes normal to the body diagonals as $\{111\}$.

2.2.2 Reciprocal Lattice and Brillouin Zone

Considering the periodicity of crystal structures in space, it should not be surprising that many useful insights can be gained by transforming the structure into the Fourier domain. Fourier analysis decomposes the periodic spatial arrangement of the crystal into its spatial frequency components. Describing a crystal in this way is particularly useful for photonic crystals because the space of spatial frequencies, i.e., k-space, is also the space used to characterize electromagnetic waves according to their wave vector $\mathbf{k}$.

The three-dimensional Fourier decomposition of the periodic dielectric permittivity that defines a photonic crystal can be expressed as

$$\varepsilon(\mathbf{r}) = \sum_{\mathbf{G}} \varepsilon(\mathbf{G}) e^{j\mathbf{G}\cdot\mathbf{r}}, \tag{2.84}$$

where $\mathbf{G}$ is an infinite set of vectors in k-space that define the spatial frequencies contributing to the expansion. The Fourier expansion is thus an expansion in a plane-wave basis, and the expansion coefficients are given by $\varepsilon(\mathbf{G})$. To ensure that the expansion preserves the periodicity of the crystal, we must constrain the set $\mathbf{G}$ by insisting that

$$\varepsilon(\mathbf{r}+\mathbf{R}) = \sum_{\mathbf{G}} \varepsilon(\mathbf{G}) e^{j\mathbf{G}\cdot\mathbf{r}} e^{j\mathbf{G}\cdot\mathbf{R}} = \varepsilon(\mathbf{r}). \tag{2.85}$$

This condition implies that $\mathbf{e}^{j\mathbf{G}\cdot\mathbf{R}} = 1$, and therefore that $\mathbf{G}\cdot\mathbf{R} = 2\pi m$, where m is an integer. Recall the definition of the lattice vectors $\mathbf{R}$, Eq. (2.82). We can satisfy Eq. (2.85) with the following form for the expansion wave vectors:

$$\mathbf{G} = n_1\mathbf{g}_1 + n_2\mathbf{g}_2 + n_3\mathbf{g}_3, \tag{2.86}$$

where the fundamental vectors **g** are chosen such that they are related to the fundamental vectors **a** according to

$$\mathbf{g}_i \cdot \mathbf{a}_j = 2\pi\delta_{ij}, \tag{2.87}$$

with $i, j = 1, 2, 3$, and δ_{ij} is the Kronecker delta function: $\delta_{ij} = 1$ if $i = j$, and 0 if $i \neq j$. The most convenient definition for the vectors **g** that meets these criteria is

$$\mathbf{g}_1 = 2\pi\frac{\mathbf{a}_2 \times \mathbf{a}_3}{V}; \qquad \mathbf{g}_2 = 2\pi\frac{\mathbf{a}_3 \times \mathbf{a}_1}{V}; \qquad \mathbf{g}_3 = 2\pi\frac{\mathbf{a}_1 \times \mathbf{a}_2}{V}, \tag{2.88}$$

with V given by Eq. (2.83). We identify the set **g** as the fundamental vectors of a lattice in k-space, which are called the *reciprocal lattice*. The volume of the primitive cell in the reciprocal lattice is $(2\pi)^3/V$. The vectors **G** are called reciprocal lattice vectors (RLVs), and they are the only vectors in k-space that contribute to the Fourier expansion of the crystal, Eq. (2.84).

Every lattice in real space has a corresponding lattice in reciprocal space (k-space). By applying the definitions in Eq. (2.88), we can make the following observations:

1. The reciprocal lattice of the simple cubic real-space lattice is a simple cubic lattice in k-space.
2. The reciprocal lattice of the body-centered-cubic real-space lattice is a face-centered-cubic lattice in k-space.
3. The reciprocal lattice of the face-centered-cubic real-space lattice is a body-centered-cubic lattice in k-space.

As we did for the real-space lattices, we can construct a Wigner-Seitz cell in k-space. This cell has special significance, and is known as the first Brillouin zone (FBZ), after Leon Brillouin (1889–1969), who developed the concept [8]. Owing to the periodicity of the crystal, all points in k-space are equivalent to points within the FBZ from which they differ by an RLV. This fact greatly simplifies the analysis of photonic crystals, as we will observe throughout this book.

Because the FCC and BCC lattices are mutually reciprocal, likewise each lattice's Wigner-Seitz cell is the other's FBZ: The FCC FBZ is a truncated octahedron, and the BCC FBZ is a rhombic dodecahedron. Trivially, the SC FBZ is a cube. The FBZs of these important lattices are pictured in Fig. 2.14. Conventionally, certain high-symmetry directions associated with each lattice are labeled by letters, which represent the k-vectors that correspond to the point on the FBZ surface along that particular direction; these points are indicated on the surfaces of the FBZs in Fig. 2.14. The origin in k-space is designated by the greek letter gamma Γ.

Also depicted for each lattice type in Fig. 2.14 is the region of k-space called the *irreducible Brillouin zone*, or IBZ. All points within the FBZ can be mapped on to points within the IBZ by symmetry operations, such as rotations or mirror inversions, which leave the crystal unchanged. This fact can be used to simplify photonic-crystal analysis further. Note, however, that the IBZs depicted apply

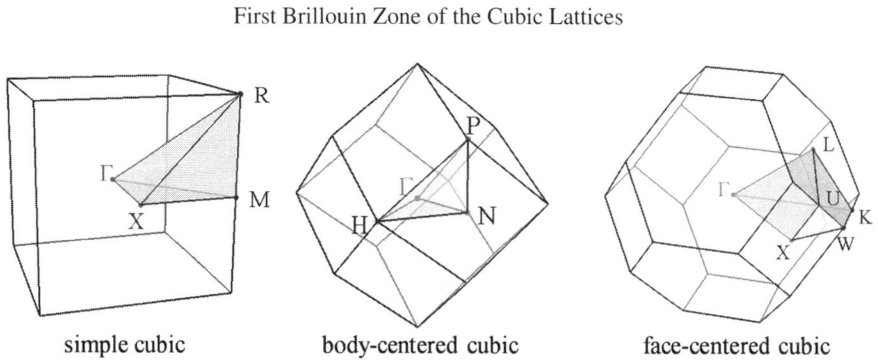

Figure 2.14 First Brillouin zones of the simple cubic, body-centered-cubic, and face-centered-cubic lattices. Symmetry points are labeled according to convention. Also indicated are the irreducible Brillouin zones; the full BZ can be obtained by acting on the irreducible zone with symmetry operations that leave the lattice unchanged.

only to those crystals where the basis possesses at least as much symmetry as the lattice itself; in many cases, the basis does not have such symmetry. Thus, the IBZ in such cases is larger than the region depicted. For example, in a diamond-like crystal, the basis introduces a unique direction among the family $\langle 111 \rangle$, namely the body diagonal on which the second atom resides. Thus, in diamond-like crystal structures, not all L points are equivalent; there are at least two distinct L-points, and they must be distinguished, say as L_1 and L_2, respectively. Likewise, all the other symmetry points that lie on hexagons centered on L_m points would become U_m, W_m, and K_m, where m represents 1 or 2. Such symmetry breaking plays a crucial role in the existence of 3D photonic bandgaps.

2.3 WAVES IN PERIODIC MEDIA

We now turn our attention to the properties of electromagnetic waves in periodic media, joining the material presented in the first two sections of this chapter.

2.3.1 Wave Equation in Periodic Dielectric Structures

The electromagnetic properties of a periodic medium, such as the one depicted in Fig. 2.15, are determined by the dielectric permittivity and magnetic permeability of the constituent materials, which reflect the translational symmetry of the medium:

$$\begin{aligned} \varepsilon(\mathbf{r}) &= \varepsilon(\mathbf{r}+\mathbf{R}) \\ \mu(\mathbf{r}) &= \mu(\mathbf{r}+\mathbf{R}) \end{aligned}, \tag{2.89}$$

where $\mathbf{r}$ is a vector that represents the position of some arbitrary point in space, and $\mathbf{R}$ is any arbitrary lattice translation vector, which along with the fundamental lattice

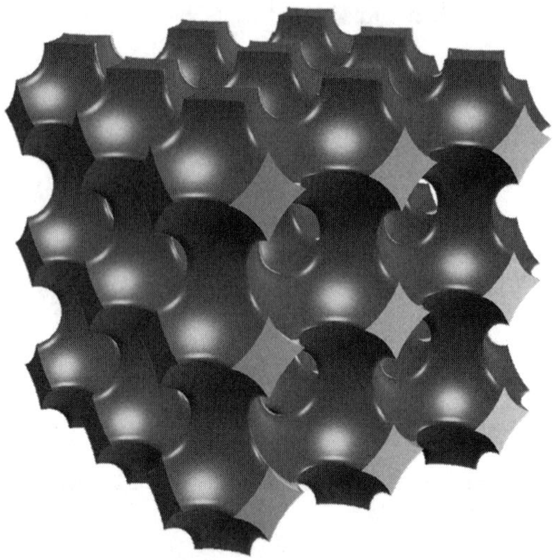

Figure 2.15 Simple cubic lattice, showing three periods in each direction, which consist of overlapping air spheres with radii of $0.55a$ in a dielectric matrix, where a is the lattice constant.

vectors defines the periodicity of the lattice such that the medium becomes invariant under translation through any vector **R**.

The propagation of an electromagnetic wave with frequency ω through a periodic dielectric medium is described by Maxwell's equations:

$$\nabla \times \mathbf{H}(\mathbf{r}) = j\omega\varepsilon(\mathbf{r})\mathbf{E}(\mathbf{r}), \tag{2.90}$$

$$\nabla \times \mathbf{E}(\mathbf{r}) = -j\omega\mu(\mathbf{r})\mathbf{H}(\mathbf{r}), \tag{2.91}$$

$$\nabla \cdot \mathbf{H}(\mathbf{r}) = 0, \tag{2.92}$$

$$\nabla \cdot [\varepsilon(\mathbf{r})\mathbf{E}(\mathbf{r})] = 0. \tag{2.93}$$

Note that ε and μ are in general tensors ($\bar{\bar{\varepsilon}}$ and $\bar{\bar{\mu}}$); however, this fact is not important to the following discussion, so we will use scalar notation for simplicity. Equations (2.90) and (2.91) remain invariant under the substitution of $(\mathbf{r} + \mathbf{R})$ for $\mathbf{r}$ in the arguments of **H**, **E**, ε, and μ. For nonmagnetic materials, the relative permeability tensor $\mu_r = \mu(\mathbf{r})/\mu_0$ becomes the identity matrix **I**, and hence, $\mu(\mathbf{r}) = \mu_0\mu_r(\mathbf{r}) = \mu_0$.

The periodic translational symmetry of the medium allows the electric and magnetic eigenmodes to be expressed as

$$\tilde{\mathbf{E}} = \mathbf{E}_\mathbf{k}(\mathbf{r})e^{-j\mathbf{k}\cdot\mathbf{r}}, \tag{2.94}$$

$$\tilde{\mathbf{H}} = \mathbf{H}_\mathbf{k}(\mathbf{r})e^{-j\mathbf{k}\cdot\mathbf{r}}, \tag{2.95}$$

where $\mathbf{E}_\mathbf{k}(\mathbf{r})$ and $\mathbf{H}_\mathbf{k}(\mathbf{r})$ are both periodic

$$\mathbf{E}_\mathbf{k}(\mathbf{r}) = \mathbf{E}_\mathbf{k}(\mathbf{r} + \mathbf{R}), \tag{2.96}$$

$$\mathbf{H}_\mathbf{k}(\mathbf{r}) = \mathbf{H}_\mathbf{k}(\mathbf{r} + \mathbf{R}). \tag{2.97}$$

Equations (2.94) and (2.95) are simply plane waves $e^{-j\mathbf{k}\cdot\mathbf{r}}$, which are modulated by a complex amplitude function, $\mathbf{E_k}(\mathbf{r})$ or $\mathbf{H_k}(\mathbf{r})$, that has the periodicity of the lattice. This result is known as the Bloch theorem [9], and the solutions of this form are called Bloch waves. **k** is known as the Bloch wave vector. A proof of the theorem is straightforward; at the end of this section, we present a proof of Eq. (2.94). Equation (2.95) can be obtained by a similar analysis.

Using Eqs. (2.90) and (2.91), we can derive the wave equation in periodic structures by taking the curl of Eq. (2.91) and by using Eq. (2.90) to eliminate $\mathbf{H}(\mathbf{r})$:

$$\nabla \times [\nabla \times \mathbf{E}(\mathbf{r})] = \omega^2 \mu_0 \varepsilon(\mathbf{r}) \mathbf{E}(\mathbf{r}). \tag{2.98}$$

A similar equation for $\mathbf{H}(\mathbf{r})$ can be also derived, yielding

$$\nabla \times \left[\frac{1}{\varepsilon(\mathbf{r})} \nabla \times \mathbf{H}(\mathbf{r})\right] = \omega^2 \mu_0 \mathbf{H}(\mathbf{r}), \tag{2.99}$$

which is an eigenvalue equation. The left-hand side of Eq. (2.98) can be further expanded using vector identities, which yield

$$-\nabla^2 \mathbf{E}(\mathbf{r}) + \nabla[\nabla \cdot \mathbf{E}(\mathbf{r})] = \omega^2 \mu_0 \varepsilon(\mathbf{r}) \mathbf{E}(\mathbf{r}). \tag{2.100}$$

If the periodic medium where the electromagnetic wave is propagating is free of sources/sinks of displacement fields, then $\nabla \cdot \mathbf{E}(\mathbf{r}) = 0$ and Eq. (2.100) can be simplified to

$$\nabla^2 \mathbf{E}(\mathbf{r}) + \omega^2 \mu_0 \varepsilon(\mathbf{r}) \mathbf{E}(\mathbf{r}) = 0. \tag{2.101}$$

In this case, each component of $\mathbf{E}(\mathbf{r})$ obeys the scalar wave equation:

$$-\nabla^2 \psi - \omega^2 \mu_0 \varepsilon(\mathbf{r}) \psi = 0, \tag{2.102}$$

where ψ can be any spatial component of the **E** field. Equation (2.102) is frequently used in optics, as opposed to the more fundamental form of Maxwell's equations. Such an approximation is valid only if the magnitude of the neglected term $\nabla \cdot \mathbf{E}(\mathbf{r})$ is small. Consider the divergence of the displacement field in the absence of free charge,

$$\nabla \cdot \mathbf{D} = \nabla \cdot (\varepsilon \mathbf{E}) = \varepsilon \nabla \cdot \mathbf{E} + \mathbf{E} \cdot \nabla \varepsilon = 0. \tag{2.103}$$

Solving for $\nabla \cdot \mathbf{E}$ and taking the gradient, we see that

$$\nabla(\nabla \cdot \mathbf{E}) = -\nabla(\mathbf{E} \cdot \nabla \ln \varepsilon). \tag{2.104}$$

The term on the right-hand side of Eq. (2.104) vanishes inside a homogenous medium, and if the variation in ε is small over distances comparable with the wavelengths of interest, it can be neglected in an inhomogeneous medium as well. As a result, the electric field components can be decoupled, and the scalar wave equation in Eq. (2.102) can be applied. However, in periodic dielectric structures of interest in this book, the unit cell occupies a size comparable with the operating wavelength. Therefore, this term cannot be neglected while analyzing a periodic arrangement of dielectric objects, and the scalar wave equation cannot be used to analyze periodic structures accurately in this case. Indeed, the first attempts to solve the scalar wave equation inside periodic dielectric structures were found to provide inaccurate and nonphysical solutions to the problem [10]. Instead, the vector nature of Maxwell's equation was incorporated for a more precise understanding of these structures. In such analysis, Eq. (2.99) seems more suitable because the differential operator on the left-hand side is Hermitian, which guarantees that its eigenvalues on the right-hand side of the same equation are real.

2.3.1.1 Proof of Bloch's Theorem. If we begin with Eq. (2.98) and substitute a Fourier integral expansion for the electric field,

$$\mathbf{E}(\mathbf{r}) = \int \mathbf{E}(\mathbf{k}) e^{j\mathbf{k}\cdot\mathbf{r}} d\mathbf{k}, \tag{2.105}$$

as well as a Fourier series expansion of the periodic permittivity function over the reciprocal lattice,

$$\varepsilon(\mathbf{r}) = \sum_{\mathbf{G}} \varepsilon(\mathbf{G}) e^{j\mathbf{G}\cdot\mathbf{r}}, \tag{2.106}$$

then we obtain the following equation:

$$-\int \mathbf{k} \times [\mathbf{k} \times \mathbf{E}(\mathbf{k})] e^{j\mathbf{k}\cdot\mathbf{r}} d\mathbf{k} = \omega^2 \mu_0 \sum_{\mathbf{G}} \varepsilon(\mathbf{G}) e^{j\mathbf{G}\cdot\mathbf{r}} \int \mathbf{E}(\mathbf{k}) e^{j\mathbf{k}\cdot\mathbf{r}} d\mathbf{k}, \tag{2.107}$$

which can be rearranged to yield

$$-\int \mathbf{k} \times [\mathbf{k} \times \mathbf{E}(\mathbf{k})] e^{j\mathbf{k}\cdot\mathbf{r}} d\mathbf{k} = \omega^2 \mu_0 \sum_{\mathbf{G}} \varepsilon(\mathbf{G}) \int \mathbf{E}(\mathbf{k}) e^{j(\mathbf{k}+\mathbf{G})\cdot\mathbf{r}} d\mathbf{k}. \tag{2.108}$$

Because this equation is valid for all $\mathbf{r}$, the equation is valid for each plane-wave component $\mathbf{k}$ individually, which implies that the coefficients of each component must be equal on both sides of the equation. This equation, with substitution of $\mathbf{k}-\mathbf{G}$ for $\mathbf{k}$ as the integration variable on the right-hand side, leads to the relation

$$\mathbf{k} \times [\mathbf{k} \times \mathbf{E}(\mathbf{k})] + \omega^2 \mu_0 \sum_{\mathbf{G}} \varepsilon(\mathbf{G}) \mathbf{E}(\mathbf{k} - \mathbf{G}) = 0, \tag{2.109}$$

which implies that for each plane-wave component, the expansion of the field depends only on those Fourier components that comprise the reciprocal lattice. Hence, the eigenmode that corresponds to $\mathbf{k}$ can be expressed as

$$\tilde{\mathbf{E}}_{\mathbf{k}}(\mathbf{r}) = \sum_{\mathbf{G}} \mathbf{E}(\mathbf{k} - \mathbf{G}) e^{j(\mathbf{k}-\mathbf{G})\cdot\mathbf{r}}. \tag{2.110}$$

If we define a periodic function $\mathbf{E}_{\mathbf{k}}(\mathbf{r})$ as

$$\mathbf{E}_{\mathbf{k}}(\mathbf{r}) = \sum_{\mathbf{G}} \mathbf{E}(\mathbf{k} - \mathbf{G}) e^{-j\mathbf{G}\cdot\mathbf{r}}, \tag{2.111}$$

then with this definition, we have shown the validity of Eq. (2.94). It also follows from this result that the solution for the field at any location is uniquely specified by the solution within the unit cell, because every point in the lattice can be mapped onto a corresponding point within the unit cell by a translation through a suitable lattice vector. The result of such a translation is

$$\tilde{\mathbf{E}}_{\mathbf{k}}(\mathbf{r} + \mathbf{R}) = \tilde{\mathbf{E}}_{\mathbf{k}}(\mathbf{r}) e^{j\mathbf{k}\cdot\mathbf{R}}, \tag{2.112}$$

which means that for the electric field of any eigenmode of a periodic dielectric system, the result of a translation through a lattice vector $\mathbf{R}$ is the mere addition of $(\mathbf{k} \cdot \mathbf{R})$ to the phase. Indeed, Eq. (2.112) is often presented as an alternative statement of the Bloch theorem itself.

2.3.1.2 Variational Principle for Eigenvalue Problem. As discussed above, in the linear dielectric with permittivity $\varepsilon(\mathbf{r})$, the source-free Maxwell's equations lead to the following equation for the magnetic field [compare Eq. (2.99)]:

$$\mathcal{L}\mathbf{H} = \frac{\omega^2}{c^2}\mathbf{H}, \tag{2.113}$$

where the operator $\mathcal{L} = \left(\nabla \times [\varepsilon_r(\mathbf{r})]^{-1} \nabla \times\right)$ is introduced, and c the is speed of light in vacuum, whereas $\varepsilon_r(\mathbf{r}) = \varepsilon(\mathbf{r})/\varepsilon_0$ is the relative permittivity. Equation (2.113) is a standard eigenvalue problem with eigenvalue[4] $\lambda = (\omega/c)^2$ and associated eigenvector $\mathbf{H}$. In most cases, this problem must be solved numerically; several numerical procedures that are used most often to find solutions for photonic crystals rely on casting the eigenvalue problem as arising from a variational principle. To observe this, note that Eq. (2.113) is a special case of a generalized eigenvalue problem

$$\mathcal{L}\Phi = \lambda\mathcal{M}\Phi, \tag{2.114}$$

[4]Note that in this discussion, we use λ to denote the eigenvalue. This should not be confused with the wavelength of an electromagnetic wave, which is also usually denoted by λ.

where $\mathcal{L}$ and $\mathcal{M}$ can be in principle any self-adjoint operators, whether differential or integral, scalar or vector. The problem is finding the eigenvalue λ and the corresponding eigenfunctions Φ.

It can be shown that Eq. (2.114) follows from a variational principle. To this end, a functional F[Φ] is defined as a ratio

$$\mathrm{F}[\Phi] = \frac{\langle \Phi, \mathcal{L}\Phi \rangle}{\langle \Phi, \mathcal{M}\Phi \rangle}, \tag{2.115}$$

where angled brackets denote the inner product

$$\langle f, g \rangle = \frac{1}{V} \int_V f^* g \, dv, \tag{2.116}$$

where V is the volume of the domain with differential element dv, and an asterisk "*" represents complex conjugation. By definition, functional F[Φ] achieves a stationary value if small, arbitrary variations δ of the field distribution Φ lead to no first-order (linear in δ) variation in F, or formally

$$\mathrm{F}[\Phi + \delta] = \mathrm{F}[\Phi] + O[\delta^2]. \tag{2.117}$$

Using the definition of F[Φ] in Eq. (2.115), we obtain

$$\mathrm{F}[\Phi + \delta] = \frac{\langle \Phi + \delta, \mathcal{L}(\Phi + \delta) \rangle}{\langle \Phi + \delta, \mathcal{M}(\Phi + \delta) \rangle}. \tag{2.118}$$

The right-hand side of the last equation is expanded by using the linearity of operators $\mathcal{L}$ and $\mathcal{M}$, as well as the sesquilinearity of the inner products in the numerator and in the denominator. By expanding the denominator as a power series in terms of δ, we obtain

$$\begin{aligned} \mathrm{F}[\Phi + \delta] = {} & (\langle \Phi, \mathcal{L}\Phi \rangle + \langle \Phi, \mathcal{L}\delta \rangle + \langle \delta, \mathcal{L}\Phi \rangle) \\ & \times \left(\frac{1}{\langle \Phi, \mathcal{M}\Phi \rangle} - \frac{\langle \Phi, \mathcal{M}\delta \rangle + \langle \delta, \mathcal{M}\Phi \rangle}{\langle \Phi, \mathcal{M}\Phi \rangle^2} \right) + O[\delta^2], \end{aligned} \tag{2.119}$$

where the terms of the order higher than linear in δ have been included $O[\delta^2]$. Multiplying out the terms in parentheses leads to

$$\begin{aligned} \mathrm{F}[\Phi + \delta] = {} & \frac{\langle \Phi, \mathcal{L}\Phi \rangle}{\langle \Phi, \mathcal{M}\Phi \rangle} + \frac{\langle \Phi, \mathcal{L}\delta \rangle + \langle \delta, \mathcal{L}\Phi \rangle}{\langle \Phi, \mathcal{M}\Phi \rangle} \\ & - \frac{\langle \Phi, \mathcal{M}\delta \rangle + \langle \delta, \mathcal{M}\Phi \rangle}{\langle \Phi, \mathcal{M}\Phi \rangle} \frac{\langle \Phi, \mathcal{L}\Phi \rangle}{\langle \Phi, \mathcal{M}\Phi \rangle} + O[\delta^2] \\ = {} & \mathrm{F}[\Phi] + \left(\frac{\langle \Phi, \mathcal{L}\delta \rangle + \langle \delta, \mathcal{L}\Phi \rangle}{\langle \Phi, \mathcal{L}\Phi \rangle} - \frac{\langle \Phi, \mathcal{M}\delta \rangle + \langle \delta, \mathcal{M}\Phi \rangle}{\langle \Phi, \mathcal{M}\Phi \rangle} \right) \frac{\langle \Phi, \mathcal{L}\Phi \rangle}{\langle \Phi, \mathcal{M}\Phi \rangle} \\ & + O[\delta^2]. \end{aligned} \tag{2.120}$$

For the linear term to vanish, the term in parentheses must be zero; i.e.,

$$\frac{\langle\Phi, \mathcal{L}\delta\rangle + \langle\delta, \mathcal{L}\Phi\rangle}{\langle\Phi, \mathcal{L}\Phi\rangle} = \frac{\langle\Phi, \mathcal{M}\delta\rangle + \langle\delta, \mathcal{M}\Phi\rangle}{\langle\Phi, \mathcal{M}\Phi\rangle}. \tag{2.121}$$

Given that δ is arbitrary and the operators $\mathcal{L}$ and $\mathcal{M}$ are self-adjoint, the only way Eq. (2.121) will hold is if

$$\frac{\langle\delta, \mathcal{L}\Phi\rangle}{\langle\Phi, \mathcal{L}\Phi\rangle} = \frac{\langle\delta, \mathcal{M}\Phi\rangle}{\langle\Phi, \mathcal{M}\Phi\rangle}, \tag{2.122}$$

for all variations δ. Because δ is arbitrary, this leads to

$$\frac{\mathcal{L}\Phi}{\langle\Phi, \mathcal{L}\Phi\rangle} = \frac{\mathcal{M}\Phi}{\langle\Phi, \mathcal{M}\Phi\rangle} \Longrightarrow \mathcal{L}\Phi = \frac{\langle\Phi, \mathcal{L}\Phi\rangle}{\langle\Phi, \mathcal{M}\Phi\rangle}\mathcal{M}\Phi, \tag{2.123}$$

which is equivalent to Eq. (2.114) with the substitution

$$\lambda = \frac{\langle\Phi, \mathcal{L}\Phi\rangle}{\langle\Phi, \mathcal{M}\Phi\rangle}. \tag{2.124}$$

This demonstrates that the generalized eigenvalue problem (2.114) is the result of a variational principle with the functional defined in Eq. (2.115). The stationary value of this functional is equal to the eigenvalue and is achieved for the field Φ equal to the eigenfunction in Eq. (2.114).

Given that Eq. (2.113), which must be satisfied for the electromagnetic field interacting with a photonic crystal, is a special case of the generalized eigenvalue problem (2.114), the field configuration for each mode in a given photonic crystal results from the respective variational principle. The appropriate functional is (proportional to) the energy held by the (normalized) mode.

2.3.2 Group Velocity in Periodic Media

In homogeneous media, the fundamental solutions to Maxwell's equations take the form of plane waves. This is not the case in periodic media, such as photonic crystals. Although plane waves still play an important role here, the fundamental objects are Bloch waves [Eqs. (2.94) and (2.95)], which solve the Maxwell's equations exactly. Thus, in periodic media, the role of plane waves is played by Bloch waves. Accordingly, an arbitrary solution to the Maxwell's equations can be expressed as a superposition of Bloch waves. Bearing this in mind, we can repeat the procedure of calculating the group velocity for homogeneous media from Section 2.1.5, except we have to use Bloch waves instead of plane waves. The strategy is the same: We choose a wave packet that is localized in k-space, where k refers this time to the index enumerating the Bloch waves, as opposed to the coordinates of the spatial frequency space in the homogeneous media case. We then express the wave packet as a product of a carrier Bloch wave and an envelope function; finally, we calculate the velocity of the envelope function and interpret it as the group velocity of a wave packet propagating in a photonic crystal.

Accordingly, in photonic crystals, we write a wave packet as a superposition of Bloch waves

$$\Psi(\mathbf{r}, t) = \int d\mathbf{k}\tilde{\Psi}(\mathbf{k})u(\mathbf{k}, \mathbf{r})\exp[j(\omega(\mathbf{k})t - \mathbf{k}\cdot\mathbf{r})], \tag{2.125}$$

where $u(\mathbf{k}, \mathbf{r})$ is a periodic function of the second argument $u(\mathbf{k}, \mathbf{r}+\mathbf{R}) = u(\mathbf{k}, \mathbf{r})$ and $\mathbf{R}$ is a lattice vector. As in the homogeneous case, we consider a wave packet with a narrow, concentrated k-space range and expand the argument of the exponential function in (2.125) with respect to $(k - k_0)$

$$\begin{aligned}\Psi(\mathbf{r}, t) &= \exp[j(\omega_0 t - \mathbf{k}_0\cdot\mathbf{r})]\int d\mathbf{k}\tilde{\Psi}(\mathbf{k})u(\mathbf{k}, \mathbf{r})\\ &\times \exp\Big[j\Big(\Big(\nabla_k\omega|_{\mathbf{k}=\mathbf{k}_0}\Big)t - \mathbf{r}\Big)\cdot(\mathbf{k}-\mathbf{k}_0) + jO(\mathbf{k}-\mathbf{k}_0)^2 t\Big].\end{aligned} \tag{2.126}$$

Furthermore, we expand the exponential function under the integral

$$\begin{aligned}\Psi(\mathbf{r}, t) &= \exp[j(\omega_0 t - \mathbf{k}_0\cdot\mathbf{r})]\int d\mathbf{k}\tilde{\Psi}(\mathbf{k})u(\mathbf{k}, \mathbf{r})\\ &\times \Big[1 + j\Big(\Big(\nabla_k\omega|_{\mathbf{k}=\mathbf{k}_0}\Big)t - \mathbf{r}\Big)\cdot(\mathbf{k}-\mathbf{k}_0) + jO(\mathbf{k}-\mathbf{k}_0)^2 t\Big],\end{aligned}$$

and multiply out the square bracket

$$\begin{aligned}\Psi(\mathbf{r}, t) &= \exp[j(\omega_0 t - \mathbf{k}_0\cdot\mathbf{r})]\int d\mathbf{k}\tilde{\Psi}(\mathbf{k})u(\mathbf{k}, \mathbf{r}) + j\exp[j(\omega_0 t - \mathbf{k}_0\cdot\mathbf{r})]\\ &\times \int d\mathbf{k}\tilde{\Psi}(\mathbf{k})u(\mathbf{k}, \mathbf{r})\Big[\Big(\Big(\nabla_k\omega|_{\mathbf{k}=\mathbf{k}_0}\Big)t - \mathbf{r}\Big)\cdot(\mathbf{k}-\mathbf{k}_0) + O(\mathbf{k}-\mathbf{k}_0)^2 t\Big].\end{aligned} \tag{2.127}$$

The function $u(\mathbf{k}, \mathbf{r})$ in the second term of the (2.127) can be Taylor-expanded around $\mathbf{k} = \mathbf{k}_0$

$$u(\mathbf{k}, \mathbf{r}) = u(\mathbf{k}_0, \mathbf{r}) + \Big(\nabla_k u|_{\mathbf{k}=\mathbf{k}_0}\Big)\cdot(\mathbf{k}-\mathbf{k}_0) + O(\mathbf{k}-\mathbf{k}_0)^2. \tag{2.128}$$

Because $u(\mathbf{k}, \mathbf{r})$ is multiplied in the second term of (2.127) by a function in square brackets, which is of at least first order in $(\mathbf{k}-\mathbf{k}_0)$, we can rewrite expression (2.127) with the help of (2.128) in the following form:

$$\begin{aligned}\Psi(\mathbf{r}, t) &= \exp[j(\omega_0 t - \mathbf{k}_0\cdot\mathbf{r})]\int d\mathbf{k}\tilde{\Psi}(\mathbf{k})u(\mathbf{k}, \mathbf{r})\\ &\quad + j\exp[j(\omega_0 t - \mathbf{k}_0\cdot\mathbf{r})]u(\mathbf{k}_0, \mathbf{r})\\ &\quad \times \int d\mathbf{k}\tilde{\Psi}(\mathbf{k})\Big[\Big(\Big(\nabla_k\omega|_{\mathbf{k}=\mathbf{k}_0}\Big)t - \mathbf{r}\Big)\cdot(\mathbf{k}-\mathbf{k}_0) + O(\mathbf{k}-\mathbf{k}_0)^2 t\Big],\end{aligned} \tag{2.129}$$

where we pulled $u(\mathbf{k}_0, \mathbf{r})$ out of the integral because it does not depend on the integration variable $\mathbf{k}$ and collected all higher order terms under $O(\mathbf{k}-\mathbf{k}_0)^2$. Now, the first term in expression (2.129) contains only high spatial frequencies, or in other words, it is rapidly varying spatially. Because it lacks the slowly varying component, it can be thought of as a high-frequency bias that does not contribute to the propagation of the wave packet; therefore, it is of no interest here. However, in the second term, we can identify a high-frequency part $\exp[j(\omega_0 t - \mathbf{k}_0 \cdot \mathbf{r})]u(\mathbf{k}_0, \mathbf{r})$ and the slowly varying envelope function

$$A(\mathbf{r}, t) = \int d\mathbf{k}\tilde{\Psi}(\mathbf{k})\left[\left(\left(\nabla_k \omega|_{\mathbf{k}=\mathbf{k}_0}\right)t - \mathbf{r}\right) \cdot (\mathbf{k}-\mathbf{k}_0) + O(\mathbf{k}-\mathbf{k}_0)^2 t\right]. \tag{2.130}$$

The envelope function of the wave packet that propagates in a photonic crystal, Eq. (2.130), has almost exactly the same form as the envelope function of a wave packet that propagates in a homogeneous medium shown in Eq. (2.79). Accordingly, we can calculate the full time derivative of the envelope function

$$\frac{dA}{dt} = (\nabla_r A)\frac{d\mathbf{r}}{dt} + \frac{\partial A}{\partial t} = \int d\mathbf{k}\tilde{\Psi}(\mathbf{k})\left[j\left(\left(\nabla_k \omega|_{\mathbf{k}=\mathbf{k}_0}\right) - \dot{\mathbf{r}}\right) \cdot (\mathbf{k}-\mathbf{k}_0) + jO(\mathbf{k}-\mathbf{k}_0)^2\right]. \tag{2.131}$$

Note that for the envelope to remain unchanged, it is necessary that the expression in square brackets vanishes. To first order in $(\mathbf{k}-\mathbf{k}_0)$, it means that the time derivative of the position $\dot{\mathbf{r}}$ is equal to the k-gradient of the frequency ω. However, $\dot{\mathbf{r}}$ is equal to the group velocity $\mathbf{v}_g$. Therefore, we can conclude that the group velocity of a wave propagating in a photonic crystal is equal to

$$\mathbf{v}_g(\mathbf{k}_0) = \nabla_k \omega|_{\mathbf{k}=\mathbf{k}_0}. \tag{2.132}$$

2.3.3 Dispersion Surfaces and Band Diagrams

We have observed in Section 2.1.3.2 that the directional dispersion, or anisotropy, of a material can be represented by the equifrequency surfaces. These surfaces contain a great deal of information about the propagating electromagnetic modes within an anisotropic material: For a given frequency, the EFS represents all the allowed wave vectors that can exist in the material at that frequency. Furthermore, it was shown in Sections 2.1.5 and 2.3.2, for homogeneous and periodic media, respectively, that if we compute how the frequency associated with a specific wave vector $\mathbf{k}_0$ changes as we change the wave vector by a small amount $d\mathbf{k}$, i.e., the gradient of the function $\omega(\mathbf{k})$, then we obtain the group velocity for wave packets centered around $\mathbf{k}_0$. In other words, an electromagnetic wave whose free-space wave vector is $\mathbf{k}$, excited by some means within in a medium, whether it be homogeneous or periodic, *will propagate along a direction normal to the EFS at that frequency.* This important property will be observed in later chapters to be of paramount importance for understanding the properties of photonic crystals. Because of this property, the EFSs are also termed *dispersion surfaces*, in the sense that $\omega(\mathbf{k})$ is the *dispersion relation*, which implicitly refers to dispersion in frequency (as opposed to directional, or spatial, dispersion, i.e., anisotropy).

Section 2.1.3.2 showed how to obtain an analytical expression for the dispersion surface in homogeneous, anisotropic dielectric media, from which one can obtain the functional form of the dispersion relation $\omega(\mathbf{k})$. The corresponding surfaces for inhomogeneous, periodic media, such as photonic crystals, must be obtained by solving Maxwell's equations numerically; the methods are described in Chapter 3. For now, we will simply point out the various ways that the dispersion relation for photonic crystals is depicted and briefly discuss how to interpret such depictions.

2.3.3.1 Interpretation of the Dispersion Surfaces and Band Diagrams of Photonic Crystals. The dispersion surfaces are the loci of all points in k-space for which the dispersion function $\omega(\mathbf{k})$ is a constant. In a periodic system, i.e., a photonic-crystal lattice, for a particular choice of $\mathbf{k}$, a discrete set of eigenfrequencies $\{\omega_{\mathbf{k}}\}$ are solutions of the eigenvalue equation, Eq. (2.113). The discrete dispersion surfaces that correspond to these eigenfrequencies are by convention enumerated in order of increasing frequency; the enumerated frequencies are labeled by a counting index n (not to be confused with the index of refraction) and are denoted by $\omega_{n\mathbf{k}}$. All solutions with a common counting index n are collectively called the nth *photonic band*; e.g., the lowest eigenfrequency obtained for each $\mathbf{k}$ within the first Brillouin zone, plotted in k-space, is called the first band. Figure 2.16a depicts a dispersion surface for a photonic-crystal structure that will be discussed in detail in a later chapter. For now, it merely serves as an example. Note that a particular dispersion surface, which is a surface of constant frequency, is a subset of its band, which spans a range of frequencies; for three-dimensional space, it is not practical to depict the entire band at once (the volume of k-space occupied by a band can be plotted, but such a plot would not indicate how the band's constituent dispersion surfaces changed with the value of the frequency). This is not the case for two dimensions. In 2D systems, the constructions analogous to dispersion surfaces are not surfaces but curves, which are called equifrequency contours. Such contours can be plotted in a three-dimensional plot over a space whose dimensions are k_x, k_y, and ω. In such a plot, the contours that belong to a given band will form continuous surfaces; thus, entire bands can be visualized at once, as can be observed in Fig. 2.16b.

The reason for referring to the grouped eigenfrequencies as "bands" is more apparent if one considers an alternative, simplified depiction of the dispersion relation, namely the dispersion diagram or "band diagram." The dispersion diagram, such as the one shown in Fig. 2.17, is a plot of several bands for $\mathbf{k}$-points that trace the perimeter of the irreducible Brillouin zone. In such a plot, the horizontal axis represents a path through k-space in any number of dimensions; note that although this path is plotted along a straight axis in the diagram, it is not straight in k-space. Rather, it typically contains several sharp turns, as can be observed in Fig. 2.14. The appearance of the several discrete curves in a dispersion diagram explains the "band" descriptive, originating from solid-state physics, which has long used dispersion diagrams to describe the eigen-energy solutions of Schrödinger's scalar-wave equation for electrons in a periodic atomic potential.

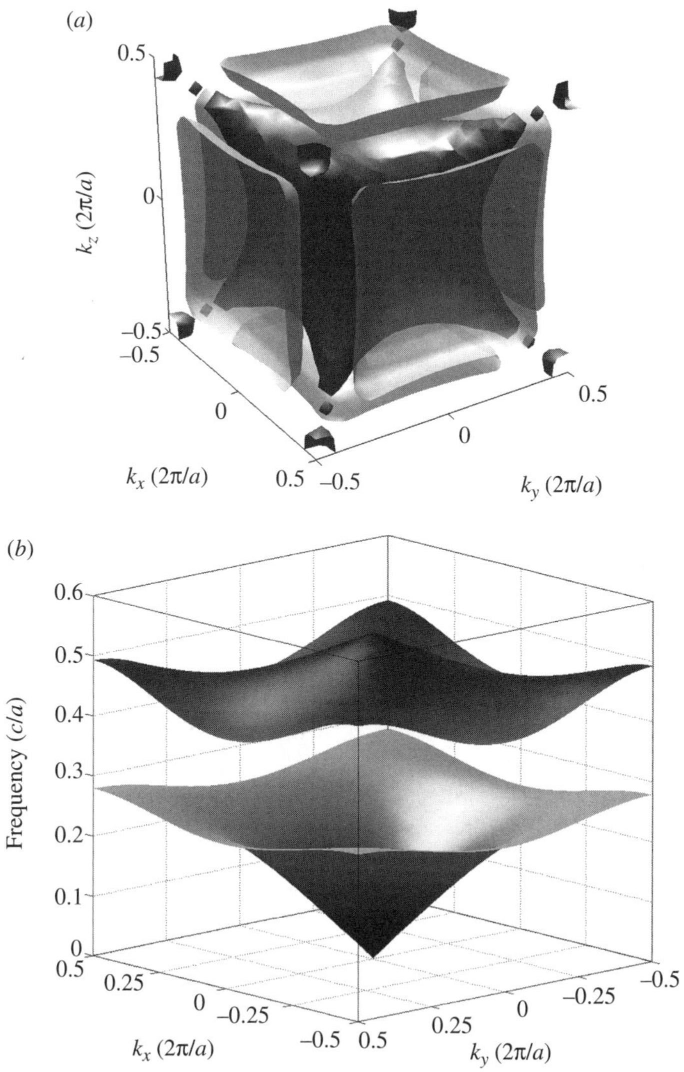

Figure 2.16 Examples of dispersion surfaces for photonic crystals. (a) Equifrequency surfaces for the third (darker) and fourth (lighter) bands of a simple cubic 3D photonic crystal. (b) Surfaces that represent the entire first (lower) and second (higher) bands for a 2D square photonic crystal are plotted in a hybrid space of k_x, k_y, and ω. Horizontal sections of these surfaces are equifrequency contours, which are the 2D analogs of the surfaces in (a).

The radial component of a point on a dispersion surface is a measure of the effective index of refraction for that combination of frequency and **k**. Specifically, the effective index is the ratio of this radial component to the radial component of a point on the spherical dispersion surface for free-space propagation along the same direction. In the normalized units typically used to plot dispersion diagrams, free space

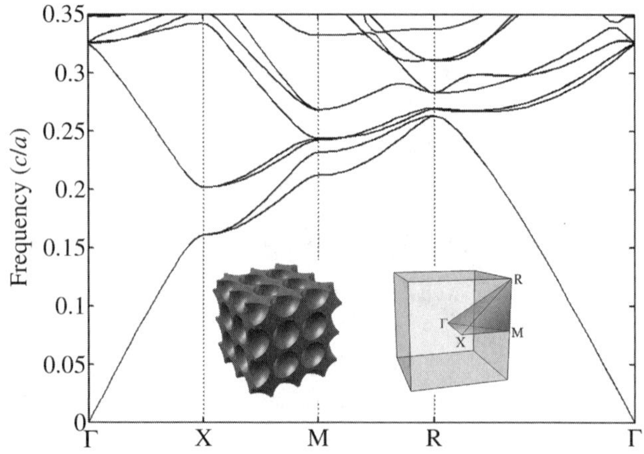

Figure 2.17 Example of a band diagram for a simple cubic 3D photonic crystal, which is depicted inset left. The curves represent the frequencies (vertical axis) of the dispersion surfaces on which a set of k-points (horizontal axis) lie. The set of k-points plotted form the perimeter of the IBZ, as indicated inset right.

propagation corresponds to $\tilde{\omega} = \tilde{k}$, so in practice the effective index is $|\tilde{\mathbf{k}}|/\tilde{\omega}$ ($\tilde{\omega} = \omega a/2\pi c$ and $\tilde{k} = ka/2\pi$, where a is the crystal lattice constant and c is the speed of light in vacuum).

When examining a band diagram, it may be the case that there are certain ranges of frequency in which none of the bands exists for any wave vector. This situation is termed a *bandgap*. More commonly, a *directional gap*, or *stop band*, may be found in the k-space vicinity of a particular direction; an example is in Fig. 2.17, in which no bands exist in the Γ-X direction (i.e., between Γ and X on the horizontal axis) for normalized frequencies between 0.16 and 0.2. Later chapters will discuss the importance of bandgaps and stop bands for many types of photonic-crystal devices and applications. Furthermore, one often finds when studying the equifrequency contours (for 2D) or surfaces (for 3D) of photonic crystals, that these contours/surfaces take on shapes that are far more complex than even the general fourth-order surface described for a homogeneous isotropic medium in Section 2.1.3.2. Examples of these unusual shapes can be observed in Fig. 2.16. We will learn in later chapters that these shapes, coupled with the important relation between the group velocity and the gradient of $\omega(\mathbf{k})$, provide powerful insight into the behavior of light in photonic crystals, particularly for frequencies far from any bandgaps.

REFERENCES

1. J. D. Jackson, *Classical Electrodynamics*, 3rd ed. New York: Wiley, 1999.
2. C. A. Balanis, *Advanced Engineering Electromagnetics*. New York: Wiley, 1989.
3. T. Inui, Y. Tanabe, and Y. Onodera, *Group Theory and its Applications in Physics*, New York: Springer-Verlag, 1996.

4. K. Sakoda, "Symmetry, Degeneracy, and Uncoupled Modes in 2-Dimensional Photonic Lattices," *Phys. Rev. B*, vol. 52, pp. 7982–7986, 1995.
5. W. L. Bragg, "The Specular Reflection of X-Rays," *Nature*, vol. 90, 410, 1912.
6. P. P. Ewald, *Fifty Years of X-Ray Diffraction*. Utrecht, The Netherlands: Published for the International Union of Crystallography by A. Oosthoek's Uitgeversmij., 1962.
7. M. von Laue, "X-radiation interferences," *Physikalische Zeitschrift*, vol. 14, pp. 1075–1079, 1913.
8. L. Brillouin, "Les Electrons Dans les Metaux et le Classement des Ondes de deBroglie Correspondantes," *Comptes Rendus Hebdomadaires des Seances de l'Academie des Sciences*, vol. 191, pp. 292, 1930.
9. N. W. Ashcroft and N. D. Mermin, *Solid State Physics*. Philadelphia: Saunders College, 1976.
10. S. Datta, C. T. Chan, K. M. Ho, and C. M. Soukoulis, "Photonic Band-Gaps in Periodic Dielectric Structures—the Scalar-Wave Approximation," *Phys. Rev. B*, vol. 46, pp. 10650–10656, 1992.

3

Numerical Methods

To understand the interaction of electromagnetic waves with periodic systems and to design practical devices that take advantage of unique material properties resulting from this interaction, it is often necessary to resort to numerical simulations because analytical treatment is restricted to highly regular patterns. Therefore, in this chapter, several numerical techniques to analyze such structures will be reviewed briefly. Two methods among them, namely the plane-wave expansion method (PWEM) and the finite-difference time-domain (FDTD) method, will be discussed in detail.

3.1 OVERVIEW

Because the modulation of the material properties in a photonic crystal is on the scale of the wavelength, a set of full vectorial simulation tools is required to carry out the task of solving for the photonic band structure. Fortunately, in periodic structures, the electromagnetic fields can be expressed in the form of Bloch waves, as described in Chapter 2, so that only a unit cell analysis of photonic crystal is necessary to determine the dispersion relations by applying appropriate Bloch boundary conditions. The dispersion relation, which is also called the *band structure* of the photonic crystal, is defined as the variation of the eigenfrequencies ω versus the wave propagation vectors $\mathbf{k}$. Because of the arbitrary inhomogeneity of material within a single unit cell, in most cases, the band structure cannot be obtained analytically, so numerical methods are required.

In the last decade, considerable effort has been put forth to develop efficient and accurate algorithms to calculate photonic-crystal band structures. Among these algorithms, the most popular are the PWEM [1–33], the FDTD method [34–46], the transfer-matrix method [21, 30, 34, 47–53], and the finite element method (FEM) [54–56]. In addition to these techniques, several other interesting approaches have been developed, which will not be covered here in detail; the interested reader is referred to Refs. 44 and 57–64.

The plane-wave expansion method described in Section 3.2 relies on a simple and straightforward (Fourier) representation of the periodic $\mathbf{E_k}$ and $\mathbf{H_k}$ fields in terms

Photonic Crystals: Theory, Applications, and Fabrication. By D. W. Prather, S. Shi, A. Sharkawy, J. Murakowski, and G. J. Schneider

of harmonic functions defined by the reciprocal lattice vectors. This simplicity, together with the development of powerful numerical procedures, has made the plane-wave expansion method the most widely used tool for finding Bloch modes and eigenfrequencies of an infinite, periodic system of scattering objects. The Fourier expansion of monochromatic fields casts Maxwell's equations into an algebraic eigenvalue problem, which can be solved numerically after truncating the infinite series to a finite sum. Although efficient, the plane-wave expansion method may not be sufficient for the simulations of complicated defects in photonic crystals, and it may not be suitable for dispersive or lossy materials such as metals.

The FDTD method [65] described in Section 3.3 is widely used, and it is considered one of the most applicable for photonic-crystal structures. It is universal, robust, methodologically simple, and intuitive. The wave that propagates through the photonic-crystal structure is found by direct integration in the time domain of Maxwell's equations in discretized, finite-difference form. Discretization is used in both time and space; the spatial discretization is done on a staggered grid, in which the electric and magnetic field components are calculated at adjacent, spatially separated grid points. Whether one defines the input signal as continuous wave (CW) or pulse, the excitation propagates through the structure by time stepping through the entire grid.

Like the FDTD method, the transfer-matrix method [34, 47, 52, 66–68] works with the discretized Maxwell's equations. However, the initial excitation is always monochromatic, and the basic structural unit is a layer of grid cells. The structure under consideration is divided into a set of layers with the same number of grid nodes in each layer. Using the discretized Maxwell's equations, the field $\mathbf{E}_i$ in the nodes of one layer is connected to the field $\mathbf{E}_{i+1}$ in the nodes of the neighboring layers via a transfer matrix

$$\mathbf{E}_{i+1} = \mathbf{T}_i \mathbf{E}_i. \tag{3.1}$$

Thus, by integrating all layers, the output field is connected to the input field by the transfer matrix $\mathbf{T} = \Pi_i \mathbf{T}_i$, which is a product of the individual layer-to-layer transfer matrices. However, the transfer-matrix method is less popular because of numerical instability during the integration even though it is more computationally efficient than the FDTD method. Also, it can be used to model infinite periodic structures to find the eigenmodes and eigenfrequencies as well as the transmission properties of finite slabs.

The FEM [54, 69, 70] is a frequency-domain approach to solve Maxwell's equations. Similar to the plane-wave expansion method, it is based on a variational principle. However, instead of using plane waves defined throughout the entire unit cell as the expansion basis, FEM uses a subdomain basis to discretize the computational unit cell. By taking into account discontinuities in the dielectric function, FEM overcomes the slow convergence of the plane-wave expansion method. Because of the large numbers of unknowns used to discretize typical electromagnetic field eigenvalue problems, particular attention must be paid to the choice and implementation of the eigensolver so as to minimize the required computational resources, both time and storage, while maintaining convergence.

In the remainder of this chapter, we focus on the PWEM and the FDTD method, which are perhaps the two most commonly employed methods of obtaining the dispersion relationships of photonic crystals. The algorithms for numerical implementation of these methods are described in detail below.

3.2 PLANE-WAVE EXPANSION METHOD

Consider a periodic arrangement of dielectric objects with a dielectric constant ε_a embedded in a background dielectric material with a dielectric constant ε_b. These objects can be arranged periodically in one, two, or three dimensions. For the sake of simplicity, we will begin by deriving the formulation of the PWEM for a one-dimensional periodic structure, proceeding from there to the case of two-dimensional structures that have planar periodicity while extending infinitely in the third dimension. Subsequent generalization of the formulation to three-dimensional structures is straightforward. But before we begin, let us first briefly consider the general principle on which the PWEM is based.

3.2.1 Preliminaries

As discussed in Chapter 2, in the linear dielectric with permittivity $\varepsilon(\mathbf{r})$, the source-free Maxwell's equations lead to the following equation for the magnetic field:

$$\mathcal{L}\mathbf{H} = \frac{\omega^2}{c^2}\mathbf{H}, \tag{3.2}$$

where operator $\mathcal{L} = (\nabla \times [\varepsilon_r(\mathbf{r})]^{-1}\nabla\times)$ is introduced and c is the speed of light in vacuum, whereas $\varepsilon_r(\mathbf{r}) = \varepsilon(\mathbf{r})/\varepsilon_0$ is the relative permittivity. Equation (3.2) is a standard eigenvalue problem with eigenvalue[1] $\lambda = (\omega/c)^2$ and associated eigenvector $\mathbf{H}$. It is a special case of a generalized eigenvalue problem

$$\mathcal{L}\Phi = \lambda\mathcal{M}\Phi, \tag{3.3}$$

where $\mathcal{L}$ and $\mathcal{M}$ can be in principle any self-adjoint differential or integral, scalar, or vector operators, and the problem is finding the eigenvalue λ and the corresponding eigenfunctions Φ.

To solve the generalized eigenvalue problem (3.3) numerically, we choose a set of basis functions $\{u_1, u_2, \ldots, u_N\}$ that satisfy the boundary conditions and assume the approximate solution $\tilde{\Phi}$ in the form

$$\tilde{\Phi} = \sum_{n=1}^{N} a_n u_n. \tag{3.4}$$

Substituting this into Eq. (3.3) gives

$$\sum_{n=1}^{N} a_n \mathcal{L} u_n = \lambda \sum_{n=1}^{N} a_n \mathcal{M} u_n. \tag{3.5}$$

[1]Note that in this discussion, we use λ to denote the eigenvalue. This should not be confused with the wavelength of an electromagnetic wave, which is also usually denoted by λ.

We choose a set of weight functions (also called testing functions) w_m and take the inner product of both sides of Eq. (3.5) with each w_m to obtain

$$\sum_{n=1}^{N} a_n[\langle w_m, \mathcal{L}u_n\rangle - \lambda\langle w_m, \mathcal{M}u_n\rangle] = 0, \quad m = 1, 2, \ldots, N, \tag{3.6}$$

which can be cast into matrix form as

$$\sum_{n=1}^{N} [A_{mn} - \lambda B_{mn}]a_n = 0 \quad (m = 1, 2, \ldots, N) \tag{3.7}$$

or

$$\mathbf{Ax} = \lambda\mathbf{Bx}, \tag{3.8}$$

where $\mathbf{x}$ is a column vector with elements $\{a_1, a_2, \ldots, a_N\}$ and $\mathbf{A}$ and $\mathbf{B}$ are matrices with elements $A_{mn}=\langle w_m, \mathcal{L}u_n\rangle$ and $B_{mn}=\langle w_m, \mathcal{M}u_n\rangle$. Solving this equation yields N approximate eigenvalues $\lambda_1, \lambda_2, \ldots, \lambda_N$ of Eq. (3.3).

The various ways of choosing a testing function w_m lead to different weighted residual techniques [71]. If the testing function w_m is the same as the expansion function u_n, it is called the Galerkin method [69]. If u_n and w_m are defined within the entire computational domain, we call it an entire-domain method. In this case, each expansion basis function has to satisfy the boundary conditions. If w_m exists only over a subdomain, then we refer to it as a subdomain method. For example, in the FEM [69], both u_n and w_m are defined over subdomains.

The choice of expansion basis functions is determined by the following criteria. First, they should form a compact representation of the problem at hand, so that a reasonable N yields good accuracy. Second, if a set of entire-domain basis functions is used to expand the unknown eigenfunction, then each basis function must satisfy the boundary conditions. For example, the Bloch wave boundary condition has to be used within a unit cell of a periodic structure. When a subdomain method is used, an additional boundary condition is required for those basis functions defined on the edge of the subdomain. Furthermore, choosing mutually orthogonal basis functions greatly simplifies the analysis.

In this framework, two algorithms have been developed to calculate the band structure of photonic-crystal structures: the plane-wave expansion method and the finite-element method. In this section, we focus on the plane-wave expansion method, which uses a set of plane waves as basis functions. We start with a one-dimensional photonic-crystal structure, which will provide the foundation for understanding the more complicated two- and three-dimensional cases.

3.2.2 One-Dimensional Plane-Wave Expansion Method

The simplest photonic-crystal structure consists of alternating layers of two materials with different dielectric constants, as shown in Fig. 3.1. The dielectric constant is periodic only along the x direction and is uniform in the lateral y and z directions.

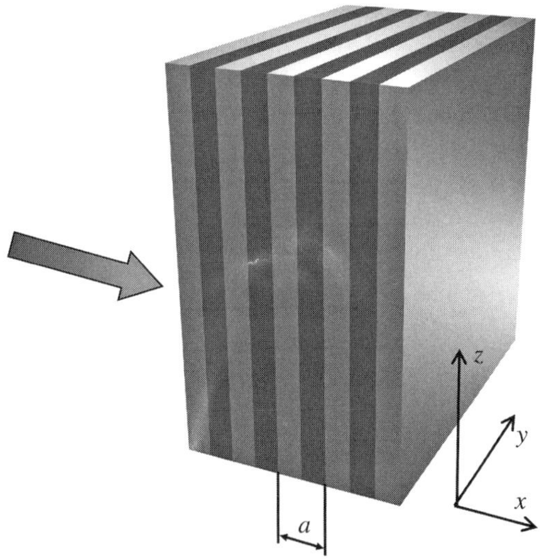

Figure 3.1 A one-dimensional photonic crystal that consists of alternating layers of materials with different dielectric constants. The periodicity extends infinitely along the x direction, and the slabs are uniform in the xy plane.

For plane waves that travel in the x direction (i.e., normal incidence), such a structure constitutes a one-dimensional optical system. Its optical properties, such as transmission and reflection, have been widely investigated for applications such as mirrors and filters. Here, we are focusing on the calculation of band structure using an approach that can be easily generalized to the more complicated two- and three-dimensional photonic crystals.

The differential forms of Maxwell's curl equations in Cartesian coordinates are

$$\begin{cases} \dfrac{\partial H_z}{\partial y} - \dfrac{\partial H_y}{\partial z} = j\omega\varepsilon_0\varepsilon_r E_x \\ \dfrac{\partial H_x}{\partial z} - \dfrac{\partial H_z}{\partial x} = j\omega\varepsilon_0\varepsilon_r E_y \\ \dfrac{\partial H_y}{\partial x} - \dfrac{\partial H_x}{\partial y} = j\omega\varepsilon_0\varepsilon_r E_z \end{cases} \quad \begin{cases} \dfrac{\partial E_y}{\partial z} - \dfrac{\partial E_z}{\partial y} = j\omega\mu_0 H_x \\ \dfrac{\partial E_z}{\partial x} - \dfrac{\partial E_x}{\partial z} = j\omega\mu_0 H_y \\ \dfrac{\partial E_x}{\partial y} - \dfrac{\partial E_y}{\partial x} = j\omega\mu_0 H_z \end{cases}. \tag{3.9}$$

The dielectric constant is uniform in y and z directions, and we assume that the E and H fields are constant within xy planes as well. This leads to vanishing spatial derivatives of the fields in the above equations,

$$\frac{\partial}{\partial y}, \frac{\partial}{\partial z} \longrightarrow 0. \tag{3.10}$$

As a result, the three-dimensional Maxwell's equations reduce to two sets of equations:

$$\begin{cases} -\dfrac{\partial H_z}{\partial x} = j\omega\varepsilon_0\varepsilon_r E_y \\ -\dfrac{\partial E_y}{\partial x} = j\omega\mu_0 H_z \end{cases} \quad \begin{cases} \dfrac{\partial E_z}{\partial x} = j\omega\mu_0 H_y \\ \dfrac{\partial H_y}{\partial x} = j\omega\varepsilon_0\varepsilon_r E_z \end{cases}. \tag{3.11}$$

Owing to the symmetry of the two sets of electromagnetic fields $\{E_z, H_y\}$ and $\{E_y, H_z\}$ only one set of field components is needed to derive the wave equations. This effect occurs because two polarization states of the wave are degenerate for normal incidence, and we arbitrarily choose the set of E_z and H_y. On eliminating either H_y or E_z, by substitution between the two equations, one-dimensional wave equations for the electric and magnetic fields are obtained as follows:

$$\begin{cases} \dfrac{1}{\varepsilon_r}\dfrac{\partial^2 E_z}{\partial x^2} + \dfrac{\omega^2}{c^2}E_z = 0 \\ \dfrac{\partial}{\partial x}\left[\dfrac{1}{\varepsilon_r}\dfrac{\partial H_y}{\partial x}\right] + \dfrac{\omega^2}{c^2}H_y = 0 \end{cases}. \tag{3.12}$$

For linear lossless media, the differential operator for the magnetic field H_y is Hermitian [72], which ensures real eigenvalues. Therefore, the second formula of Eq. (3.12) is preferred to calculate the dispersion relations.

Because only the wave propagation in the x direction is considered for now, the lateral wave vector components vanish, $k_y = k_z = 0$, and k_x is the only component that must be investigated in detail. In a one-dimensional periodic system, the primitive lattice vector $\mathbf{a}$ and the primitive reciprocal lattice vector $\mathbf{G}$ can be expressed as

$$\begin{cases} \mathbf{a} = a\hat{\mathbf{x}} \\ \mathbf{G} = g\hat{\mathbf{x}} = \dfrac{2\pi}{a}\hat{\mathbf{x}} \end{cases}, \tag{3.13}$$

where a is the period of the photonic crystal. In reciprocal vector space, the Brillouin zone and irreducible Brillouin zone are the intervals $[-\pi/a,\ \pi/a]$ and $[0,\ \pi/a]$, respectively. Because of the symmetry and periodicity in the reciprocal vector space, the vector component k_x need only be considered within the irreducible zone. Once the dispersion relation within the irreducible zone has been determined, it can be extended to the entire reciprocal vector space by using symmetry operations that leave the lattice unchanged, such as translation, mirror reflection, and in the case of more than one dimension, rotation.

To solve the magnetic field wave equation of Eq. (3.12), the procedure described in Section 3.2.1 to construct the eigensystems is applied. First, a set of plane waves is chosen as a basis in terms of which the sought field and the reciprocal of the dielectric constant $1/\varepsilon_r$ are expanded within the entire unit cell. This is possible thanks to the periodicity of the dielectric constant $\varepsilon_r(x) = \varepsilon_r(x + a)$, and Bloch's theorem that allows expressing the field as a product of a plane wave and a periodic function. Thus,

$$\begin{cases} \dfrac{1}{\varepsilon_r(x)} = \sum\limits_{n=-\infty}^{+\infty} \kappa(n) e^{jngx} \\ H_y(k_x, x) = \sum\limits_{m=-\infty}^{+\infty} h(k_x, m) e^{j(k_x+mg)x}, \end{cases} \tag{3.14}$$

where $h(k_x, m)$ are expansion coefficients, which can be also interpreted as the amplitudes of particular plane waves $e^{j(k_x+mg)x}$; g is defined in Eq. (3.13). The Fourier coefficients $\kappa(n)$ in the expansion of the reciprocal permittivity can be found using the inverse Fourier transform,

$$\kappa(n) = \frac{1}{a} \int_0^a \frac{1}{\varepsilon_r(x)} e^{-jngx} dx. \tag{3.15}$$

The higher order coefficients represent higher spatial frequencies and, hence, smaller size features or sharp changes of the dielectric constant. If the one-dimensional photonic crystal is formed by two alternating materials with relative dielectric constants ε_1 and ε_2, as shown in Fig. 3.2, then expansion coefficients can be calculated analytically,

$$\begin{aligned} \kappa(n) &= \frac{1}{a} \left(\int_0^b \frac{1}{\varepsilon_1} e^{jngx} dx + \int_b^a \frac{1}{\varepsilon_2} e^{jngx} dx \right) \\ &= \begin{cases} \dfrac{1}{jnga} \left(\dfrac{e^{jngb} - 1}{\varepsilon_1} + \dfrac{e^{jnga} - e^{jngb}}{\varepsilon_2} \right), & n \neq 0 \\ \dfrac{b}{a\varepsilon_1} + \dfrac{a-b}{a\varepsilon_2}, & n = 0. \end{cases} \end{aligned} \tag{3.16}$$

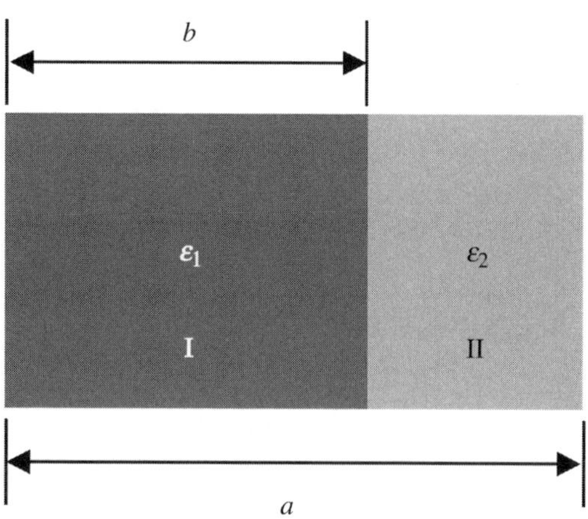

Figure 3.2 A unit cell of a one-dimensional photonic-crystal structure. It consists of two layers of dielectric material, which are labeled as layer I and layer II, with thicknesses b and $(a-b)$, and dielectric constants ε_1 and ε_2, respectively. The structure extends to infinity in the vertical direction.

To write the wave equation (3.12) in terms of expressions (3.14), we first calculate

$$\begin{aligned}\frac{1}{\varepsilon_r(x)}\frac{\partial}{\partial x}H_y &= j\sum_n \kappa(n)e^{jngx}\sum_m (k_x+mg)h(k_x,m)e^{j(k_x+mg)x}\\ &= j\sum_n\sum_m \kappa(n)(k_x+mg)h(k_x,m)e^{j[k_x+(m+n)g]x}\\ &= j\sum_n\sum_m \kappa(n-m)(k_x+mg)h(k_x,m)e^{j(k_x+ng)x},\end{aligned} \tag{3.17}$$

where in the last equality, summation index n was replaced with $n-m$. The wave equation thus becomes

$$\sum_n\sum_m \kappa(n-m)(k_x+mg)(k_x+ng)h(k_x,m)e^{j(k_x+ng)x} = \frac{\omega^2}{c^2}\sum_n h(k_x,n)e^{j(k_x+ng)x}. \tag{3.18}$$

To assemble an eigensystem, the Galerkin method is employed, in which the testing function is chosen to be the same as the expansion function of the electromagnetic field component $w_{n'}(x) = e^{j(k_x+n'g)x}$. Following the procedure outlined in Section 3.2.1, the inner product of the testing function with the terms in Eq. (3.18) is calculated

$$\begin{cases}\text{LHS} = \displaystyle\sum_n\sum_m \kappa(n-m)(k_x+mg)(k_x+ng)h(k_x,m)\int_0^a e^{j(n-n')gx}dx\\ \quad = a\displaystyle\sum_m \kappa(n'-m)(k_x+mg)(k_x+n'g)h(k_x,m),\\ \text{RHS} = \dfrac{\omega^2}{c^2}\displaystyle\sum_n h(k_x,n)\int_0^a e^{j(n-n')gx}dx\\ \quad = \dfrac{\omega^2}{c^2}ah(k_x,n'),\end{cases} \tag{3.19}$$

where LHS and RHS, respectively, refer to the left-hand side and right-hand side of Eq. (3.18). This leads to the standard eigenvalue problem

$$\sum_m (k_x+mg)\kappa(n-m)(k_x+ng)h(k_x,m) = \frac{\omega^2}{c^2}h(k_x,n), \tag{3.20}$$

which can be written in a matrix form as

$$\mathbf{Ah} = \lambda\mathbf{h}, \tag{3.21}$$

where the eigenvalues are $\lambda = \omega^2/c^2$. From the expression for the matrix elements in Eq. (3.20), it is easy to see that matrix **A** is symmetric and real, and therefore it is Hermitian with real eigenvalues.

The sum in Eq. (3.20) contains, in principle, an infinite number of terms. Thus, to solve the eigenvalue problem numerically, series expansions of both the field and the reciprocal permittivity must be truncated. However, Fourier series converge poorly for discontinuous functions, which means that the basis is ill suited to handle discontinuities of field and material present in the case analyzed here. Therefore, to calculate the band structure accurately, a large number terms in the series, i.e., many plane waves, are required. In general, the more complicated the geometry, the more plane waves are needed.

The procedure discussed above is summarized using a flowchart presented in Fig. 3.3. First, the Fourier coefficients of the reciprocal permittivity are computed in advance, either analytically or by fast Fourier transform (FFT). Then, the wave vector in the irreducible Brillouin zone is specified, and matrix **A** is constructed.

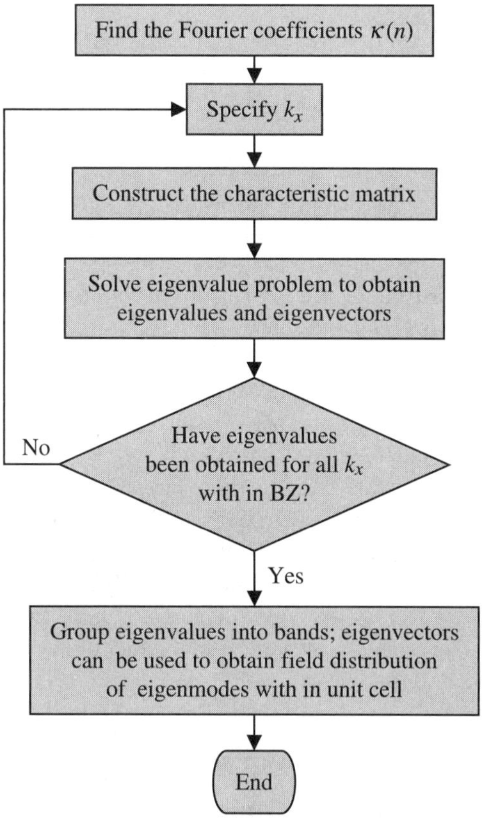

Figure 3.3 Flowchart for the calculation of the band structure of one-dimensional photonic crystals by using the plane-wave expansion method. BZ = Brillouin Zone.

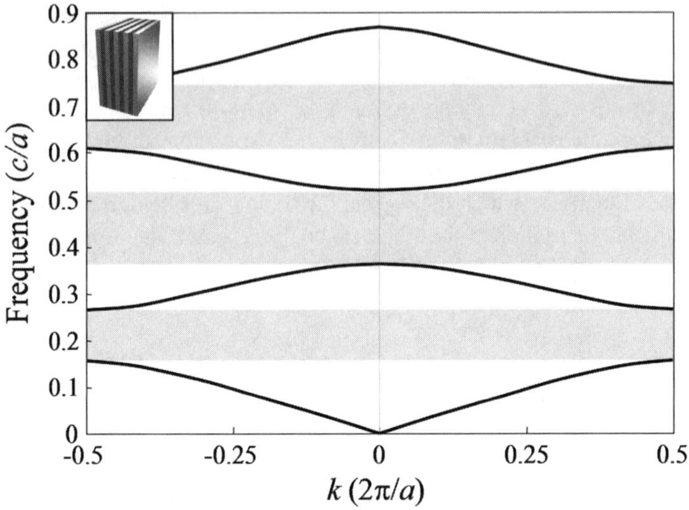

Figure 3.4 The band structure for a one-dimension photonic crystal with two layers per unit cell, with all layers having the same width of $0.5a$. The alternating dielectric constants are 12.25 and 1. Three bandgaps are shown as shaded rectangles: $0.1554 \leq a/\lambda \leq 0.2635$, $0.3626 \leq a/\lambda \leq 0.5188$, and $0.6087 \leq a/\lambda \leq 0.7487$.

The standard eigenvalue problem that results is solved by using any standard matrix algorithm. A series of discretized eigenvalues and corresponding eigenvectors is thus obtained. By repeating this procedure, all eigenvalues for all wave vectors within the discretized irreducible Brillouin zone are found. Once the band structure within the irreducible Brillouin zone is calculated, the band structure of the entire Brillouin zone is obtained by symmetry operations.

As an example of results obtained this way, consider a one-dimensional photonic crystal with two layers per unit cell, as shown in Fig. 3.1. The dielectric constants of the two materials are ε_1 and ε_2, and the layers are $0.5a$ thick, where a is the period. In Fig. 3.4, the photonic band structure for $\varepsilon_1 = 1$, $\varepsilon_2 = 12.25$ is presented, which is obtained using a total of 256 plane waves. According to this figure, in many bands of frequency, solutions to Maxwell's equations do not exist for any wave vector inside the Brillouin zone. Hence, the propagation of electromagnetic waves is prohibited in these bands, which are commonly referred to as *photonic bandgaps*. The higher the contrast of the dielectric constant, i.e., the greater $\varepsilon_>/\varepsilon_<$ (where "$>$" "$<$", respectively indicate the larger and smaller of the two permittivities), the wider the photonic bandgaps.

3.2.3 Two-Dimensional Plane-Wave Expansion Method

A two-dimensional photonic crystal is periodic along two directions and uniform in the third direction. Examples of two-dimensional photonic crystals, which are shown in Fig. 3.5, consist of arrays of parallel dielectric rods that are periodic along

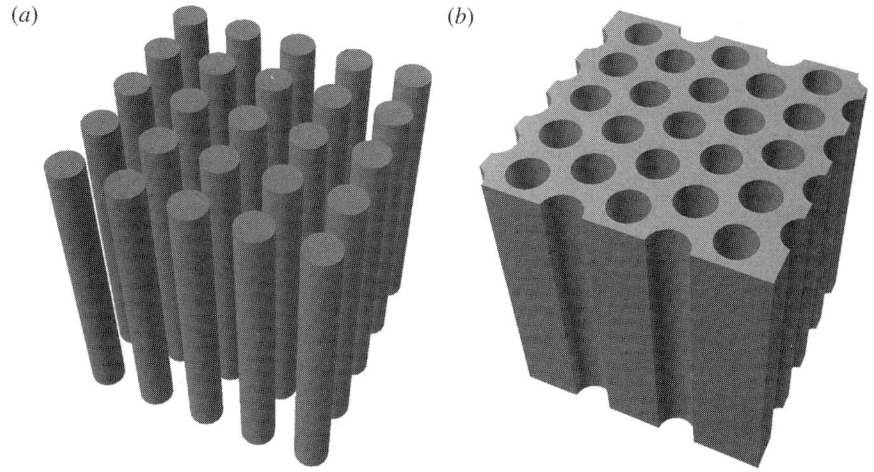

Figure 3.5 Two-dimensional photonic-crystal structures. (a) Square lattice of dielectric rods embedded in air background. (b) Triangular lattice of air holes embedded within high dielectric constant background.

x and y directions and are homogeneous along the z direction where they extend indefinitely.

In general, a two-dimensional periodic structure is defined through the set of lattice vectors $\mathbf{R}$ that are generated by primitive lattice vectors $\mathbf{a}_1$ and $\mathbf{a}_2$

$$\mathbf{R} = m\mathbf{a}_1 + n\mathbf{a}_2, \tag{3.22}$$

where m and n are integers. With the substitution of all possible integer pairs for m and n, Eq. (3.22) generates a Bravais lattice that consists of an infinite number of points periodically distributed throughout the plane. The lattice of $\mathbf{R}$ vectors thus obtained is also referred to as the primitive lattice. The reciprocal lattice is formed by a set of vectors $\mathbf{G}$ such that

$$e^{j\mathbf{G}\cdot\mathbf{R}} = 1, \tag{3.23}$$

for all $\mathbf{R}$. Equation (3.23) can also be expressed as

$$\mathbf{G} \cdot \mathbf{R} = 2\pi l, \tag{3.24}$$

where l is an integer. The reciprocal lattice of vectors $\mathbf{G}$ is generated by vectors $\mathbf{g}_1$ and $\mathbf{g}_2$

$$\mathbf{G} = m'\mathbf{g}_1 + n'\mathbf{g}_2, \tag{3.25}$$

where m' and n' are integers, whereas $\mathbf{g}_1$ and $\mathbf{g}_2$ are primitive reciprocal lattice vectors defined by the relation

$$\mathbf{a}_i \cdot \mathbf{g}_j = 2\pi\delta_{ij}, \tag{3.26}$$

where δ_{ij} is the Kronecker delta

$$\delta_{ij} = \begin{cases} 1, & i = j \\ 0, & i \neq j \end{cases}. \tag{3.27}$$

In two dimensions, two canonical highly symmetric lattice configurations exist: a square lattice and a triangular lattice. As shown in Fig. 3.6a, the primitive lattice vectors of the square lattice can be written as $\mathbf{a}_1 = a\hat{\mathbf{x}}$ and $\mathbf{a}_2 = a\hat{\mathbf{y}}$, where a is the lattice constant. The unit cell of this periodic structure is the region enclosed by the square. Using Eq. (3.26), the reciprocal lattice vectors are found as $\mathbf{g}_1 = (2\pi/a)\hat{\mathbf{x}}$ and $\mathbf{g}_2 = (2\pi/a)\hat{\mathbf{y}}$, as shown in Fig. 3.6b. In the reciprocal space, the Brillouin zone is constructed as outlined by a solid line in Fig. 3.6b. Because of the symmetry of the square lattice, one only needs to calculate the band structure within 1/8 of this Brillouin zone, which in this case consists of the triangular region $\triangle\Gamma$XM shaded in Fig. 3.6; all other points in the full Brillouin zone are mapped onto points within the irreducible zone by symmetry operations that include 90° rotations and mirror reflections. The wave vectors of the corner points of the irreducible zone are by convention assigned labels Γ, X, and M, where $\Gamma = \mathbf{0}$, $\mathrm{X} = (\pi/a)\hat{\mathbf{x}}$, and $\mathrm{M} = (\pi/a)(\hat{\mathbf{x}} + \hat{\mathbf{y}})$.

For the triangular lattice, a suitable choice of the two primitive lattice vectors is $\mathbf{a}_1 = a\hat{\mathbf{x}}$ and $\mathbf{a}_2 = (a/2)(\hat{\mathbf{x}} + \sqrt{3}\hat{\mathbf{y}})$, where a is the lattice period. As shown in Fig. 3.7a, the unit cell is the region enclosed by the rhombus. From Eq. (3.26), the primitive reciprocal lattice vectors are given by $\mathbf{g}_1 = (2\pi/a)(\hat{\mathbf{x}} - \hat{\mathbf{y}}/\sqrt{3})$ and $\mathbf{g}_2 = (4\pi/\sqrt{3}a)\hat{\mathbf{y}}$, as shown in Fig. 3.7b. The corresponding Brillouin zone is outlined by a solid line in Fig. 3.7b. Because of the symmetry of the lattice, one only needs to calculate the band structure within 1/12 of this Brillouin zone, which consists of the triangular region $\triangle\Gamma$MK shaded in Fig. 3.7. The wave vectors of the corner points are given by $\Gamma = \mathbf{0}$, $\mathrm{M} = (\pi/a)(\hat{\mathbf{x}} + \hat{\mathbf{y}}/\sqrt{3})$, and $\mathrm{K} = (4\pi/\sqrt{3}a)\hat{\mathbf{x}}$.

Although the square lattice and the triangular lattice described above are often shown as representative examples of 2D photonic crystals, it should be noted that in

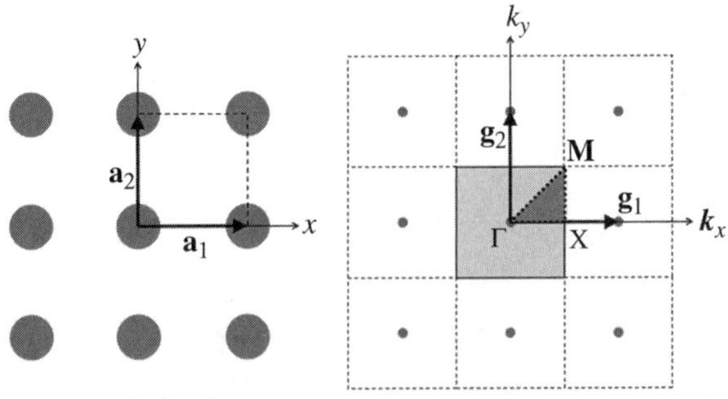

Figure 3.6 Square lattice. (a) Real-space arrangement with two primitive lattice vectors $\mathbf{a}_1$ and $\mathbf{a}_2$. (b) Reciprocal space with two primitive lattice vectors $\mathbf{g}_1$ and $\mathbf{g}_2$; the Brillouin zone is outlined by a solid line and shaded gray, and the darker shaded triangular region $\triangle\Gamma$XM represents the irreducible 1/8 Brillouin zone.

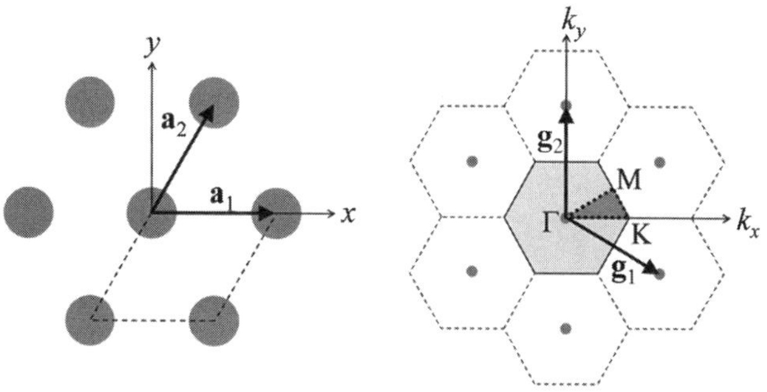

Figure 3.7 Triangular lattice. (a) Real-space arrangement with two primitive lattice vectors $\mathbf{a}_1$ and $\mathbf{a}_2$. (b) Reciprocal wave vector space with two primitive reciprocal lattice vectors $\mathbf{g}_1$ and $\mathbf{g}_2$; the Brillouin zone is outlined by a solid line and shaded gray, and the darker shaded triangular region $\triangle\Gamma$MK represents irreducible 1/12 Brillouin zone.

general, a Bravais lattice in two dimensions is generated by two vectors $\mathbf{a}_1$ and $\mathbf{a}_2$ that are neither of equal length nor perpendicular to one another.

If the dielectric function $\varepsilon(\mathbf{r})$ satisfies the relation $\varepsilon(\mathbf{r}+\mathbf{R}) = \varepsilon(\mathbf{r})$ for each $\mathbf{R}$ in the primitive lattice, i.e., $\varepsilon_r(\mathbf{r})$ is periodic in 2D, then so is the reciprocal of the dielectric function, $\varepsilon^{-1}(\mathbf{r}+\mathbf{R}) = \varepsilon^{-1}(\mathbf{r})$. Accordingly, it can be expressed as a Fourier series:

$$\varepsilon^{-1}(\mathbf{r}) = \sum_{\mathbf{G}'} \kappa(\mathbf{G}')e^{j(\mathbf{G}'\cdot\mathbf{r})}, \tag{3.28}$$

where $\kappa(\mathbf{G})$ is the Fourier coefficient, given by

$$\kappa(\mathbf{G}) = \frac{1}{S}\int_S \varepsilon_r^{-1}(\mathbf{r})e^{-j\mathbf{G}\cdot\mathbf{r}}ds, \tag{3.29}$$

where S is the area of the unit cell.

For arbitrary variation of the dielectric constant within the unit cell, numerical integration is required to obtain the Fourier coefficients $\kappa(\mathbf{G})$, wherein the continuous integral in Eq. (3.29) is replaced by an approximation in the form of a discrete finite sum. To this end, a unit cell is divided into $N_1 \times N_2$ small cells, as shown in Fig. 3.8, where the two primitive lattice vectors $\mathbf{a}_1$ and $\mathbf{a}_2$ define a skewed coordinate system with skew angle θ. The 1-small-cell increments $\Delta\xi = a_1/N_1$ and $\Delta\eta = a_2/N_2$ in the ξ and η directions are enumerated with integer labels $n_1 = 1, 2, \ldots, N_1$ and $n_2 = 1, 2, \ldots, N_2$. As a result, the position vector in the skewed coordinate system can be written as

$$\mathbf{r} = \xi\frac{\mathbf{a}_1}{a_1} + \eta\frac{\mathbf{a}_2}{a_2} = n_1\Delta\xi\frac{\mathbf{a}_1}{a_1} + n_2\Delta\eta\frac{\mathbf{a}_2}{a_2}. \tag{3.30}$$

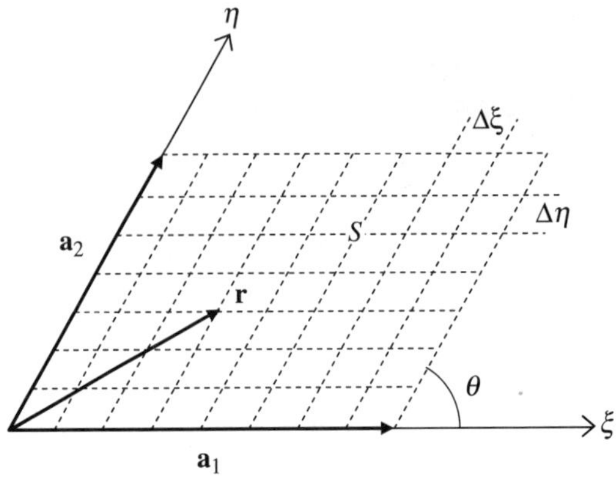

Figure 3.8 Diagram of the unit cell mesh for numerical integration in the determination of Fourier coefficients. The primitive lattice vectors are $\mathbf{a}_1$ and $\mathbf{a}_2$. The unit cell is divided into an $N_1 \times N_2$ array of small cells in the skewed coordinate system.

With $\mathbf{G}=m_1\mathbf{g}_1 + m_2\mathbf{g}_2$, Eq. (3.30) is substituted into Eq. (3.29) to yield

$$\begin{aligned}\kappa(m_1, m_2) &= \frac{1}{S}\int_0^{a_1}\int_0^{a_2} \frac{1}{\varepsilon_r(\xi, \eta)} e^{-j[m_1\mathbf{g}_1+m_2\mathbf{g}_2]\left[\xi\frac{\mathbf{a}_1}{a_1}+\eta\frac{\mathbf{a}_2}{a_2}\right]} \sin\theta \, d\xi \, d\eta \\ &= \frac{1}{S}\int_0^{a_1}\int_0^{a_2} \frac{1}{\varepsilon_r(\xi, \eta)} e^{-j2\pi\left[\frac{m_1 n_1}{N_1}+\frac{m_2 n_2}{N_2}\right]} \sin\theta \, d\xi \, d\eta \\ &\cong \frac{1}{N_1 N_2}\sum_{n_1}\sum_{n_2} \frac{1}{\varepsilon_r(n_1, n_2)} e^{-j2\pi\left[\frac{m_1 n_1}{N_1}+\frac{m_2 n_2}{N_2}\right]},\end{aligned} \tag{3.31}$$

where in the last step, the integrals are approximated by discrete sums. The Fourier coefficients $\kappa(m_1, m_2)$ can be obtained by using the FFT, with computational complexity of $O(N \log N)$, where $N = N_1 N_2$ is the total number of plane waves.

For a circular rod of radius r and dielectric constant ε_a embedded in a background material with dielectric constant ε_b, the Fourier coefficients can be calculated analytically

$$\kappa(\mathbf{G}) = \begin{cases} \dfrac{1}{\varepsilon_a} f + \dfrac{1}{\varepsilon_b}(1-f), & |\mathbf{G}| = 0 \\ \left(\dfrac{1}{\varepsilon_a} - \dfrac{1}{\varepsilon_b}\right) f \, \dfrac{J_1(|\mathbf{G}|R)}{|\mathbf{G}|R}, & |\mathbf{G}| \neq 0, \end{cases} \tag{3.32}$$

where $f = \pi r^2/(a^2 \sin\theta)$ is the fill factor and $J_1(x)$ is the first-order Bessel function.

Now we turn to the calculation of the photonic band structure for electromagnetic waves that propagate in such a structure in a direction perpendicular to the dielectric rods. Because rods are homogeneous in the z direction, derivatives of the fields with respect to z should be zero; i.e., $\partial/\partial z \to 0$. As a result, Maxwell's equations in the Cartesian coordinate system, Eq. (3.9), reduce to two sets of decoupled equations as follows:

$$\begin{cases} \dfrac{\partial H_z}{\partial y} = j\omega\varepsilon_0\varepsilon_r E_x \\ -\dfrac{\partial H_z}{\partial x} = j\omega\varepsilon_0\varepsilon_r E_y \\ \dfrac{\partial E_x}{\partial y} - \dfrac{\partial E_y}{\partial x} = j\omega\mu_0 H_z \end{cases} \quad \begin{cases} -\dfrac{\partial E_z}{\partial y} = j\omega\mu_0 H_x \\ \dfrac{\partial E_z}{\partial x} = j\omega\mu_0 H_y \\ \dfrac{\partial H_y}{\partial x} - \dfrac{\partial H_x}{\partial y} = j\omega\varepsilon_0\varepsilon_r E_z \end{cases}. \tag{3.33}$$

The first set of equations has only electric field components within the xy plane; they describe what we refer to as the transverse-electric (TE) mode, or TE-polarized wave. The second set of equations has only magnetic field components within the xy plane, describing what we refer to as the transverse-magnetic (TM) mode, or TM-polarized wave.

By eliminating the electric field components E_x and E_y from the first set of equations, we obtain the wave equation for the magnetic field component H_z for the TE mode

$$\frac{\partial}{\partial x}\left(\frac{1}{\varepsilon(\mathbf{r})}\frac{\partial}{\partial x}H_z(\mathbf{r})\right) + \frac{\partial}{\partial y}\left(\frac{1}{\varepsilon(\mathbf{r})}\frac{\partial}{\partial y}H_z(\mathbf{r})\right) + \left(\frac{\omega}{c}\right)^2 H_z(\mathbf{r}) = 0. \tag{3.34}$$

Using the same procedure, we can eliminate the magnetic field components H_x and H_y from the second set of equations to obtain the wave equation for the electric field component E_z for the TM mode:

$$\frac{1}{\varepsilon(\mathbf{r})}\left[\frac{\partial^2}{\partial x^2} + \frac{\partial^2}{\partial y^2}\right]E_z(\mathbf{r}) + \left(\frac{\omega}{c}\right)^2 E_z(\mathbf{r}) = 0. \tag{3.35}$$

In the following sections, the analysis and field solutions for each class of eigenmodes, namely TE and TM polarizations, are presented separately.

3.2.3.1 TE Polarization. A "TE-polarized" wave is a wave in which the electric field is polarized in the xy plane, i.e., a wave with only E_x, E_y, and H_z nonvanishing field components. In a periodic medium, the **H** field can be decomposed into Bloch waves, i.e., plane waves with their amplitudes modulated with periodic functions

$$H_z(\mathbf{r}) = u(\mathbf{k}, \mathbf{r})e^{j(\mathbf{k}\cdot\mathbf{r})}, \tag{3.36}$$

where the function $u(\mathbf{k}, \mathbf{r})$ has the periodicity of the lattice $u(\mathbf{k}, \mathbf{r} + \mathbf{R}) = u(\mathbf{k}, \mathbf{r})$ and therefore can be expanded in a Fourier series

$$u(\mathbf{k}, \mathbf{r}) = \sum_{\mathbf{G}} h(\mathbf{k}, \mathbf{G})e^{j(\mathbf{G}\cdot\mathbf{r})}, \tag{3.37}$$

which leads to

$$H_z(\mathbf{r}) = \sum_{\mathbf{G}} h(\mathbf{k}, \mathbf{G}) e^{j(\mathbf{k}+\mathbf{G})\cdot\mathbf{r}}, \tag{3.38}$$

where $\mathbf{k} = k_x\hat{\mathbf{x}} + k_y\hat{\mathbf{y}}$ is the two-dimensional wave vector.

Substituting the field and $1/\varepsilon_r(\mathbf{r})$ expansions, Eqs. (3.38) and (3.28), respectively, into the first term of the wave Eq. (3.34) yields

$$\begin{aligned}
\frac{\partial}{\partial x}\left(\frac{1}{\varepsilon_r}\frac{\partial H_z}{\partial x}\right) &= \frac{\partial}{\partial x}\left[\sum_{\mathbf{G}'} \kappa(\mathbf{G}') e^{j\mathbf{G}'\cdot\mathbf{r}} \cdot \frac{\partial}{\partial x}\sum_{\mathbf{G}} h(\mathbf{k}, \mathbf{G}) e^{j(\mathbf{k}+\mathbf{G})\cdot\mathbf{r}}\right] \\
&= \frac{\partial}{\partial x}\left[\sum_{\mathbf{G}'} \kappa(\mathbf{G}') e^{-j\mathbf{G}'\cdot\mathbf{r}} \cdot j(k_x + G_x)\sum_{\mathbf{G}} h(\mathbf{k}, \mathbf{G}) e^{j(\mathbf{k}+\mathbf{G})\cdot\mathbf{r}}\right] \\
&= j\frac{\partial}{\partial x}\left[\sum_{\mathbf{G}',\mathbf{G}} (k_x + G_x)\kappa(\mathbf{G}') h(\mathbf{k}, \mathbf{G}) e^{j(\mathbf{k}+\mathbf{G}+\mathbf{G}')\cdot\mathbf{r}}\right] \\
&= j\sum_{\mathbf{G}'',\mathbf{G}} \frac{\partial}{\partial x}\left[(k_x + G_x)\kappa(\mathbf{G}'' - \mathbf{G}) h(\mathbf{k}, \mathbf{G}) e^{j(\mathbf{k}+\mathbf{G}'')\cdot\mathbf{r}}\right] \\
&= -\sum_{\mathbf{G}'',\mathbf{G}} (k_x + G_x)(k_x + G''_x)\kappa(\mathbf{G}'' - \mathbf{G}) h(\mathbf{k}, \mathbf{G}) e^{j(\mathbf{k}+\mathbf{G}')\cdot\mathbf{r}},
\end{aligned} \tag{3.39}$$

where $\mathbf{G}''$ is defined as $\mathbf{G}'' = \mathbf{G} + \mathbf{G}'$. Similarly, the second term can be written as

$$\begin{aligned}
\frac{\partial}{\partial y}\left(\frac{1}{\varepsilon_r}\frac{\partial H_z}{\partial y}\right) &= j\sum_{\mathbf{G}'',\mathbf{G}} \frac{\partial}{\partial y}\left[(k_y + G_y)\kappa(\mathbf{G}'' - \mathbf{G}) h(\mathbf{k}, \mathbf{G}) e^{j(\mathbf{k}+\mathbf{G}'')\cdot\mathbf{r}}\right] \\
&= -\sum_{\mathbf{G}'',\mathbf{G}} (k_y + G_y)(k_y + G''_y)\kappa(\mathbf{G}'' - \mathbf{G}) h(\mathbf{k}, \mathbf{G}) e^{j(\mathbf{k}+\mathbf{G}'')\cdot\mathbf{r}}.
\end{aligned} \tag{3.40}$$

Assembling two terms together, the wave equation can be written as

$$\begin{aligned}
&\sum_{\mathbf{G},\mathbf{G}''} (\mathbf{k} + \mathbf{G}) \cdot (\mathbf{k} + \mathbf{G}'')\kappa(\mathbf{G}'' - \mathbf{G}) h(\mathbf{k}, \mathbf{G}) e^{j(\mathbf{k}+\mathbf{G}'')\cdot\mathbf{r}} \\
&\quad = \frac{\omega^2}{c^2}\sum_{\mathbf{G}} h(\mathbf{k}, \mathbf{G}) e^{j(\mathbf{k}+\mathbf{G})\cdot\mathbf{r}}.
\end{aligned} \tag{3.41}$$

Choosing the testing function to be the same as the expansion basis function, i.e., a plane wave $\exp(j\mathbf{G}' \cdot \mathbf{r})$, and taking the inner product of both sides with each testing function, which includes integration over the unit cell, yields

$$\begin{aligned}
&\sum_{\mathbf{G},\mathbf{G}''} (\mathbf{k} + \mathbf{G}) \cdot (\mathbf{k} + \mathbf{G}'')\kappa(\mathbf{G}'' - \mathbf{G}) h(\mathbf{k}, \mathbf{G}) \int_s e^{j(\mathbf{G}''-\mathbf{G}')\cdot\mathbf{r}} ds \\
&\qquad = \frac{\omega^2}{c^2}\sum_{\mathbf{G}} h(\mathbf{k}, \mathbf{G}) \int_s e^{j(\mathbf{G}-\mathbf{G}')\cdot\mathbf{r}} ds.
\end{aligned} \tag{3.42}$$

The integral over the unit cell can be carried out analytically

$$\int_s e^{j(\mathbf{G}''-\mathbf{G}')\cdot\mathbf{r}}ds = \begin{cases} S, & \mathbf{G}'' = \mathbf{G}' \\ 0, & \mathbf{G}'' \neq \mathbf{G}' \end{cases}. \tag{3.43}$$

This leads to a standard eigenvalue problem,

$$\sum_{\mathbf{G}} (\mathbf{k}+\mathbf{G}) \cdot (\mathbf{k}+\mathbf{G}')\kappa(\mathbf{G}'-\mathbf{G})h(\mathbf{k}, \mathbf{G}) = \frac{\omega^2}{c^2} h(\mathbf{k}, \mathbf{G}'), \tag{3.44}$$

or in matrix form

$$\mathbf{Ah} = \lambda\mathbf{h}, \tag{3.45}$$

where the eigenvalues are $\lambda = \omega^2/c^2$. It is noted that matrix $\mathbf{A}$ is self-adjoint, which means that this eigensystem has real eigenvalues.

3.2.3.2 TM Polarization. A "TM-polarized" wave is a wave in which the magnetic field components are polarized in the transverse direction (xy plane), i.e., a wave with only H_x, H_y, and E_z field components. The electric field can be then expressed as

$$E_z(\mathbf{r}) = \sum_{\mathbf{G}} \eta(\mathbf{k}, \mathbf{G}) e^{j(\mathbf{k}+\mathbf{G})\cdot\mathbf{r}}. \tag{3.46}$$

Substituting the expanded reciprocal permittivity, Eq. (3.28), and electric field, Eq. (3.46), into the wave equation, Eq. (3.35), yields

$$\sum_{\mathbf{G}'} \kappa(\mathbf{G}')e^{j\mathbf{G}'\cdot\mathbf{r}} \sum_{\mathbf{G}} [(k_x+G_x)^2 + (k_y+G_y)^2]\eta(\mathbf{k}, \mathbf{G})e^{j(\mathbf{k}+\mathbf{G})\cdot\mathbf{r}}$$

$$= \frac{\omega^2}{c^2} \sum_{\mathbf{G}} \eta(\mathbf{k}, \mathbf{G})e^{j(\mathbf{k}+\mathbf{G})\cdot\mathbf{r}},$$

which, on substitution of $\mathbf{G}''$ for $\mathbf{G}+\mathbf{G}'$, simplifies to

$$\sum_{\mathbf{G}'',\mathbf{G}} \kappa(\mathbf{G}''-\mathbf{G})(\mathbf{k}+\mathbf{G})^2\eta(\mathbf{k}, \mathbf{G})e^{j(\mathbf{k}+\mathbf{G}'')\cdot\mathbf{r}} = \frac{\omega^2}{c^2} \sum_{\mathbf{G}''} \eta(\mathbf{k}, \mathbf{G}'')e^{j(\mathbf{k}+\mathbf{G}'')\cdot\mathbf{r}}. \tag{3.47}$$

Taking the inner product of this equation with each testing function and using Eq. (3.43), yields

$$\sum_{\mathbf{G}} \kappa(\mathbf{G}'-\mathbf{G})(\mathbf{k}+\mathbf{G})^2\eta(\mathbf{k}, \mathbf{G}) = \frac{\omega^2}{c^2} \eta(\mathbf{k}, \mathbf{G}'), \tag{3.48}$$

which also has the form standard eigenvalue equation, albeit for a nonsymmetric matrix. However, the substitution

$$b(\mathbf{k}, \mathbf{G}) = |\mathbf{k}+\mathbf{G}|\eta(\mathbf{k}, \mathbf{G}) \tag{3.49}$$

yields an eigenvalue problem for a symmetric matrix. Multiplying both sides of Eq. (3.48) by $|\mathbf{k}+\mathbf{G}'|$ leads to

$$\sum_{\mathbf{G}} |\mathbf{k}+\mathbf{G}'|\kappa(\mathbf{G}'-\mathbf{G})(\mathbf{k}+\mathbf{G})^2\eta(\mathbf{k},\mathbf{G}) = \frac{\omega^2}{c^2}|\mathbf{k}+\mathbf{G}'|\eta(\mathbf{k},\mathbf{G}'),$$

which, using Eq. (3.49) to substitute for $\eta(\mathbf{k},\mathbf{G})$, can be rearranged to yield

$$\sum_{\mathbf{G}} |\mathbf{k}+\mathbf{G}'|\kappa(\mathbf{G}'-\mathbf{G})|\mathbf{k}+\mathbf{G}|b(\mathbf{k},\mathbf{G}) = \frac{\omega^2}{c^2}b(\mathbf{k},\mathbf{G}'). \tag{3.50}$$

This result can also be written in matrix form,

$$\mathbf{A}\mathbf{b} = \lambda\mathbf{b}, \tag{3.51}$$

where the eigenvalues are again $\lambda=\omega^2/c^2$.

In both Eqs. (3.44) and (3.50), the reciprocal lattice sum is truncated to obtain a numerical solution to the eigenvalue problem. Once Eq. (3.44) or Eq. (3.50) has been solved for both the eigenvalues $(\omega/c)^2$ and the respective eigenvectors $h(\mathbf{k},\mathbf{G})$ or $b(\mathbf{k},\mathbf{G})$, the electric and magnetic fields that correspond to a given eigenfrequency $\omega(\mathbf{k})$ can be computed using Eq. (3.38), or Eqs. (3.46) and (3.49), respectively. The solution of Eq. (3.44) or Eq. (3.50) for a set of wave vectors $\{\mathbf{k}\}$ that span the irreducible Brillouin zone results in a dispersion relation for the eigenmodes that are supported by a particular photonic crystal.

A flowchart is presented in Fig. 3.9 to illustrate the procedure discussed above. In summary, before beginning to calculate the band structure, the Fourier coefficients are computed in advance, either analytically, or numerically by FFT. By specifying the wave vector in the irreducible Brillouin zone, matrix $\mathbf{A}$ is constructed for both TE and TM modes. Then, the eigenvalue problem is solved by a standard matrix algorithm. A series of discretized eigenfrequencies $\omega(\mathbf{k})$ and corresponding eigenvectors is thus obtained. By repeating this procedure, eigenmodes are found for a set of wave vectors within the irreducible Brillouin zone. These modes are enumerated from the lowest frequency to the highest, where the number is the band index. The set of all solutions with the same band index n, $\omega_n(\mathbf{k})$, is called the nth band. Once we have the band structure within the irreducible Brillouin zone, the rest of the Brillouin zone is obtained by symmetry operations. The dispersion surfaces, which express the eigenfrequencies as a function of the wave vector for each band, are obtained by connecting the nearest points in the ω-$\mathbf{k}$-space within each band.

This procedure was used for a square lattice with dielectric rods in air background shown in Fig. 3.5a and in the inset of Fig. 3.10a. The rods have a circular cross section with radius $r = 0.2a$, where a is the lattice constant and the dielectric constant is $\varepsilon_r = 12.25$. A total of 256 plane waves are used to obtain the results described below. The dispersion relation along the edge of the irreducible Brillouin zone is shown in Fig. 3.10a. The in-plane wave vector $\mathbf{k}$ goes from Γ to X to M and back to Γ, as shown by the dashed triangle in Fig. 3.6b. Solid and dashed curves represent the TM and TE modes, respectively. Examining the dispersion diagram, we observe

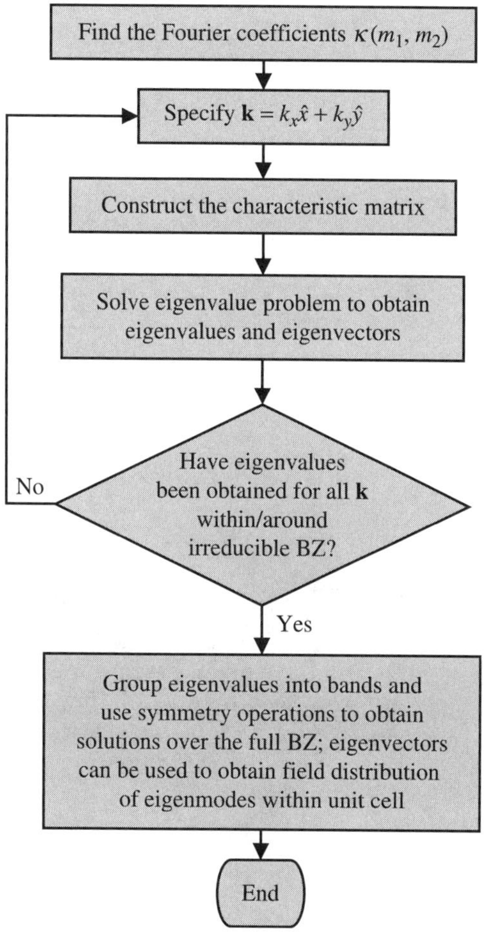

Figure 3.9 Flowchart for the calculation of the band structure of two-dimensional photonic crystals by using the plane-wave expansion method.

that there are ranges of frequency over which solutions to the eigenvalue equations for TM modes do not exist for any wave vector inside the Brillouin zone, i.e., in all directions. This means that propagation of TM-polarized electromagnetic waves is prohibited within these frequency ranges, which are referred to as *photonic bandgaps* and in Fig. 3.10a are indicated as shaded rectangles. The wider bandgap spans the frequency range from $0.2783c/a$ to $0.4158c/a$. However, there is no photonic band for TE modes.

A three-dimensional plot of dispersion surfaces for the first two bands of TM modes is shown in Fig. 3.10b, where the photonic bandgap between the first and the second bands is evident.

Now consider a triangular lattice with air cylinders embedded in dielectric background with dielectric constant $\varepsilon_r = 12.25$, as depicted in Fig. 3.5b. The rods are of

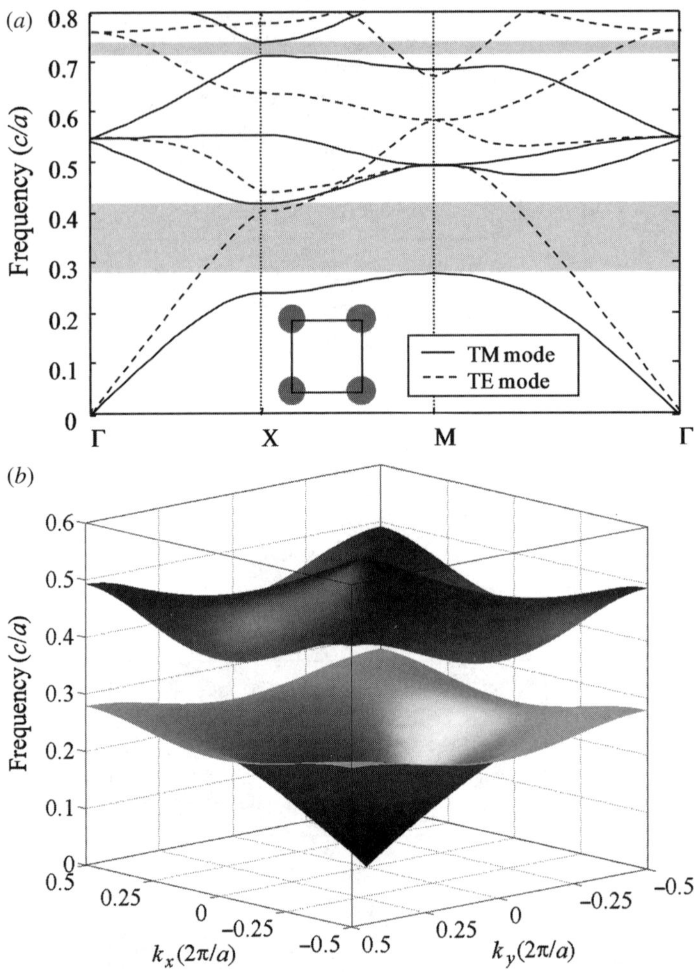

Figure 3.10 The band structure for a square lattice with dielectric rods embedded in air background. (a) Dispersion diagram along the perimeter of the irreducible Brillouin zone. (b) Dispersion surfaces of first and second bands.

a circular cross section with radius $r = 0.4a$. Again, a total of 256 plane waves are used to obtain the results described below. The dispersion relation along the perimeter of the irreducible Brillouin zone is shown in Fig. 3.11a. The in-plane wave vector **k** goes from Γ to M to K and back to Γ, as shown in Fig. 3.7b. Solid and dashed lines again represent the TM and TE modes, respectively.

Examining the dispersion diagram, one can see that for this lattice, a TE photonic bandgap between the first and the second bands is depicted as a gray rectangle in Fig. 3.11a. A three-dimensional plot of the dispersion surfaces of the first two TE bands is shown in Fig. 3.11b, which confirms the presence of a photonic bandgap. In contrast, this photonic crystal displays no photonic bandgap for the TM mode.

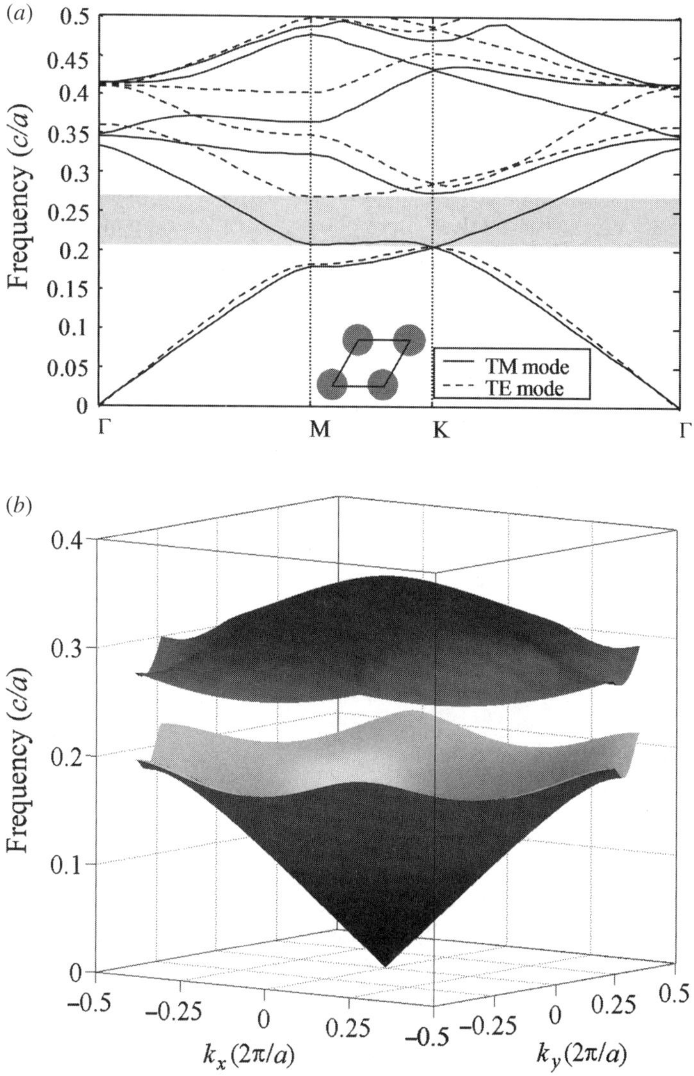

Figure 3.11 The band structure for a triangular lattice with dielectric rods embedded in air background. (a) Dispersion diagram along the edge of irreducible Brillouin zone. (b) Dispersion surfaces of first and second bands.

The presence of photonic bandgaps has been exploited in the design of many useful photonic devices as discussed in Chapter 4. However, the unique dispersion within the transmission bands leads to unusual phenomena such as superprism [73–76] and self-collimation [77–79], which are discussed in Chapter 5. In either case, the plane-wave expansion method described above is instrumental both as a tool of discovery and design.

3.2.4 Three-Dimensional Plane-Wave Expansion Method

The PWEM described above for two dimensions can be extended to three-dimensional (3D) periodic structures, in which the periodic dielectric function $\varepsilon(\mathbf{r})$ defines a 3D lattice. Because of the lack of translational invariance along z (or any direction) in the 3D case, there is no decoupling of the Maxwell equations for the Cartesian field components analogous Eq. (3.9). In other words, solutions cannot in general be separated into TE or TM modes where one scalar field component is sufficient to describe the problem fully. In 3D, a full vector problem must be solved.

As pointed out in Chapter 2, the Maxwell operator based on the magnetic field, which appears on the left-hand side of equation

$$\nabla \times \left(\frac{1}{\varepsilon_r(\mathbf{r})} \nabla \times \mathbf{H}(\mathbf{r}) \right) = \left(\frac{\omega}{c} \right)^2 \mathbf{H}(\mathbf{r}) \tag{3.52}$$

is Hermitian, whereas the operator based on the electric field is not. Therefore, we choose to work with the wave equation for the magnetic field as it guarantees real eigenvalues. Because the dielectric function in three-dimensional space is periodic, it can be expanded in the form of a Fourier series

$$\frac{1}{\varepsilon_r(\mathbf{r})} = \sum_{\mathbf{G}'} \kappa(\mathbf{G}') e^{j(\mathbf{G}' \cdot \mathbf{r})}, \tag{3.53}$$

where $\kappa(\mathbf{G})$ is the Fourier coefficient

$$\kappa(\mathbf{G}) = \frac{1}{V} \int_V \frac{1}{\varepsilon_r(\mathbf{r})} e^{-j\mathbf{G} \cdot \mathbf{r}} dv, \tag{3.54}$$

and V is the volume of a unit cell. As in the two-dimensional case, by discretizing the unit cell as an array of $N_1 \times N_2 \times N_3$ uniform cells, we can obtain the Fourier coefficient in the form of a discretized Fourier transform (DFT):

$$\kappa(m_1, m_2, m_3) = \frac{1}{N_1 N_2 N_3} \sum_{n_1} \sum_{n_2} \sum_{n_3} \frac{1}{\varepsilon_r(n_1, n_2, n_3)} e^{-j2\pi \left[\frac{m_1 n_1}{N_1} + \frac{m_2 n_2}{N_2} + \frac{m_3 n_3}{N_3} \right]}, \tag{3.55}$$

where m_1, m_2, m_3 are integer indices that specify a reciprocal lattice vector $\mathbf{G}$ as $\mathbf{G} = m_1 \mathbf{g}_1 + m_2 \mathbf{g}_2 + m_3 \mathbf{g}_3$, and $\mathbf{g}_i$ ($i = 1, 2, 3$) are primitive reciprocal lattice vectors. The Fourier coefficient can be efficiently calculated by a 3D FFT algorithm, which is built into commercial mathematical software suites and is available in public-domain computational libraries.

Similar to the 2D case examined previously, the magnetic field can be expanded in a plane-wave basis that consists of all plane waves with wave vectors given by $\mathbf{k} + \mathbf{G}$, where $\mathbf{k} = k_x \hat{\mathbf{x}} + k_y \hat{\mathbf{y}} + k_z \hat{\mathbf{z}}$ lies within the first Brillouin zone and $\mathbf{G}$ is a reciprocal lattice vector. But because the field cannot be decomposed into TE and

TM modes, in 3D, our treatment must be generalized to include vector fields. Because the amplitude of any plane wave $\sim e^{j(\mathbf{k}+\mathbf{G})\cdot\mathbf{r}}$ is perpendicular to its propagation wave vector $\mathbf{k}+\mathbf{G}$, the former can be expressed as a linear combination of a pair of unit vectors $\hat{\mathbf{e}}_{\mathbf{G}}^{1}$ and $\hat{\mathbf{e}}_{\mathbf{G}}^{2}$, such that $\{\hat{\mathbf{e}}_{\mathbf{G}}^{1}, \hat{\mathbf{e}}_{\mathbf{G}}^{2}, \mathbf{k}+\mathbf{G}\}$ forms a mutually orthogonal right-handed triple. Thus, in 3D, the expansion of a Bloch wave $\mathbf{H}_{\mathbf{k}}(\mathbf{r})$ becomes

$$\mathbf{H}_{\mathbf{k}}(\mathbf{r}) = \sum_{\mathbf{G}}\sum_{\lambda=1}^{2} h_{\mathbf{G}}^{\lambda}\hat{\mathbf{e}}_{\mathbf{G}}^{\lambda} e^{j(\mathbf{k}+\mathbf{G})\cdot\mathbf{r}}. \tag{3.56}$$

Before deriving the formulation for the calculation of the band structure for three dimensions, consider a curl operator acting on any plane wave:

$$\begin{aligned}
\nabla\times\hat{\mathbf{e}}_{\mathbf{G}}^{\lambda}e^{j\mathbf{K}\cdot\mathbf{r}} &= \nabla e^{j\mathbf{K}\cdot\mathbf{r}}\times\hat{\mathbf{e}}_{\mathbf{G}}^{\lambda}\\
&= \left[\left(\frac{\partial}{\partial x}\hat{\mathbf{x}}+\frac{\partial}{\partial y}\hat{\mathbf{y}}+\frac{\partial}{\partial z}\hat{\mathbf{z}}\right)e^{j(K_x x+K_y y+K_z z)}\right]\times\hat{\mathbf{e}}_{\mathbf{G}}^{\lambda}\\
&= j(K_x\hat{\mathbf{x}}+K_y\hat{\mathbf{y}}+K_z\hat{\mathbf{z}})e^{j(K_x x+K_y y+K_z z)}\times\hat{\mathbf{e}}_{\mathbf{G}}^{\lambda}\\
&= j\mathbf{K}\times\hat{\mathbf{e}}_{\mathbf{G}}^{\lambda}e^{j\mathbf{K}\cdot\mathbf{r}}.
\end{aligned} \tag{3.57}$$

Combining this result with Eqs. (3.53) and (3.56) yields

$$\begin{aligned}
\frac{1}{\varepsilon_r(\mathbf{r})}\nabla\times\mathbf{H}(\mathbf{r}) &= \left[\sum_{\mathbf{G}'}\kappa(\mathbf{G}')e^{j\mathbf{G}'\cdot\mathbf{r}}\right]\nabla\times\sum_{\mathbf{G}}\sum_{\lambda}h_{\mathbf{G}}^{\lambda}\hat{\mathbf{e}}_{\mathbf{G}}^{\lambda}e^{j(\mathbf{k}+\mathbf{G})\cdot\mathbf{r}}\\
&= \sum_{\mathbf{G}'}\kappa(\mathbf{G}')e^{j\mathbf{G}'\cdot\mathbf{r}}\sum_{\mathbf{G}}\sum_{\lambda=1}^{2}h_{\mathbf{G}}^{\lambda}j[(\mathbf{k}+\mathbf{G})\times\hat{\mathbf{e}}_{\mathbf{G}}^{\lambda}]e^{j(\mathbf{k}+\mathbf{G})\cdot\mathbf{r}}\\
&= j\sum_{\mathbf{G}}\sum_{\mathbf{G}'}\sum_{\lambda=1}^{2}h_{\mathbf{G}}^{\lambda}\kappa(\mathbf{G}')[(\mathbf{k}+\mathbf{G})\times\hat{\mathbf{e}}_{\mathbf{G}}^{\lambda}]e^{j(\mathbf{k}+\mathbf{G}+\mathbf{G}')\cdot\mathbf{r}}\\
&= j\sum_{\mathbf{G}}\sum_{\mathbf{G}''}\sum_{\lambda=1}^{2}h_{\mathbf{G}}^{\lambda}\kappa(\mathbf{G}''-\mathbf{G})[(\mathbf{k}+\mathbf{G})\times\hat{\mathbf{e}}_{\mathbf{G}}^{\lambda}]e^{j(\mathbf{k}+\mathbf{G}'')\cdot\mathbf{r}},
\end{aligned} \tag{3.58}$$

where we have again introduced $\mathbf{G}''=\mathbf{G}+\mathbf{G}'$. Taking the curl of both sides of Eq. (3.58) yields, with another application of Eq. (3.57),

$$\begin{aligned}
\nabla\times\frac{1}{\varepsilon_r(\mathbf{r})}\nabla\times\mathbf{H}_{\mathbf{k}}(\mathbf{r}) &= j\sum_{\mathbf{G}}\sum_{\mathbf{G}''}\sum_{\lambda=1}^{2}h_{\mathbf{G}}^{\lambda}\kappa(\mathbf{G}''-\mathbf{G})\nabla\times[(\mathbf{k}+\mathbf{G})\times\hat{\mathbf{e}}_{\mathbf{G}}^{\lambda}]e^{j(\mathbf{k}+\mathbf{G}'')\cdot\mathbf{r}}\\
&= -\sum_{\mathbf{G}}\sum_{\mathbf{G}''}\sum_{\lambda=1}^{2}h_{\mathbf{G}}^{\lambda}\kappa(\mathbf{G}''-\mathbf{G})(\mathbf{k}+\mathbf{G}'')\times[(\mathbf{k}+\mathbf{G})\times\hat{\mathbf{e}}_{\mathbf{G}}^{\lambda}]e^{j(\mathbf{k}+\mathbf{G}'')\cdot\mathbf{r}},
\end{aligned} \tag{3.59}$$

and the wave equation can be written as

$$-\sum_{\mathbf{G}}\sum_{\mathbf{G}''}\sum_{\lambda=1}^{2}h_{\mathbf{G}}^{\lambda}\kappa(\mathbf{G}''-\mathbf{G})(\mathbf{k}+\mathbf{G}'')\times[(\mathbf{k}+\mathbf{G})\times\hat{\mathbf{e}}_{\mathbf{G}}^{\lambda}]e^{j(\mathbf{k}+\mathbf{G}'')\cdot\mathbf{r}}$$
$$=\frac{\omega^2}{c^2}\sum_{\mathbf{G}}\sum_{\lambda=1}^{2}h_{\mathbf{G}}^{\lambda}\hat{\mathbf{e}}_{\mathbf{G}}^{\lambda}e^{j(\mathbf{k}+\mathbf{G})\cdot\mathbf{r}}. \tag{3.60}$$

Multiplying both sides of Eq. (3.60) by each of the testing functions $e^{-j(\mathbf{k}+\mathbf{G}')\cdot\mathbf{r}}$, integrating over the unit cell, and generalizing to 3D the delta-function integral result we have observed previously in the two-dimensional case

$$\int_V e^{j(\mathbf{G}-\mathbf{G}')\cdot\mathbf{r}}dv=\begin{cases}V, & \mathbf{G}=\mathbf{G}'\\ 0, & \mathbf{G}\neq\mathbf{G}'\end{cases}, \tag{3.61}$$

yields

$$-\sum_{\mathbf{G}'}\sum_{\lambda=1}^{2}h_{\mathbf{G}'}^{\lambda}\kappa(\mathbf{G}-\mathbf{G}')(\mathbf{k}+\mathbf{G})\times[(\mathbf{k}+\mathbf{G}')\times\hat{\mathbf{e}}_{\mathbf{G}'}^{\lambda}]=\frac{\omega^2}{c^2}\sum_{\lambda=1}^{2}h_{\mathbf{G}}^{\lambda}\hat{\mathbf{e}}_{\mathbf{G}}^{\lambda}. \tag{3.62}$$

Then, we take a dot product of both sides of Eq. (3.62) by $\hat{\mathbf{e}}_{\mathbf{G}}^{1}$ to obtain

$$-\sum_{\mathbf{G}'}\sum_{\lambda=1}^{2}h_{\mathbf{G}'}^{\lambda}\kappa(\mathbf{G}-\mathbf{G}')\hat{\mathbf{e}}_{\mathbf{G}}^{1}\cdot\{(\mathbf{k}+\mathbf{G})\times[(\mathbf{k}+\mathbf{G}')\times\hat{\mathbf{e}}_{\mathbf{G}'}^{\lambda}]\}=\frac{\omega^2}{c^2}\sum_{\lambda=1}^{2}h_{\mathbf{G}}^{\lambda}\hat{\mathbf{e}}_{\mathbf{G}}^{1}\cdot\hat{\mathbf{e}}_{\mathbf{G}}^{\lambda},$$

which, with the application of vector identity $\mathbf{A}\cdot(\mathbf{B}\times\mathbf{C})=-\mathbf{C}\cdot(\mathbf{B}\times\mathbf{A})$, simplifies to

$$\sum_{\mathbf{G}'}\sum_{\lambda=1}^{2}h_{\mathbf{G}'}^{\lambda}\kappa(\mathbf{G}-\mathbf{G}')[(\mathbf{k}+\mathbf{G}')\times\hat{\mathbf{e}}_{\mathbf{G}'}^{\lambda}]\cdot[(\mathbf{k}+\mathbf{G})\times\hat{\mathbf{e}}_{\mathbf{G}}^{1}]=\frac{\omega^2}{c^2}h_{\mathbf{G}}^{1}. \tag{3.63}$$

Because $\{\hat{\mathbf{e}}_{\mathbf{G}}^{1},\hat{\mathbf{e}}_{\mathbf{G}}^{2},\mathbf{k}+\mathbf{G}\}$ is a right-handed triple,

$$\begin{cases}(\mathbf{k}+\mathbf{G})\times\hat{\mathbf{e}}_{\mathbf{G}}^{1}=|\mathbf{k}+\mathbf{G}|\hat{\mathbf{e}}_{\mathbf{G}}^{2}\\ (\mathbf{k}+\mathbf{G})\times\hat{\mathbf{e}}_{\mathbf{G}}^{2}=-|\mathbf{k}+\mathbf{G}|\hat{\mathbf{e}}_{\mathbf{G}}^{1},\end{cases} \tag{3.64}$$

and Eq. (3.63) can be further simplified to

$$\sum_{\mathbf{G}'}\kappa(\mathbf{G}-\mathbf{G}')|\mathbf{k}+\mathbf{G}||\mathbf{k}+\mathbf{G}'|[h_{\mathbf{G}'}^{1}(\hat{\mathbf{e}}_{\mathbf{G}'}^{2}\cdot\hat{\mathbf{e}}_{\mathbf{G}}^{2})-h_{\mathbf{G}'}^{2}(\hat{\mathbf{e}}_{\mathbf{G}'}^{1}\cdot\hat{\mathbf{e}}_{\mathbf{G}}^{2})]=\frac{\omega^2}{c^2}h_{\mathbf{G}}^{1}. \tag{3.65}$$

Similarly, if we take a dot product by $\hat{\mathbf{e}}^2_{\mathbf{G}}$, we obtain

$$\sum_{\mathbf{G}'} \kappa(\mathbf{G}-\mathbf{G}')|\mathbf{k}+\mathbf{G}||\mathbf{k}+\mathbf{G}'|[-h^1_{\mathbf{G}'}(\hat{\mathbf{e}}^2_{\mathbf{G}'}\cdot\hat{\mathbf{e}}^1_{\mathbf{G}}) + h^2_{\mathbf{G}'}(\hat{\mathbf{e}}^1_{\mathbf{G}'}\cdot\hat{\mathbf{e}}^1_{\mathbf{G}})] = \frac{\omega^2}{c^2} h^2_{\mathbf{G}}. \quad (3.66)$$

Combining Eqs. (3.65) and (3.66), we finally obtain a standard eigenvalue problem,

$$\sum_{\mathbf{G}'}\sum_{\lambda'=1}^{2} H^{\lambda,\lambda'}_{\mathbf{G},\mathbf{G}'} h^{\lambda'}_{\mathbf{G}'} = \frac{\omega^2}{c^2} h^{\lambda}_{\mathbf{G}}, \quad (3.67)$$

where $H^{\lambda,\lambda'}_{\mathbf{G},\mathbf{G}'}$ is an element of a 2×2 matrix

$$H_{\mathbf{G},\mathbf{G}'} = |\mathbf{k}+\mathbf{G}||\mathbf{k}+\mathbf{G}'|\kappa(\mathbf{G}-\mathbf{G}')\begin{pmatrix} \hat{\mathbf{e}}^2_{\mathbf{G}}\cdot\hat{\mathbf{e}}^2_{\mathbf{G}'} & -\hat{\mathbf{e}}^2_{\mathbf{G}}\cdot\hat{\mathbf{e}}^1_{\mathbf{G}'} \\ -\hat{\mathbf{e}}^1_{\mathbf{G}}\cdot\hat{\mathbf{e}}^2_{\mathbf{G}'} & \hat{\mathbf{e}}^1_{\mathbf{G}}\cdot\hat{\mathbf{e}}^1_{\mathbf{G}'} \end{pmatrix}, \quad (3.68)$$

which itself is an element of the full matrix **H**. Because the matrix **H** is real and symmetric, the dispersion relation of electromagnetic waves can be found using standard matrix diagonalization techniques applied to the linear system of Eq. (3.67).

In summary, the flowchart to calculate the band structure of photonic crystals is the same as the two-dimensional case. The Fourier coefficients are computed analytically or by FFT in advance. After specifying the wave vector in the irreducible Brillouin zone, matrix **H** is constructed. Then the eigenvalue problem is solved by standard matrix algorithms, and a series of discretized eigenmodes $\omega(\mathbf{k})$ and corresponding eigenvector are obtained. By repeating this procedure, all eigenmodes for all eigenvectors within the irreducible Brillouin zone are found. Once we have the band structure within the irreducible Brillouin zone, the rest of the Brillouin zone is obtained by symmetry operations.

An example of a full three-dimensionally periodic structure is shown in Fig. 3.12a, which is a simple cubic structure composed of air spheres in a high-index dielectric material. The air spheres have a radius of $0.45a$ and are embedded within a medium with a refractive index of 3.5. The dispersion diagram, which is represented as bands of eigenfrequecies plotted versus wave vectors tracing the perimeter of the irreducible Brillouin zone, is illustrated in Fig. 3.12b. The first Brillouin zone and irreducible zone are shown in the inset of Fig. 3.12b.

3.2.5 Practical Considerations in the Implementation of the Plane-Wave Expansion Method

The computational cost of the 3D plane-wave expansion method is of order N^3, where N is the number of plane waves, because the method involves the diagonalization of a matrix. As the number of plane waves increases, the required computer resources, such as memory and computation time, increase rapidly, and the method quickly becomes impractical. It is less of a problem in 2D where a few hundred plane waves are often sufficient to simulate photonic crystals accurately. To achieve

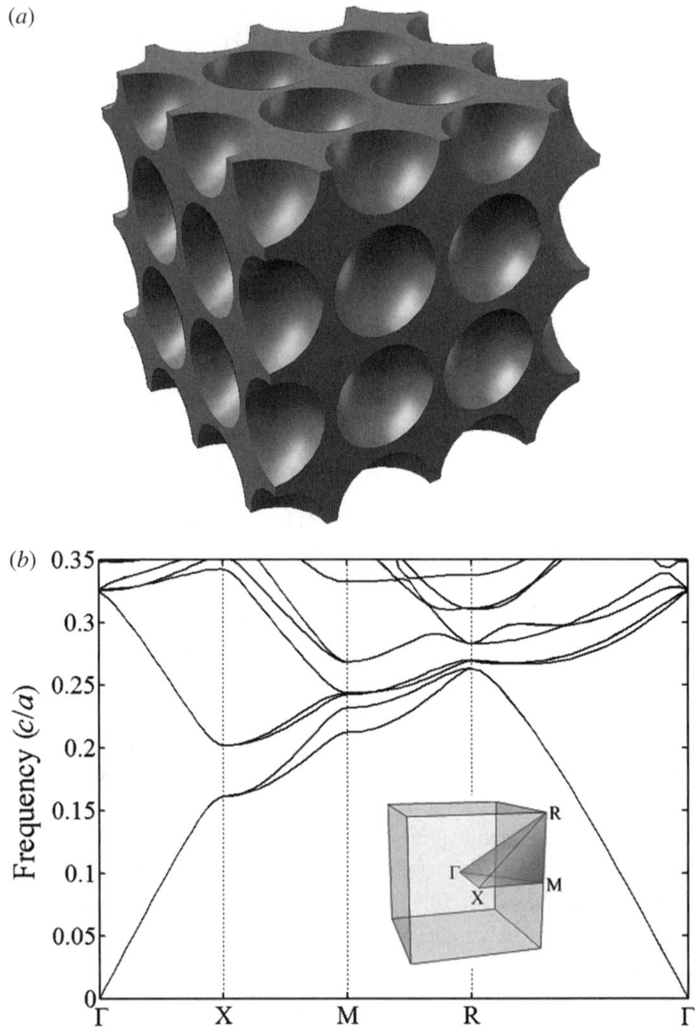

Figure 3.12 Three-dimensional simple cubic structure. (a) Structure. (b) Band diagram. Inset: Brillouin zone labeled with symmetry points (irreducible zone is shaded gray).

comparable accuracy in 3D, many thousands of plane waves are required. Fortunately, it is often the case that only some of the lowest bands are of interest in real problems. In these situations, iterative methods can be applied to compute the lowest p eigenvalues with storage requirements of $O(pN)$ [80] and time requirements of $O(pN \log N)$.

To observe how this is possible, recall the eigenvalue problem in Eq. (3.62)

$$-\sum_{\mathbf{G}'} \kappa(\mathbf{G} - \mathbf{G}')(\mathbf{k} + \mathbf{G}) \times [(\mathbf{k} + \mathbf{G}') \times \mathbf{h}_{\mathbf{G}'}] = \frac{\omega^2}{c^2} \mathbf{h}_{\mathbf{G}}, \tag{3.69}$$

where $\mathbf{h}_\mathbf{G} = \sum_{\lambda=1}^{2} h_\mathbf{G}^\lambda \hat{\mathbf{e}}_\mathbf{G}^\lambda$. For $\mathbf{G} = m_1\mathbf{g}_1 + m_2\mathbf{g}_2 + m_3\mathbf{g}_3$, $\mathbf{G}' = m_1'\mathbf{g}_1 + m_2'\mathbf{g}_2 + m_3'\mathbf{g}_3$, and with $\kappa(\mathbf{G} - \mathbf{G}')$ in the DFT form, Eq. (3.55), we obtain

$$-\frac{(\mathbf{k}+\mathbf{G})}{N_1N_2N_3}\times\sum_{n_1,n_2,n_3}\left\{\frac{e^{-j2\pi\sum_{j=1}^{3}\frac{m_jn_j}{N_j}}}{\varepsilon_r(n_1,n_2,n_3)}\left[\sum_{m_1',m_2',m_3'}[(\mathbf{k}+\mathbf{G}')\times\mathbf{h}_{\mathbf{G}'}]e^{j2\pi\sum_{i=1}^{3}\frac{m_i'n_i}{N_i}}\right]\right\}=\frac{\omega^2}{c^2}\mathbf{h}_\mathbf{G}$$

$$-(\mathbf{k}+\mathbf{G})\times\mathrm{IFFT}\left\{\frac{1}{\varepsilon_r}\mathrm{FFT}[(\mathbf{k}+\mathbf{G}')\times\mathbf{h}_{\mathbf{G}'}]\right\}=\frac{\omega^2}{c^2}\mathbf{h}_\mathbf{G}$$

$$\mathbf{A}\mathbf{h}=\frac{\omega^2}{c^2}\mathbf{h}, \quad (3.70)$$

where FFT and IFFT indicate a fast Fourier transform operation and the inverse fast Fourier transform, respectively. Thus, **Ah** can be calculated without storing the entire characteristic matrix **A**, which alleviates the computer storage requirements. In addition, **Ah** can be efficiently evaluated by two FFTs, with a computational complexity of $O(N \log N)$.

To find the p smallest eigenvalues of a matrix **A** while knowing only the results of its action on selected vectors, and not its explicit form, recall from Chapter 2 that the solution of a generalized eigenvalue problem

$$\mathbf{A}\mathbf{h} = \lambda\mathbf{B}\mathbf{h} \quad (3.71)$$

is obtained by minimizing the functional

$$\frac{\langle\mathbf{h},\,\mathbf{A}\mathbf{h}\rangle}{\langle\mathbf{h},\,\mathbf{B}\mathbf{h}\rangle}. \quad (3.72)$$

The minimum value of the functional (3.72) is equal to the smallest eigenvalue λ of the problem (3.71), whereas the vector **h** that minimizes it is the corresponding eigenvector. To find the minimum of functional (3.72), any of the standard iterative numerical methods for finding minimum values of multivariable functions can be employed. The convergence to the solution can be hastened by using as the initial guess a result from a neighboring **k** in the irreducible Brillouin zone, because the field configurations and the mode frequencies vary continuously as a function of **k**.

The next smallest eigenvalue and its corresponding eigenvector is found by modifying the functional (3.72) so as to limit the search for minimum to a space orthogonal to the first eigenvector. This can be done, for example, by sandwiching matrix **B** between two operators that project **h** onto a space orthogonal to the first found eigenvector. Such a procedure is then repeated until all of the required p eigenvalues and eigenvectors are found.

The sequential process of finding the smallest p eigenvalues and their corresponding eigenvectors described above is known as "deflation." More efficient algorithms exist that allow finding all of the smallest p eigenvalues/eigenvectors in a single shot. The reader interested in such methods is referred to Refs. 59, 60, and 80.

3.2.6 Photonic-Crystal Slab by Plane-Wave Expansion Method

One of the most promising and interesting classes of photonic-crystal structures is the photonic-crystal slab (PCS), which has two-dimensional (in-plane) periodicity, whereas its height is finite and comparable with the wavelength of light (Fig. 3.13). The PCS is attractive because of its relative ease of fabrication compared with three-dimensional photonic crystals and the associated ease of chip-level integration [81–89]. Being finite in height requires an alternative mechanism to the photonic bandgap for light confinement in the third dimension, namely total internal reflection (TIR), which is the same mechanism that is responsible for the light confinement in optical fibers. It is then the combination of these two phenomena: the in-plane light confinement through either coherent Bragg reflections (bandgap guiding, Chapter 4) or engineered dispersion (self-collimation guiding, Chapter 5), with the finite height of the structure with TIR at the slab/cladding interface, that enables the use of photonic-crystal slabs in photonic integrated circuits.

Because of the practical importance of these types of structures, specialized methods have been developed to model efficiently the behavior of light interacting with them. The methods based on PWEM are described in this section. As a PCS is not a photonic crystal in the strict sense of this term as it lacks periodicity (or uniformity) in the direction normal to the slab, certain modifications of the structure or the method must be made to determine its dispersive properties in the spirit of PWEM. In the sections below, several such modifications, which exhibit varied levels of accuracy and computational efficiencies, are described. Although inspired by the study of PCSs, these methods also find applications in other areas of photonic-crystal research and beyond.

3.2.6.1 3D Plane-Wave Expansion Method in Photonic-Crystal Slabs. The 3D plane-wave expansion method described in Section 3.2.4 was developed for general three-dimensional structures that extend periodically in all three dimensions; as

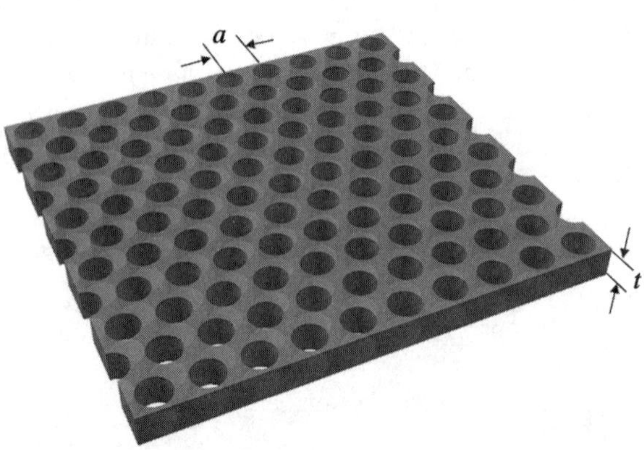

Figure 3.13 Photonic-crystal slab that consists of a perforated slab with periodically distributed air holes. The thickness of the slab is t, and the lattice constant is a.

such, it is not suited for structures whose periodicity does not extend infinitely in all directions. However, although the photonic-crystal slab has two-dimensional periodicity, it is neither periodic nor invariant along the third dimension. To make the PWEM applicable to this structure, artificial periodicity is introduced in the third dimension by periodically distributing identical slabs separated by a certain amount of background material, as shown in Fig. 3.14. The guided modes, which are of primary practical interest, are strongly localized within the slab(s), and therefore, the additional periodicity of an infinity of slabs separated by large intervals does not affect their properties. The nonguided or radiation modes extend into and propagate through the regions between the slabs and, therefore, *are* affected by the artificially imposed periodicity.

The "light line" or "light cone" represents the dispersion relation for the background medium, and as such it comprises a boundary between radiation modes, which propagate into the cladding region outside the slab, and guided modes, which are localized within the plane of the slab, as shown in Fig. 3.15. States that lie below the light line in the band diagram cannot couple with modes that propagate in the bulk background. Thus, the discrete bands below the light line are confined. Mathematically, if we express the wave vector $\mathbf{k}$ as

$$\mathbf{k} = \mathbf{k}_{\parallel} + \mathbf{k}_z, \tag{3.73}$$

where $\mathbf{k}_{\parallel} = \mathbf{k}_x + \mathbf{k}_y$ is the in-plane wave vector, restricted to the first Brillouin zone, and $\mathbf{k}_z$ is the out-of-plane wave vector; then the light line can be represented by the simple equation

$$k_{\parallel} = \frac{\omega}{c}\sqrt{\varepsilon_b}, \tag{3.74}$$

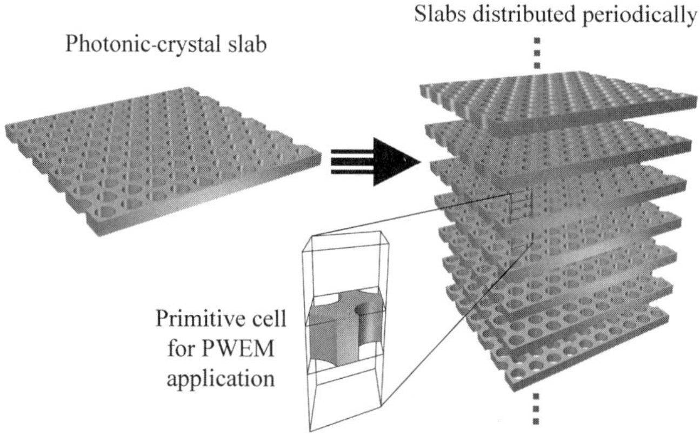

Figure 3.14 Construction of primitive cell for the application of PWEM to the analysis of a photonic-crystal slab. First, the system is made periodic by replacing a single slab with an array distributed periodically in the z direction. From such a stack, a primitive cell is selected. The stack of perforated slabs is equivalent to a 3D array of identical primitive cells.

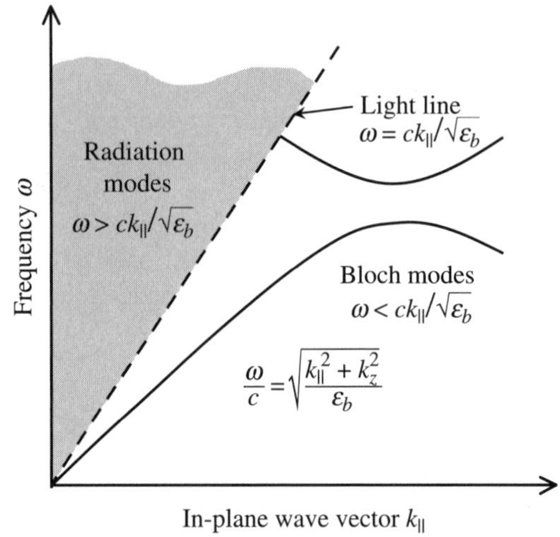

Figure 3.15 Dispersion diagram for in-plane wave vector of a periodic structure. Modes in the shaded region are radiation modes; i.e., they are not confined to the slab.

where ε_b is the dielectric permittivity of the background material of the cladding region. Accordingly, for the guided modes that can only propagate in the plane of periodicity, we have

$$k_{\|} > \frac{\omega}{c}\sqrt{\varepsilon_b}. \tag{3.75}$$

In other words, radiation modes have a real k_z component, whereas the guided modes have an imaginary k_z, which makes them evanescent as they extend into the cladding region. Therefore, modes that lie below the light line ($\omega < ck_{\|}/\sqrt{\varepsilon_b}$) cannot couple to radiation modes that lie above the light line ($\omega > ck_{\|}/\sqrt{\varepsilon_b}$).

Another distinguishing property of dielectric slabs is that eigenmodes can no longer be classified into pure TE or pure TM modes, as is the case for 2D photonic crystals. However, for a perforated dielectric slab (core) sandwiched symmetrically between the cladding layers of the identical (lower) index, the eigenmodes of the slab can be classified into even and odd modes. For the even mode, the in-plane electric field components and the magnetic field component in the z direction (H_z) are even functions of z, with respect to the symmetry plane that bisects the slab at $z = 0$; for the odd mode, the in-plane magnetic field components and electric field component in the z direction (E_z) are odd functions of z. In fact, these even and odd modes are similar to TE and TM modes in two-dimensional systems, and on the symmetry plane $z = 0$, they are purely TE and TM. Thus, these modes are also called TE-like and TM-like modes, respectively.

In calculating the properties of guided modes, the z-component k_z of the wave vector $\mathbf{k}$ is imaginary, so that the real part, which is responsible for propagation, reduces to $\Re(\mathbf{k}) = k_x\hat{\mathbf{x}} + k_y\hat{\mathbf{y}} = \mathbf{k}_{xy}$ ($\Re$ denotes the real part). If in addition the slab has even symmetry in the out-of-plane direction, i.e., $\varepsilon(z) = \varepsilon(-z)$, the computational cost of the three-dimensional eigenvalue problem of Eq. (3.67) reduces considerably thanks to the separation of all modes into even and odd. Below, this separation is calculated explicitly.

Recall the expansion of magnetic field in a Bloch wave

$$\mathbf{H}_{\mathbf{k}}(\mathbf{r}) = \sum_{\mathbf{G}} \sum_{\lambda} h_{\mathbf{G}}^{\lambda} \hat{\mathbf{e}}_{\mathbf{G}}^{\lambda} e^{j(\mathbf{k}+\mathbf{G})\cdot\mathbf{r}}, \tag{3.56}$$

where we now set $\mathbf{G} = (m\mathbf{g}_1 + n\mathbf{g}_2 + lg_3\hat{\mathbf{z}}) = \mathbf{G}_{xy} + lg_3\hat{\mathbf{z}}$ and $g_3 = 2\pi/\zeta$, in which ζ is the period of the artificial z-periodicity and m, n, and l are integers. Now consider two plane waves with wave vectors $\mathbf{k}_{xy} + (m\mathbf{g}_1 + n\mathbf{g}_2 + lg_3\hat{\mathbf{z}})$ and $\mathbf{k}_{xy} + (m\mathbf{g}_1 + n\mathbf{g}_2 - lg_3\hat{\mathbf{z}})$, as shown in Fig. 3.16. Decomposing the unit vector $\hat{\mathbf{e}}_{\mathbf{G}}^{\lambda}$ from Eq. (3.56) into its components along the z direction and in the xy plane,

$$\hat{\mathbf{e}}_{\mathbf{G}}^{\lambda} = \mathbf{e}_{z,\mathbf{G}}^{\lambda} + \mathbf{e}_{xy,\mathbf{G}}^{\lambda} = \mathbf{e}_{z,mnl}^{\lambda} + \mathbf{e}_{xy,mnl}^{\lambda}, \tag{3.76}$$

and carefully choosing the two sets of unit vectors (corresponding to $\lambda = 1, 2$), yields the following relations:

$$\begin{cases} \mathbf{e}_{z,mnl}^{1} = -\mathbf{e}_{z,mn(-l)}^{1} \\ \mathbf{e}_{xy,mnl}^{1} = \mathbf{e}_{xy,mn(-l)}^{1} \end{cases}, \quad \begin{cases} \mathbf{e}_{z,mnl}^{2} = \mathbf{e}_{z,mn(-l)}^{2} \\ \mathbf{e}_{xy,mnl}^{2} = -\mathbf{e}_{xy,mn(-l)}^{2} \end{cases}. \tag{3.77}$$

If we assume that the field components of the two plane waves satisfy

$$\begin{cases} h_{mnl}^{1} = -h_{mn(-l)}^{1} \\ h_{mn0}^{1} = 0 \\ h_{mnl}^{2} = h_{mn(-l)}^{2} \end{cases}, \tag{3.78}$$

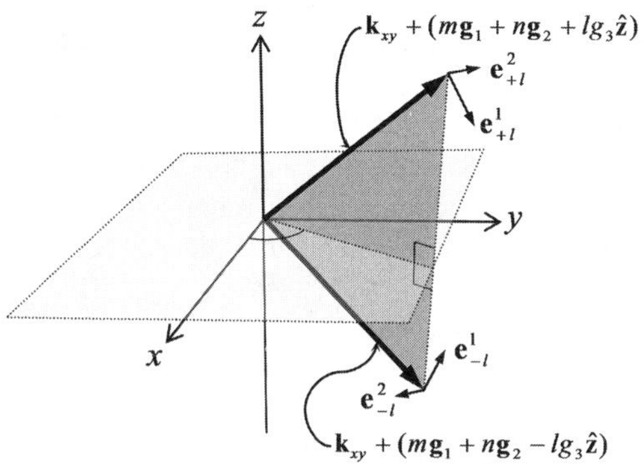

Figure 3.16 Two plane waves and their expansion vectors are symmetric with respect to the middle (xy) plane of the slab. For simplicity, the repeated subscript indices m,n have been omitted from the vectors **e**.

then with the help of relations (3.77), we have

$$\mathbf{H}_{\mathbf{k}}(\mathbf{r}) = \sum_{mn} e^{j(\mathbf{G}_{xy}+\mathbf{k}_{xy})\cdot\mathbf{r}_{xy}} \Bigg[h^2_{mn0}\mathbf{e}^2_{z,mn0} + \sum_{l=1}^{\infty} 2(h^1_{mnl}\mathbf{e}^1_{z,mnl} + h^2_{mnl}\mathbf{e}^2_{z,mnl})\cos(lg_3 z) + \sum_{l=1}^{\infty} 2j(h^1_{mnl}\mathbf{e}^1_{xy,mnl} + h^2_{mnl}\mathbf{e}^2_{xy,mnl})\sin(lg_3 z) \Bigg]. \quad (3.79)$$

According to these expressions, in the symmetry plane ($z = 0$), the last term vanishes; hence, the in-plane field components h^λ (i.e., those that multiply $\mathbf{e}^\lambda_{xy}$) do not contribute to the field $\mathbf{H}(x, y, 0)$. The resulting mode is even or TE-like because the electric field at $z = 0$ is oriented perpendicular to the slab normal.

If, however, we assume that

$$\begin{cases} h^1_{mnl} = h^1_{mn(-l)} \\ h^2_{mnl} = h^2_{mn(-l)}, \\ h^2_{mn0} = 0 \end{cases} \quad (3.80)$$

then with the relations of Eq. (3.77), we have

$$\mathbf{H}_{\mathbf{k}}(\mathbf{r}) = \sum_{mn} e^{j(\mathbf{G}_{xy}+\mathbf{k}_{xy})\cdot\mathbf{r}_{xy}} \Bigg[h^1_{mn0}\mathbf{e}^1_{xy,mn0} + \sum_{l=1}^{\infty} 2j(h^1_{mnl}\mathbf{e}^1_{z,mnl} + h^2_{mnl}\mathbf{e}^2_{z,mnl})\sin(lg_3 z) + \sum_{l=1}^{\infty} 2(h^1_{mnl}\mathbf{e}^1_{xy,mnl} + h^2_{mnl}\mathbf{e}^2_{xy,mnl})\cos(lg_3 z) \Bigg]. \quad (3.81)$$

In this expression, it is the middle term that vanishes in the symmetry plane; hence, the out-of-plane field components h^λ (i.e., those that multiply $\mathbf{e}^\lambda_z$) do not contribute to the field $\mathbf{H}(x, y, 0)$. Therefore, under the conditions of Eq. (3.80), the solution is referred to as the odd or TM-like mode because the magnetic field at $z = 0$ is oriented perpendicular to the slab normal.

Thus, because of the symmetry of the slab in the out-of-plane (z) direction, the eigenmodes can be decomposed into even and odd modes. As a result, the relations described above allow the reduction of the computational cost of the eigenvalue problem. Combining the $+l$ and $-l$ elements reduces the size of the matrix by a factor of two.

As an example, consider a photonic-crystal slab with thickness $0.5a$ and dielectric constant $\varepsilon_r = 12$ suspended in air. The perforations form a triangular lattice of air holes with radii of $0.5a$, as is depicted in the inset of Fig. 3.17. The primitive cell height has been chosen to be $\zeta = 4a$, and a total of 1053 plane waves have been used to calculate the band structure shown in Fig. 3.17, where the gray region represents the radiation modes above light cone. According to this figure, there

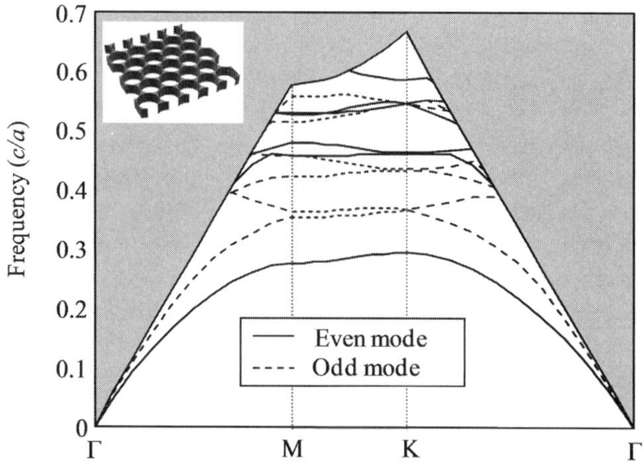

Figure 3.17 Dispersion diagram for even and odd modes along the edge of the irreducible Brillouin zone. The triangular lattice of air holes with radii 0.5a is shown in the inset.

seems to be a bandgap for TE-like (even) modes below the light cone, similar to the one obtained in the case of infinitely long or truly two-dimensional structures. However, strictly speaking, this is not a true bandgap but rather a pseudo-bandgap because several modes exist with frequencies inside this pseudo-gap range, except they lie above the light cone. In other words, waves with these frequencies are not prohibited to propagate in *all* directions, only in those defined by the light cone, i.e., confined to the slab.

For an asymmetric structure, such as when the slab is sandwiched between two different cladding materials of a lower index than the slab itself (e.g., air and glass or SiO_2), the symmetry no longer exists. Therefore, modes can no longer be classified into even or odd; instead, they will be more loosely classified in terms of having the predominant field components either in-plane or out-of-plane. Accordingly, modes with a magnetic field predominantly out-of-plane, i.e., with fields dominated by H_z, E_x, and E_y, are designated as the TE-like modes, and those with a magnetic field predominantly in-plane, i.e., with fields dominated by E_z, H_x, and H_y, are designated as the TM-like modes.

3.2.6.2 Plane-Wave Expansion Method with Perfectly Matched Layers. The previous section discussed an approach to calculating the band structure of the photonic-crystal slabs based on artificially introduced periodicity in the third dimension where the slabs were separated by a sufficient amount of background material to ensure electromagnetic isolation between modes bound within the slabs. Such isolation between slabs allows calculating dispersion properties of the bound or guided modes lying below the light line, which extend only evanescently into the cladding region. However, in addition to the guided modes, the structure also supports leaky modes lying above the light line, which propagate in the cladding region. For these modes, no amount of separation is sufficient to provide electromagnetic isolation between the slabs. As a result, the PWEM as described in the previous section is inadequate for modeling leaky modes in photonic-crystal slabs. To remedy this situation and allow the application of the PWEM to analyze leaky modes, it is necessary that the flow of energy carried by the electromagnetic wave propagating in the cladding region be interrupted. The interruption must be such that it does not affect the field distribution in the vicinity of the photonic-crystal slab; i.e., no reflection back into the slab is allowed. This requirement is satisfied by placing an absorber with impedance perfectly matched to the cladding material. A slab of such absorber is referred to as a perfectly matched layer (PML) and has been investigated in the context of electromagnetic simulation both in real space and in frequency domain [90–95]. For the purpose of simulating periodically distributed photonic-crystal slabs, the PML based on anisotropic material proposed by Sacks et al. [95] seems most suitable because it requires no modification of Maxwell's equations, which makes it easy to implement in frequency-domain methods such as PWEM considered here. Furthermore, the dielectric materials of the photonic-crystal slab and of the cladding layer may be encoded in a way similar to the PML, thus providing for a uniform treatment of all objects present in the computational region.

Figure 3.18a shows a photonic-crystal slab in which the perforated dielectric layer has a thickness t and the array of air holes with radius r has a period a. The unit cell for band structure calculations is shown in Fig. 3.18b. It was constructed in a manner similar to that presented in Fig. 3.14. It consists of a two-dimensional photonic-crystal slab with relative dielectric constant ε_r sandwiched between two cladding layers that have a lower dielectric constant. The primitive cell with height ζ is terminated with two homogeneous PMLs, each t_{PML} thick.

The electromagnetic properties of the materials in the unit cell are described using a general constitutive tensor $\bar{\bar{\Lambda}}$. This tensor provides for both the lossless, isotropic medium in the primary computational region and the anisotropic PML absorbers at the upper and lower unit-cell boundaries for suppressing waves propagating out

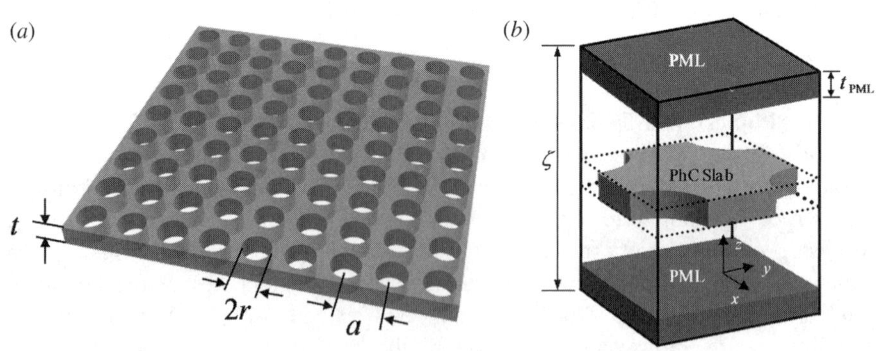

Figure 3.18 (a) Photonic-crystal slab. (b) Unit cell for the band structure calculations using PWEM.

of plane. The PML is made of a uniaxial anisotropic material as described in Ref. 95, such that

$$\bar{\bar{\Lambda}} = \begin{pmatrix} \alpha - j\beta & 0 & 0 \\ 0 & \alpha - j\beta & 0 \\ 0 & 0 & (\alpha - j\beta)^{-1} \end{pmatrix}, \tag{3.82}$$

where $\alpha, \beta > 0$ to ensure absorption of the wave propagating in the PML. The choice (3.82) of tensor $\bar{\bar{\Lambda}}$ makes the impedance of such PML perfectly matched to the adjacent cladding layer, which guarantees zero reflection at the boundary of the PML for an arbitrary angle of incidence. In Eq. (3.82), term α determines the wavelength inside the PML material, whereas β is responsible for absorption, or wave decay, therein. Outside the PML region, $\bar{\bar{\Lambda}}$ is simply an identity matrix.

With the use of $\bar{\bar{\Lambda}}$, the time-harmonic Maxwell's curl equations can be written as [95, 96]

$$\begin{cases} \nabla \times \mathbf{H} = j\omega\varepsilon_0\varepsilon_r\bar{\bar{\Lambda}}\mathbf{E} \\ \nabla \times \mathbf{E} = -j\omega\mu_0\bar{\bar{\Lambda}}\mathbf{H} \end{cases}. \tag{3.83}$$

The loss produced by β in tensor $\bar{\bar{\Lambda}}$ results in a decay of waves that propagate in such a medium. As a result, the mode propagating inside the cladding is efficiently absorbed without any reflection back to the slab. In other words, the photonic-crystal slabs are effectively isolated from one another, and each appear as if in an unbounded homogeneous space. Thus, the introduction of PMLs in the unit cell allows the calculation of not only the truly guided modes but also the leaky modes.

The PWEM with PMLs proceeds in a manner similar to the general PWEM approach. First, the electric field is eliminated from Eq. (3.83) to obtain the vector wave equation for the magnetic field,

$$\nabla \times \left[(\bar{\bar{\varepsilon}}_r(\mathbf{r}))^{-1}\nabla \times \mathbf{H}(\mathbf{r})\right] = \left(\frac{\omega}{c}\right)^2 \bar{\bar{\mu}}_r(\mathbf{r})\mathbf{H}(\mathbf{r}), \tag{3.84}$$

where $\bar{\bar{\varepsilon}}_r = \varepsilon_r\bar{\bar{\Lambda}}$ and $\bar{\bar{\mu}}_r = \mu_r\bar{\bar{\Lambda}}$. To improve the convergence of the PWEM, which is compromised by permittivity discontinuities at dielectric interfaces, the model of Fig. 3.18 is modified to smooth out these interfaces [6, 80, 97]. Thus, following Ref. 80, in the vicinity of the dielectric interfaces, the inverse of the effective dielectric tensor is assumed in the following form:

$$\overline{\varepsilon_r^{-1}}P + \overline{\varepsilon_r}^{-1}(1 - P), \tag{3.85}$$

where P is the projection operator on the direction normal to the dielectric boundary $P_{lm} = n_l n_m$, with the unit vector normal to said boundary defined as $\hat{\mathbf{n}} = n_x\hat{\mathbf{x}} + n_y\hat{\mathbf{y}} + n_z\hat{\mathbf{z}}$; $\overline{\varepsilon_r^{-1}}$ denotes the average of the inverse, whereas $\overline{\varepsilon_r}^{-1}$ is the inverse of the average dielectric constant. The particular form of the smoothed dielectric constant inverse given in Eq. (3.85) has been shown in Ref. 80 to yield the fastest convergence. It should be noted that such smoothing of interfaces is not necessary at the PML boundary, which as the name implies, is perfectly matched to the adjacent cladding material.

Using the anisotropic permittivity tensor defined above, the PWEM with PMLs can now be employed in the calculation of the complete band structure of photonic-crystal slabs. The PMLs efficiently absorb the outgoing waves, thereby suppressing the spurious Bloch modes that develop from the artificially imposed periodicity. Therefore, the material in three-dimensional space can be safely treated as periodic, as required for 3D PWEM, and the inverse dielectric and permeability functions can be expanded in Fourier series,

$$\begin{cases} \left(\bar{\bar{\varepsilon}}_r(\mathbf{r})\right)^{-1} = \sum_{\mathbf{G}} \bar{\bar{\kappa}}(\mathbf{G}) e^{j\mathbf{G}\cdot\mathbf{r}} \\ \bar{\bar{\mu}}_r(\mathbf{r}) = \sum_{\mathbf{G}} \bar{\bar{\tau}}(\mathbf{G}) e^{j\mathbf{G}\cdot\mathbf{r}} \end{cases}, \tag{3.86}$$

where $\bar{\bar{\kappa}}(\mathbf{G})$ and $\bar{\bar{\tau}}(\mathbf{G})$ are the Fourier coefficients given by

$$\begin{cases} \bar{\bar{\kappa}}(\mathbf{G}) = \dfrac{1}{V} \displaystyle\int_V \left(\bar{\bar{\varepsilon}}_r(\mathbf{r})\right)^{-1} e^{-j\mathbf{G}\cdot\mathbf{r}} dv \\ \bar{\bar{\tau}}(\mathbf{G}) = \dfrac{1}{V} \displaystyle\int_V \bar{\bar{\mu}}_r(\mathbf{r}) e^{-j\mathbf{G}\cdot\mathbf{r}} dv \end{cases}, \tag{3.87}$$

and V is the volume of the unit cell. $\mathbf{G} = m_1\mathbf{g}_1 + m_2\mathbf{g}_2 + m_3\mathbf{g}_3$ is the reciprocal lattice vector, and $\mathbf{g}_i$ $(i = 1, 2, 3)$ are the primitive reciprocal lattice vectors. The Fourier coefficients can be efficiently calculated by 3D FFT.

As described in previous sections, the magnetic field can be expanded as

$$\mathbf{H}(\mathbf{r}) = \sum_{\mathbf{G}} \sum_{\lambda} h_{\mathbf{G}}^{\lambda} \hat{\mathbf{e}}_{\mathbf{G}}^{\lambda} e^{j(\mathbf{k}+\mathbf{G})\cdot\mathbf{r}}, \tag{3.88}$$

where as before $\hat{\mathbf{e}}_{\mathbf{G}}^{\lambda}$ $(\lambda = 1, 2)$ denotes a pair of unit vectors such that $\{\hat{\mathbf{e}}_{\mathbf{G}}^{1}, \hat{\mathbf{e}}_{\mathbf{G}}^{2}, \mathbf{k}+\mathbf{G}\}$ forms a right-handed triplet of mutually orthogonal vectors and $\mathbf{k} = k_x\hat{\mathbf{x}} + k_y\hat{\mathbf{y}}$ is the in-plane wave vector.

Substituting Eqs. (3.86), (3.87), and (3.88) into Eq. (3.84) yields the following eigenvalue problem [98]:

$$\sum_{\mathbf{G}'} \sum_{\lambda'=1}^{2} H_{\mathbf{G},\mathbf{G}'}^{\lambda,\lambda'} h_{\mathbf{G}'}^{\lambda'} = \frac{\omega^2}{c^2} \sum_{\mathbf{G}'} \sum_{\lambda'=1}^{2} B_{\mathbf{G},\mathbf{G}'}^{\lambda,\lambda'} h_{\mathbf{G}'}^{\lambda'}, \tag{3.89}$$

where each of the matrix elements $H_{\mathbf{G},\mathbf{G}'}$ and $B_{\mathbf{G},\mathbf{G}'}$ is a 2×2 matrix of tensors, given by

$$\begin{cases} H_{\mathbf{G},\mathbf{G}'} = |\mathbf{k}+\mathbf{G}||\mathbf{k}+\mathbf{G}'| \begin{pmatrix} \hat{\mathbf{e}}_{\mathbf{G}}^{2} \bar{\bar{\kappa}}_{\mathbf{G},\mathbf{G}'} \hat{\mathbf{e}}_{\mathbf{G}'}^{2} & -\hat{\mathbf{e}}_{\mathbf{G}}^{2} \bar{\bar{\kappa}}_{\mathbf{G},\mathbf{G}'} \hat{\mathbf{e}}_{\mathbf{G}'}^{1} \\ -\hat{\mathbf{e}}_{\mathbf{G}}^{1} \bar{\bar{\kappa}}_{\mathbf{G},\mathbf{G}'} \hat{\mathbf{e}}_{\mathbf{G}'}^{2} & \hat{\mathbf{e}}_{\mathbf{G}}^{1} \bar{\bar{\kappa}}_{\mathbf{G},\mathbf{G}'} \hat{\mathbf{e}}_{\mathbf{G}'}^{1} \end{pmatrix} \\ B_{\mathbf{G},\mathbf{G}'} = \begin{pmatrix} \hat{\mathbf{e}}_{\mathbf{G}}^{1} \bar{\bar{\tau}}_{\mathbf{G},\mathbf{G}'} \hat{\mathbf{e}}_{\mathbf{G}'}^{1} & \hat{\mathbf{e}}_{\mathbf{G}}^{1} \bar{\bar{\tau}}_{\mathbf{G},\mathbf{G}'} \hat{\mathbf{e}}_{\mathbf{G}'}^{2} \\ \hat{\mathbf{e}}_{\mathbf{G}}^{2} \bar{\bar{\tau}}_{\mathbf{G},\mathbf{G}'} \hat{\mathbf{e}}_{\mathbf{G}'}^{1} & \hat{\mathbf{e}}_{\mathbf{G}}^{2} \bar{\bar{\tau}}_{\mathbf{G},\mathbf{G}'} \hat{\mathbf{e}}_{\mathbf{G}'}^{2} \end{pmatrix} \end{cases}, \tag{3.90}$$

with $\bar{\bar{\kappa}}_{\mathbf{G},\mathbf{G}'} = \bar{\bar{\kappa}}(\mathbf{G}-\mathbf{G}')$ and $\bar{\bar{\tau}}_{\mathbf{G},\mathbf{G}'} = \bar{\bar{\tau}}(\mathbf{G}-\mathbf{G}')$.

Since lossy material was introduced into the system in the form of the PMLs to absorb the radiation modes, the matrices $H^{\lambda,\lambda'}_{\mathbf{G},\mathbf{G}'}$ and $B^{\lambda,\lambda'}_{\mathbf{G},\mathbf{G}'}$ are no longer Hermitian; this leads to a generalized complex eigenvalue problem, in which complex frequencies are obtained as eigenvalues. The real part of the complex frequency represents the eigenfrequency of interest, whereas the imaginary part is associated with the lifetime of leaky modes in the slab or, equivalently, with the rate of energy absorption of the mode by the PMLs. Once the eigenvalue and the associated eigenvector are found, the mode profile for each eigenmode is calculated by inserting the eigenvector into the magnetic-field expansion, Eq. (3.88).

As was the case in the absence of PMLs, the symmetry of a photonic-crystal slab in the z direction means that the eigenmodes can be classified into two categories: even and odd modes, or TE-like and TM-like modes, respectively. This symmetry can be used to reduce the size of matrices $H^{\lambda,\lambda'}_{\mathbf{G},\mathbf{G}'}$ and $B^{\lambda,\lambda'}_{\mathbf{G},\mathbf{G}'}$ by a factor of two, as before.

Introducing the PMLs suppresses spurious modes and allows taking the leaky modes into account. However, undesired so-called "PML modes" are generated in their stead. In other words, the PML only modifies the undesired spurious modes and does not really eliminate them. To distinguish among the guided modes, the leaky modes, and the PML modes, additional tools have been developed for examining the eigenfunctions and for identifying modes of interest. One such tool relies on calculating the quality factor of the eigenmodes and their change as a function of the PML parameters α and β. Modes that are unaffected by the change of PML parameters are the true physical modes.

To observe how this method of distinguishing among different eigenmodes works in practice, consider a square lattice of air holes embedded in a slab with a high dielectric constant of 12.25 and a thickness of $0.6a$, as shown in Fig. 3.18a. The air holes are cylindrical with radius $r = 0.3a$. The computational region consists of a box with height $\zeta = 4a$ and PML absorbing layers with thickness $t_{\mathrm{PML}} = 0.5a$. As an example, consider even modes at a wave vector $\mathbf{k} = 0.125(2\pi/a)\hat{\mathbf{x}}$. The complex eigenfrequencies are calculated for two different absorbing material parameters of PMLs, $\alpha - j\beta = 1 - 3.2j$ and $\alpha - j\beta = 1 - 6.4j$. The results are shown in Fig. 3.19, where five modes appear unchanged when the absorption of PML varies, and therefore, can be identified as the guided modes of interest. The remaining modes are the PML modes because their eigenfrequencies depend heavily on the particular choice of PML parameters.

Because the increased absorption of the PML region ($\beta = 6.4$) produces more undesirable PML modes within the normalized (real) frequency range [0, 0.5] than the lower absorption ($\beta = 3.2$) does, the latter is preferred for the problem studied here. Therefore, in the rest of this section, the parameters of the PML are chosen as $\alpha - j\beta = 1 - 3.2j$ and $t_{\mathrm{PML}} = 0.5a$. In general, the choice of PML parameters is the result of a compromise because lower absorption requires a thicker layer for the same electromagnetic isolation and, therefore, is computationally more expensive, whereas higher absorption introduces more spurious modes as in the example studied here.

One may also observe from the above example that the normalized frequencies of the guided modes have a relatively small imaginary part. The imaginary part of the

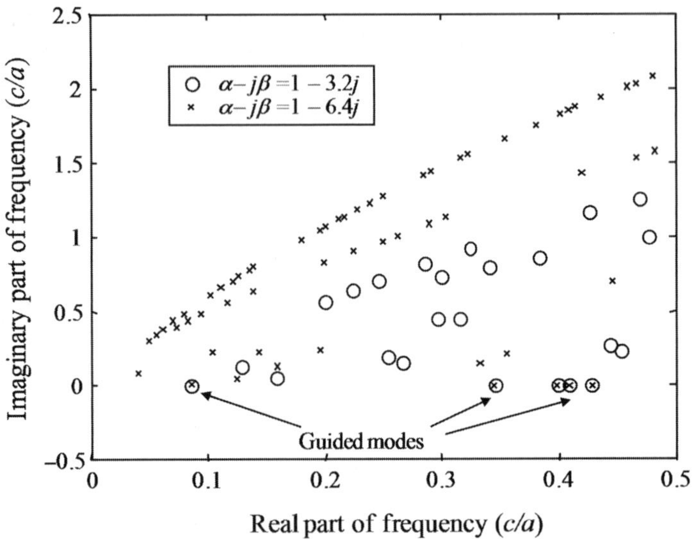

Figure 3.19 Complex frequencies of the eigenmodes of a slab PhC with PMLs bounding the computational unit cell along the z direction. Modes were calculated at a single wave vector for two different values of the PML loss β. Eigenfrequencies of the guided modes have small imaginary parts and are unchanged when the loss is varied; other modes are spurious PML modes.

complex frequency is related to the lifetime of the mode and to its quality factor Q defined as

$$Q = \frac{\Re(f)}{2\Im(f)}, \tag{3.91}$$

where $f = \omega a/2\pi c$ is the normalized frequency and $\Re$ and $\Im$ denote its real and imaginary parts, respectively.

The Q factors of the modes that correspond to the propagation vector $\mathbf{k} = 0.125(2\pi/a)\hat{\mathbf{x}}$ are shown in Fig. 3.20a. Comparison of the graphs in Fig. 3.20a and b (see explanation below) indicates that the guided modes have higher Q factors than PML modes. This fact is understandable because the PML consists of a lossy medium, and consequently, the wave that propagates within the PML has a shorter lifetime. Thus, PML modes generally have a low Q factor.

In this way, the Q factor can be used as an initial criterion to filter out some PML modes. However, the Q values can vary significantly for different physical modes. For example, the Q values of truly guided modes (with an unbounded cladding region) are infinite because the wave propagation is completely confined within the slab. However, leaky modes have a finite Q, and the propagation losses are different for different modes. Therefore, another criterion is required to help distinguish the physical modes from the PML modes. Such a criterion is obtained by noting that guided modes are concentrated mostly in the photonic-crystal slab,

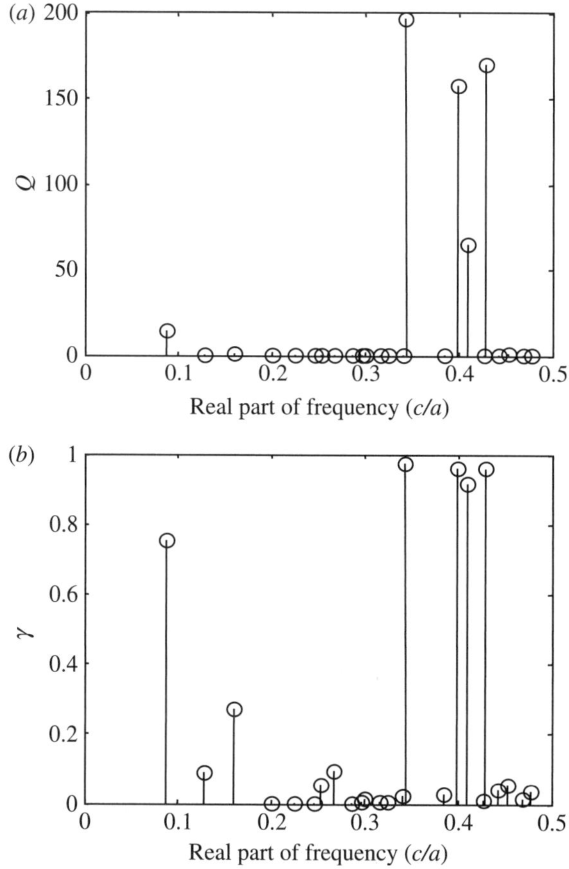

Figure 3.20 (a) Quality factor Q versus frequency (real part) for modes of a square-lattice slab PhC with wave vector $\mathbf{k} = k_x\hat{\mathbf{x}} = 0.125(2\pi/a)\hat{\mathbf{x}}$ obtained by PWEM. The slab (+cladding) was sandwiched between PMLs with $\alpha - j\beta = 1 - 3.2j$. (b) Fraction of energy within the slab γ for the same modes. Note the general correspondence between high-Q modes and high-γ modes. The five modes with the highest Q and γ values are the same modes indicated as guided modes in Fig. 3.19.

whereas PML modes are concentrated mostly in the PML region. Therefore, if we calculate the energy U of each mode within the slab, the PML, and the entire computational region

$$\begin{cases} U_{\text{slab}} = \dfrac{1}{4} \displaystyle\int\limits_{V_{\text{slab}}} (\mu|\mathbf{H}|^2 + \varepsilon|\mathbf{E}|^2)\, dv \\ U_{\text{PML}} = \dfrac{1}{4} \displaystyle\int\limits_{V_{\text{PML}}} (\mu|\mathbf{H}|^2 + \varepsilon|\mathbf{E}|^2)\, dv, \\ U = U_{\text{slab}} + U_{\text{PML}} \end{cases} \tag{3.92}$$

we can use γ, defined as the fraction of energy stored in the slab

$$\gamma \equiv \frac{U_{\text{slab}}}{U}, \tag{3.93}$$

to identify the physical modes.

The plot of the fraction of energy within the slab for modes with the propagation vector $\mathbf{k} = 0.125(2\pi/a)\hat{\mathbf{x}}$ is shown in Fig. 3.20b. According to this figure, for guided modes with higher frequencies, the fraction of energy within the slab exceeds 80% and is smaller for lower frequencies, e.g., less than $0.1c/a$. Because the wavelength of the low-frequency modes in the lowest order band is much larger than the thickness of slab, the evanescent tail of the guided mode is very long; i.e., the mode extends beyond the entire computational region, which reduces the fraction of energy within the slab. This finding indicates that the computational method described in this section is not suitable for calculating guided modes with very low frequencies as it requires commensurate increase of the computational region in the z direction to accommodate the spatial extent of the mode. However, for higher frequencies, and higher order bands, the guided modes are well confined within the slab and the method works well.

The profiles of various modes are illustrated in Fig. 3.21, which shows the three-dimensional distribution of the field amplitude H for three modes: a truly guided mode (a) as well as two PML modes (b) and (c). The real parts of their corresponding normalized eigenfrequencies are 0.0867, 0.159, and 0.2241, respectively. As indicated, the guided mode with lower frequency, Fig. 3.21a, has a long evanescent tail that extends into the PML region, which thereby artificially suppresses its Q value. From Fig. 3.21b and c, it is also clear that the energy of the PML modes is concentrated within the PML layers.

Although not entirely bulletproof, the criteria presented above provide a useful means to determine whether a particular mode is physical or an artifact of the structure modification with PMLs.

3.2.7 Revised Plane-Wave Method for Dispersive Material and its Application to Band-Structure Calculations of Photonic-Crystal Slabs

The practical importance of photonic-crystal slabs creates a need for efficient and accurate methods to analyze their dispersive properties. Although the methods discussed in the section above provide accurate results, the need to use three-dimensional analysis makes them computationally expensive. However, the planar nature of such structures, as is evident in Fig. 3.13a, offers an irresistible temptation to apply 2D tools in their analysis. Before such an approach is tried, one must be aware of certain complications that stem from the finite thickness of the slab. The confinement in the z direction is the result of total internal reflection at the slab boundaries. This gives rise to evanescent tails that extend beyond the slab physical boundaries and consequently to the reduction of the refractive index that the wave confined to the slab

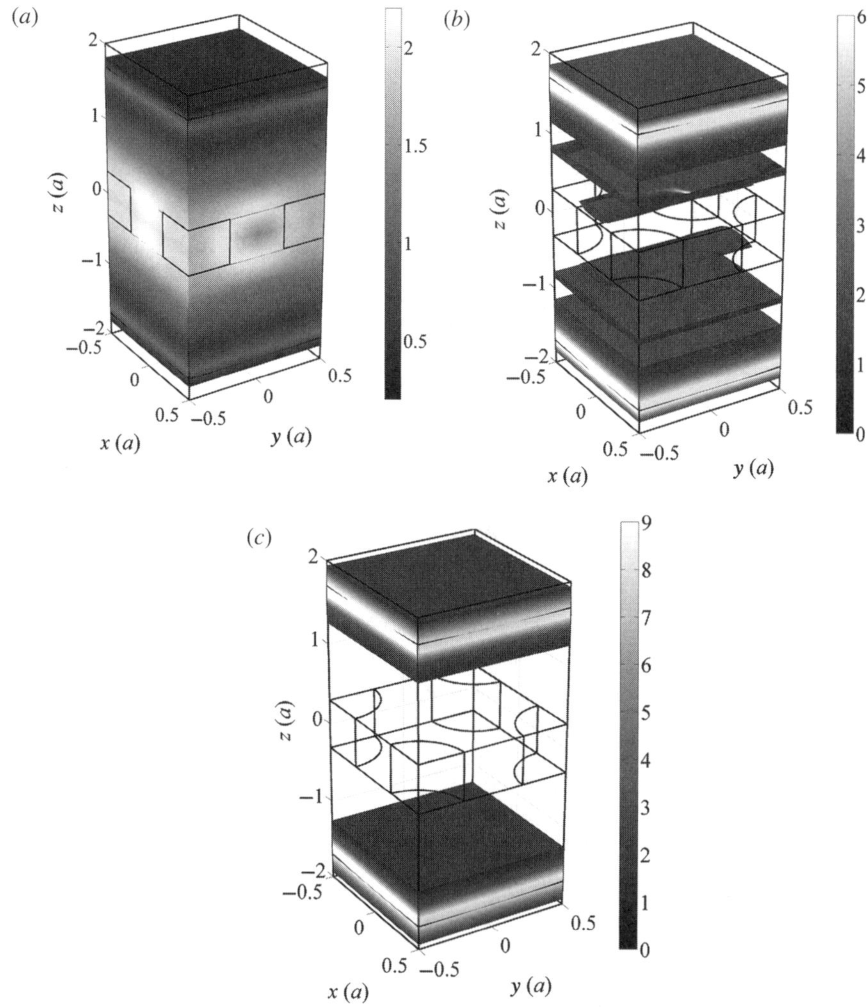

Figure 3.21 Three-dimensional H field amplitude for three typical modes: (a) guided mode as well as (b) and (c) PML modes.

experiences. Therefore, instead of using the refractive index of the slab material in bulk, one must use the effective index of the slab, which is generally lower than the bulk index. Furthermore, for a given material, the slab effective index is a function of frequency and the slab thickness, and it depends on which of the modes supported by the slab is excited. Assuming that the slab is thin enough that it supports only a single mode in the frequency range of interest, the frequency dependence of the slab effective index makes it highly dispersive for electromagnetic waves that propagate within.

The difficulty in employing the PWEM to photonic crystals made of highly dispersive materials lies in the frequency dependence of the matrix **A** in the eigenvalue equation

$$\mathbf{Ah} = \left(\frac{\omega}{c}\right)^2 \mathbf{h} \tag{3.94}$$

that the PWEM reduces to, as discussed in Chapter 2, and in more detail in Sections 3.2.3 and 3.2.4. Because in general, ω-dependence of **A** is nontrivial and indeed nonlinear, Eq. (3.94) does not represent a standard eigenvalue problem as **A** is not a fixed matrix but one whose entries depend on the "eigenvalues" through ω. This dependence precludes the use of standard, highly efficient algorithms for solving eigenvalue problems in this case. Instead, when solving Eq. (3.94), one must take into account the variability of **A** as ω changes with every step of the iteration process converging to the eigensystem, or alternately one can use a piecewise effective-index method [99].

However, an alternative approach restores the ability to use standard eigensolvers. To see how it comes about, let us back off a little and examine what is the goal of these calculations. In analyzing the dispersive properties of a photonic crystal, the object is the dispersion diagram, i.e., a set of dispersion surfaces in the **k**-ω space, such as the one shown in Fig. 3.10. In this figure, each point on the surface is determined by three numbers, k_x, k_y, and ω, two of which can be chosen freely, whereas the third one is determined by the first two. In standard PWEM, ω is calculated for each choice of the pair (k_x, k_y) as schematically presented in Fig. 3.22a, which in the case of dispersive materials leads to the difficulties discussed above. However, if instead we start with the pair (ω, k_x) and calculate k_y, as schematically shown in Fig. 3.22b, the problem of dispersion goes away because ω is fixed for each round of calculations.

The remaining task consists of casting Maxwell's equations in periodic media in the form of an eigenvalue problem, where k_x or k_y rather than ω is the sought quantity. To begin, in a 2D photonic crystal, the problem of calculating dispersion relations is decomposed into two problems for the two different polarizations,

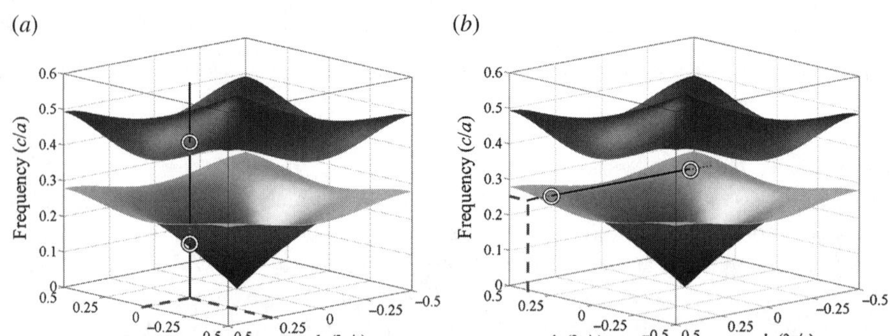

Figure 3.22 Construction of the photonic-crystal dispersion surfaces in (a) conventional PWEM and (b) revised PWEM. In both (a) and (b), dashed lines indicate a particular pair of givens and circles indicate the eigensolutions found by the respective methods. In each case, two parameters are specified to solve for the third among the set $\{k_x, k_y, \omega\}$.

TE and TM. The time-harmonic form of Maxwell's equations for TE polarization is

$$\begin{cases} \dfrac{\partial H_z}{\partial y} = j\omega\varepsilon_0\varepsilon_r(\omega)E_x \\ -\dfrac{\partial H_z}{\partial x} = j\omega\varepsilon_0\varepsilon_r(\omega)E_y, \\ \dfrac{\partial E_x}{\partial y} - \dfrac{\partial E_y}{\partial x} = j\omega\mu_0 H_z \end{cases} \tag{3.95}$$

whereas for TM polarization, it is

$$\begin{cases} \dfrac{\partial E_z}{\partial y} = -j\omega\mu_0 H_x \\ \dfrac{\partial E_z}{\partial x} = j\omega\mu_0 H_y \\ \dfrac{\partial H_y}{\partial x} - \dfrac{\partial H_x}{\partial y} = j\omega\varepsilon_0\varepsilon_r(\omega)E_z \end{cases} . \tag{3.96}$$

Note that the permittivity is shown explicitly to be a function of frequency. Then, a rectangular unit cell is defined by a pair of perpendicular lattice vectors with lengths a and b.

The periodicity of the permittivity function ε_r allows its expansion in a Fourier series, and the expansion of the electromagnetic field components into Bloch modes. (Note that we now expand the permittivity itself, not its reciprocal, as we had in the conventional PWEM.) For TE polarization, these expansions take the form

$$\begin{cases} \varepsilon_r(\omega, \mathbf{r}) = \sum_{\mathbf{n}} \kappa_{\mathbf{n}}(\omega) \exp(j\mathbf{G}_{\mathbf{n}} \cdot \mathbf{r}) \\ \mathbf{H}(\mathbf{r}) = \sum_{\mathbf{n}} h_{z,\mathbf{n}} \exp[j(\mathbf{G}_{\mathbf{n}} + \mathbf{k}) \cdot \mathbf{r}]\hat{\mathbf{z}} \\ \mathbf{E}(\mathbf{r}) = \sum_{\mathbf{n}} (e_{x,\mathbf{n}}\hat{\mathbf{x}} + e_{y,\mathbf{n}}\hat{\mathbf{y}}) \exp[j(\mathbf{G}_{\mathbf{n}} + \mathbf{k}) \cdot \mathbf{r}] \end{cases} , \tag{3.97}$$

where $\mathbf{r} = x\hat{\mathbf{x}} + y\hat{\mathbf{y}}$ is the position vector, the summation index $\mathbf{n}$ is two-dimensional $\mathbf{n} = n_x\hat{\mathbf{x}} + n_y\hat{\mathbf{y}}$ with integer n_x and n_y, $\mathbf{G}_{\mathbf{n}} = n_xG_x\hat{\mathbf{x}} + n_yG_y\hat{\mathbf{y}}$ is the reciprocal lattice vector with $G_x = 2\pi/a$, $G_y = 2\pi/b$, and $\mathbf{k} = k_x\hat{\mathbf{x}} + k_y\hat{\mathbf{y}}$ is the wave vector in the first Brillouin zone. In computer simulations, these expansions are truncated to a finite number M of plane waves, and the Fourier coefficients $\kappa_{\mathbf{n}}$ of the permittivity are efficiently calculated by 2D FFT.

By substituting expressions (3.97) into Eq. (3.95), we obtain

$$\begin{cases} \sum_{\mathbf{n}} h_{z,\mathbf{n}}(n_y G_y + k_y)\exp[j(\mathbf{G_n}+\mathbf{k})\cdot\mathbf{r}] = \omega\varepsilon_0 \sum_{\mathbf{n}'\mathbf{n}''} \kappa_{\mathbf{n}''} e_{x,\mathbf{n}'} \exp[j(\mathbf{G}_{\mathbf{n}'}+\mathbf{G}_{\mathbf{n}''}+\mathbf{k})\cdot\mathbf{r}] \\ -\sum_{\mathbf{n}} h_{z,\mathbf{n}}(n_x G_x + k_x)\exp[j(\mathbf{G_n}+\mathbf{k})\cdot\mathbf{r}] = \omega\varepsilon_0 \sum_{\mathbf{n}'\mathbf{n}''} \kappa_{\mathbf{n}''} e_{y,\mathbf{n}'} \exp[j(\mathbf{G}_{\mathbf{n}'}+\mathbf{G}_{\mathbf{n}''}+\mathbf{k})\cdot\mathbf{r}], \\ \sum_{\mathbf{n}} [e_{x,\mathbf{n}}(n_y G_y + k_y) - e_{y,\mathbf{n}}(n_x G_x + k_x)]\exp[j(\mathbf{G_n}+\mathbf{k})\cdot\mathbf{r}] = \omega\mu_0 \sum_{\mathbf{n}} h_{z,\mathbf{n}} \exp[j(\mathbf{G_n}+\mathbf{k})\cdot\mathbf{r}] \end{cases} \tag{3.98}$$

which by redefining the summation indexes on the right-hand sides of the first two equations according to $\mathbf{n} = \mathbf{n}' + \mathbf{n}''$, $\mathbf{n}'' = \mathbf{n} - \mathbf{n}'$, become

$$\begin{cases} \sum_{\mathbf{n}} h_{z,\mathbf{n}}(n_y G_y + k_y)\exp[j(\mathbf{G_n}+\mathbf{k})\cdot\mathbf{r}] = \omega\varepsilon_0 \sum_{\mathbf{n}\mathbf{n}'} \kappa_{\mathbf{n}-\mathbf{n}'} e_{x,\mathbf{n}'} \exp[j(\mathbf{G_n}+\mathbf{k})\cdot\mathbf{r}] \\ -\sum_{\mathbf{n}} h_{z,\mathbf{n}}(n_x G_x + k_x)\exp[j(\mathbf{G_n}+\mathbf{k})\cdot\mathbf{r}] = \omega\varepsilon_0 \sum_{\mathbf{n}\mathbf{n}'} \kappa_{\mathbf{n}-\mathbf{n}'} e_{y,\mathbf{n}'} \exp[j(\mathbf{G_n}+\mathbf{k})\cdot\mathbf{r}]. \\ \sum_{\mathbf{n}} [e_{x,\mathbf{n}}(n_y G_y + k_y) - e_{y,\mathbf{n}}(n_x G_x + k_x)]\exp[j(\mathbf{G_n}+\mathbf{k})\cdot\mathbf{r}] = \omega\mu_0 \sum_{\mathbf{n}} h_{z,\mathbf{n}} \exp[j(\mathbf{G_n}+\mathbf{k})\cdot\mathbf{r}] \end{cases} \tag{3.99}$$

For the series equations (3.99) to be identically true at any position **r**, the individual terms of the **n**-series must be equal, and thus, we obtain

$$\begin{cases} h_{z,\mathbf{n}}(n_y G_y + k_y) = \omega\varepsilon_0 \sum_{\mathbf{n}'} \kappa_{\mathbf{n}-\mathbf{n}'} e_{x,\mathbf{n}'} \\ -h_{z,\mathbf{n}}(n_x G_x + k_x) = \omega\varepsilon_0 \sum_{\mathbf{n}'} \kappa_{\mathbf{n}-\mathbf{n}'} e_{y,\mathbf{n}'} , \\ e_{x,\mathbf{n}}(n_y G_y + k_y) - e_{y,\mathbf{n}}(n_x G_x + k_x) = \omega\mu_0 h_{z,\mathbf{n}} \end{cases} \tag{3.100}$$

which can be cast in the form of a system of matrix equations[2]

$$\begin{cases} ([G_y]+k_y)[h_z] = k_0[\kappa][e_x] \\ -([G_x]+k_x)[h_z] = k_0[\kappa][e_y], \\ ([G_y]+k_y)[e_x] - ([G_x]+k_x)[e_y] = k_0[h_z] \end{cases} \tag{3.101}$$

with suitable definitions of $M \times M$ matrices $[G_x]$, $[G_y]$, $[\kappa]$, and column M-vectors $[e_x]$, $[e_y]$, $[h_z]$, and with $k_0 = \omega/c$. Note that in the transition from Eq. (3.100) to Eq. (3.101), the Fourier coefficients of the magnetic field h_z are rescaled by the vacuum impedance $\sqrt{\mu_0/\varepsilon_0}$.

[2]In the formulas that follow, we use square brackets to denote $M \times M$ matrices and column M-vectors, rather than boldface, to distinguish them from Cartesian vectors.

By eliminating the field component $[e_y]$ using the second of Eqs. (3.101), the following eigenvalue problem is obtained:

$$\frac{1}{k_0}\begin{pmatrix} -k_0[G_y] & k_0^2 - (k_x + [G_x])[\kappa]^{-1}(k_x + [G_x]) \\ k_0^2[\kappa] & -k_0[G_y] \end{pmatrix}\begin{pmatrix} [e_x] \\ [h_z] \end{pmatrix} = k_y \begin{pmatrix} [e_x] \\ [h_z] \end{pmatrix}. \quad (3.102)$$

In a similar fashion, the eigenvalue equation for TM polarization is obtained

$$-\frac{1}{k_0}\begin{pmatrix} k_0[G_y] & k_0^2[\kappa] - (k_x + [G_x])^2 \\ k_0^2 & k_0[G_y] \end{pmatrix}\begin{pmatrix} [h_x] \\ [e_z] \end{pmatrix} = k_y \begin{pmatrix} [h_x] \\ [e_z] \end{pmatrix}. \quad (3.103)$$

Thus, Maxwell's equations in periodic media again lead to an eigenvalue problem, as in the standard PWEM, except that this time the sought eigenvalue is the y-component of the propagation vector $\mathbf{k}$ rather than the frequency ω. The matrices, whose eigenvalues and eigenvectors are calculated, are independent of the eigenvalues themselves, which allows the application of standard numerical techniques in this case. As a result, this formulation is not sensitive to the constituent-material dispersion, and it can be readily applied to photonic-crystal slabs approximated as 2D photonic crystals via the effective-index method.

The effectiveness of this revised PWEM is best demonstrated by applying it to a specific example and by comparing the results with those obtained using a 3D FDTD simulation (discussed in detail in Section 3.3). The slab PhC considered for this purpose consists of a square lattice of circular air holes with lattice constant a and hole radius $0.3a$, in a slab of permittivity $\varepsilon_r = 12.25$ and thickness $0.6a$.

Figure 3.23 shows a comparison of the dispersion diagrams for TE mode obtained by the FDTD, which are represented as circles, and by the revised 2D PWEM

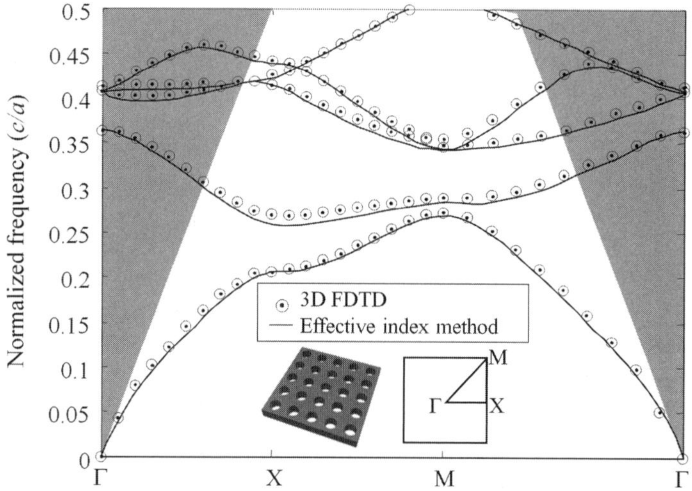

Figure 3.23 Comparison of dispersion diagrams for TE modes obtained using FDTD and the modified PWEM. The slab has thickness $0.6a$ and dielectric constant 12.25. Air holes distributed on a square lattice have radii $0.3a$, where a is the lattice constant.

method, which is shown as solid lines. The gray region indicates the modes that lie above the light line, which are leaky; modes below the light line are guided. This figure shows excellent agreement between the two methods within the normalized frequency range under consideration, $0 \leq c/a \leq 0.5$. Also notable is the absence of the spurious modes often observed in 3D PWEM calculations as discussed in Section 3.2.6, and both the leaky and guided modes are determined with comparable accuracy.

The power of the revised PWEM method can be appreciated by noting that it took 128 minutes to obtain the eigenfrequencies for each of the considered wave vectors by using the FDTD method. However, it took only 0.05 seconds to calculate the eigen-wave vectors for a given frequency by using the revised PWEM algorithm on the same desktop computer.

3.3 FINITE-DIFFERENCE TIME-DOMAIN (FDTD) METHOD

Since the initial work of Yee [100], the FDTD method [65] has been one of the most widely used numerical techniques for solving electromagnetic boundary value problems. The essence of the method is the direct computational implementation of the Maxwell curl equations, which approximate all derivatives, both spatial and temporal, as second-order-accurate finite differences. The computational space is sampled at intervals that are small compared with all wavelengths under consideration, and material properties are specified at each sample point. The fields are calculated directly for all space points at successive time steps, where the time increment between steps is small compared with the reciprocals of all frequencies under consideration. Sources are implemented by defining the fields at certain points to have specified values at specified times, and they can be continuous or transient; continuous sources are typically sinusoidal in time and hence monochromatic, whereas transient sources possess a finite bandwidth.

The method has two primary advantages over frequency-domain techniques such as the FEM and boundary integral methods. First and foremost, the FDTD method's computational requirements scale linearly with the size of the problem, which permits the analysis of larger, irregular structures in addition to perfectly periodic structures such as PhCs. Second, the method is simple to implement, even for complicated scattering structures, because arbitrary dielectric parameters can be assigned to each lattice point. Consequently, structures that contain inhomogeneous, lossy, dispersive, active, or anisotropic materials can be analyzed easily [65]. In addition, the FDTD algorithm lends itself to parallel implementation over multiple processors [65] or hardware accelerated boards [101, 102] because of the inherent data independencies of the difference equations from which it is derived.

In the remainder of this section, we first describe the FDTD method for both the two-dimensional and full three-dimensional cases. Next, we provide brief descriptions of the conditions for ensuring stability and accuracy, followed by discussions of techniques for simulating transient and CW system excitation and response extraction. Then, absorbing and Bloch boundary conditions for terminating the computational region are discussed. The section ends with examples of solutions of photonic-crystal problems obtained using FDTD, which include dispersion relations, and transmission/reflection across finite PhC slabs.

3.3.1 Central-Difference Expressions of Maxwell's Equations

First, consider the Maxwell's time-dependent curl equations for a linear, isotropic, and source-free region,

$$\begin{cases} \mu \dfrac{\partial \mathbf{H}}{\partial t} = -\nabla \times \mathbf{E} \\ \varepsilon \dfrac{\partial \mathbf{E}}{\partial t} = \nabla \times \mathbf{H} - \sigma \mathbf{E} \end{cases}, \tag{3.104}$$

where **E** and **H** are time-variable electric and magnetic fields, whereas μ, ε, and σ characterize the permittivity, permeability, and conductivity, respectively. The FDTD method derives its name from a direct, central finite-difference approximation to these time-dependent curl equations. Let $f(u,v,w,t)$ designate any field component, either electric or magnetic, in an orthogonal 3D coordinate system with points specified by coordinates (u,v,w). We label discrete points in space as $(i\Delta u, j\Delta v, k\Delta w)$ and discrete points in time as $n\Delta t$, where i, j, k, and n are all integers, whereas Δu, Δv, and Δw are the spatial intervals and Δt is the time increment. Consequently, at the discrete points in space-time, $f(u,v,w,t)$ can be expressed as

$$f(u,v,w,t) = f(i\Delta u, j\Delta v, k\Delta w, n\Delta t) \equiv f^n_{i,j,k}, \tag{3.105}$$

where the notation uses subscripts to indicate spatial indices (in any number of dimensions up to three) and the superscript to indicate the time index.

Central finite-difference expressions that are second-order accurate in both time and space increments are used for the space and time derivatives. For example, consider the first partial derivative of f with respect to u, calculated at a fixed time $t_n = n\Delta t$,

$$\frac{\partial f(i\Delta u, j\Delta v, k\Delta w, n\Delta t)}{\partial u} = \frac{f^n_{i+1/2,j,k} - f^n_{i-1/2,j,k}}{\Delta u} + O[(\Delta u)^2]. \tag{3.106}$$

Note the $\pm(1/2)$ increment in the index i of the u coordinate to indicate a finite difference of f over interval $\pm(1/2)\Delta u$. This result is obtained by taking the difference between the Taylor expansions of f at points $u = u_0 + \Delta u/2 = i\Delta u + \Delta u/2$ and $u = u_0 - \Delta u/2 = i\Delta u - \Delta u/2$, and by solving for the first derivative, while gathering terms of second order and higher in Δu in the last term. Note the use of half-integer increments in the index, and that the points used to approximate the derivative at point i are separated by Δu. This notation is adopted to separate the grid points spatially where the electric and magnetic fields are computed; in other words, the electric and magnetic field grid points are interleaved in the space lattice at intervals $\Delta u/2$. Hence, the value of a particular component of E is updated using the difference between two adjacent values of H, which are separated from each other by Δu and located at distances of $\pm(1/2)\Delta u$ on either side of the E-field grid point.

Likewise, the expression for the first partial derivative of f with respect to time, evaluated at a point (i, j, k), is

$$\frac{\partial f(i\Delta u, j\Delta v, k\Delta w, n\Delta t)}{\partial t} = \frac{f_{i,j,k}^{n+1/2} - f_{i,j,k}^{n-1/2}}{\Delta t} + O[(\Delta t)^2]. \qquad (3.107)$$

Here, the $\pm(1/2)$ increments are in the time index n, which denotes a finite difference over $\pm(1/2)\Delta t$. This expression reflects the fact that the algorithm computes the electric and magnetic fields at alternating time steps that are separated by intervals of $\Delta t/2$. In other words, the E and H fields are updated alternately in half time steps, such that *both* the electric and the magnetic fields are updated during each full time step Δt. For this reason, this algorithm is often called a "leapfrog" algorithm.

In the next two subsections, the FDTD formulations are presented for TM and TE polarizations in two dimensions and for the full three-dimensional case—all based on Cartesian coordinates.

3.3.2 Two-Dimensional FDTD Method

As in Section 3.2.3, in the two-dimensional case, it is assumed that neither the fields nor the modeled material geometry has any variation in the z direction; therefore, all partial derivatives with respect to z are equal to zero. In this case, Maxwell's equations in the rectangular coordinate system, which in general involve coupled differential equations for six field components, Eq. (3.104), can be separated to two independent sets of coupled differential equations, each with only three field components:

$$\begin{cases} -\dfrac{\partial E_z}{\partial y} = \mu_0 \dfrac{\partial H_x}{\partial t} \\ \dfrac{\partial E_z}{\partial x} = \mu_0 \dfrac{\partial H_y}{\partial t} \\ \dfrac{\partial H_y}{\partial x} - \dfrac{\partial H_x}{\partial y} = \varepsilon_0 \varepsilon_r \dfrac{\partial E_z}{\partial t} + \sigma E_z \end{cases}, \qquad (3.108)$$

$$\begin{cases} \dfrac{\partial H_z}{\partial y} = \varepsilon_0 \varepsilon_r \dfrac{\partial E_x}{\partial t} + \sigma E_x \\ -\dfrac{\partial H_z}{\partial x} = \varepsilon_0 \varepsilon_r \dfrac{\partial E_y}{\partial t} + \sigma E_y, \\ \dfrac{\partial E_x}{\partial y} - \dfrac{\partial E_y}{\partial x} = \mu_0 \dfrac{\partial H_z}{\partial t} \end{cases} \qquad (3.109)$$

where we make the additional simplifying assumption that all simulated materials are nonmagnetic, i.e., $\mu = \mu_0$ everywhere, which is reasonable for most photonic-crystal applications at optical wavelengths. (Note, however, that the FDTD method has no difficulty handling magnetic materials.) The first set, which involves only magnetic field components H_x and H_y in the xy plane and E_z as the sole field component in the z direction, is designated as the TM mode. Likewise, the second set includes only E_x and E_y in the xy plane and H_z as the sole component in the z direction, and it is designated as the TE mode.

Applying the central finite-difference expressions of the last section to Eq. (3.108), the following set of difference equations for the TM mode is obtained:

$$\left\{\begin{aligned} H_x\big|_{i,j}^{n+1/2} &= H_x\big|_{i,j}^{n-1/2} - \frac{1}{\mu_0}\frac{\Delta t}{\Delta y}\left(E_z\big|_{i,j+1/2}^{n} - E_z\big|_{i,j-1/2}^{n}\right) \\ H_y\big|_{i,j}^{n+1/2} &= H_y\big|_{i,j}^{n-1/2} + \frac{1}{\mu_0}\frac{\Delta t}{\Delta x}\left(E_z\big|_{i+1/2,j}^{n} - E_z\big|_{i-1/2,j}^{n}\right) \\ E_z\big|_{i,j}^{n+1} &= \frac{\varepsilon_{i,j} - \sigma_{i,j}\Delta t/2}{\varepsilon_{i,j} + \sigma_{i,j}\Delta t/2} E_z\big|_{i,j}^{n} \\ &\quad + \frac{\Delta t}{\varepsilon_{i,j} + \sigma_{i,j}\Delta t/2}\left(\frac{H_y\big|_{i+1/2,j}^{n+1/2} - H_y\big|_{i-1/2,j}^{n+1/2}}{\Delta x} - \frac{H_x\big|_{i,j+1/2}^{n+1/2} - H_x\big|_{i,j-1/2}^{n+1/2}}{\Delta y}\right) \end{aligned}\right. , \tag{3.110}$$

where the electric field component is updated at times $t = n\Delta t$ and the magnetic field components are updated at times $t = (n + 1/2)\Delta t$. The difference expressions for TE mode can be determined in a similar fashion and are given by

$$\left\{\begin{aligned} E_x\big|_{i,j}^{n+1} &= \frac{\varepsilon_{i,j} - \sigma_{i,j}\Delta t/2}{\varepsilon_{i,j} + \sigma_{i,j}\Delta t/2} E_x\big|_{i,j}^{n} + \frac{\Delta t}{\varepsilon_{i,j} + \sigma_{i,j}\Delta t/2}\left(\frac{H_z\big|_{i,j+1/2}^{n+1/2} - H_z\big|_{i,j-1/2}^{n+1/2}}{\Delta y}\right) \\ E_y\big|_{i,j}^{n+1} &= \frac{\varepsilon_{i,j} - \sigma_{i,j}\Delta t/2}{\varepsilon_{i,j} + \sigma_{i,j}\Delta t/2} E_y\big|_{i,j}^{n} - \frac{\Delta t}{\varepsilon_{i,j} + \sigma_{i,j}\Delta t/2}\left(\frac{H_z\big|_{i+1/2,j}^{n+1/2} - H_z\big|_{i-1/2,j}^{n+1/2}}{\Delta y}\right) \\ H_z\big|_{i,j}^{n+1/2} &= H_z\big|_{i,j}^{n-1/2} + \frac{\Delta t}{\mu_0}\left(\frac{E_x\big|_{i,j+1/2}^{n} - E_x\big|_{i,j-1/2}^{n}}{\Delta y} - \frac{E_y\big|_{i+1/2,j}^{n} - E_y\big|_{i-1/2,j}^{n}}{\Delta x}\right) \end{aligned}\right. . \tag{3.111}$$

The spatial field components are evaluated on an interleaved grid, as shown in Fig. 3.24 for both TE and TM modes. Examining the figure, we can observe more clearly how the interleaved spatial grid works. The central-difference equations take the separation between adjacent components to be $\Delta x/2$ or $\Delta y/2$. Thus, in Eqs. (3.110) and (3.111), index $i + 1/2$ $(j + 1/2)$ indicates that the component being referenced is located to the right (above) of the component being updated at index i, whereas $i-1/2$ $(j-1/2)$ indicates that it is located to the left (below). Using the grouping convention indicated in Fig. 3.24 by the dashed boxes allows the algorithm to be programmed using only integer indices: For the TM mode, the

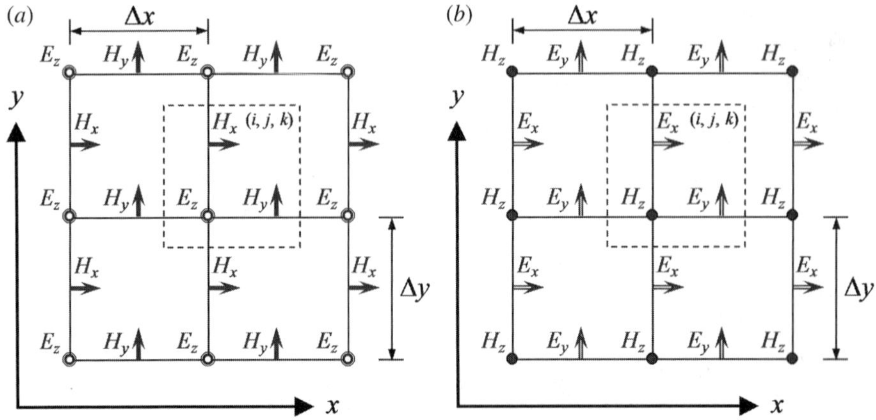

Figure 3.24 Two-dimensional FDTD meshes for (a) TM mode and (b) TE mode. Magnetic (electric) field components are indicated by filled (open) symbols; arrows indicate in-plane components, and dots indicate out-of-plane components. Note that the field points are interleaved with a spacing of $\Delta x/2$ or $\Delta y/2$. The dashed boxes indicate a set of components that could be associated with a single set of grid indices (i, j, k).

fields are updated according to

$$\begin{cases} H_x|_{i,j}^n = H_x|_{i,j}^{n-1} - \dfrac{1}{\mu_0}\dfrac{\Delta t}{\Delta y}(E_z|_{i,j+1}^n - E_z|_{i,j}^n) \\ H_y|_{i,j}^n = H_y|_{i,j}^{n-1} + \dfrac{1}{\mu_0}\dfrac{\Delta t}{\Delta x}(E_z|_{i+1,j}^n - E_z|_{i,j}^n) \\ E_z|_{i,j}^{n+1} = \dfrac{\varepsilon_{i,j} - \sigma_{i,j}\Delta t/2}{\varepsilon_{i,j} + \sigma_{i,j}\Delta t/2} E_z|_{i,j}^n + \dfrac{\Delta t}{\varepsilon_{i,j} + \sigma_{i,j}\Delta t/2}\left(\dfrac{H_y|_{i,j}^n - H_y|_{i-1,j}^n}{\Delta x} - \dfrac{H_x|_{i,j}^n - H_x|_{i,j-1}^n}{\Delta y}\right) \end{cases}; \tag{3.112}$$

likewise, a similar result is obtained for the TE mode. Note, too, that we can eliminate the half-integer time index as well: We merely must be certain to update *all* in-plane field components based on the most recent prior values of the out-of-plane field before updating the out-of-plane field values based on the just-obtained in-plane fields, according to Yee's leapfrog scheme.

3.3.3 Three-Dimensional FDTD Method

In a three-dimensional Cartesian coordinate system, Maxwell's curl equations, Eq. (3.104), can be written as a set of coupled differential equations

$$\begin{cases} \dfrac{\partial E_y}{\partial z} - \dfrac{\partial E_z}{\partial y} = \mu_0 \dfrac{\partial H_x}{\partial t} \\ \dfrac{\partial E_z}{\partial x} - \dfrac{\partial E_x}{\partial z} = \mu_0 \dfrac{\partial H_y}{\partial t} \\ \dfrac{\partial E_x}{\partial y} - \dfrac{\partial E_y}{\partial x} = \mu_0 \dfrac{\partial H_z}{\partial t} \end{cases}, \quad \begin{cases} \dfrac{\partial H_z}{\partial y} - \dfrac{\partial H_y}{\partial z} = \varepsilon_0\varepsilon_r \dfrac{\partial E_x}{\partial t} + \sigma E_x \\ \dfrac{\partial H_x}{\partial z} - \dfrac{\partial H_z}{\partial x} = \varepsilon_0\varepsilon_r \dfrac{\partial E_y}{\partial t} + \sigma E_y \\ \dfrac{\partial H_y}{\partial x} - \dfrac{\partial H_x}{\partial y} = \varepsilon_0\varepsilon_r \dfrac{\partial E_z}{\partial t} + \sigma E_z \end{cases}. \tag{3.113}$$

In accordance with the results of Section 3.3.1, we can approximate each partial derivative as a second-order finite-difference equation, in both space and time, to obtain

$$\begin{aligned}
E_x|_{i,j,k}^{n+1} &= C_{i,j,k}^{(1)} E_x|_{i,j,k}^{n} + C_{i,j,k}^{(2)} \left(\frac{H_z|_{i,j+1/2,k}^{n+1/2} - H_z|_{i,j-1/2,k}^{n+1/2}}{\Delta y} - \frac{H_y|_{i,j,k+1/2}^{n+1/2} - H_y|_{i,j,k-1/2}^{n+1/2}}{\Delta z} \right) \\
E_y|_{i,j,k}^{n+1} &= C_{i,j,k}^{(1)} E_y|_{i,j,k}^{n} + C_{i,j,k}^{(2)} \left(\frac{H_x|_{i,j,k+1/2}^{n+1/2} - H_x|_{i,j,k-1/2}^{n+1/2}}{\Delta z} - \frac{H_z|_{i+1/2,j,k}^{n+1/2} - H_z|_{i-1/2,j,k}^{n+1/2}}{\Delta x} \right), \\
E_z|_{i,j,k}^{n+1} &= C_{i,j,k}^{(1)} E_z|_{i,j,k}^{n} + C_{i,j,k}^{(2)} \left(\frac{H_y|_{i+1/2,j,k}^{n+1/2} - H_y|_{i-1/2,j,k}^{n+1/2}}{\Delta x} - \frac{H_x|_{i,j+1/2,k}^{n+1/2} - H_x|_{i,j-1/2,k}^{n+1/2}}{\Delta y} \right)
\end{aligned} \tag{3.114}$$

with coefficients $C^{(1)}$ and $C^{(2)}$, which are functions only of the material properties at each point in the simulation grid and do not change with time

$$C_{i,j,k}^{(1)} = \frac{\varepsilon_{i,j,k} - \sigma_{i,j,k}\Delta t/2}{\varepsilon_{i,j,k} + \sigma_{i,j,k}\Delta t/2}, \quad C_{i,j,k}^{(2)} = \frac{\Delta t}{\varepsilon_{i,j,k} + \sigma_{i,j,k}\Delta t/2}. \tag{3.115}$$

Likewise, the magnetic field components at time $t = (n + 1/2)\Delta t$ are

$$\begin{aligned}
H_x|_{i,j,k}^{n+1/2} &= H_x|_{i,j,k}^{n-1/2} + \frac{\Delta t}{\mu_0} \left(\frac{E_y|_{i,j,k+1/2}^{n} - E_y|_{i,j,k-1/2}^{n}}{\Delta z} - \frac{E_z|_{i,j+1/2,k}^{n} - E_z|_{i,j-1/2,k}^{n}}{\Delta y} \right) \\
H_y|_{i,j,k}^{n+1/2} &= H_y|_{i,j,k}^{n-1/2} + \frac{\Delta t}{\mu_0} \left(\frac{E_z|_{i+1/2,j,k}^{n} - E_z|_{i-1/2,j,k}^{n}}{\Delta x} - \frac{E_x|_{i,j,k+1/2}^{n} - E_x|_{i,j,k-1/2}^{n}}{\Delta z} \right). \\
H_z|_{i,j,k}^{n+1/2} &= H_z|_{i,j,k}^{n-1/2} + \frac{\Delta t}{\mu_0} \left(\frac{E_x|_{i,j+1/2,k}^{n} - E_x|_{i,j-1/2,k}^{n}}{\Delta y} - \frac{E_y|_{i+1/2,j,k}^{n} - E_y|_{i-1/2,j,k}^{n}}{\Delta x} \right)
\end{aligned} \tag{3.116}$$

Note that each electric field component in three-dimensional space is surrounded by four circulating magnetic field components and, similarly, that each magnetic field component is surrounded by four circulating electric field components, as shown in Fig. 3.25. Thus, the numerical solution of Maxwell's equations using the FDTD method provides a nearly perfect—albeit discretized—correspondence with the physical propagation of electromagnetic waves.

The FDTD procedure is summarized in the flowchart of Fig. 3.26. The process begins by defining a grid in the computational region and assigning electromagnetic properties, such as permittivity, permeability, and conductivity, to each node as well as initial field values at time step $n = 0$. The new values of the electric and magnetic fields are then calculated using Eqs. (3.114) and (3.116), respectively, at all

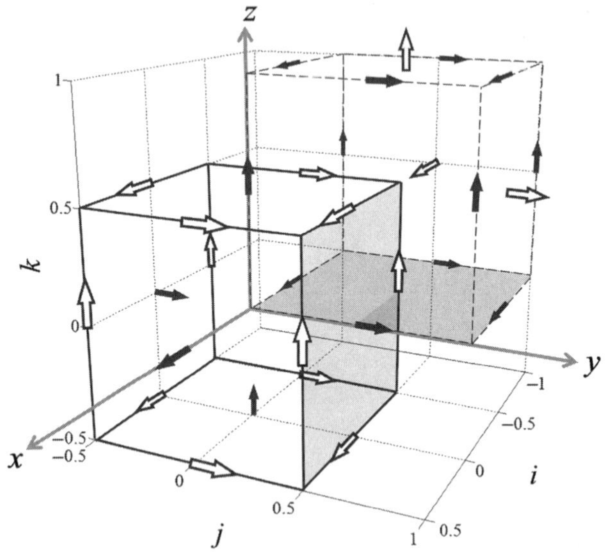

Figure 3.25 Three-dimensional FDTD mesh. Open (filled) arrows indicate electric (magnetic) field components, with the arrow's direction indicating the Cartesian component of the field computed at its location. The solid-edged cube is a single Yee cell centered at the origin; electric field components are located at the center of each edge, and magnetic field components are located at the center of each face. The cube with dashed edges, which is centered at $(-1/2, 1/2, 1/2)$, illustrates how the extended mesh is composed of interpenetrating cells with complementarily located electric and magnetic field components. Each face of the solid-edged cube can be viewed as a Faraday loop, with E components surrounding and normal to the H component centered thereupon. Likewise, each face of the dashed cube can be viewed as an Ampère loop. Two such interlocking loops are shaded in the figure, as an indication of the natural correspondence between the Yee algorithm and the propagating-wave solutions to Maxwell's equations. Inspection of this figure shows that, for this choice of origin, in the (i, j, k)th Yee cell, the E_q component is located at $([i+1/2]\Delta x, [j+1/2]\Delta y, [k+1/2]\Delta z) - (\Delta q/2)\hat{\mathbf{e}}_q$, where q is any Cartesian component x, y, or z, and $\hat{\mathbf{e}}_q$ is a unit vector in the q-direction. Similarly, the H_q component is found at $(i\Delta x, j\Delta y, k\Delta z) + (\Delta q/2)\hat{\mathbf{e}}_q$.

points of the lattice and are modified according to the sources present in the computational region. This application of sources takes place only at grid points where they are defined. The time step is incremented by one, and the field-evaluation process is repeated in a leapfrog fashion.

In the above sections, the discretizations of Maxwell's curl equations in terms of central difference equations was discussed. In the implementation of the FDTD algorithm, several other key issues must be addressed, namely the stability and dispersion properties of the method as well as the boundary conditions. These topics are discussed in the following two subsections.

3.3.4 Numerical Stability and Dispersion

When implementing any numerical algorithm, it is necessary to establish the conditions under which the stability of the technique can be ensured. For the FDTD method, stability means that as time marching continues, the electric and magnetic

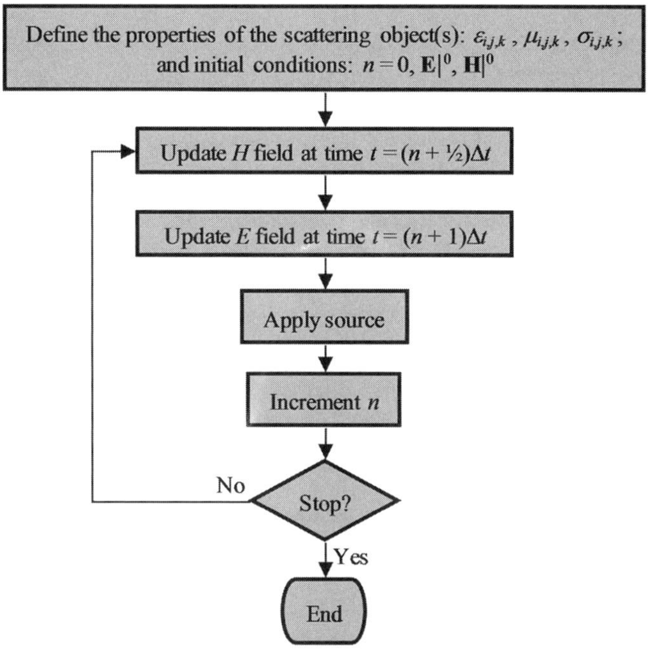

Figure 3.26 Flowchart of the basic FDTD computational method.

field values do not grow without bound. Detailed discussion of stability can be found in Ref. 65; here we present only the criteria that ensure stability of the FDTD method.

The stability condition is dependent on both the spatial interval Δ and the time increment Δt. For a chosen spatial increment, the upper bound of the time increment that ensures stability is

$$\begin{cases} c\Delta t \leq \dfrac{1}{\sqrt{1/\Delta x^2 + 1/\Delta y^2}} & \text{for two dimensions} \\ c\Delta t \leq \dfrac{1}{\sqrt{1/\Delta x^2 + 1/\Delta y^2 + 1/\Delta z^2}} & \text{for two dimensions} \end{cases}. \tag{3.117}$$

If $\Delta = \Delta x = \Delta y = \Delta z$, then the stability condition can be simplified to

$$c\Delta t \leq \Delta/\sqrt{D}, \tag{3.118}$$

where $D = 1, 2, 3$ is the dimensionality of the system being simulated.

Another factor that affects the accuracy of the FDTD method is numerical dispersion, which is an artifact that develops from the discretization of the physical space-time. In three-dimensions, it is governed by the following expression:

$$\frac{\sin^2(\omega\Delta t/2)}{c^2\Delta t^2} = \frac{\sin^2(\tilde{k}_x\Delta x/2)}{\Delta x^2} + \frac{\sin^2(\tilde{k}_y\Delta y/2)}{\Delta y^2} + \frac{\sin^2(\tilde{k}_z\Delta z/2)}{\Delta z^2}, \tag{3.119}$$

where $\tilde{k}_x$, $\tilde{k}_y$, and $\tilde{k}_z$ are the x, y, and z components of the wave vector, respectively, and ω is the angular frequency. Numerical dispersion manifests itself as phase errors that result from the deviation of the calculated phase velocity from its physical counterpart. It can be alleviated by reducing the space and time increments. In particular, it is noted that as Δt, Δx, Δy, and Δz all go to zero, the numerical dispersion, Eq. (3.119), reduces to the familiar expression for dispersion of a continuous lossless medium

$$\frac{\omega^2}{c^2} = k_x^2 + k_y^2 + k_z^2. \tag{3.120}$$

The increased accuracy is obtained at the cost of increased memory requirements, and $\Delta \leq \lambda_{\min}/20$ is a good rule of thumb for typical FDTD simulations, where $\lambda_{\min}$ is the minimal wavelength excited by the source.

3.3.5 Simulating Transient and Steady-State System Response

One of the most significant advantages of the FDTD method is its ability to perform both steady-state and transient analysis. If we are interested in system response to a single-frequency excitation, we can simply launch a sinusoidal wave $A\sin(\omega t)$ and obtain a steady-state condition by extracting the complex (amplitude and phase) field components over a complete cycle of the sinusoidal incident field. The key component in this approach is to allow the simulation to run for a sufficient number of time steps to ensure that the transient fields have decayed out. Typically, this requires two to five transient periods T, where T is defined as Δt times the number of time steps required for one cycle of the incident field to traverse the maximum extent of the scatterers [103]. If the FDTD calculation has reached a steady state after N time steps, then the field at any lattice point within the solution space can then be described by

$$E_1 = E(\mathbf{r})\sin[\omega N\Delta t + \varphi(\mathbf{r})]. \tag{3.121}$$

The steady-state magnitude $E(\mathbf{r})$ and phase $\varphi(\mathbf{r})$ can be extracted by continuing the simulation for a small number of additional time steps Δn, such that

$$\omega\Delta n\Delta t = \pi/2, \tag{3.122}$$

after which we have

$$\begin{aligned} E_2 &= E(\mathbf{r})\sin[\omega N\Delta t + \omega\,\Delta n\Delta t + \varphi(\mathbf{r})] \\ &= E(\mathbf{r})\cos[\omega N\Delta t + \varphi(\mathbf{r})]. \end{aligned} \tag{3.123}$$

Combining Eqs. (3.121) and (3.123), the magnitude $E(\mathbf{r})$ and phase $\varphi(\mathbf{r})$ are determined from

$$\begin{cases} E(\mathbf{r}) = \sqrt{E_2^2 + E_1^2} \\ \varphi(\mathbf{r}) = \tan^{-1}(E_1/E_2) \end{cases}. \tag{3.124}$$

In a transient analysis, one is generally interested in the system response over a certain bandwidth. This can be achieved by using a pulsed source

$$E^{inc}(t) = E_0 g(t). \tag{3.125}$$

For example, a Gaussian pulse $g(t) = e^{-\pi(2F_b t)^2}$ has a spectrum given by $G(f) \propto \exp[-\pi(f/2F_b)^2]$. Modulating a sine-wave with such a pulse, as in $g(t) = e^{2\pi j f_0 t} e^{-\pi(2F_b t)^2}$, shifts the maximum of the frequency spectrum to f_0, $G(f) \propto \exp[-\pi([f - f_0]/2F_b)^2]$. Thus, the wider the frequency range we are interested in, the shorter the pulse needs to be, and the center of the excited bandwidth is easily controlled by multiplying the temporal profile of the pulse by a sine wave of chosen frequency.

After launching a pulse of finite duration, the frequency response of the system at position $\mathbf{r}$ is obtained by Fourier transformation (FT)

$$\tilde{E}(\mathbf{r}, f) = \mathrm{FT}[E(\mathbf{r}, n\Delta t)], \tag{3.126}$$

where $E(\mathbf{r}, n\Delta t)$ is the temporal transient response. Because the field values need to be recorded for all time steps, Eq. (3.126) is useful only when a small number of positions $\mathbf{r}$ are monitored—otherwise, the amount of collected data would quickly overwhelm the available computer memory. Therefore, Eq. (3.126) is used when only several discrete "detectors" are placed throughout the computational region. To observe spectral response over an extended region, such as a plane or volume, a different approach is employed. In this case, instead of collecting all temporal data and performing Fourier transformation when the time evolution of the fields has run its course, the spectral content of the electromagnetic field is calculated on the fly, while the FDTD time marching is taking place. Thus, at a time step N and position $\mathbf{r}$, the field amplitude at frequency f is obtained using recursive formula

$$\tilde{E}_N(\mathbf{r}, f) = \tilde{E}_{N-1}(\mathbf{r}, f) + \Delta t\, E(\mathbf{r}, N\Delta t) e^{-2\pi j f N \Delta t}, \tag{3.127}$$

which can be easily confirmed to produce the desired Fourier transformation of the field $E(\mathbf{r}, t)$ from $t = 0$ up to time $t = N\Delta t$

$$\tilde{E}_N(\mathbf{r}, f) = \Delta t \sum_{n=1}^{N} E(\mathbf{r}, n\Delta t) e^{-2\pi j f n \Delta t}. \tag{3.128}$$

Thus, at each of the monitored positions $\mathbf{r}$, in addition to the instantaneous field values required for FDTD simulations, only a single number, which is calculated with Eq. (3.127), needs to be stored, which reduces the memory requirements if larger regions of the computational space are monitored. The price to pay for this reduction of computer resources is a limit on the number of different frequencies that can be monitored at the same time. Indeed, according to Eqs. (3.127) and (3.128), a separate number needs to be stored for each monitored frequency; thus, only a relatively small number of frequencies can be monitored at the same time. Accordingly, this approach is referred to as discrete Fourier transformation (DFT).

Therefore, Eqs. (3.126) and (3.127) provide two complementary approaches to analyzing the spectral response of the system under consideration. The former provides high-resolution spectral data at few discrete points in space, whereas the latter allows monitoring larger spatial regions for the presence of few discrete frequencies.

3.3.6 Absorbing Boundary Conditions

Many electromagnetic-wave interaction problems involve geometries defined in an open region where the space domain of computed fields is unbounded in one or more directions. However, because computer memory and computation time are finite, it is always necessary to truncate the domain that encloses the analyzed structure. To avoid unphysical reflections from artificial boundaries, suitable boundary conditions that simulate the extension of the computational region to infinity are required. As the goal is to absorb all incident radiation, endowing the outer layer of the computational space with such a property is referred to as an absorbing boundary condition, or ABC.

Because of its importance in realistic modeling of electromagnetic wave scattering, one of the most active areas of research in FDTD development over the last few years has been in the construction of efficient ABCs. As stated, the goal of an ABC is to absorb all outwardly traveling energy without producing any reflections—the lower the reflection, the better the ABC. The most commonly used ABCs are the Mur absorbing boundary condition [104] and PMLs [90, 95]. Because the reflections from the PML ABCs are typically several orders of magnitude lower than for other ABCs, we focus on PML ABCs in this section.

A PML for FDTD is based on the splitting of electric and magnetic field components in the absorbing region, with the possibility of assigning losses to the individual split-field components. The net effect is the creation of a nonphysical absorbing medium adjacent to the outer FDTD mesh boundary that has wave impedance independent of the incident angle or frequency of outgoing scattered waves.

For the TM mode in two dimensions, Maxwell's equations modified by the presence of PML take the form

$$\begin{cases} \mu \dfrac{\partial H_x}{\partial t} + \sigma_x^* H_x = -\dfrac{\partial E_z}{\partial y} \\ \mu \dfrac{\partial H_x}{\partial t} + \sigma_y^* H_y = \dfrac{\partial E_z}{\partial x} \\ \varepsilon \dfrac{\partial E_{zx}}{\partial t} + \sigma_x E_{zx} = \dfrac{\partial H_y}{\partial x} \\ \varepsilon \dfrac{\partial E_{zy}}{\partial t} + \sigma_x E_{zy} = -\dfrac{\partial H_x}{\partial y} \\ E_{zx} + E_{zy} = E_z \end{cases}, \tag{3.129}$$

where the parameters σ_x, σ_y, σ_x^*, and σ_y^* are electric and magnetic conductivities and satisfy relations

$$\begin{cases} \dfrac{\sigma_x}{\varepsilon} = \dfrac{\sigma_x^*}{\mu} \\ \dfrac{\sigma_y}{\varepsilon} = \dfrac{\sigma_y^*}{\mu} \end{cases}, \tag{3.130}$$

which ensure the same impedance for a wave propagating in the PMLs as in the homogeneous space for any incident angle. Note that the PML represents a generalization of normally modeled physical media: If all electric and magnetic conductivities go to zero, the original Maxwell's equations in lossless media are recovered. Furthermore, the formulas in Eq. (3.129) can be easily discretized by the central difference method discussed in Section 3.3.1.

Thus, the electromagnetic wave enters the PML region with virtually no reflection, as its impedance is the same as that of free space and decays quickly because of the loss medium. The PML region has to be thick enough to absorb and extinguish the incident field completely by the time it reaches the outer boundary, which usually requires 8–16 PML layers.

In three dimensions, all six Cartesian field vector components are split, and conductivities σ_x, σ_y, σ_z, σ_x^*, σ_y^*, and σ_z^* that obey the same matching conditions as in Eq. (3.130) are introduced.

Since Berenger first introduced the split-field PML [90], several other PML formalisms have been proposed, which include the unaxial perfectly matched layer (UPML) [105] and the complex coordinate stretching method [106]. The former performs as well as Berenger's PML while avoiding the nonphysical field splitting. It is defined by a general constitutive tensor $\bar{\bar{s}}$

$$\bar{\bar{s}} = \begin{bmatrix} s_y s_z s_x^{-1} & 0 & 0 \\ 0 & s_x s_z s_y^{-1} & 0 \\ 0 & 0 & s_x s_y s_z^{-1} \end{bmatrix}, \tag{3.131}$$

which enters the time-harmonic Maxwell's equations as follows:

$$\begin{cases} \nabla \times \mathbf{H} = j\omega\varepsilon\bar{\bar{s}}\,\mathbf{E} \\ \nabla \times \mathbf{E} = -j\omega\mu\bar{\bar{s}}\,\mathbf{H} \end{cases}. \tag{3.132}$$

The multiplicative components of the diagonal elements of $\bar{\bar{s}}$ are

$$\begin{cases} s_x = \kappa_x + \dfrac{\sigma_x}{j\omega\varepsilon} \\ s_y = \kappa_y + \dfrac{\sigma_y}{j\omega\varepsilon} \\ s_z = \kappa_z + \dfrac{\sigma_z}{j\omega\varepsilon} \end{cases}. \tag{3.133}$$

To use the frequency-dependent $\bar{\bar{s}}$ in FDTD, which, as the name implies, works in the time domain, the FDTD time marching is modified to introduce an intermediate step wherein D-fields (B-fields) are calculated prior to updating the values of E-fields (H-fields). This process makes $j\omega$ a multiplicative factor in all formulas, which allows the substitution of $\partial/\partial t$ per Fourier transformation [105] and then the use of central difference equations. The procedure essentially amounts to acknowledging that the (frequency-dependent) s-components, Eq. (3.133), are dynamic variables that undergo time evolution driven by electromagnetic fields.

Although the UPML as described above works well when the absorbing boundary is adjacent to uniform material, such as free space, it is not always possible to have the scattering structure separated from the boundary. For example, when analyzing a photonic-crystal waveguide, it is desirable to continue the photonic-crystal (PhC) pattern to the edge of the computational region to avoid reflections that result from the impedance mismatch at the boundary between the photonic crystal and free space. Fortunately, the UPML described above can be easily modified to accommodate material inhomogeneities by introducing appropriate spatial variation of the dielectric constant in Eq. (3.132). An example of this approach is presented in Fig. 3.27, where the PhC-based inhomogeneous PML covers several periods of the photonic crystal near the boundary of the computational region [107]. As a result, the UPML becomes a natural, albeit lossy, extension of the analyzed and, in principle, infinitely periodic photonic crystal. Because no boundary exists between the photonic-crystal waveguide and free space, no impedance mismatch occurs and back-reflection at the end of the waveguide is suppressed.

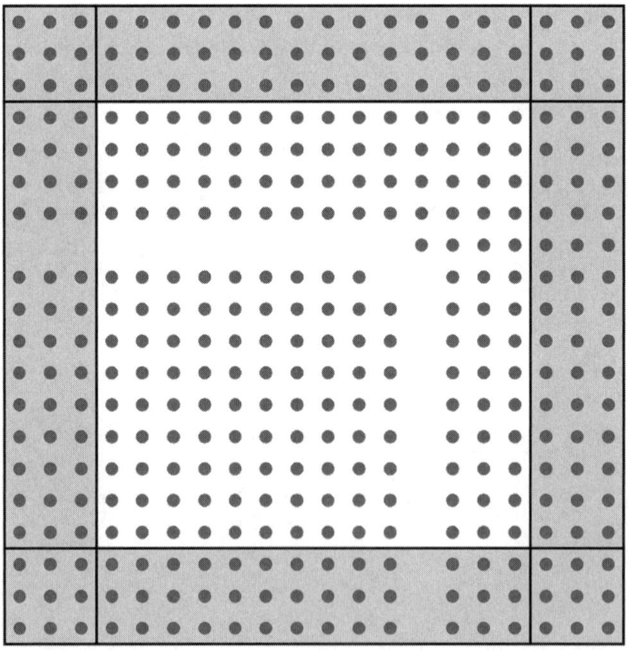

Figure 3.27 Inhomogeneous PMLs (shaded) used to terminate the photonic crystals to minimize the computational region while suppressing reflections.

The advantage of this approach over homogeneous UPMLs can be ascertained by directly comparing the simulation results obtained using these two methods. Such a comparison is shown in Fig. 3.28, in which homogeneous UPML has been used at the boundary of the computational region to obtain the steady-state result of Fig. 3.28a, and inhomogeneous UPML that covers five lattice periods has been used in the case presented in Fig. 3.28b. In both cases, the electromagnetic wave is launched at the left end of the waveguide. The presence of a standing-wave pattern with nodes and antinodes along the waveguide in Fig. 3.28a indicates the existence of a reflected wave that interferes with the launched beam. The reflection, which is estimated at about 20–30%, is the result of discontinuity between the photonic-crystal region of the waveguide and the homogeneous material of the UPML. In Fig. 3.28b, the high uniformity of the field along the waveguide direction demonstrates

(*a*)

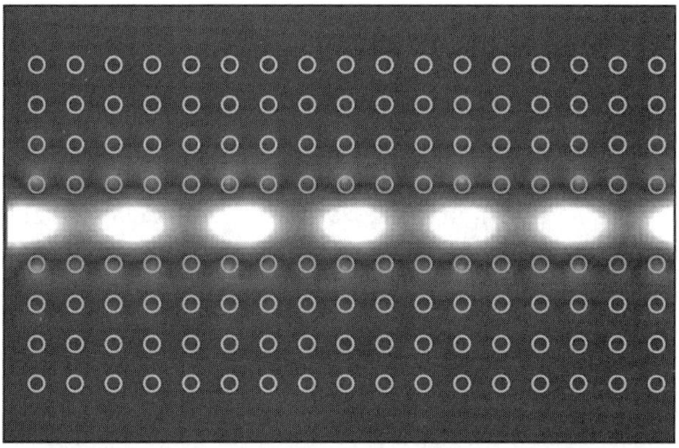

(*b*)

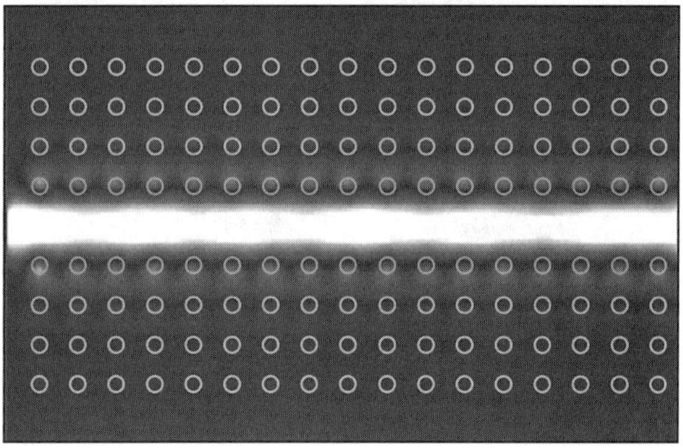

Figure 3.28 Comparison of steady-state FDTD results with a conventional PML (a) and inhomogeneous PML (b) used to terminate a single-row photonic-crystal waveguide; the E_z field amplitude is shown. The source is at the left edge of the computational region, and the PhC–PML region lies beyond the right edge (not shown). Note the absence of a standing wave pattern in (b), which indicates that a reflection from the waveguide end has been suppressed, thereby effectively simulating indefinite continuation of the waveguide.

the unidirectional nature of light propagation in this case and, therefore, efficient suppression of the reflected wave. In this case, the reflection from the end of the waveguide is estimated at -30 dB, or 0.1%, within a wide frequency range.

3.3.7 FDTD for Photonic Crystals

3.3.7.1 Bloch Boundary Conditions. Periodicity, which is the defining characteristic of photonic crystals, enables certain efficiencies in the time-domain simulations. In this section, the FDTD method is used in combination with Bloch boundary conditions and absorbing boundary conditions to calculate the band structure and the attenuation produced by a finite number of layers of photonic-bandgap material.

In photonic crystals, the Bloch boundary conditions in time domain take the form

$$\begin{cases} \mathbf{H}(\mathbf{r}+\mathbf{R}, t) = e^{j\mathbf{k}\cdot\mathbf{a}}\mathbf{H}(\mathbf{r}, t) \\ \mathbf{E}(\mathbf{r}+\mathbf{R}, t) = e^{j\mathbf{k}\cdot\mathbf{a}}\mathbf{E}(\mathbf{r}, t) \end{cases}, \tag{3.134}$$

where $\mathbf{R}$ is a lattice vector. Such boundary conditions are implemented as follows. For example, consider a photonic crystal with primitive lattice vectors $\mathbf{a}_1 = a_1\hat{\mathbf{x}}$ and $\mathbf{a}_2 = a_1\hat{\mathbf{y}}$ in a two-dimensional coordinate system.[3] The unit cell is divided into $N_x \times N_y$ grid cells, as shown in Fig. 3.29. The electromagnetic fields are located on staggered grids according to the Yee scheme (see Section 3.3.2).

The time marching for TM polarization proceeds as follows. At time $t = (n+1/2)\Delta t$, the magnetic field components within the unit cell are updated using Eq. (3.110), with the exception of points at the boundaries; i.e., we update

$$\begin{cases} H_x|_{i,j} & \text{for } i = 1, \ldots, N_x; j = 1+1/2, \ldots, N_y - 1/2 \\ H_y|_{i,j} & \text{for } i = 1+1/2, \ldots, N_x - 1/2; j = 1, \ldots, N_y \end{cases}. \tag{3.135}$$

Then, Bloch boundary conditions are imposed

$$\begin{cases} H_x\Big|_{i,N_y+1/2}^{n+1/2} = e^{jk_y a_2} H_x\Big|_{i,1+1/2}^{n+1/2} \\ H_y\Big|_{N_x+1/2,j}^{n+1/2} = e^{jk_x a_1} H_y\Big|_{1+1/2,j}^{n+1/2} \end{cases}, \tag{3.136}$$

where $\mathbf{k} = k_x\hat{\mathbf{x}} + k_y\hat{\mathbf{y}}$ is a wave vector in the first Brillouin zone. (Note the distinction between the index j in the y-component of the field and the imaginary unit that appears in the exponentials, which is also denoted by j.) At the time $t = (n+1)\Delta t$, the electric field at points inside the unit cell; i.e.,

$$E_z|_{i,j} \quad \text{for } i = 2, \ldots, N_x; j = 2, \ldots, N_y \tag{3.137}$$

[3] The approach can be generalized to nonrectangular lattices; see Refs. 65 and 108.

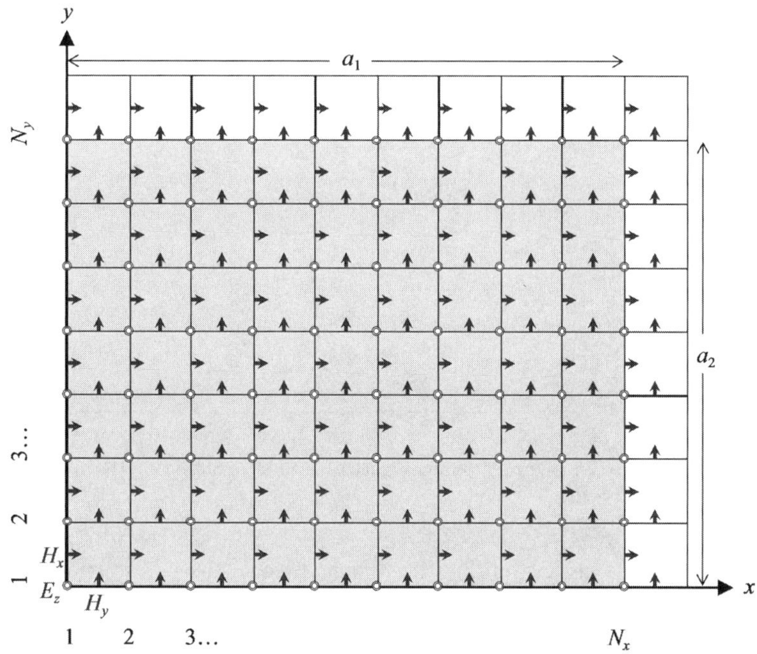

Figure 3.29 FDTD computational region of a unit cell. The shaded area is the unit cell with dimensions $a_1 \times a_2$.

is updated according to the standard FDTD equation (3.110) and along the boundaries by

$$\begin{cases} E_z\big|_{i,1}^{n+1} = e^{-jk_y a_2} E_z\big|_{i,N_y}^{n+1} \\ E_z\big|_{1,j}^{n+1} = e^{-jk_x a_1} E_z\big|_{N_x,j}^{n+1} \end{cases}. \tag{3.138}$$

As a result, the electromagnetic fields inside the unit cell are updated completely according to the Yee leapfrog algorithm. The simulation of the TE mode can be conducted in a similar manner.

This procedure allows for restricting the FDTD computational region to a single-unit cell defined by primitive lattice vectors, thereby realizing the efficiencies mentioned above. The band structure is obtained by performing FDTD simulations for wave vectors **k** within the first Brillouin zone and by extracting resonances from the Fourier transforms of the temporal responses. To this end, for a given wave vector **k**, a pulse with a bandwidth that covers the frequency range of interest is launched. To excite all modes in the PhC, multiple point sources are randomly distributed throughout the cell, and the temporal response is captured by several randomly placed detectors. Modes that remain after transient decay are the resonances; they are easily identified as peaks in the frequency domain because of their relatively high quality factor. Once the eigenstates are found, the corresponding mode profiles are studied individually by using a long excitation pulse with the spectral peak at the eigenstate frequency and DFT, as discussed in Section 3.3.5.

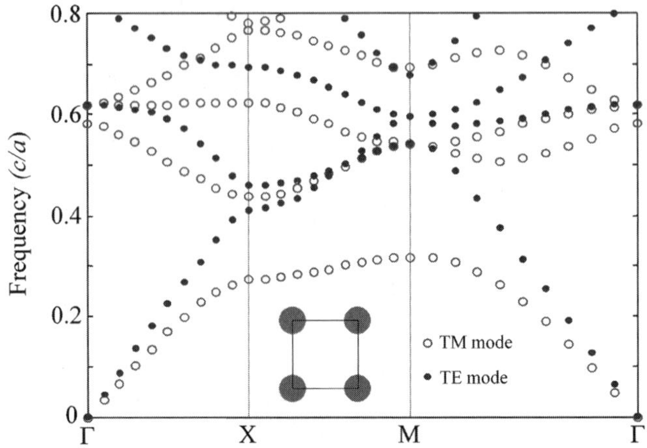

Figure 3.30 Band structure of two-dimensional square lattice obtained by FDTD calculation. Open circles: TM mode; filled circles: TE mode.

An example of a dispersion diagram obtained using this procedure for both TE and TM modes is shown in Fig. 3.30. In this case, a square lattice with dielectric rods embedded in air was analyzed. The radius of the dielectric rods was $0.2a$, and their dielectric constant was 8.9.

3.3.7.2 Analyzing Finite Sections of Photonic Crystals. In addition to analyzing dispersive properties of photonic crystals, the FDTD method can be used to calculate the transmission through and reflection from a finite number of photonic-crystal layers. This is an advantage over the PWEM, which in principle is applicable only to infinitely periodic structures. As an example of such use in two-dimensions, consider the structure shown in Fig. 3.31, in which a system infinitely periodic in

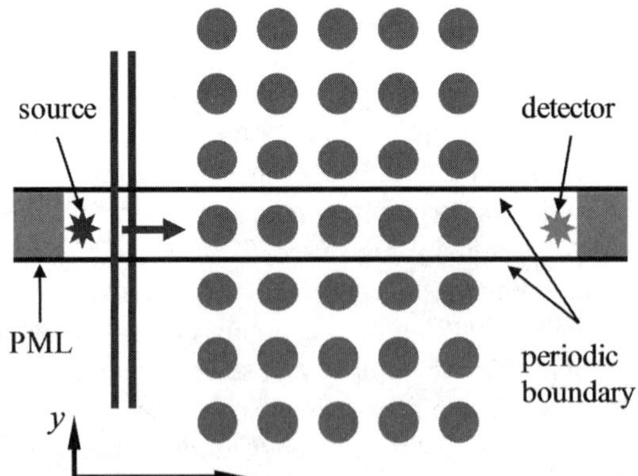

Figure 3.31 Schematic diagram for the calculation of transmission accross a finite slab of photonic-bandgap material.

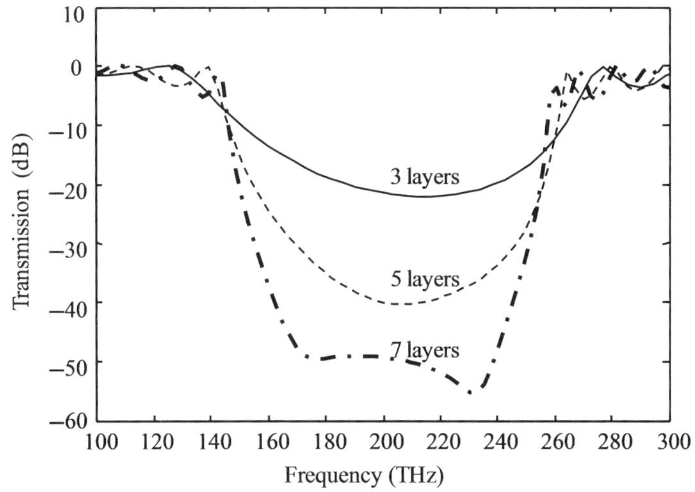

Figure 3.32 Transmission through photonic-bandgap material with a different number of layers.

the *y* direction and having a finite number of layers in the *x* direction is examined. To simulate infinite extension of the homogeneous material outside the photonic crystal, the computational region was terminated with PMLs at the left and right boundaries. At the same time, on the top and bottom boundaries, Bloch periodic boundary conditions were used to account for an infinitely long crystal in the *y* direction. A pulse with a wide frequency profile launched at normal incidence yields the broadband transmission spectrum shown in Fig. 3.32—as captured by a detector placed on the far end of the crystal. To obtain the transmission as a function of frequency, the temporal variation of the field captured by the detector was Fourier transformed, and the amplitude was normalized to the incident field. In Fig. 3.32, stop bands, which are the frequency regions where the transmission is lower than −10 dB, and pass bands, which are located where the transmission is higher than −10 dB, can be easily identified.

Such studies of stop bands and pass bands for finite sections of PhC lattices are important prior to using a particular PhC structure in a specific application. They provide critical design information, such as the number of layers necessary to achieve a certain transmission or attenuation for optical isolation. The FDTD method is, therefore, an invaluable tool in the design of PhC-based devices.

REFERENCES

1. S. Datta, C. T. Chan, K. M. Ho, and C. M. Soukoulis, "Photonic Band-Gaps in Periodic Dielectric Structures—the Scalar-Wave Approximation," *Phys. Rev. B*, vol. 46, pp. 10650–10656, 1992.
2. K. M. Ho, C. T. Chan, and C. M. Soukoulis, "Existence of a Photonic Gap in Periodic Dielectric Structures," *Phys. Rev. Lett.*, vol. 65, pp. 3152–3155, 1990.
3. R. D. Meade, K. D. Brommer, A. M. Rappe, and J. D. Joannopoulos, "Photonic Bound-States in Periodic Dielectric Materials," *Phys. Rev. B*, vol. 44, pp. 13772–13774, 1991.

4. E. Yablonovitch, T. J. Gmitter, K. M. Leung, R. D. Meade, A. M. Rappe, K. D. Brommer, and J. D. Joannopoulos, "3-Dimensional Photonic Band-Structure," *Opt. Quantum Electron.*, vol. 24, pp. S273–S283, 1992.
5. R. D. Meade, K. D. Brommer, A. M. Rappe, and J. D. Joannopoulos, "Existence of a Photonic Band-Gap in 2 Dimensions," *Appl. Phys. Lett.*, vol. 61, pp. 495–497, 1992.
6. R. D. Meade, A. M. Rappe, K. D. Brommer, J. D. Joannopoulos, and O. L. Alerhand, "Accurate Theoretical-Analysis of Photonic Band-Gap Materials," *Phys. Rev. B*, vol. 48, pp. 8434–8437, 1993.
7. R. D. Meade, A. Devenyi, J. D. Joannopoulos, O. L. Alerhand, D. A. Smith, and K. Kash, "Novel Applications of Photonic Band-Gap Materials—Low-Loss Bends and High Q-Cavities," *J. Appl. Phys.*, vol. 75, pp. 4753–4755, 1994.
8. M. L. Povinelli, S. G. Johnson, S. H. Fan, and J. D. Joannopoulos, "Emulation of Two-Dimensional Photonic Crystal Defect Modes in a Photonic Crystal with a Three-Dimensional Photonic Band Gap," *Phys. Rev. B*, vol. 6407, art. no.-075313, 2001.
9. S. G. Johnson, A. Mekis, S. H. Fan, and J. D. Joannopoulos, "Molding the Flow of Light," *Comput. Sci. Eng.*, vol. 3, pp. 38–47, 2001.
10. L. F. Shen, S. L. He, and L. Wu, "The Application of Effective-Medium Theory in the Plane-Wave Expansion Method for Analyzing Photonic Crystals," *Acta Physica Sinica*, vol. 51, pp. 1133–1138, 2002.
11. L. F. Shen and S. L. He, "Analysis for the Convergence Problem of the Plane-Wave Expansion Method for Photonic Crystals," *J. Opt. Soc. Am. A*, vol. 19, pp. 1021–1024, 2002.
12. S. S. Xiao, L. F. Shen, and S. L. He, "A Plane-Wave Expansion Method Based on the Effective Medium Theory for Calculating the Band Structure of a Two-Dimensional Photonic Crystal," *Phys. Lett. A*, vol. 313, pp. 132–138, 2003.
13. K. M. Leung and Y. F. Liu, "Photon Band Structures—the Plane-Wave Method," *Phys. Rev. B*, vol. 41, pp. 10188–10190, 1990.
14. K. M. Leung and Y. F. Liu, "Full Vector Wave Calculation of Photonic Band Structures in Face-Centered-Cubic Dielectric Media," *Phys. Rev. Lett.*, vol. 65, pp. 2646–2649, 1990.
15. M. Plihal and A. A. Maradudin, "Photonic Band-Structure of 2-Dimensional Systems - the Triangular Lattice," *Phys. Rev. B*, vol. 44, pp. 8565–8571, 1991.
16. M. Plihal, A. Shambrook, A. A. Maradudin, and S. Ping, "2-Dimensional Photonic Band Structures," *Opt. Commun.*, vol. 80, pp. 199–204, 1991.
17. A. A. Maradudin and A. R. McGurn, "Photonic Band Structures of Two-Dimensional Dielectric Media," *NATO ASI Series, Series B: Physics*, vol. 308, pp. 247–268, 1993.
18. A. R. McGurn and A. A. Maradudin, "Photonic Band Structures of 2-Dimensional and 3-Dimensional Periodic Metal or Semiconductor Arrays," *Phys. Rev. B*, vol. 48, pp. 17576–17579, 1993.
19. I. H. H. Zabel and D. Stroud, "Photonic Band Structures of Optically Anisotropic Periodic Arrays," *Phys. Rev. B*, vol. 48, pp. 5004–5012, 1993.
20. J. W. Haus, "A Brief Review of Theoretical Results for Photonic Band Structures," *J. Modern Opt.*, vol. 41, pp. 195–207, 1994.
21. J. B. Pendry, "Photonic Band Structures," *J. Modern Opt.*, vol. 41, pp. 209–229, 1994.
22. T. Suzuki and P. K. L. Yu, "Tunneling in Photonic Band Structures," *J. Opt. Soc. Am. B*, vol. 12, pp. 804–820, 1995.

23. T. Suzuki and P. K. L. Yu, "Complex Photonic Band Structures of a Conductive Metal Lattice by a Quadratic Eigensystem," *Opt. Lett.*, vol. 20, pp. 2520–2522, 1995.
24. S. Simeonov, U. Bass, and A. R. McGurn, "Photonic Band Structure of Zinc Blende Type Periodic Dielectric Media," *Physica B*, vol. 228, pp. 245–250, 1996.
25. W. Y. Zhang, A. Hu, X. Y. Lei, N. Xu, and N. B. Ming, "Photonic Band Structures of a Two-Dimensional Ionic Dielectric Medium," *Phys. Rev. B*, vol. 54, pp. 10280–10283, 1996.
26. P. Lalanne, "Effective Properties and Band Structures of Lamellar Subwavelength Crystals: Plane-Wave Method Revisited," *Phys. Rev. B*, vol. 58, pp. 9801–9807, 1998.
27. T. Suzuki and P. K. L. Yu, "Method of Projection Operators for Photonic Band Structures with Perfectly Conducting Elements," *Phys. Rev. B*, vol. 57, pp. 2229–2241, 1998.
28. L. Liu and J. T. Liu, "Photonic Band Structure in the Nearly Plane Wave Approximation," *Eur. Phys. J. B*, vol. 9, pp. 381–388, 1999.
29. F. Zhuang, L. Wu, S. S. Xiao, and S. L. He, "Band Structure of a Two-Dimensional Photonic Crystal with a Triangular Lattice of Anisotropic Elliptic Cylinders," *Chinese Phys. Lett.*, vol. 19, pp. 73–75, 2002.
30. Z. Y. Li and L. L. Lin, "Photonic Band Structures Solved by a Plane-Wave-Based Transfer-Matrix Method," *Phys. Rev. E*, vol. 67, pp. 761–764, 2003.
31. S. P. Guo and S. Albin, "Simple Plane Wave Implementation for Photonic Crystal Calculations," *Opt. Express*, vol. 11, pp. 167–175, 2003.
32. W. C. Sailor, F. M. Mueller, and P. R. Villeneuve, "Augmented-Plane-Wave Method for Photonic Band-Gap Materials," *Phys. Rev. B*, vol. 57, pp. 8819–8822, 1998.
33. H. S. Sözüer, J. W. Haus, and R. Inguva, "Photonic Bands - Convergence Problems with the Plane-Wave Method," *Phys. Rev. B*, vol. 45, pp. 13962–13972, 1992.
34. J. B. Pendry, "Calculating Photonic Band Structure," *J. Phys. Condens. Matter*, vol. 8, pp. 1085–1108, 1996.
35. C. T. Chan, Q. L. Yu, and K. M. Ho, "Order-N Spectral Method for Electromagnetic-Waves," *Phys. Rev. B*, vol. 51, pp. 16635–16642, 1995.
36. A. Kamli, S. Abboudy, and M. S. Al-Assiri, "Effect of Impurities on Dipole Emission Rate in Doped Photonic Structures," *Phys. Low-Dimensional Struc.*, vol. 1–2, pp. 231–236, 1998.
37. M. Qiu and S. L. He, "A Nonorthogonal Finite-Difference Time-Domain Method for Computing the Band Structure of a Two-Dimensional Photonic Crystal with Dielectric and Metallic Inclusions," *J. Appl. Phys.*, vol. 87, pp. 8268–8275, 2000.
38. M. Qiu, K. Azizi, A. Karlsson, M. Swillo, and B. Jaskorzynska, "Numerical Studies of Mode Gaps and Coupling Efficiency for Line-Defect Waveguides in Two-Dimensional Photonic Crystals," *Phys. Rev. B*, vol. 64, art no. 155113, 2001.
39. J. P. He, L. F. Shen, Q. Zhang, and S. L. He, "A Pseudospectral Time-Domain Algorithm for Calculating the Band Structure of a Two-Dimensional Photonic Crystal," *Chinese Phys. Lett.*, vol. 19, pp. 507–510, 2002.
40. L. F. Shen, S. L. He, and S. S. Xiao, "A Finite-Difference Eigenvalue Algorithm for Calculating the Band Structure of a Photonic Crystal," *Comp. Phys. Commun.*, vol. 143, pp. 213–221, 2002.
41. Y. Tanaka and S. Tamura, "Band Structures of Acoustic Waves in Phononic Lattices," *Physica B*, vol. 316, pp. 237–239, 2002.

42. L. Wu and S. L. He, "Revised Finite-Difference Time-Domain Algorithm in a Nonorthogonal Coordinate System and its Application to the Computation of the Band Structure of a Photonic Crystal," *J. Appl. Phys.*, vol. 91, pp. 6499–6506, 2002.
43. S. S. Xiao and S. L. He, "FDTD Method for Computing the Off-Plane Band Structure in a Two-Dimensional Photonic Crystal Consisting of Nearly Free-Electron Metals," *Physica B*, vol. 324, pp. 403–408, 2002.
44. Y. Ohtera, K. Kurokawa, and S. Kawakami, "Variational Expression for the Analysis of Photonic Crystal Devices," *IEEE J. Quantum Electron.*, vol. 38, pp. 919–926, 2002.
45. A. Reineix and B. Jecko, "A New Photonic Band Gap Equivalent Model Using Finite Difference Time Domain Method," *Ann. Telecommun. An.*, vol. 51, pp. 656–662, 1996.
46. A. J. Ward and J. B. Pendry, "Calculating Photonic Green's Functions Using a Nonorthogonal Finite-Difference Time-Domain Method," *Phys. Rev. B*, vol. 58, pp. 7252–7259, 1998.
47. P. M. Bell, J. B. Pendry, L. M. Moreno, and A. J. Ward, "A Program for Calculating Photonic Band Structures and Transmission Coefficients of Complex Structures," *Comp. Phys. Commun.*, vol. 85, pp. 306–322, 1995.
48. C. J. Jin, B. Qin, M. Yang, and R. H. Qin, "Two-Dimensional Photonic Band Structure in the Chiral Medium-Transfer Matrix Method," *Opt. Commun.*, vol. 142, pp. 179–183, 1997.
49. V. Kuzmiak and A. A. Maradudin, "Photonic Band Structures of One—and Two—Dimensional Periodic Systems with Metallic Components in the Presence of Dissipation," *Phys. Rev. B*, vol. 55, pp. 7427–7444, 1997.
50. S. Khorasani and B. Rashidian, "Modified Transfer Matrix Method for Conducting Interfaces," *J. Opt. A*, vol. 4, pp. 251–256, 2002.
51. J. P. Albert, C. Jouanin, D. Cassagne, and D. Monge, "Photonic Crystal Modelling Using a Tight-Binding Wannier Function Method," *Opt. Quantum Electron.*, vol. 34, pp. 251–263, 2002.
52. Y. C. Tsai, K. W. K. Shung, and J. B. Pendry, "Three-Dimensional Photonic Band Gaps in Woven Structures," *J. Phys. Condens. Matter*, vol. 10, pp. 753–760, 1998.
53. M. M. Sigalas, R. Biswas, K. M. Ho, and C. M. Soukoulis, "Theoretical Investigation of Off-Plane Propagation of Electromagnetic Waves in Two-Dimensional Photonic Crystals," *Phys. Rev. B*, vol. 58, pp. 6791–6794, 1998.
54. W. Axmann and P. Kuchment, "An Efficient Finite Element Method for Computing Spectra of Photonic and Acoustic Band-Gap Materials—I. Scalar Case," *J. Comput. Phys.*, vol. 150, pp. 468–481, 1999.
55. M. Koshiba, "Full-Vector Analysis of Photonic Crystal Fibers Using the Finite Element Method," *IEICE Trans. Electron.*, vol. E85C, pp. 881–888, 2002.
56. C. Mias, J. P. Webb, and R. L. Ferrari, "Finite Element Modelling of Electromagnetic Waves in Doubly and Triply Periodic Structures," *IEE Proc. Optoelectro.*, vol. 146, pp. 111–118, 1999.
57. W. M. Robertson, "Transmission-Line Matrix Modeling of Superluminal Electromagnetic-Pulse Tunneling Through the Forbidden Gap in Two-Dimensional Photonic Band Structures," *J. Opt. Soc. Am. B*, vol. 14, pp. 1066–1073, 1997.
58. W. C. Tan, T. W. Preist, J. R. Sambles, M. B. Sobnack, and N. P. Wanstall, "Calculation of Photonic Band Structures of Periodic Multilayer Grating Systems by Use of a Curvilinear Coordinate Transformation," *J. Opt. Soc. Am. A*, vol. 15, pp. 2365–2372, 1998.

59. D. C. Dobson, "An Efficient Method for Band Structure Calculations in 2D Photonic Crystals," *J. Computat. Phys.*, vol. 149, pp. 363–376, 1999.
60. D. C. Dobson, J. Gopalakrishnan, and J. E. Pasciak, "An Efficient Method for Band Structure Calaculations in 3D Photonic Crystals," *J. Computat. Phys.*, vol. 161, pp. 668–679, 2000.
61. S. L. He, M. Qiu, and C. R. Simovski, "Obtaining the Band Structure of a Two-Dimensional Photonic Crystal by an Averaged Field Approach," *Ch. Phys. Lett.*, vol. 17, pp. 352–354, 2000.
62. L. C. Botten, N. A. Nicorovici, R. C. McPhedran, C. M. de Sterke, and A. A. Asatryan, "Photonic Band Structure Calculations Using Scattering Matrices," *Phys. Rev. E*, vol. 64, art. no. 046603, 2001.
63. H. T. Miyazaki, H. Miyazaki, K. Ohtaka, and T. Sato, "Photonic Band in Two-Dimensional Lattices of Micrometer-Sized Spheres Mechanically Arranged Under a Scanning Electron Microscope," *J. Appl. Phys.*, vol. 87, pp. 7152–7158, 2000.
64. D. Hermann, M. Frank, K. Busch, and P. Wolfle, "Photonic Band Structure Computations," *Opt. Express*, vol. 8, pp. 167–172, 2001.
65. A. Taflove and S. C. Hagness, *Computational Electrodynamics: The Finite-Difference Time-Domain Method*, 3rd ed. Boston: Artech House, 2005.
66. C. Barnes and J. B. Pendry, "Multiple-Scattering of Waves in Random-Media—a Transfer-Matrix Approach," *Proceedings of the Royal Society of London Series A*, vol. 435, pp. 185–196, 1991.
67. A. S. Mclean and J. B. Pendry, "A Polarized Transfer-Matrix for Electromagnetic-Waves in Structured Media," *J. Modern Opt.*, vol. 41, pp. 1781–1802, 1994.
68. A. J. Ward, J. B. Pendry, and W. J. Stewart, "Photonic Dispersion Surfaces," *J. Phys. Condens. Matter*, vol. 7, pp. 2217–2224, 1995.
69. J. Jin, *The Finite Element Method in Electromagnetics*, New York: Wiley, 1993.
70. D. C. Dobson, J. Gopalakrishnan, and J. E. Pasciak, "An Efficient Method for Band Structure Calculations in 3D Photonic Crystals," *J. Computat. Phys.*, vol. 161, pp. 668–679, 2000.
71. M. N. Sadiku, *Numerical Techniques in Electromagnetics*, Boca Raton, FL: CRC Press, 1992.
72. J. D. Joannopoulos, R. D. Meade, and J. N. Winn, *Photonic Crystals: Molding the Flow of Light*. Princeton, NJ: Princeton University Press, 1995.
73. H. Kosaka, T. Kawashima, A. Tomita, M. Notomi, T. Tamamura, T. Sato, and S. Kawakami, "Superprism Phenomena in Photonic Crystals," *Phys. Rev. B*, vol. 58, pp. R10096–R10099, 1998.
74. T. Baba and T. Matsumoto, "Resolution of Photonic Crystal Superprism," *Appl. Phys. Lett.*, vol. 81, pp. 2325–2327, 2002.
75. L. Wu, M. Mazilu, and T. F. Krauss, "Beam Steering in Planar-Photonic Crystals: From Superprism to Supercollimator," *J. Lightwave Technol.*, vol. 21, pp. 561–566, 2003.
76. A. Di Falco, C. Conti, and G. Assanto, "Three-Dimensional Superprism Effect in Photonic-Crystal Slabs," *J. Lightwave Technol.*, vol. 22, pp. 1748–1753, 2004.
77. D. W. Prather, S. Y. Shi, D. M. Pustai, C. H. Chen, S. Venkataraman, A. Sharkawy, G. J. Schneider, and J. Murakowski, "Dispersion-Based Optical Routing in Photonic Crystals," *Opt. Lett.*, vol. 29, pp. 50–52, 2004.

78. Z. L. Lu, S. Y. Shi, J. A. Murakowski, G. J. Schneider, C. A. Schuetz, and D. W. Prather, "Experimental Demonstration of Self-Collimation Inside a Three-Dimensional Photonic Crystal," *Phys. Rev. Lett.*, vol. 96, p. 173902, 2006.
79. D. W. Prather, S. Shi, S. Venkataraman, Z. Lu, J. Murakowski, and G. J. Schneider, "Self-Collimation in Three-Dimensional Photonic Crystals," *Proc. SPIE*, vol. 5733, pp. 84–93, 2005.
80. S. G. Johnson and J. D. Joannopoulos, "Block-Iterative Frequency-Domain Methods for Maxwell's Equations in a Planewave Basis," *Opt. Express*, vol. 8, pp. 173–190, 2001.
81. H. Y. Ryu, H. G. Park, and Y. H. Lee, "Two-Dimensional Photonic Crystal Semiconductor Lasers: Computational Design, Fabrication, and Characterization," *IEEE J. Sel. Topics Quantum Electron.*, vol. 8, pp. 891–908, 2002.
82. R. Shimada, A. L. Yablonskii, S. G. Tikhodeev, and T. Ishihara, "Transmission Properties of a Two-Dimensional Photonic Crystal Slab with an Excitonic Resonance," *IEEE J. Quantum Electron.*, vol. 38, pp. 872–879, 2002.
83. T. Baba, A. Motegi, T. Iwai, N. Fukaya, Y. Watanabe, and A. Sakai, "Light Propagation Characteristics of Straight Single-Line-Defect Waveguides in Photonic Crystal Slabs Fabricated into a Silicon-on-Insulator Substrate," *IEEE J. Quantum Electron.*, vol. 38, pp. 743–752, 2002.
84. A. Chutinan and S. Noda, "Waveguides and Waveguide Bends in Two-Dimensional Photonic Crystal Slabs," *Phys. Rev. B*, vol. 62, pp. 4488–4492, 2000.
85. M. Notomi, A. Shinya, E. Kuramochi, I. Yokohama, C. Takahashi, K. Yamada, J. Takahashi, T. Kawashima, and S. Kawakami, "Si-Based Photonic Crystals and Photonic-Bandgap Waveguides," *IEICE Trans. Electron.*, vol. E85C, pp. 1025–1032, 2002.
86. H. G. Park, J. K. Hwang, J. Huh, H. Y. Ryu, S. H. Kim, J. S. Kim, and Y. H. Lee, "Characteristics of Modified Single-Defect Two-Dimensional Photonic Crystal Lasers," *IEEE J. Quantum Electron.*, vol. 38, pp. 1353–1365, 2002.
87. S. G. Johnson, S. H. Fan, P. R. Villeneuve, J. D. Joannopoulos, and L. A. Kolodziejski, "Guided Modes in Photonic Crystal Slabs," *Phys. Rev. B*, vol. 60, pp. 5751–5758, 1999.
88. A. Scherer, O. Painter, J. Vučković, M. Lončar, and T. Yoshie, "Photonic Crystals for Confining, Guiding, and Emitting Light," *IEEE Trans. Nanotechnol.*, vol. 1, pp. 4–11, 2002.
89. M. Lončar, T. Doll, J. Vučković, and A. Scherer, "Design and Fabrication of Silicon Photonic Crystal Optical Waveguides," *J. Lightwave Technol.*, vol. 18, pp. 1402–1411, 2000.
90. J. P. Berenger, "A Perfectly Matched Layer for the Absorption of Electromagnetic Waves," *J. Comp. Physics*, vol. 114, pp. 185–200, 1994.
91. H. Rogier and D. D. Zutter, "Berenger and Leaky Modes in Microstrip Substrates Terminated by a Perfectly Match Layer," *IEEE Trans. Microwave Theory Tech.*, vol. 49, pp. 712–715, 2001.
92. Q. Cao, P. Lalanne, and J.-P. Hugonin, "Stable and Efficient Bloch-Mode Computational Method for One-Dimensional Grating Waveguides," *J. Opt. Soc. Am. A*, vol. 19, pp. 335–338, 2002.
93. T. Tischler and W. Heinrich, "The Perfectly Matched Layer as Lateral Boundary in Finite-Difference Transmission-Line Analysis," *IEEE Trans. Microwave Theory Techn.*, vol. 48, pp. 2249–2253, 2000.

94. H. Derudder, F. Olyslager, D. D. Zutter, and S. V. D. Berghe, "Efficient Mode-Matching Analysis of Discontinuities in Finite Planar Substrates Using Perfectly Matched Layers," *IEEE Tran. Antennas Propagat.*, vol. 49, pp. 185–195, 2001.
95. Z. S. Sacks, D. M. Kingsland, R. Lee, and J. Lee, "A Perfectly Matched Anisotropic Absorber for Use as an Absorbing Boundary Condition," *IEEE Trans. Antennas Propagat.*, vol. AP-43, pp. 14601463, 1995.
96. A. Taflove and S. C. Hagness, *Computational Electrodynamics*: *The Finite-Difference Time-Domain Method*, 2nd ed. Boston: Artech House, 2000.
97. D. E. Aspnes, "Effective Medium Theory," *Am. J. Phys.*, vol. 50, 1982.
98. Z. Y. Li, J. Wang, and B. Y. Gu, "Creation of Partial Band Gaps in Anisotropic Photonic-Band-Gap Structures," *Phys. Rev. B*, vol. 58, pp. 3721–3729, 1998.
99. C. Chen, S. Shi, and D. Prather, "Stepwise Effective Index Method in Modeling of High Index Contrast Planar Photonic Crystals," *Proc. SPIE*, vol. 5360, pp. 390, 2004.
100. K. S. Yee, "Numerical Solution of Initial Boundary Value Problems Involving Maxwell's Equations in Isotropic Media," *IEEE Trans. Antennas Propagat.*, vol. AP-14, pp. 302–307, 1966.
101. J. P. Durbano, J. R. Humphrey, F. E. Ortiz, P. F. Curt, D. W. Prather, and M. S. Mirotznik, "Hardware Acceleration of the 3D Finite-Difference Time-Domain Method," presented at *IEEE Antennas and Propagation Society Symposium 2004 Digest held in Conjunction with: USNC/URSI National Radio Science Meeting, Jun 20–25 2004*, Monterey, CA, vol. 1, pp. 77–80, 2004.
102. P. F. Curt, J. P. Durbano, M. R. Bodnar, S. Shi, and M. S. Mirotznik, "Enhanced Functionality for Hardware-Based FDTD Accelerators," *Appl. Computat. Electromagn. Soc. J.*, vol. 22, pp. 39–46, 2007.
103. A. Taflove, "Application of the Finite-Difference Time-Domain Method to Sinusoidal Steady-State Electromagnetic-Penetration Problems," *IEEE Trans. Electromagn. Comp.*, vol. 22, pp. 191–202, 1980.
104. G. Mur, "Absorbing Boundary-Conditions for the Finite-Difference Approximation of the Time-Domain Electromagnetic-Field Equations," *IEEE Trans. Electromagn. Comp.*, vol. 23, pp. 377–382, 1981.
105. S. D. Gedney, "An Anistropic Perfectly Matched Layer-Absorbing Medium for the Truncation of Finite-Difference Time-Domain Lattices," *IEEE Trans. Antenn. Prop.*, vol. 44, pp. 1630–1639, 1996.
106. W. C. Chew and W. H. Weedon, "A 3D Perfectly Matched Medium from Modified Maxwell's Equations with Stretched Coordinates," *Microwave Opt. Tech. Lett.*, vol. 7, pp. 599–604, 1994.
107. M. Koshiba, Y. Tsuji, and S. Sasaki, "High-Performance Absorbing Boundary Conditions for Photonic Crystal Waveguide Simulations," *IEEE Microwave Wireless Components Lett.*, vol. 11, pp. 152–154, 2001.
108. M. Qiu and S. He, "A Nonorthogonal Finite-Difference Time-Domain Method for Computing the Band Structure of a Two-Dimensional Photonic Crystal with Dielectric and Metallic Inclusions," *J. Appl. Phys.*, vol. 87, pp. 8268–8275, 2000.

4

Devices and Applications Based on Photonic Bandgaps

4.1 INTRODUCTION

Once a certain periodic structure is analyzed and bandgaps are identified for specific polarizations using the numerical techniques presented in Sections 3.3 and 3.4, the next challenge is to design structures, such as those based on two-dimensional (2D), three-dimensional (3D), or slab photonic crystals (PhCs), for a particular application. The goal is the implementation and optimization of basic signal-processing operations at scales comparable with the wavelength of light inside the periodic structures.

Implementation of photonic-crystal devices in planar dielectric slabs, such as silicon-on-insulator (SOI), not only opens many exciting opportunities in integrated optics and high-density optical interconnects but also provides the basic building blocks for nanophotonic circuits (NPCs). Just as the development of nanophotonic semiconductor materials led to the ongoing electronic revolution, photonic-crystal device integration in planar dielectric slabs holds the promise of achieving the long-sought goal of large-scale integrated photonic circuits (LSIPCs). Semiconductor materials are engineered by adding a concentration of donor or acceptor atoms, which results in a modification of the electrical properties of the atomic crystal by making the material n-type or p-type, respectively. In a similar fashion, the optical properties of a photonic crystal can be altered by introducing defects, that is by adding, removing, or substituting a certain amount of dielectric material at particular locations in the lattice. Such defects can be of one of two types: a donor type or an acceptor type [1]. A donor-type defect corresponds to a local increase of the effective refractive index, whereas an acceptor-type defect corresponds to a local decrease of the effective refractive index [2–5].

In the presence of a bandgap, such perturbations of the regular photonic-crystal array lead to the localization of optical modes, which become confined as a result of coherent scattering and interference of waves from the surrounding periodic structure. The defects often take the form of a point, a line, or a plane, where a single lattice site, a row of lattice sites, or a plane of lattice sites is modified, respectively. The efficient confinement of light in the vicinity of defects leads to their use

Photonic Crystals: Theory, Applications, and Fabrication. By D. W. Prather, S. Shi, A. Sharkawy, J. Murakowski, and G. J. Schneider

as resonant cavities in the case of point defects, and as waveguides in the case of line defects.

Looking from the *k*-space perspective, by locally adding or removing a certain amount of dielectric material to the photonic crystal, we disrupt the symmetry of the photonic lattice. In so doing, we allow a single state or a group of closely spaced states to exist within the bandgap.

In this chapter, we investigate the effects such changes have on the properties of a perfectly periodic photonic crystal, and we use thus-introduced defects to engineer structures for various photonic applications. The next section discusses point defects, and Section 4.3 examines the properties of line defects in photonic crystals. Finally, Section 4.4 discusses a range of applications that use the strong light confinement provided by a photonic bandgap.

4.2 POINT DEFECTS

As stated, a resonant cavity can be made by introducing a point defect in a photonic crystal [6–9]. The defect is obtained by locally modifying the shape, size, or dielectric constant of the PhC features. By varying these parameters, the number of modes supported by the cavity and their resonant frequencies can be tuned. Consider, for example, a square lattice of cylindrical rods with a sufficiently high dielectric contrast (i.e., greater than 2.0) [1] between the host and the rod materials. In such a PhC, a point defect can be introduced by changing one parameter of a single rod within the crystal. For example, reducing the radius of one dielectric rod results in a point defect that supports a single localized mode. Alternatively, if we increase the radius of one rod, then we will introduce multiple modes localized within such a cavity. The resonant frequencies supported by such cavities are determined by how much the rod radius is changed.

To study the optical properties of such cavities, various numerical techniques are used. Two techniques that have been found to be particularly useful are the plane-wave expansion method (PWEM), which works in frequency domain, and the finite-difference time-domain method (FDTD), which, as the name implies, solves electromagnetic systems in the time domain. These methods were discussed in detail in Chapter 3. Here, we apply them to analyze defects in photonic crystals.

4.2.1 Numerical Analysis of Point Defects

In Chapter 3, the standard eigenvalue problem was derived starting from the wave equations, which were obtained from Maxwell's equations, for one-, two-, and three-dimensional photonic crystals, as well as for a photonic-crystal slab. The solution to the eigenvalue problem yielded the dispersion properties of photonic crystals. It was found that by optimizing the lattice type, pitch, fill-factor, or index of refraction, an omnidirectional bandgap can be obtained.

In this section, we introduce a point defect into such a lattice by locally perturbing the periodicity of the photonic crystal and we calculate modes localized within the point-defect region. As discussed in Chapter 3, the application of the PWEM requires periodicity in all directions. However, a single point defect occupies a finite region in the photonic crystal and thereby breaks its periodicity, which renders PWEM no longer applicable. To remedy this situation and allow for the use of PWEM, identical defects are distributed periodically across the entire photonic crystal as shown in Fig. 4.1. If the defect mode is well confined within the cavity, it can be isolated from the neighboring defects by introducing a sufficient number of PhC layers around it. Because the electromagnetic field of the defect modes decays rapidly as we depart the defect region in any direction, the cavities can be considered decoupled from one another, each behaving like a single cavity inside a photonic crystal. To analyze the structure using PWEM, a supercell is defined as indicated in Fig. 4.1, which consists of several periods of the photonic crystal that surrounds a point defect located in the center of a square region. The supercell is treated as the unit cell in the PWEM, which repeats periodically in all directions and constitutes a single period of the new lattice. Based on this period, the band structure is calculated. By careful examination of the band structure, one may find that the defect modes have frequencies within the bandgap and that the resonant frequencies are independent of the wave vectors.

As an example, consider a point defect in a two-dimensional supercell that consists of a 7×7 square array of dielectric rods with one rod missing at the center. The dielectric rods are embedded within air background, have a dielectric constant of 12.25, and have a circular cross section with a radius of $0.2a$, where a is the lattice constant. The dispersion surfaces of different bands in the case of transverse-magnetic (TM) polarization are shown in Fig. 4.2a, projected onto the ωk_x plane. They are obtained by solving the respective eigenequation for wave vectors within the first Brillouin zone (recall that the unit cell is the $7a \times 7a$ supercell, and hence, the first

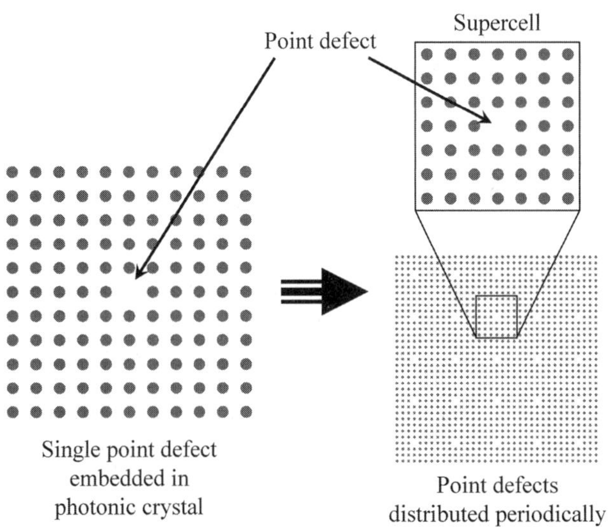

Figure 4.1 Construction of a supercell for point defect analysis using PWEM. Introducing a single defect inside a regular photonic-crystal lattice breaks its periodicity, which thereby renders PWEM inapplicable. By distributing the defects periodically across the photonic crystal, we reclaim the ability to employ PWEM albeit at the cost of an enlarged computational region.

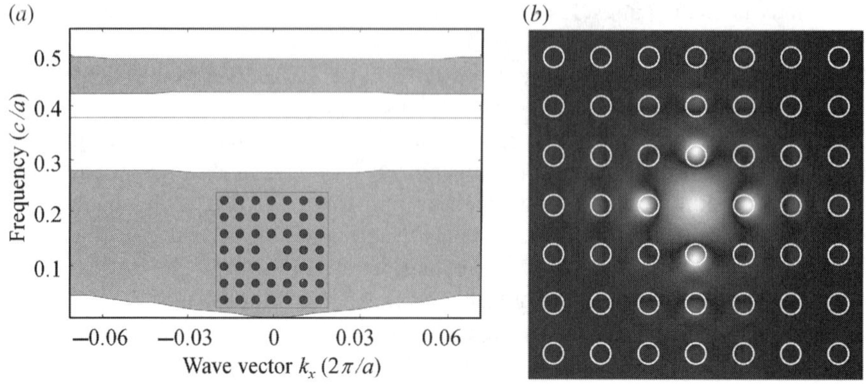

Figure 4.2 Supercell analysis of point defect. (a) Dispersion diagram. (b) Electric field distribution for wave vector $(0.4\hat{\mathbf{x}} + 0.3\hat{\mathbf{y}})2\pi/a$.

Brillouin zone is much smaller compared with that of the unperturbed lattice). In this figure, notice that a planar dispersion surface appears within the photonic bandgap and that the eigenfrequencies of the corresponding modes are independent of the wave vector **k**. These additional modes are the consequence of the introduction of the defect into the photonic crystal. Note that it is conventional to group these defect modes into "bands" in order of frequency, just as the propagating modes of the defect-free PhC are grouped in this way. Each "band" of defect modes is often referred to as just one defect mode. The resonant frequency of the cavity mode within the bandgap is $0.383c/a$. The shaded regions in Fig. 4.2a represent the slab-band continuum. Once the cavity mode is known, the field distribution can be obtained from the associated eigenvector calculated using PWEM. Figure 4.2b shows the amplitude distribution of the electric field for an arbitrarily chosen wave vector $(0.4\hat{\mathbf{x}} + 0.3\hat{\mathbf{y}})2\pi/a$; the electromagnetic wave is well confined within the cavity, and the intensity drops rapidly in all directions. Similar defect-mode confinement can be observed at other wave vectors.

To study the impact of the number of photonic-crystal layers that surround a point defect on the resonant mode properties, consider the photonic-crystal cavity analyzed in Figs. 4.1 and 4.2. To determine additional information that cannot be obtained using the PWEM approach, such as the quality factor, the FDTD method was used. Perfectly matched layers (PMLs) that minimize backreflections, which thereby mimics an open surrounding, are used for the computational region boundary.

All modes are excited by placing multiple dipole sources at quasirandom positions within the defect region and launching a short pulse that covers the bandwidth of interest. Multiple detectors placed in the cavity detect the transient response. Resonant modes are identified as peaks in the frequency response obtained by performing a Fourier transform of the temporal response.

To see how the number of layers confining the point defect affects the properties of the resonant mode, consider the same structure of Fig. 4.1 with a single missing dielectric rod. Figure 4.3 shows the cavity mode as the number of surrounding photonic-crystal layers is increased from two to four. Just as before, when the

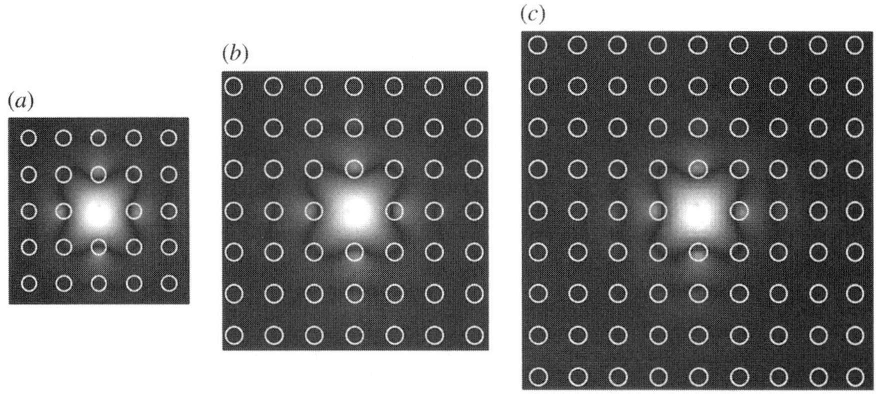

Figure 4.3 Field amplitude for cavity modes of photonic-crystal cavities with different numbers of confining photonic-crystal layers. (a) Two layers, $Q = 148$. (b) Three layers, $Q = 461$. (c) Four layers, $Q = 1636$.

system was analyzed using PWEM, the electromagnetic field decays rapidly as it enters the photonic crystal that surrounds the defect, which is a consequence of the resonant frequency being located in the middle of a wide photonic bandgap. It seems that as few as two layers of the photonic crystal are sufficient to confine the resonant mode (Fig. 4.3a). However, such a visual assessment of the mode can be deceiving. In fact, the quality factor Q (see Section 4.2.2) changes significantly with the number of surrounding layers, increasing from 148 to 461 to 1636 for two, three, and four surrounding layers, respectively, which indicates that for a small number of isolation layers, the energy leaks out of the cavity substantially.

Another way to introduce a simple point defect into a photonic crystal is by locally adding dielectric material instead of removing it. This technique produces an acceptor-type defect. As an example, consider a triangular lattice of circular air holes drilled in a silicon slab with a refractive index of 3.5. The air holes have a radius $0.4a$, where a is the lattice constant. This PhC lattice possesses a wide bandgap that opens for transverse-electric (TE) polarization. A defect is introduced by plugging an air hole in the center of a hexagonal arrangement of the photonic crystal, as illustrated in Fig. 4.4a. The PWEM using the supercell technique produces the dispersion diagram shown in Fig. 4.4b, in which four cavity modes are identified at frequencies $0.3832c/a$, $0.3681c/a$, $0.3453c/a$, and $0.2748c/a$. Then, FDTD simulations are used to examine mode profiles that correspond to these resonant frequencies. The four cavity modes, which are shown in Fig. 4.4c through f, are identified as monopole, dipole, quadrupole, and hexapole modes, respectively. The associated quality factors are 633, 660, 1736, and 739, respectively.

The ability to obtain high Q in small volumes of point defects using only dielectric materials is an attractive feature of photonic-crystal–based cavities [10].

4.2.2 Design Criteria for Photonic-Crystal Cavities

In the previous section, we presented different ways to introduce point defects into a photonic-crystal structure. Those point defects trapped electromagnetic modes

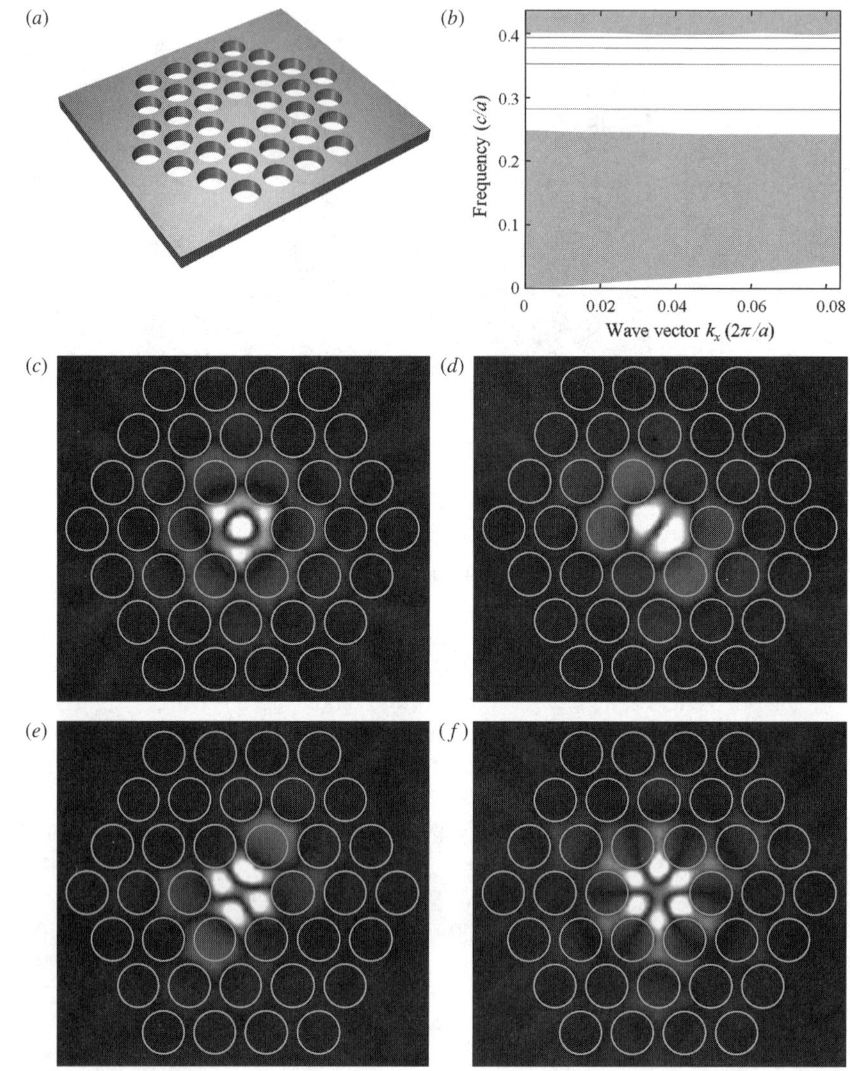

Figure 4.4 (a) Point defect that consists of a missing air hole in perforated silicon slab, (b) dispersion relation (projected on to the ωk_x plane) obtained with PWEM using the supercell technique, (c) monopole mode ($f = 0.3681c/a$), (d) dipole mode ($f = 0.2748c/a$), (e) quadrupole ($f = 0.3453c/a$), and (f) hexapole mode ($f = 0.3832c/a$).

and, hence, formed optical cavities. The light-trapping efficiency of an optical cavity is characterized by the quality factor Q, which is a measure of how long the electromagnetic wave stays inside the cavity and can be defined using the following formula [11]:

$$Q = \frac{\omega E}{P}, \tag{4.1}$$

where E is the energy stored inside the cavity, ω is the resonant frequency, and $P = -dE/dt$ is the power dissipated outside the cavity [12, 13]. In other words, the quality factor is, up to a factor of 2π, the inverse of the fraction of energy lost in a single period of the oscillating electromagnetic field. The quality factor of a cavity plays a major role in designing high-density wavelength filtering and multiplexing systems, as a high Q value translates to more channels that can be multiplexed. To see this, we consider an alternative way of calculating Q,

$$Q = \frac{f_{\text{center}}}{\Delta f}, \tag{4.2}$$

where f_{center} is the center frequency and Δf is the full width at half maximum (FWHM) of the resonance peak [14]. Thus, high Q corresponds to a narrow spectral width of the resonance peak.

In two-dimensional photonic crystals, the quality factor is mainly a function of the number of layers that surround the defect [15–20]. It has been shown that the spectral widths of the defect modes decrease rapidly with an increasing number of lattice layers [11, 14], which is desirable in high-density multiplexing systems because it improves the selectivity of the available bandwidth.

High-Q microcavities can also be realized in a 2D photonic-crystal slab if light scattering in the vertical direction and out of the slab is minimized. This was demonstrated using FDTD simulations of an optical microcavity defined in a three-layer slab waveguide and 2D photonic-crystal mirrors. It was shown that the Q factor of the cavity mode depends strongly on the depth of the holes that define the photonic crystal and on the refractive index of the material that surrounds the waveguide core [21].

Another parameter often taken into account when designing an optical cavity, and particularly a laser cavity, is the β factor. For a given mode m, β_m is the fraction of spontaneous emission that is channeled into mode m. Typically, in laser applications, we would prefer to have all spontaneous emission end up in a single desired mode; hence, ideally β is equal to one.

Recall that the spontaneous emission rate into a single-cavity mode is inversely proportional to the mode volume [22], in which the latter is defined as a product of the cross-sectional area A_b of the mode and the physical cavity length L_c. Often, the mode volume is expressed in units of $0.5(\lambda/n)^3$, where n is the refractive index of the background material. To enhance the β factor, the cavity needs to support one or at most a few high-Q modes. Hence, the true challenge in designing a photonic-crystal-based resonant laser cavity lies in realizing one that has a small mode volume, high Q, and near-unity β factor.

4.3 LINE DEFECTS

When a line defect is introduced into a photonic crystal that exhibits a bandgap, waveguides may be created. Such waveguides can guide light at optical wavelengths with minimal propagation losses [23–27]. A waveguide is an important element in any optical integrated circuit [28, 29]. It is used to transmit electromagnetic waves between various devices and components in an interconnected system. Historically, dielectric waveguides have been used to carry light between various components on a single chip, whereas optical fiber cables were used between systems. Although these architectures were successful in satisfying the bandwidth requirements for an optically interconnected system, the physical size of these links is considerable. However, large lateral dimensions of data pipes are unacceptable for chip-level optical system integration, in which the minimum feature size shrinks rapidly with every new generation of chip-manufacturing technology. As a result, nanoscale optical interconnections are now required to satisfy future interconnect needs in terms of throughput and latency, as well as size. Here, photonic crystals seem to be a good candidate for the miniaturization and large-scale integration of high-density optical interconnects [30]. This is mainly because of their unique ability to control the propagation of light in one, two, and three dimensions at a scale comparable with the wavelength.

4.3.1 Photonic-Crystal Line Defects for Waveguiding

To demonstrate the concept of waveguiding in photonic crystals, consider a two-dimensional system periodic in the xy plane and extending infinitely in the z direction. For this example, the system consists of an array of high-index dielectric rods arranged on a square lattice embedded in a low-index dielectric background. If the index contrast is high enough, a photonic bandgap may open for TM-polarized waves. A line defect introduced into such a photonic-crystal lattice breaks its symmetry along one of its main crystalline directions, which may result in the appearance of single or multiple modes in the otherwise vacant bandgap [31–36]. Depending on their spatial and temporal distribution within the photonic-crystal line-defect structure, these modes can be classified into *evanescent*, *standing*, or *traveling* modes. The traveling modes can be used to carry optical signals between components in an integrated system, which thereby turns linear defects into bona fide waveguides.

In the following sections, we present two distinct numerical methods used to examine such modes, which emerge as a result of the introduction of a line defect into a photonic-crystal structure.

4.3.1.1 Plane-Wave Expansion Analysis. Dispersion properties for photonic-crystal waveguides are often calculated by the plane-wave expansion method, which solves Maxwell's equations in the frequency domain for a given photonic-crystal configuration. In a manner similar to point-defect analysis, we define a supercell for the line defect, as shown in Fig. 4.5. It consists of a short column of rods (seven cells of the original photonic crystal) in the photonic-crystal region, with a single rod missing in the middle; see Fig. 4.5. When such supercells are distributed periodically throughout the xy plane, they form a set of parallel, evenly spaced line defects in a uniform photonic crystal. The dimensions of the supercell are, in this case, $(1 \times 7)a$, where a is the lattice constant. The width of the supercell ($7a$ in this case) must be chosen to ensure minimal coupling between the parallel line defects of the periodic

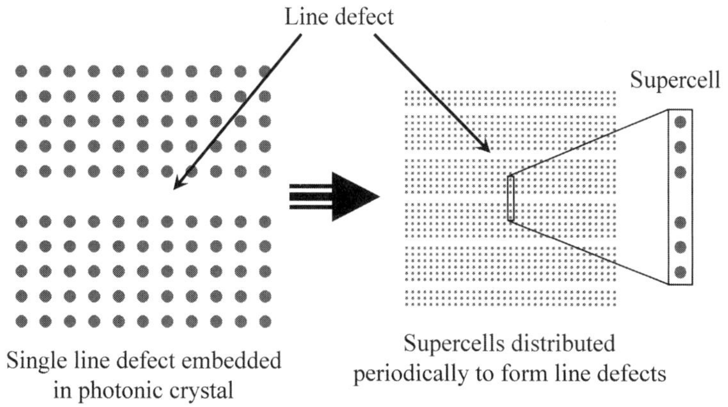

Figure 4.5 Construction of a supercell for PWEM analysis of a line defect. Introducing a single defect inside a regular photonic-crystal lattice breaks its periodicity, which thereby renders PWEM inapplicable. By distributing the defects periodically across the photonic crystal, we reclaim the ability to employ PWEM albeit at the cost of an enlarged computational region.

system construct. A rule of thumb is that the amplitude of the electromagnetic field at the boundary of the supercell should be less than 10% of its peak value within the defect region. Thus, it can only be verified in hindsight whether the number of isolation layers was sufficient for a particular line defect.

Having constructed a periodic system, we can now apply the PWEM to the problem of finding and characterizing modes supported by a particular line defect in a photonic crystal. In other words, the band structure, which describes the functional dependence of frequency ω on the wave vector $\mathbf{k}$, can be calculated.

To illustrate the application of the PWEM in this situation, consider the structure of Fig. 4.5. The rods are assumed to be made of silicon with a dielectric constant 12.25 and have a circular cross section with a radius $0.2a$, where a is the lattice constant. The dispersion diagram that corresponds to this structure is shown in Fig. 4.6. It was obtained by projecting the eigenfrequencies along the k_y direction in the spatial frequency space onto the ωk_x plane. The shaded regions in Fig. 4.6a represent the band continuum, whereas the solitary curve within the bandgap corresponds to a mode that results from the introduction of the line defect. The electromagnetic-field distribution of particular modes can be recovered from the eigenvectors associated with the respective eigenvalues. Figure 4.6b shows the amplitude of electric-field distribution within the line-defect region for the wave vector $k_x = 0.318(2\pi/a)$.

As discussed, the use of PWEM requires the introduction of artificial periodic boundary conditions, which effectively extend the structure to infinity in all directions. However, in some situations, we need to consider a genuinely finite number of lattice periods. In this case, alternative numerical techniques must be sought. An application of FDTD to this situation is discussed in the next section.

4.3.1.2 Finite-Difference Time-Domain Analysis. To analyze a straight photonic-crystal waveguide using the FDTD method, a point source is used to launch a short

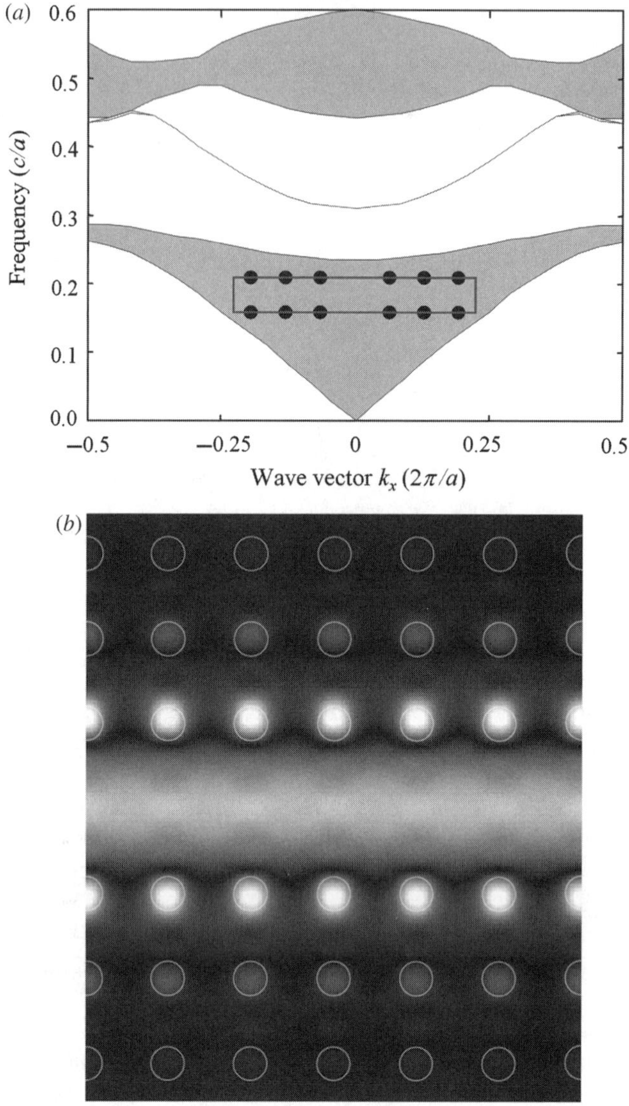

Figure 4.6 (a) Dispersion diagram for a two-dimensional square array of dielectric rods with $r = 0.2a$. (b) Electric-field distribution for $k_x = 0.318(2\pi/a)$.

pulse into one of its ends. The frequency response of the structure is then extracted from the Fourier transform of the transient field measured at the other end of the waveguide. Although the approach is conceptually simple, an important issue must be taken into consideration to obtain realistic results. If we define a finite length of a photonic-crystal structure by simply truncating the line defect, we introduce modal discontinuities between confined line-defect modes and free-space modes outside the photonic-crystal region, and therefore, backreflection from the waveguide ends. This leads to the formation of standing waves, instead of traveling waves, within the line-defect region, which contaminate the results. Although the reflection from the ends of the line defect can be significantly reduced by applying inhomogeneous PML-absorbing boundaries, as described in Chapter 3, the computational cost is increased. Therefore, below we present a modified FDTD scheme to solve for the eigenmodes of photonic-crystal line defects, which alleviates these problems.

Consider a two-dimensional photonic-crystal channel created by introducing a line defect in the photonic-crystal lattice of dielectric rods in a dielectric background, as shown in Fig. 4.7. In the two-dimensional case, the electromagnetic field components can be decoupled into TM and TE polarizations. Because the straight photonic-crystal channels are formed by periodically repeating the supercell in the x direction, one period of crystal lattices is sufficient for the computation region. By using the Bloch theory, as discussed in Chapter 3.5, the transverse field of a wave that travels in $+x$ direction may be written in the form

$$\begin{cases} E_z(x, y) = U(x, y) \exp(jk_x x) \\ H_y(x, y) = V(x, y) \exp(jk_x x) \end{cases}, \tag{4.3}$$

where k_x is a Bloch wave number and x,y are spatial coordinates. U and V are periodic Bloch functions that satisfy

$$\begin{cases} U(x, y) = U(x + a, y) \\ V(x, y) = V(x + a, y) \end{cases}. \tag{4.4}$$

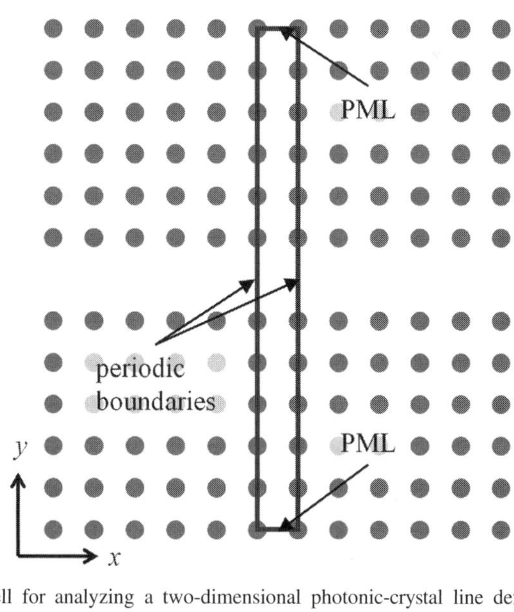

Figure 4.7 Supercell for analyzing a two-dimensional photonic-crystal line defect using the FDTD method.

By applying Yee's formulation (Section 3.4), a set of two-dimensional FDTD equations is then derived. Because only a finite number of crystal-lattice periods in the y direction is used to form the cladding, a nonreflecting boundary is needed to truncate the computation region, and simulate free-space extension. Here, a PML is applied to minimize backreflections.

A Gaussian (temporal) pulse produced by a dipole source excites a wide range of frequencies. To ensure that all eigenmodes are excited, multiple dipole sources are arbitrarily distributed throughout the waveguide channel, and transient data from several detectors placed randomly in the defect region are captured. In the frequency domain, the line-defect resonances can then be easily distinguished from the guided modes as the latter have a relatively high-quality factor. Each guided state can then be studied individually by using a long, narrowband excitation pulse. The electromagnetic-field components of the mode are extracted by using discretized Fourier transform at a given resonant frequency during the evolution of the FDTD time marching. Once the electromagnetic components for a guided mode are known, the energy flow in the direction of propagation can be obtained by taking the real part of the complex Poynting vector,

$$\mathbf{P} = \frac{1}{2}\Re(\mathbf{E} \times \mathbf{H}^*) = P_x\hat{\mathbf{x}}, \tag{4.5}$$

where "*" denotes complex conjugation, and $\Re$ denotes the real part.

4.3.1.3 Comparison between FDTD and PWEM. By implementing PWEM and FDTD for the analysis of a linear photonic-crystal defect, as described in Sections 4.3.1.1 and 4.3.1.2, dispersion relations are obtained for the structure shown in Figs. 4.5 and 4.7. The guided mode appears as an isolated line inside the photonic bandgap (Fig. 4.8). The results of PWEM and FDTD simulations overlap, which demonstrates an excellent agreement between the two methods and indicates that the introduction of a linear defect in a photonic crystal indeed creates an additional band of states that was absent in the uniform photonic crystal. The guiding properties of this structure can be inferred from an analysis of the dispersion diagram. For example, the group velocity for a particular wave vector is the slope of the curve in Fig. 4.8.

4.3.2 Line Defects in Photonic-Crystal Slabs

Two-dimensional photonic crystals are conceptually simple and require less computational resources to analyze than their three-dimensional counterparts. However, in principle, they extend to infinity in the z direction (or the direction perpendicular to the crystal plane). Although in practice, structures several wavelengths tall can often be thought of as "infinite" with little loss in accuracy, when the z extension becomes comparable with or smaller than the wavelength of light in the material, this approximation breaks down. In this case, the height of the structure must be taken into account, and full three-dimensional analysis must be applied to obtain

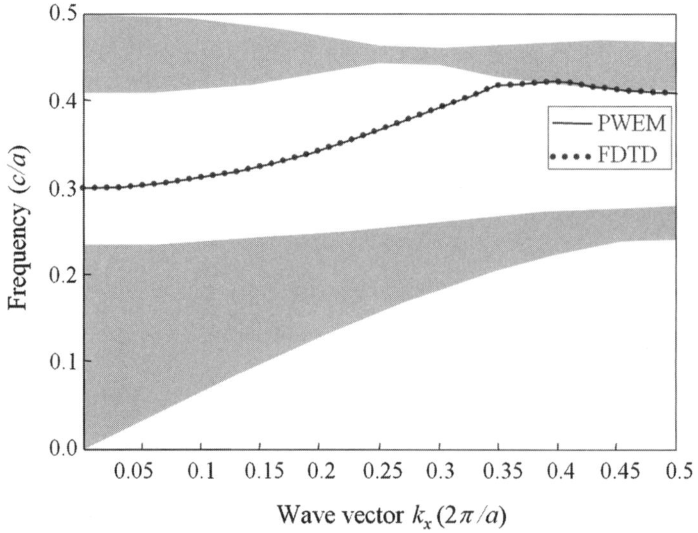

Figure 4.8 Dispersion diagram of a line defect inside a two-dimensional photonic crystal. The solid line indicates the results of PWEM calculations, whereas FDTD is shown as a dotted line.

trustworthy results. Furthermore, to take advantage of the confinement provided by total internal reflection (TIR), the typical planar photonic crystal will consist of a high-index slab with voids, or holes, drilled through it in a periodic arrangement, as opposed to high-index rods in a low-index background considered thus far. Fortunately, in this case, the methods described above can still be used for a linear defect to minimize the computational region.

Consider a structure shown in Fig. 4.9a where a row of holes is filled with the dielectric material in a slab photonic crystal, which produces a donor-type waveguide.

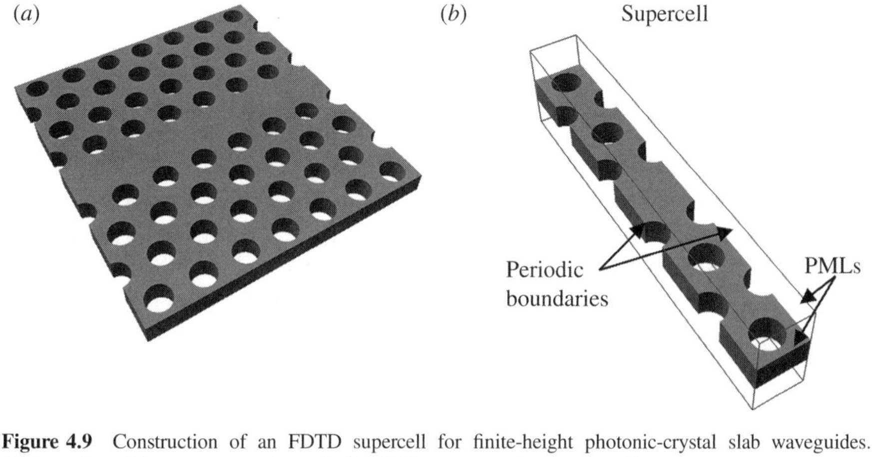

Figure 4.9 Construction of an FDTD supercell for finite-height photonic-crystal slab waveguides. (a) A section of the structure under analysis. (b) The computational supercell.

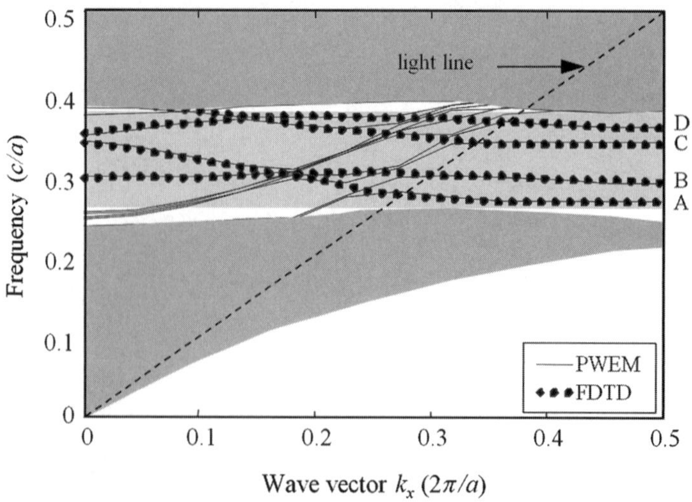

Figure 4.10 Comparison of a dispersion diagram obtained using PWEM and FDTD for photonic-crystal slab waveguide. Photonic bandgap of the photonic crystal is shaded in light gray. Shaded in dark gray are the continuum bands. Waveguide modes are indicated as isolated lines inside the bandgap. The diagonal dashed line represents the light line. The region above the light line contains radiation modes that are not confined to the slab.

For FDTD analysis, PML-absorbing boundary conditions are used in both the y (lateral) and z (out-of-plane) directions, whereas the Bloch periodic boundary condition is used in the x, or propagation, direction, as shown in Fig. 4.9b. For PWEM analysis, the supercell is repeated periodically in all three directions.

The photonic crystal is defined as follows: The slab has a dielectric constant $\varepsilon_r = 12.56$ and thickness $0.6a$, where a is the lattice constant of the crystal; the air holes have a radius of $r = 0.34a$. With these design parameters, the photonic-crystal structure is found to exhibit a bandgap below the light line in the spectral range $(0.273 \leq a/\lambda \leq 0.374)$ for TE-like polarization. The dispersion diagram for the photonic crystal with a line defect of Fig. 4.9a is shown in Fig. 4.10.

The region in the dispersion diagram above the light line corresponds to out-of-plane radiation.[1] Below the light line, four guided modes within the bandgap can be identified. Although excellent agreement is found between the PWEM and FDTD results below the light line, the former produced some spurious modes above the light line, which are absent in FDTD analysis. This discrepancy is the result of different boundary conditions in the z direction used in these two approaches. Whereas in FDTD, the absorbing boundary was used, PWEM required periodicity in the z direction, which allows for coupling between the periodically arranged photonic-crystal slabs even at considerable separations. This shows that one must be careful when analyzing slab photonic crystals using PWEM, as unphysical modes may

[1]Note, however, that because of the symmetry of the eigenmodes' field distributions, certain modes above the light line cannot couple to radiating modes in the cladding material. See Ref. 37, Ch. 3.

appear in the radiative portion of the band diagram. Fortunately, FDTD allows for easy identification of such spurious modes. If these modes are eliminated, the agreement between FDTD and PWEM is restored for all wave vectors in the Brillouin zone.

The mode profiles in the center of slab for the four guided modes are shown in Fig. 4.11. To illustrate the differences between guided and radiative modes, the profiles of modes that correspond to two different wave vectors ($k_x = 0.5 \times 2\pi/a$ and $k_x = 0$) are shown. Modes B and C are odd, whereas modes A and D are even, with respect to the vertical plane along the wave propagation direction.

Because of the discrete translation symmetry of the photonic-crystal waveguide in the propagation direction, the field components of each waveguide mode can be expressed as

$$F(x, y, z) = u(x, y, z)e^{jkx} = \sum_K U_k(y, z)e^{j(K+k)x}, \tag{4.6}$$

where $u(x, y, z) = u(x + a, y, z)$ and $F(x, y, z)$ represents any one of the six electromagnetic field components. k is the (Bloch) wave vector, a is the periodicity in the x direction (the waveguiding direction), and K is a reciprocal-space lattice vector ($K = 0, \ \pm 2\pi/a, \pm 4\pi/a, \ldots$). Note that a may or may not be the lattice constant, depending on the lattice geometry. In the last equality, we used the periodicity of u in the x direction to expand it as a Fourier series. Casting the mode this way shows how the interference of the different Fourier components form a guided photonic-crystal mode that has nonuniform amplitude distribution within one period of the photonic crystal. This distinguishes the photonic-crystal waveguide from its conventional dielectric counterpart, in which the amplitude distribution is uniform along the waveguide.

4.3.3 Extracting Dispersion Properties Using a Single-Frequency Source

The dispersion properties of a photonic-crystal waveguide can also be extracted from its response to a single-frequency signal. As opposed to the FDTD and PWEM methods presented above, this approach does not require periodic boundary conditions. Instead, the single-frequency signal is launched into a finite section of the waveguide. The steady-state field configuration is captured and subjected to spatial Fourier transform. The peak value of the field represented in the spatial frequency domain marks the wave vector that corresponds to the launched frequency.

To illustrate the method, consider the photonic-crystal waveguide shown in Fig. 4.12a. The parameters of the photonic crystal are the same as in Section 4.3.2. The length of the waveguide section is 40 times the photonic-crystal period, with four layers of photonic-crystal cladding flanking it on either side. We probe the waveguide using a continuous-wave signal with normalized frequency 0.3195 to obtain the steady-state results shown in Fig. 4.12b and c. The spatial Fourier transform of this field distribution along the direction of propagation yields the graph depicted in Fig. 4.13. It shows three distinct peaks at spatial frequencies of -0.9058, -0.113, and 0.9055, with the last peak being the tallest. Folding the spatial frequency spectrum into the irreducible Brillouin zone yields the wave vector, or propagation constant,

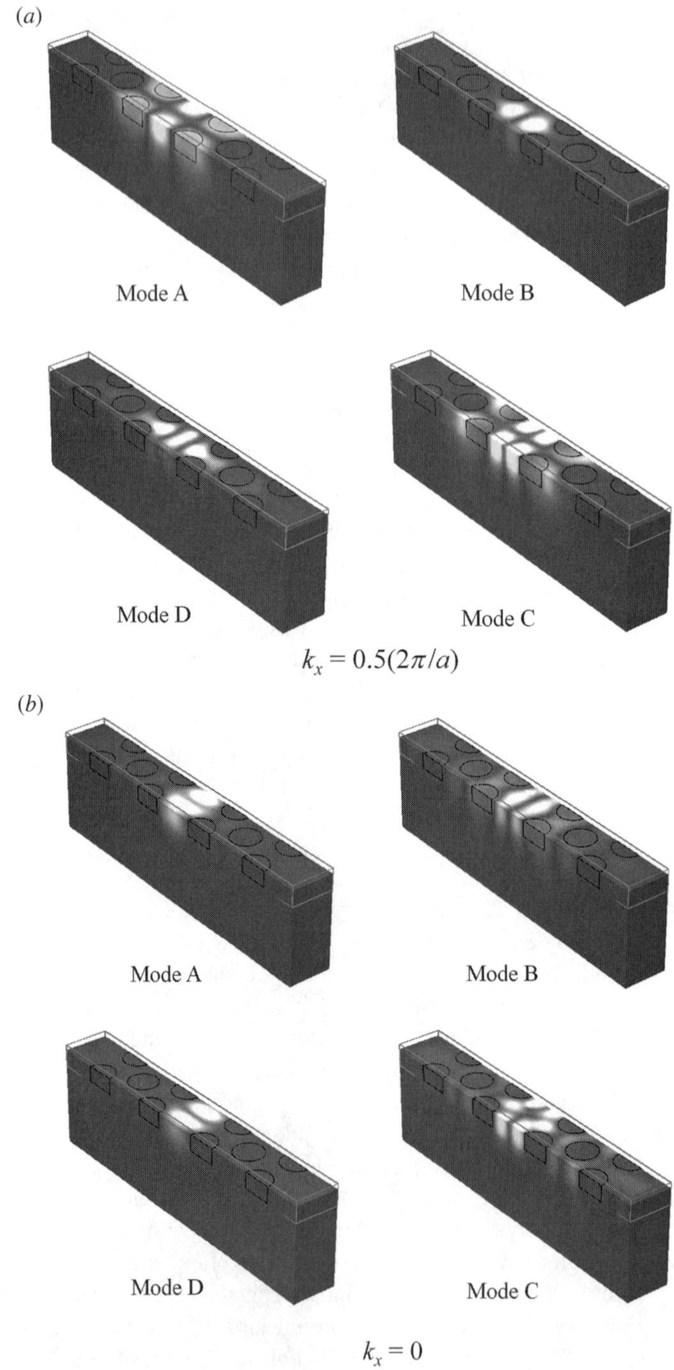

Figure 4.11 Optical mode profiles for two different wave vectors: (a) $k_x = 0.5 \times 2\pi/a$ and (b) $k_x = 0$. The modes at $k_x = 0.5 \times 2\pi/a$ for A, B, C, and D have normalized frequencies of 0.27, 0.285, 0.345, and 0.358, respectively.

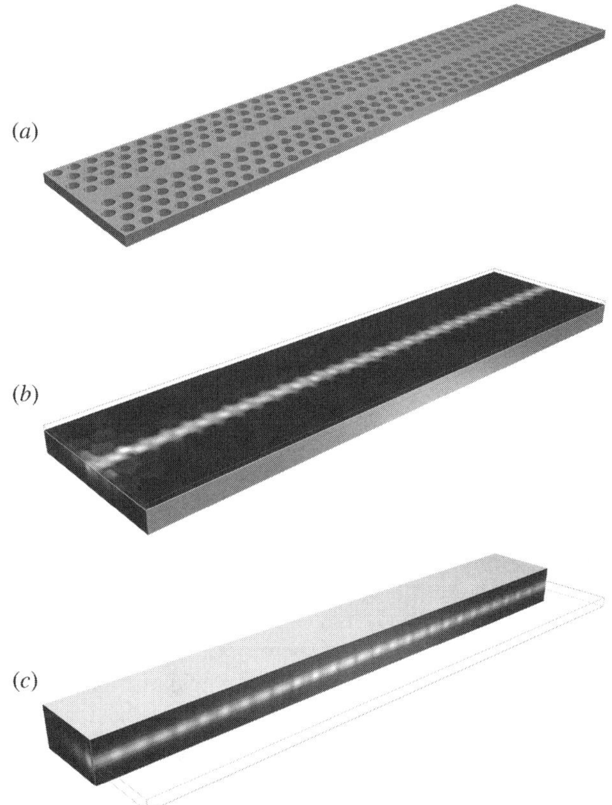

Figure 4.12 FDTD simulation of a straight photonic-crystal waveguide. (a) Line-defect waveguide formed by filling a row of air holes in perforated slab; (b) and (c) show the cross sections of guided mode profiles cut along *xy* and *xz* planes, respectively.

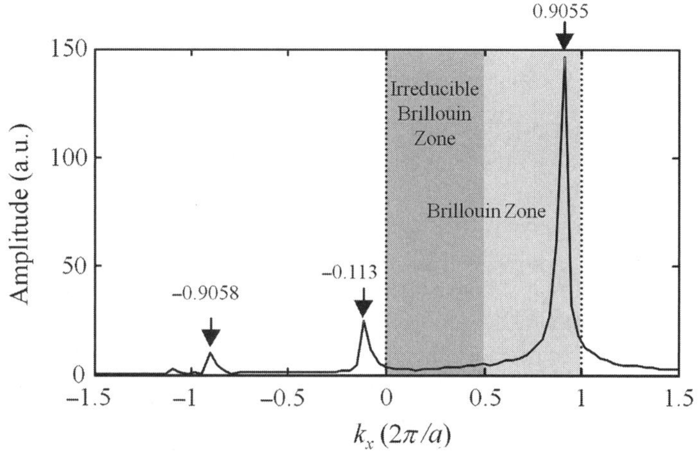

Figure 4.13 Spectral response for a slab photonic-crystal line defect obtained using FDTD followed by a spatial Fourier transform.

$0.0945 \times 2\pi/a$, which agrees (within computational accuracy) with $k = 0.09 \times 2\pi/a$ obtained using methods described in Section 4.3.2 for the same frequency.

Photonic crystals exhibit several properties that are desirable for photonic integrated circuits [30, 38], mainly strong confinement and irregular dispersion [39]. The strong confinement can be exploited for compact channel waveguides, sharp waveguide bends, switches, and high-quality-factor nanocavities, whereas the irregular dispersive properties can be exploited for superprism and self-collimation-based applications.

Numerous applications and devices that use both the confinement of photonic crystals and their dispersive properties have been presented in the literature. In the remaining sections of this chapter, we summarize the design and analysis of several confinement-based devices and their applications. A similar treatment of dispersion-based devices can be found in Chapter 5.

4.4 APPLICATIONS THAT USE STRONG CONFINEMENT IN PHC

In the sections above, we described elementary building blocks of photonic-crystal devices that take advantage of the presence of a photonic bandgap to confine light spatially. We also described methods frequently used to analyze them. Below, we provide a catalog, with brief explanations, of a variety of applications that take advantage of such strong confinement of the electromagnetic field. Although a complete (omni directional) photonic bandgap provides full spatial confinement, a partial/directional bandgap or stop band allows for tuning the degree of confinement, which may be exploited in some applications.

Although the reader will find that many concepts from conventional optical materials are simply extended to the field of photonic crystals, other concepts are unique to this field in that they are difficult or impossible to achieve otherwise. We have attempted to make the list as exhaustive as possible. However, given the high activity in the field, some applications might not have found their way here.

4.4.1 Waveguide Bends

Section 4.3 presented the use of a line defect in a photonic crystal exhibiting a photonic bandgap as a waveguide. For such a device to have practical value, it must allow arbitrary routing of optical signals between devices of an optical integrated circuit. In particular, any waveguiding structure must allow for bends in addition to the straight lines analyzed in Section 4.3. Although a circuit that contains a single discontinuity may tolerate some backreflection or loss in the propagating signal, realistic photonic devices may contain waveguides with multiple bends. Therefore, it is important to engineer such structures with the maximization of throughput across the operational bandwidth in mind. Thus, in this section, we analyze how the bends and discontinuities in waveguides defined in a photonic crystal affect the propagation of light through them, and we present some approaches to maximize the throughput.

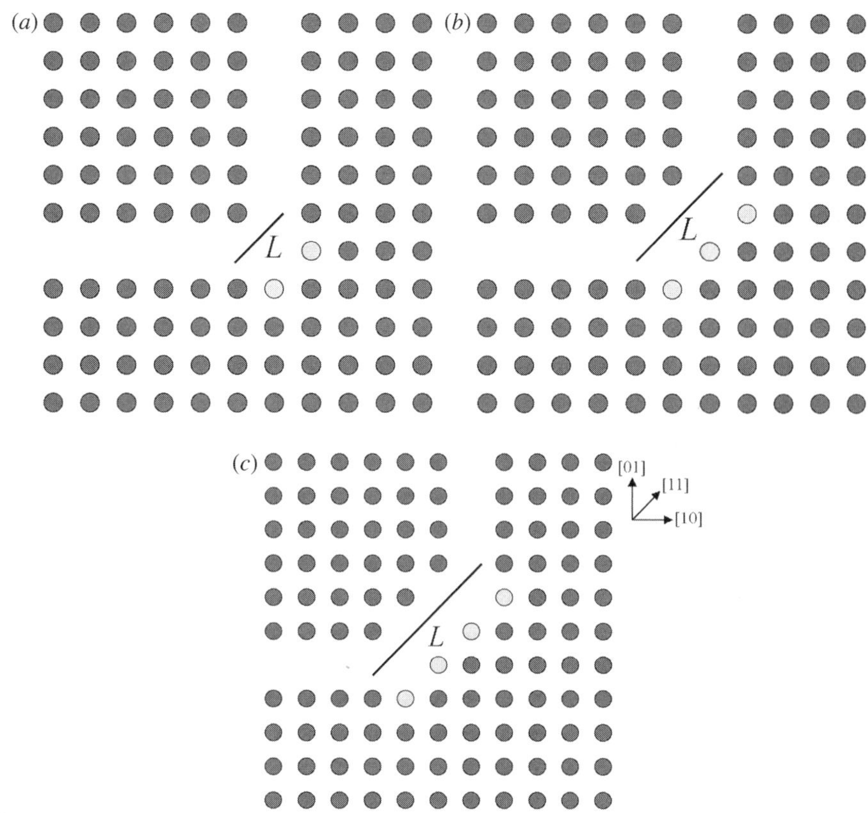

Figure 4.14 A 90° photonic-crystal waveguide bend created in a square lattice of high-index dielectric rods in air background. Such a bend can be realized by simply overlapping the ends of two straight waveguide sections (a) or by connecting them with a section of diagonal waveguide (b), (c). Note that even in (a), there is still an effective length of diagonal waveguide.

Consider a simple 90° bend in a 2D photonic crystal made of high-index dielectric rods on a square grid as shown in Fig. 4.14a–c. Investigations of such a structure have been reported in Refs. 14, 21, and 40. Theoretically, it was shown by scattering theory that nearly 100% transmission is possible: A simple one-dimensional scattering problem was solved to obtain reflection coefficient [14]

$$R(\omega) = \left[1 + \left(\frac{2k_1(\omega)k_2(\omega)}{[k_1^2(\omega) - k_2^2(\omega)]\sin[k_2(\omega)L]}\right)^2\right]^{-1}, \tag{4.7}$$

where $k_1(\omega)$ is the wave vector for a mode that propagates through the straight waveguide sections (along [10] and [01] directions) with no discontinuity; $k_2(\omega)$ is the wave vector for a mode that propagates through the waveguide bend section (along the [11] direction), which represents the discontinuity; and L is the effective length of the discontinuity. L is determined by finding the zeros of Eq. (4.7) at a frequency that was found to show zero reflection in a 2D FDTD simulation of the structure shown in Fig. 4.14b. Note that even in the case where the bend region seems to have zero length (Fig. 4.14a), the effective length L is nonzero; this can be interpreted as resulting from the two highlighted rods in Fig. 4.14a that act as a small 45° PhC mirror.

Experimentally, over 80% transmission was demonstrated at a frequency of 100 GHz in a square array of circular alumina rods that have a dielectric constant of 8.9 and a radius of $0.2a$, where a is the lattice constant of the array. For $a = 1.27$ mm, this crystal had a large bandgap between 76 GHz and 105 GHz. A line defect and several bends like those in Fig. 4.14 were created inside the crystal by removing rods. The guided modes produced by the defect extended over a large bandwidth, which extends over the entire bandgap [40].

If a similar discontinuity is introduced in a perforated dielectric substrate with air holes arranged on a triangular lattice, as shown in Fig. 4.15, the frequency bandwidth for high transmission through the discontinuity is narrow [41, 42] (less than 1% of the bandwidth available for the straight waveguide). The waveguide is defined in a dielectric substrate perforated with an array of circular, 380-nm-wide air holes arranged on a triangular lattice with a lattice constant $a = 450$ nm.

To reduce computational cost, the inherently three-dimensional problem of a slab photonic crystal is reduced to two dimensions by employing an effective index of refraction equal to the ratio of the speed of light and the phase velocity of a mode propagating in the slab (as explained in Chapter 2),

$$n_{eff} = \frac{c}{v_p} = \frac{c}{\omega/k}, \tag{4.8}$$

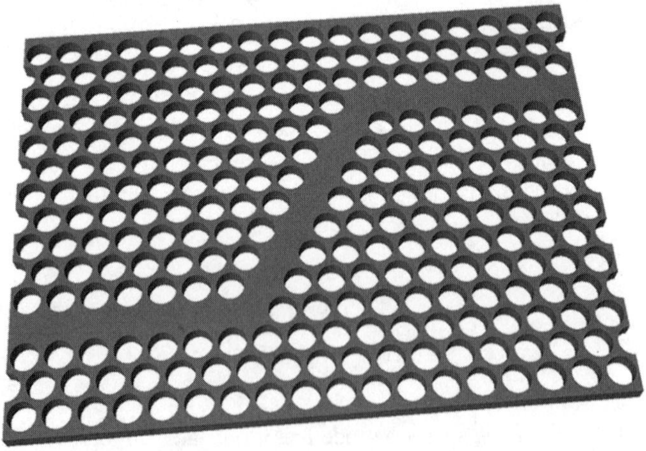

Figure 4.15 Two 60° photonic-crystal waveguide bends created in a hexagonal lattice of air holes in a high-index dielectric background.

where ω and k are the frequency and the wave vector of the slab mode, respectively. In our case, the effective index of the slab waveguide, $n_{eff} = 2.8$, was calculated by solving a transcendental equation [43]

$$\left(\frac{t}{\lambda}\right)_{\mathrm{TE}} = \frac{1}{2\pi\sqrt{n_2^2 - n_3^2}}\left[m\pi + \tan^{-1}\left(\frac{n_3^2 - n_1^2}{n_2^2 - n_3^2}\right)^{1/2}\right], \tag{4.9}$$

where m is an integer ($m = 1, 2, 3, \ldots$) that refers to the mth confined mode, t is the cutoff thickness of the TE_m mode, n_1 is the cladding index, n_2 is the core-layer index, and n_3 is the substrate index (Fig. 4.16). The derivation of this result can be found in Chapter 2. The effective-index value used in the analysis to follow is the refractive index of the fundamental mode ($m = 1$), extracted at the cut-off thickness of the slab waveguide structure, which guarantees a single mode of operation in the sense that the boundary conditions in the z direction (perpendicular to the slab plane) allow only one solution for the z dependence of the transverse mode profile. The corresponding cutoff thickness was also calculated using Eq. (4.9) and was found to be $t = 200$ nm at $\lambda = 1.3$ μm. Using this effective index ($n_{eff} = 2.8$), we can perform FDTD simulations of the perforated slab in two dimensions instead of three and still obtain fairly accurate results over a sufficiently narrow frequency band [44–46].

Such a photonic crystal exhibits a two-dimensional bandgap for a TE-polarized electromagnetic wave between $\lambda = 1.16$ μm and $\lambda = 1.45$ μm. To determine numerically the transmission of the optical power through the double-bend discontinuity, detectors are placed at various points within both the straight and the

Figure 4.16 Slab waveguide. The high-index core layer is flanked by lower index substrate and cladding, $n_2 \geq n_1$ and $n_2 \geq n_3$. The effective index of such a waveguide is lower than the bulk index of the core material.

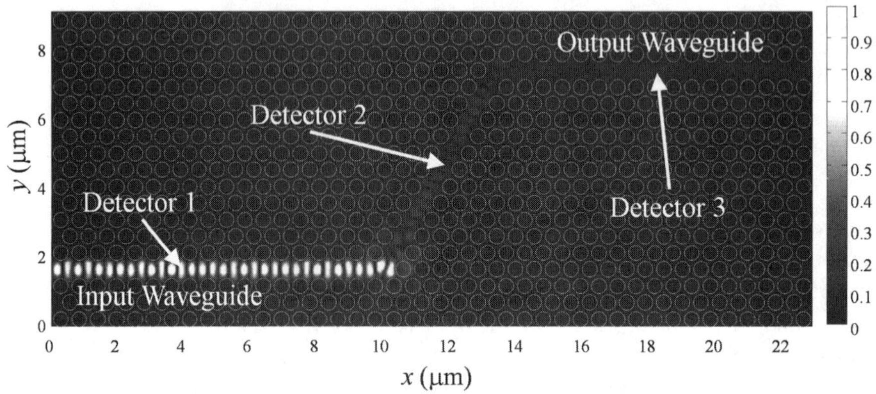

Figure 4.17 Transient electric-field magnitude of a broadband pulse launched into the input waveguide. Connections between the waveguide sections are conventional 60° bends.

angled waveguide sections. A broadband pulse with a center wavelength of $\lambda = 1.16$ μm and a pulse width of $\Delta\lambda = 290$ μm is launched at the lower left input into the waveguide, and the transient response is captured with the aforementioned detectors. A snapshot of the transient electric-field magnitude is shown in Fig. 4.17.

The transmission spectrum through each of the three sections is calculated by taking the Fourier transform of the squared amplitude of the field captured by each of the three detectors. The results show that a considerable loss of up to −40 dB is incurred because of backreflections at the waveguide discontinuities. According to these transmission spectra, this structure allows only a narrow band of frequencies to propagate through [45, 47–52].

To alleviate the problem of low transmission through waveguide bends, various solutions have been proposed. A resonant cavity placed at the discontinuity enhances transmission for the narrow band of the cavity resonance [53, 54]. However, wide-band transmission is achieved by carefully engineering the junction region, e.g., by using a mirror-like structure at the corner to redirect the incoming light toward the subsequent section of the waveguide [55–59].

4.4.2 Zero-Cross-Talk Waveguide Crossing

Thanks to the linearity of Maxwell's equations, optical beams that cross in free space (or any homogeneous linear media) do not interact; i.e., a beam of light does not feel the presence of other beams in the same region. However, if the beams are guided by crossing geometrical structures patterned in the medium, light can be scattered away from its intended path and couple into other waveguides—cross talk ensues. Therefore, if two different paths for optical beams are patterned in the same region, such as when two waveguides cross, care must be taken to avoid scattering of one beam into the path of another. The figure of merit in this situation is cross talk *CT*, defined as

$$CT = \frac{\text{Output power in undesired waveguide}}{\text{Output power in desired waveguide}}. \tag{4.10}$$

In conventional dielectric waveguides, the cross talk increases as the width of the waveguide decreases and becomes comparable with the guided wavelength. This limits the utility of such waveguides, as consequently they require larger real estate for reliable operation. However, photonic crystals allow for a low-cross-talk operation of dense embedded arrays of waveguides. A hint of how this could be possible was provided by the structure analyzed in the previous section, in which light launched in a straight waveguide "refused" to follow a bend in the triangular-lattice slab photonic crystal. What seemed to be a challenge to address for arbitrary light routing is an advantage in the case of crossing waveguides, because it is reasonable to expect that light launched in one waveguide would not readily enter another waveguide that intersects it.

To put the discussion on a firm footing, consider the structure of Fig. 4.18 in which two linear defects in a photonic crystal intersect at a 60° angle. The photonic crystal consists of an array of circular voids patterned in a high-index dielectric slab. The lattice constant is $a = 1.0$ μm, and the hole radius is $r = 0.3$ μm. The effective dielectric constant of the dielectric slab is set to $n_{eff} = 2.8$, which is the same as the effective index of the slab considered in the previous section. The structure exhibits a 2D photonic bandgap for TE-polarized waves in the spectral range between $\lambda = 3.125$ μm and $\lambda = 3.924$ μm. A system of two crossing waveguides is defined by filling the respective air holes.

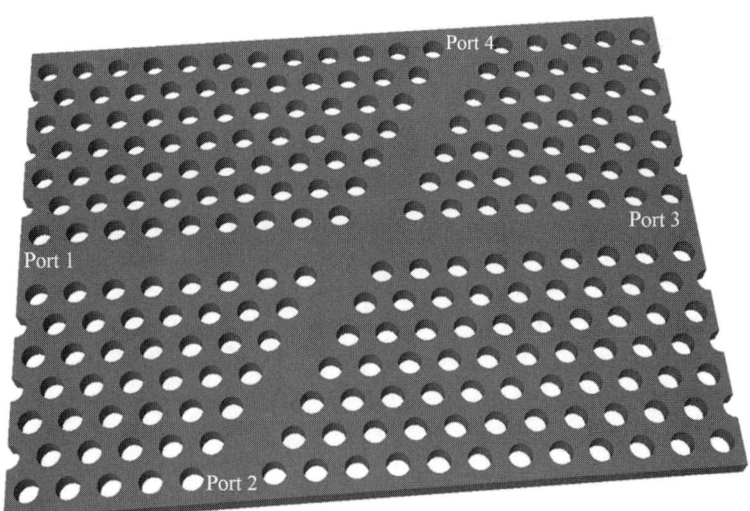

Figure 4.18 Two intersecting linear defects in a 2D photonic crystal.

Because of the symmetry of the waveguide-crossing geometry shown in Fig. 4.18, the four input/output ports are equivalent. The cross-talk figure of merit according to Eq. (4.10) is obtained using FDTD simulation by launching a broadband pulse into port 1 and observing the electromagnetic field in the remaining ports. We obtain $CT_{12} = P(2)/P(3) = -14$ dB and $CT_{14} = P(4)/P(3) = -10$ dB, which shows that, as anticipated, even this simplistic approach produces high suppression of cross talk. Additional refinements to the crossing region have shown that cross talk can be reduced by 8 orders of magnitude in 2D photonic crystals as compared with a conventional high-index-contrast waveguide crossing [60]. It has been conjectured that the proposed design principles could also apply to 3D systems and are not restricted to photonic crystals [60].

4.4.3 Narrow-Band Beam Splitter

The ability to split an optical signal into two or more paths is essential to realize highly functional photonic circuits. In bench-top optics, beam splitting is achieved by semi-reflective mirrors or prisms with suitable coating of the diagonal facet. Such beam splitting and waveguide branching have also been demonstrated in photonic crystals [61–63] where, additionally, the splitting can be made frequency dependent by employing resonant cavities or point defects in the junction region. The operational frequency range of such photonic-crystal beam splitters depends on the size, type, and material properties of the point defect, and is expected to be narrow based on the high-Q resonances of such point defects, as discussed in Section 4.2.1. Figure 4.19 shows an example of a narrow-band splitter formed by introducing two smaller dielectric rods in a simple T-junction.

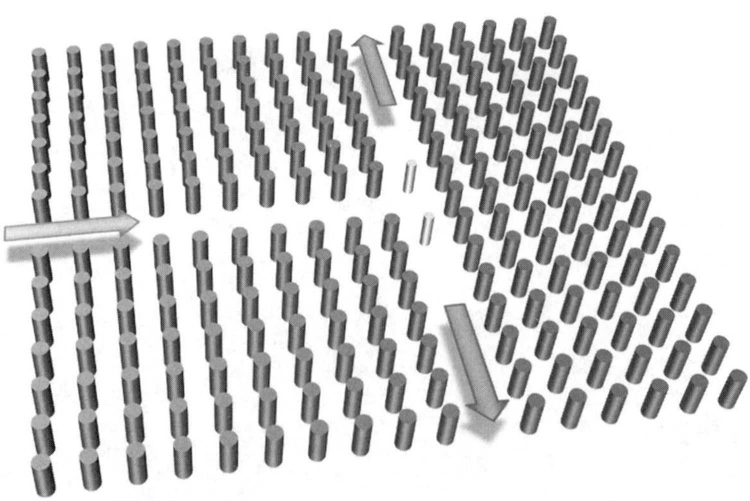

Figure 4.19 A narrow-band beam splitter formed by a system of two point defects (lighter rods), which act as resonant cavities, where the dielectric rod diameter was set at $r = 0.14a$; in the host photonic crystal, $r = 0.2a$.

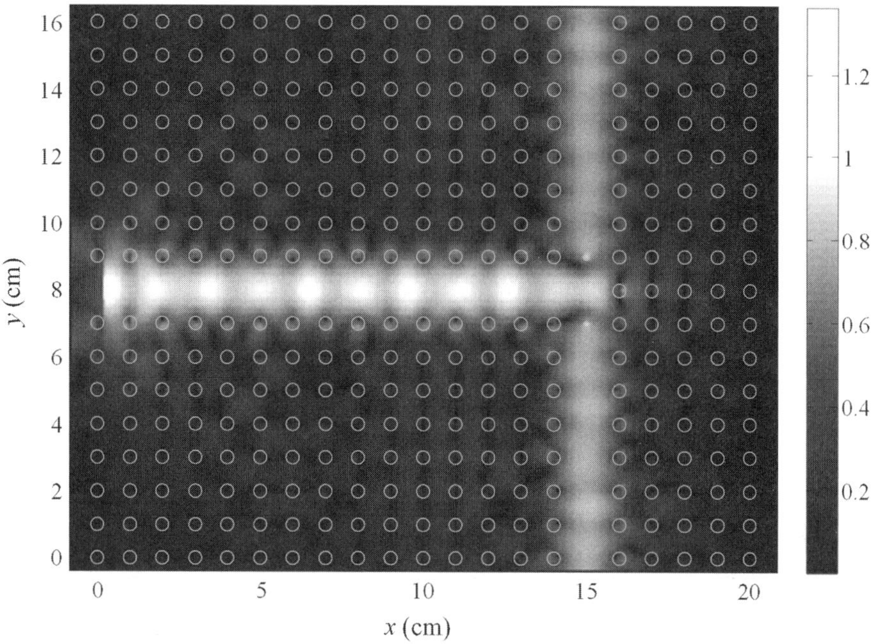

Figure 4.20 Steady-state field amplitude for the structure of Fig. 4.19.

The performance of such a device is illustrated in Fig. 4.20, which shows the steady-state field amplitude for a microwave wavelength of 3 cm. The concept has been demonstrated experimentally in the microwave region, $f \sim 10$ GHz, where square-shaped alumina rods with a dielectric constant $\varepsilon_r = 10.24$ were used [63].

4.4.4 Air-Bridge Microcavity

An interesting type of high-Q microcavity that consists of a channel waveguide and a one-dimensional photonic crystal was investigated in Refs. 15, 19, and 64. The structure essentially consists of a perforated air bridge as in Fig. 4.21. Holes drilled in a high-index suspended dielectric waveguide act like distributed Bragg mirrors on two sides of a resonant cavity. Total internal reflection, or index guiding, provides the lateral confinement. Alternatively, the structure can be thought of as a 1D photonic crystal surrounding a point defect. Numerical analysis of the eigenstates shows that strong field confinement of the defect state can be achieved with a modal volume less than $\frac{1}{2}$ of a cubic half-wavelength.

To illustrate the performance of such a cavity as a filter, we launch a broadband pulse at the input end and measure the spectrum at the output. Results of this numerical experiment are presented in Fig. 4.22. The transmission is suppressed between $\lambda = 1.45$ μm and $\lambda = 1.8$ μm, except for a sharp peak at $\lambda = 1.62$ μm. The suppression of transmission is caused by the photonic bandgap of the periodic arrangement of holes in the waveguide, whereas the peak results from the presence

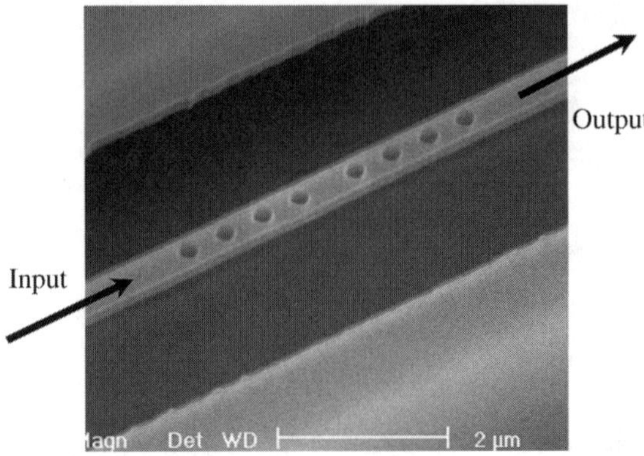

Figure 4.21 Air-bridge microcavity. Regularly spaced holes are etched in a suspended high-index dielectric waveguide to form a one-dimensional distributed Bragg reflector. A point defect in the middle supports a resonant mode.

of a defect mode supported by the disturbance of this periodicity. Equivalently, the several periods of the photonic crystal form high-efficiency dielectric mirrors that flank a linear cavity, similar to a Fabry-Perot etalon. Accordingly, the position of the peak can be tuned by changing the size (length) of the defect, whereas its width is determined by the efficiency of the "mirrors" and can be modified by increasing or decreasing the number of periodically arranged holes.

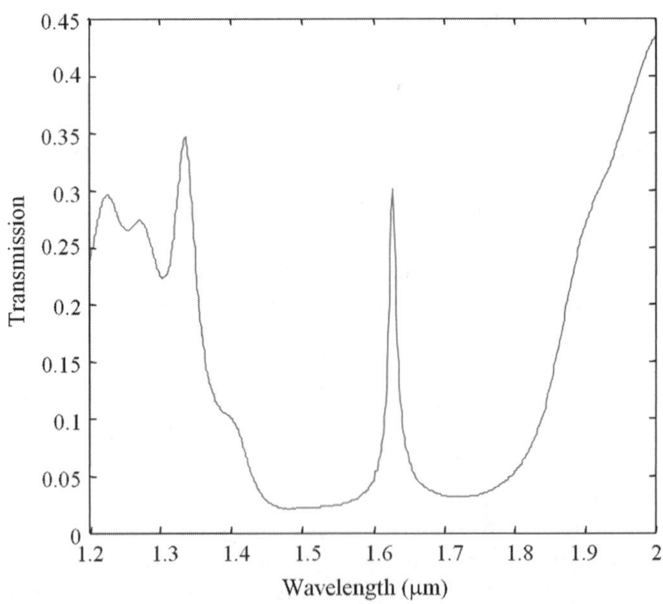

Figure 4.22 Transmission spectrum of a one dimensional air-bridge microcavity. A sharp transmission peak in the middle of the bandgap emerges as a result of the presence of a point defect.

4.4.5 Channel-Drop Filters in Photonic Crystals

The ability to transmit multiple wavelengths, each modulated with a separate data signal, in a single optical fiber enabled the realization of high-throughput connections spanning the globe, which in turn led to the Internet revolution. This multichannel approach relies on combining different wavelengths at the fiber input, and then separating them at the output. For optical fiber communication, this is accomplished using precise dielectric-stack filters that guarantee high-efficiency channel-add and channel-drop functions. Similar capability on the microscopic scale would enable high-throughput multichannel links in integrated photonic devices. Photonic-crystal line defects for guiding, in combination with point-defect resonators, appear to be ideal for this purpose [65–69].

A two-cavity solution is presented in Fig. 4.23. A two-dimensional photonic crystal with a lattice constant $a = 1.0$ μm is made of dielectric rods with $\varepsilon_r = 11.56$ and a radius of $r = 0.2$ μm. To enhance the coupling between the waveguides and the two cavities, the dielectric constant of the four dielectric rods adjacent to the waveguides was reduced to $\varepsilon_{rc} = 9.5$; these are indicated in Fig. 4.23 by dark rods. The two cavities were formed by reducing the radius at two lattice sites to $r_d = 0.05$ μm, and the dielectric constant of the altered point defects is $\varepsilon_{rd} = 6.6$. In an FDTD simulation of this device, a broadband pulse with a center wavelength of $\lambda = 2.71$ μm was launched into the bus channel. The spectral response of the system, shown in Fig. 4.24, is obtained by taking the Fourier transform of the transient response captured by the detectors placed at several locations within the system, one of which is indicated in the figure. Maximum transfer between the two waveguides occurs by creating resonant states of different symmetry and by forcing an accidental degeneracy between them. This approach produces a highly selective drop filter with a 6-nm line width and transmission approaching unity.

Although such a filter exhibits excellent performance, the use of two *identical* cavities places stringent requirements on the fabrication process to realize it. Tight fabrication tolerances are an issue of general importance in the field, which needs

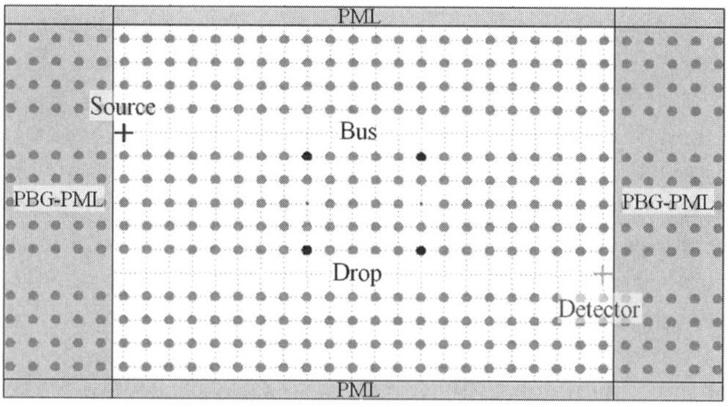

Figure 4.23 Photonic-crystal channel-drop/add filter using two waveguides (a bus and a drop) coupled through a resonator that consists of two point defects.

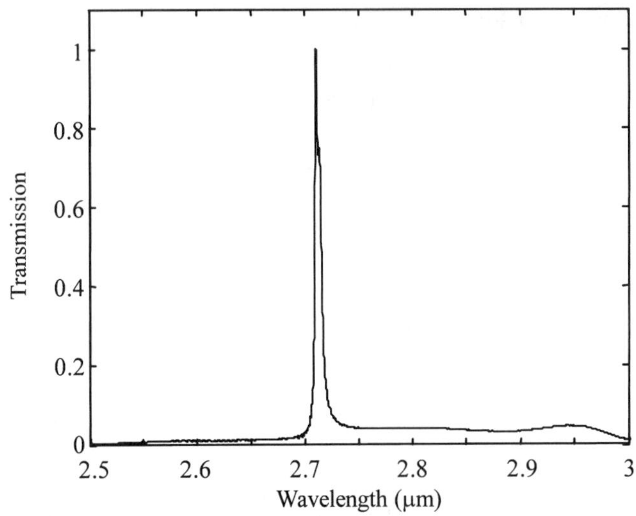

Figure 4.24 Spectral response of a drop filter. The graph shows the spectrum captured by the detector in Fig. 4.23, which corresponds to a drop channel. High selectivity of the order of 6 nm and high transmission approaching unity are achieved.

to be resolved before the potential of photonic crystals can be realized on a large scale. The operation of a single-channel drop filter was experimentally validated using a single point defect for a photonic crystal made of air holes in Si background [70]. A resonant wavelength was observed at $\lambda = 1563$ nm, with a line width of 4.1 nm.

4.4.6 Optical Spectrometer

The concept of a channel-drop filter can be extended to include extraction of multiple wavelengths. In this case, the device is referred to as an optical spectrometer, whose role is to separate optical signals spatially, based on their respective wavelengths. This characteristic is used in wavelength-division multiplexing (WDM)/demultiplexing (WDD) systems to transmit many optical signals through a single physical communication channel, such as an optical fiber. In such systems, each wavelength represents an optical channel, which is analogous to a radio or television channel carried by a specific frequency. In dense WDM (DWDM), wavelengths are closely spaced at intervals as small as 0.4 or 0.8 nm in the main telecommunications band near 1550 nm. The International Telecommunications Union (ITU) has specified a grid of standard frequencies separated by increments of 100 GHz (approximately 0.8 nm), referenced to a frequency 193.1 THz, which corresponds to a wavelength 1552.52 nm. These wavelengths are in the "conventional" or C band of the erbium-doped fiber amplifier (EDFA) operating between 1530 nm and 1570 nm. Other bands of interest are the "long" or L band (approximately 1570 nm to 1610 nm) and the "short" or S band (approximately 1490 to 1530 nm).

Although optical spectrometers that conform to the telecommunication standard have been realized using several different technologies, such as array waveguide gratings (AWGs) [71, 72] or fiber gratings [73], these devices often require dimensions in the range of centimeters to meters to support many densely spaced channels. In contrast, photonic crystals can achieve such channel separation with a much smaller footprint [66, 74–76]. Early designs were based on the superprism phenomenon [77–81], which is discussed in Chapter 5, and on using channel-drop filters [66–68, 82], which are similar to that described in Section 4.4.5, in which two waveguides are coupled through a resonant system of two or four cavities. Although the latter system works adequately for single-channel add/drop functionality, it lacks the scalability to multiple channels. Therefore, a scalable design of a photonic-crystal-based optical spectrometer that incorporates a bank of high-Q cavities tuned to different frequencies was explored in Ref. 76.

To understand how such a spectrometer operates, consider first a single-channel version, which is obtained by bringing a point-defect cavity in close proximity to a line defect waveguide, such that the waveguide and the cavity are evanescently coupled. If the frequency of the guided mode overlaps with the resonant frequency of the cavity, energy transfer from the linear defect to the point defect will occur. On the other hand, no such energy transfer will occur when the frequency of the propagating wave differs from the resonant frequency.

The device is illustrated in Fig. 4.25, in which the photonic crystal, made of dielectric rods 140 nm in diameter, with $\varepsilon_r = 11.56$, distributed in a square lattice with pitch $a = 350$ nm, exhibits an omnidirectional photonic bandgap for λ between 833 nm and 1250 nm. The acceptor-type line-defect waveguide is formed by removing

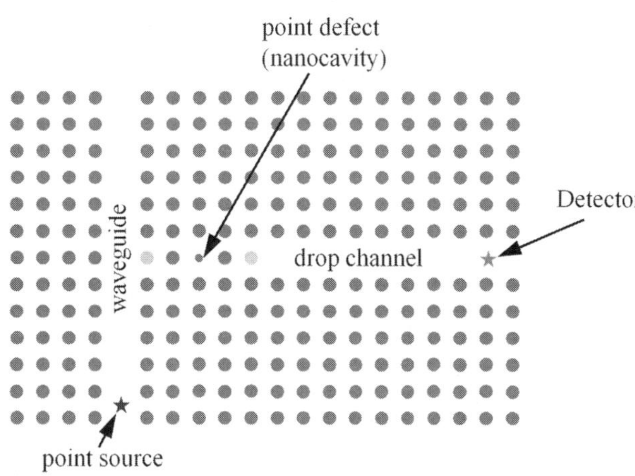

Figure 4.25 Single-channel optical spectrometer that comprises a point defect cavity ($r = 52.5$ nm and $\varepsilon_r = 7.0$) and an acceptor line defect waveguide created by removing a column of dielectric rods. The photonic-crystal lattice consists of dielectric rods with a radius $r = 70.0$ nm and a dielectric constant $\varepsilon_r = 11.56$, which are distributed on a square lattice with a pitch $a = 350$ nm. The two rods ($r = 70.0$ nm) of light gray color have a dielectric constant $\varepsilon_r = 8.0$ to enhance coupling between the cavity and the waveguides.

one column of dielectric rods along one of the cardinal directions of the crystal. As determined in Section 4.3, such a waveguide supports a single broadband mode in the PhC bandgap. The point-defect cavity is formed by reducing the diameter of a single dielectric rod to $2r = 105$ nm and by reducing its dielectric constant to $\varepsilon_r = 7.0$. Decreasing or increasing the diameter of a single lattice site was previously shown to affect the number of modes a cavity supports, as well as their resonant wavelengths.

To analyze this structure, a pulse with a center wavelength of $\lambda_0 = 1.0$ μm and a width of $\Delta\lambda_0 = 0.6$ μm is launched into the waveguide. The pulse propagates along the waveguide channel, confined to the defect region by the virtual walls formed by the photonic-crystal layers. Similar to the dielectric waveguide case, the propagating mode has evanescent tails that extend several lattice constants beyond the defect region. Hence, if the resonant cavity is close enough, the evanescent tail of the propagating mode will excite a single mode or multiple modes within the cavity. Then, the electromagnetic field in the cavity evanescently couples to a nearby output or "drop" channel (see Fig. 4.25). Coupling to and from the point defect cavity is enhanced by decreasing the dielectric constant of the two rods both at the cavity/waveguide interface and at the cavity/channel interface to $\varepsilon_r = 8.0$.

A detector placed in the output channel captures the time-variable field, which is then Fourier transformed to obtain its spectral content. Figure 4.26 shows the resulting drop-off spectrum overlaid on the spectrum of the incident pulse. It shows a line width of the transmitted signal in a 2-nm range with a center wavelength of $\lambda = 1025$ nm, which is in the ballpark of the 0.8 nm required for DWDM.

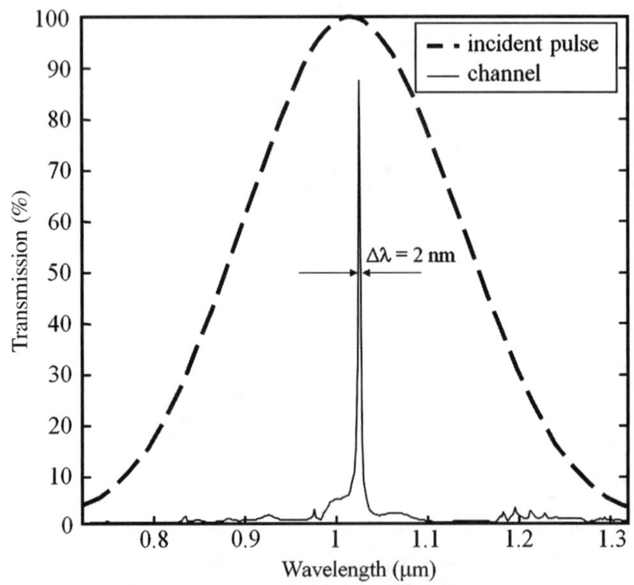

Figure 4.26 Single-channel optical spectrometer simulation. Wavelength spectra are shown for an incident broadband pulse (dashed curve) and the pulse measured at the detector placed in the filtered channel (solid curve), normalized to the input pulse.

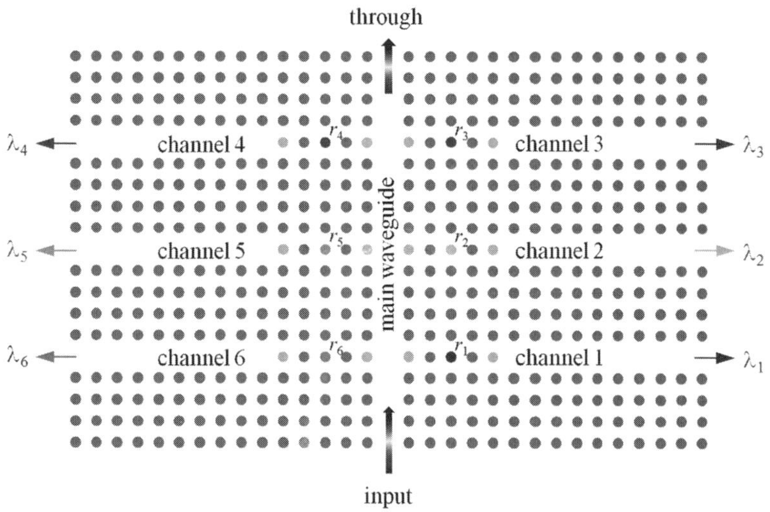

Figure 4.27 Multichannel optical spectrometer using cavities with different point defect sizes: $r_1 = 8.75$ nm, $r_2 = 17.5$ nm, $r_3 = 26.25$ nm, $r_4 = 35$ nm, $r_5 = 43.75$ nm, and $r_6 = 52.5$ nm. Each cavity couples to a separate output channel.

The single-channel device analyzed above serves as a building block to realize a multichannel spectrometer. Placing resonator cavities tuned to different wavelengths along a single-line defect allows extracting individual wavelengths from a broad-band optical signal propagating in the waveguide. Figure 4.27 shows an example where six cavities are included, each having a different defect size and its own drop channel. The radii of the rods in the point defects are 8.75 nm, 17.5 nm, 26.25 nm, 35 nm, and 43.75 nm, in addition to the 52.5 nm analyzed above, and the dielectric constant is maintained at $\varepsilon_r = 7.0$ for all of them. The resonant wavelengths that correspond to these defects are $\lambda_1 = 1.025$ μm, $\lambda_2 = 0.875$ μm, $\lambda_3 = 0.895$ μm, $\lambda_4 = 0.925$ μm, $\lambda_5 = 0.949$ μm, and $\lambda_6 = 0.985$ μm.

The spectral response of this device is shown in Fig. 4.28, in which the individual peaks are labeled using the corresponding channel numbers. With the exception of Channel 5, the relative heights of the peaks generally follow the spectral content of the incident pulse shown with a dashed curve in Fig. 4.26.

The spectrometer design described in this section provides an avenue to realizing DWDM on a chip scale. It offers a scalable platform for high-channel-density systems with hundreds of individual optical wavelengths that carry separate signals.

4.4.7 Hybrid Photonic-Crystal Structures

The applications considered so far use a single type of photonic-crystal lattice with local modifications to realize various functionalities. In this section, different types of lattices, such as triangular and square, are combined to enhance the functional properties of the devices. Such structures, which combine two or more types of photonic lattices, are referred to as heterostructure photonic crystals [83].

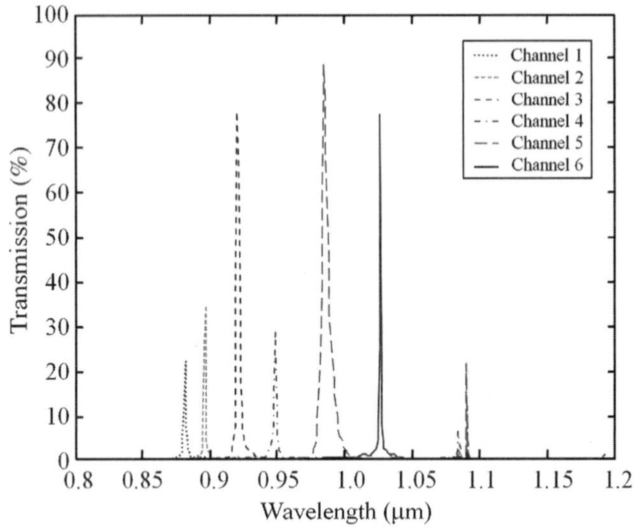

Figure 4.28 Spectral response of the six-channel device shown in Fig. 4.27. Peaks correspond to the resonant wavelengths of the respective cavities.

When different photonic crystals are brought together, discontinuities in their respective energy bands are expected. Consider, for example, triangular and square photonic crystals of dielectric rods made of the same material, with identical rod diameters $2r$ and identical lattice constants a. Their respective band diagrams are shown in Fig. 4.29. Because the band edges of the two photonic crystals do not line up, a

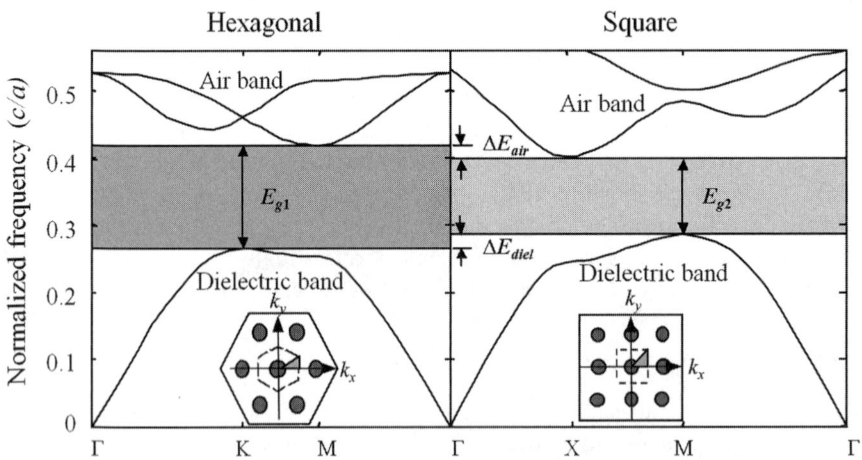

Figure 4.29 Band diagram of hexagonal PhC lattice ($r/a = 0.2$) with bandgap E_{g1} and square PhC lattice ($r/a = 0.2$) with bandgap E_{g2}. As the lattices are brought together to form a heterostructure, the PhC band edge discontinuities ΔE_{diel}, and ΔE_{air} appear at the dielectric and air bands.

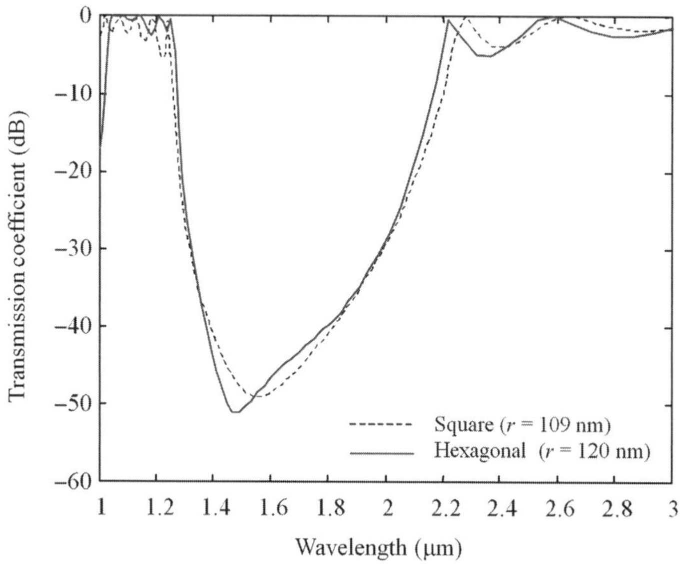

Figure 4.30 Transmission spectra at normal incidence for square-lattice and triangular-lattice photonic crystals. The bandgaps that correspond to the two lattices are matched for specific rod radii: $r = 109$ nm for the square lattice and $r = 120$ nm for the triangular lattice. The PhC period is $a = 543$ nm in both cases.

band-energy discontinuity exists at the interface between them by definition. Such discontinuities can be exploited to control photon transport in a fashion similar to that by which electron transport is controlled in semiconductor heterostructures. For example, by sandwiching a layer of wide-bandgap material between two narrow-bandgap structures, an analog of a semiconductor quantum well is formed, which in this case can be called a photonic quantum well. On the other hand, by carefully designing the triangular and the square lattices, the discontinuities in the dielectric the air bands shown in Fig. 4.29 can be minimized if not completely eliminated. Figure 4.30 illustrates this possibility in the transmission spectra of finite triangular and square lattices. In this case, heterostructure PhCs can be used as matching elements.

Heterostructure photonic crystals can also be used to optimize the directional bandgap size obtained from a single photonic-bandgap structure, such that a bandgap size of 94% can be obtained from a heterostructure of a square photonic-crystal lattice (bandgap size of 55%) and a triangular photonic-crystal lattice (bandgap size of 39% and 19%), as shown in Fig. 4.31. Wide-directional-bandgap PhCs are advantageous for applications such as wide-band optical mirrors, wide-band optical matching elements, and wide-band optical couplers.

A particularly interesting application where PhC heterostructures considerably improve device performance is optical beam splitting. Shown in Fig. 4.32 is an optical beam splitter implemented using a single photonic-crystal structure (square). Such a device achieves an overall throughput efficiency of 50%, with most of the loss in transmission caused by the backreflections in the splitting region.

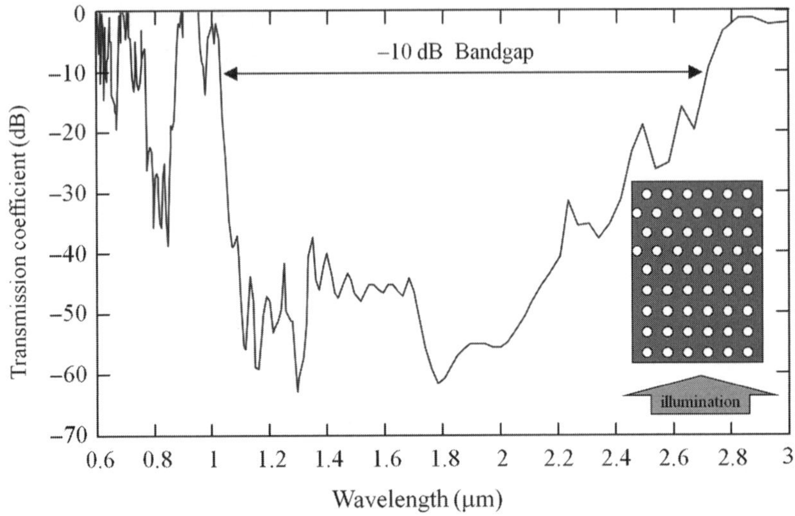

Figure 4.31 Transmission spectrum (normal incidence) of a heterostructure PhC with a directional gap between $\lambda = 1.0\ \mu m$ and $\lambda = 2.7\ \mu m$ for TM polarization.

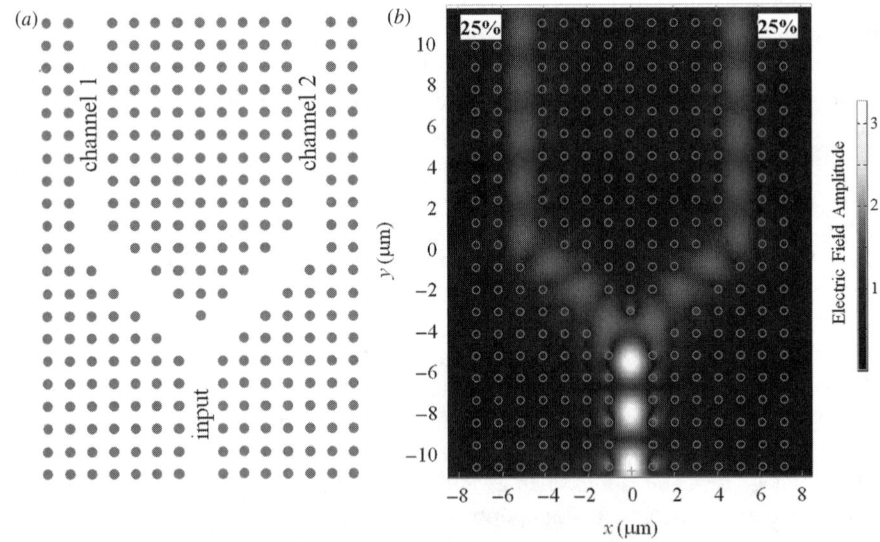

Figure 4.32 (a) Beam splitter built using a square PhC lattice. (b) Snapshot of FDTD simulation of the structure shown in (a). Throughput is 50%, with each of the branches capturing 25% of the input power. The remaining 50% is reflected back toward the input waveguide. The PhC is defined on a square grid with $a = 543$ nm and dielectric rod radius $r = 109$ nm. It has been found to exhibit a photonic bandgap for $1.234\ \mu m < \lambda < 2.172\ \mu m$.

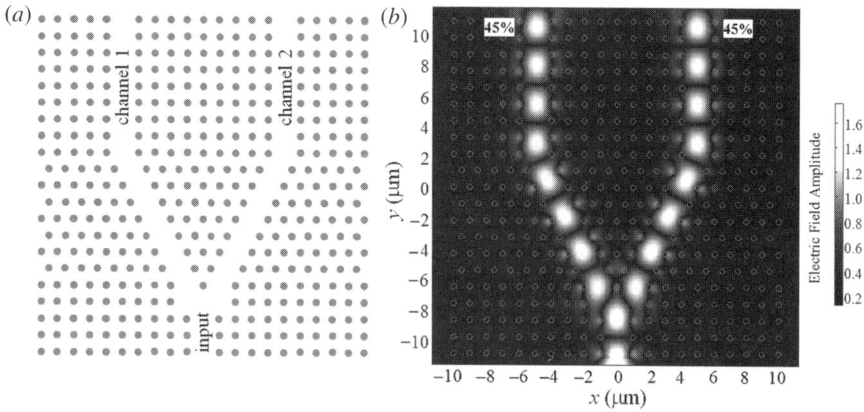

Figure 4.33 (a) A beam splitter formed in a photonic-crystal heterostructure by removing one column of dielectric rods to form a waveguide in the square part of the crystal, and a diagonal row of rods on both sides of the waveguide in the hexagonal part to form a splitter section. (b) A snapshot of FDTD simulations. Throughput is 90% with the input optical power equally divided between the output waveguides.

Judicious use of a PhC heterostructure improves the performance of the splitter from 50% to 90%, as shown in Fig. 4.33. Other examples that use PhC heterostructures include a 1-to-4 beam splitter shown in Fig. 4.34 and a beam splitter/combiner shown in Fig. 4.35.

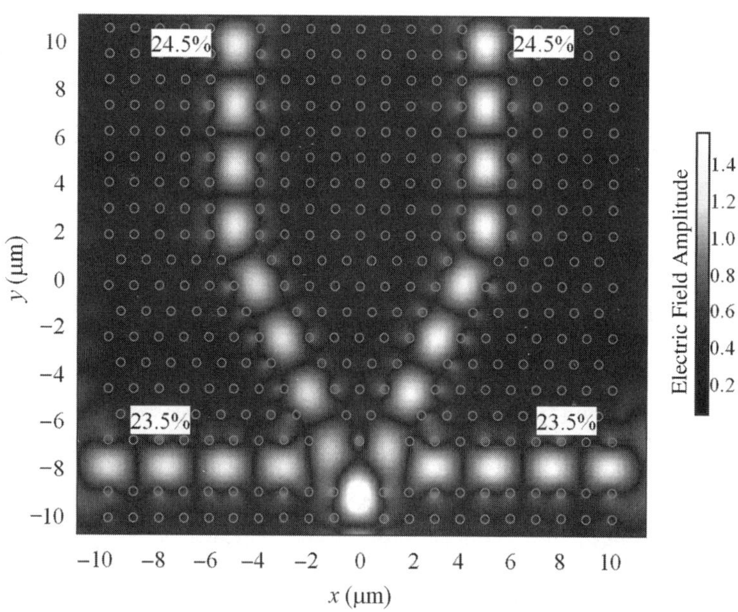

Figure 4.34 One-to-four optical beam splitter using a PhC heterostructure. Total device throughput is 94% with well-balanced outputs. Light enters at the bottom and exits through the top and side branches.

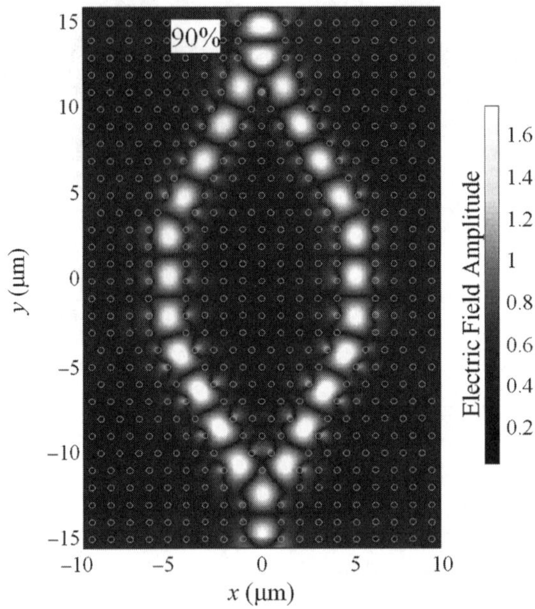

Figure 4.35 Optical beam splitter/combiner in heterostructure PhC lattice.

4.4.8 Electrically and Thermally Tunable Photonic Crystals

Photonic crystals and PhC-based devices discussed thus far are static: Once they are fabricated, their properties are determined by the geometry of the structures and by the electromagnetic properties of the materials used. As a result, the routing of optical signals within any waveguides and the resonant frequencies of any cavities are fixed. An electronic analog of such structures would be the wiring within an integrated circuit. However, in the latter, signal processing can be reconfigured by the use of active elements such as transistors. Reconfiguring optical signal processing within photonic crystals requires the ability to change dynamically either the geometry of the structures or the electromagnetic properties of the materials used to realize them. In this section, the latter possibility is discussed.

Optical properties of semiconductors are known to be affected by free-carrier injection or by the application of the external electrostatic field. For example, the presence of free charges in silicon leads to free-carrier absorption of an optical beam in the material, whereas the application of an electric field to intrinsic gallium arsenide affects its index of refraction due to the electro-optic effect. Thus, if photonic crystals are defined in these materials, then their band structures can be dynamically modified by the injection of free charge carriers or by the application of an electric field, respectively. Such tunability of photonic crystals has been proposed in Ref. 84, where the authors modeled the dependence of the dielectric constant on the temperature and impurity concentration through the plasma frequency ω_p as in $\varepsilon(\omega) = \varepsilon_0(1 - \omega_p^2/\omega^2)$. The disadvantage here is that the presence of free carriers leads to absorption, or optical losses, which cannot be avoided and can only be mitigated by careful selection of materials.

Another class of materials known to change their optical properties under the application of electric field is liquid crystals. Because liquid crystals are liquid by their very nature, using them to make a free-standing PhC structure is difficult if not impossible. However, a photonic crystal made of another (solid) material can be infiltrated with a liquid crystal [85, 86]. In this case, the latter becomes one of the PhC constituent materials, and therefore, its optical properties can be modified by externally applied electric fields.

The optical properties of liquid crystals are also a strong function of the temperature. Therefore, the properties of a photonic crystal infiltrated with this material will exhibit strong temperature dependence. For example, a macroporous silicon photonic crystal with a triangular lattice pitch of 1.58 μm exhibits a photonic bandgap in the range 3.3–5.7 μm. Experimental results presented in Ref. 85 show that after infiltration with liquid crystal E7^{TM}, the bandgap for the TE mode shifted to 4.4–6.0 μm, whereas that of the TM mode collapsed. Heating the sample above the nematic-isotropic phase transition temperature of the liquid crystal, 59°C, shifted the short-wavelength band edge of the TE gap by 70 nm without affecting the long-wavelength edge.

4.4.9 Photonic-Crystal Optical Networks

The bottleneck in high-performance computing has shifted from the switching speed of individual gates to data delivery. Increasingly, the bandwidth of electrical on-chip interconnects limits the speed of modern processors. At a time when human society is becoming increasingly dependent on automatic data processing, this difficulty threatens the continued economic expansion and improvement of productivity afforded by the heretofore unstoppable march of Moore's law. In this context, the ability to route optical signals on the chip scale in a manner akin to electrical signal routing would open a wide range of possibilities in ultra-fast computing. Because optical interconnects provide orders of magnitude of improvement in data bandwidth compared with their electrical counterparts (as demonstrated on the global scale where switching from copper wires to optical fiber enabled the explosive growth of the Internet), such a conversion on the chip scale has a potential to revolutionize computers as we know them. Thus, high-density optical interconnects using PhCs were recently proposed as a technique for three-dimensional optical signal distribution and routing through multiple planes and in different directions [87]. These optical networks offer the ability to guide light in a way analogous to electrical printed circuit boards (PCBs), which transport electric current through electrical networks.

A schematic diagram showing cross sections (xz plane and xy plane) of a conceptual two-layer integrated optical network is presented in Fig. 4.36. Here, the bottom layer consists of a photonic-crystal slab, which is used to guide, split, and process optical waves between different points within that layer. In the top layer, another

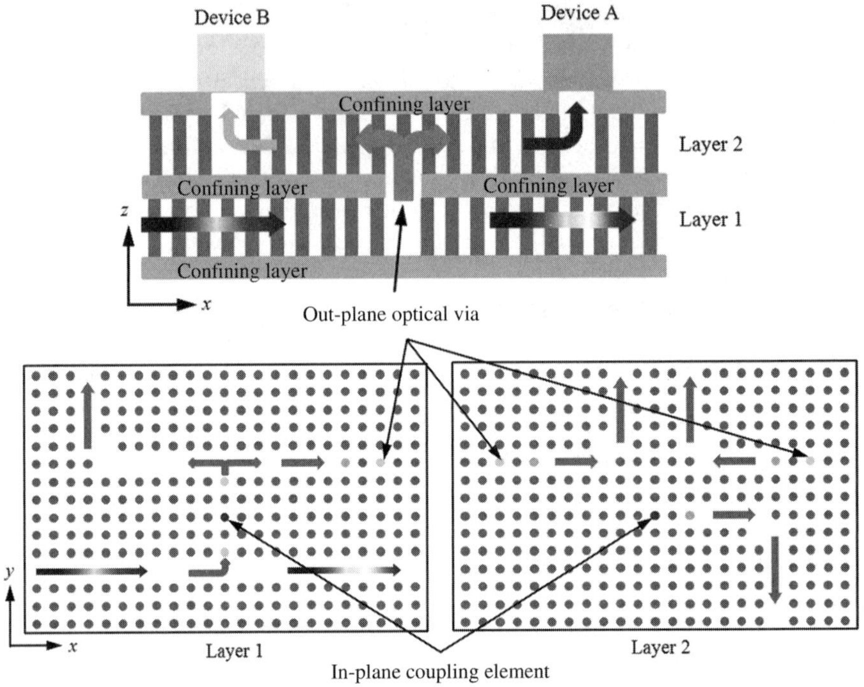

Figure 4.36 Two-layer photonic-crystal optical network. Coupling between different waveguides in the same plane is achieved through point defects or nanocavities. Optical vias provide for coupling between PhC layers. A confining layer is used to both isolate the two PhC layers from one another and to provide vertical confinement. Note that the layers depicted are merely conceptual examples; the figure does not represent any specific device or function.

photonic-crystal circuit is used for different optical guiding and processing. Optical signals of specific wavelengths are sent to their destinations at photodetectors, optoelectronic on-chip devices, or via optical fibers to other chips. The top and bottom PhC layers are isolated from one another with a confining layer. For the purpose of simulations, a perfect electric conductor is used as the confining layer for a photonic crystal made of dielectric rods in low-index material; for perforated high-index dielectric slabs, a layer of SiO_2 may be used for isolation. Optical vias are created to couple the electromagnetic waves efficiently between the PhC layers.

Although an electrical via can be as simple as a vertical section of wire penetrating across an isolation layer, optical vias are somewhat more demanding because the light beam must be redirected from its original in-plane path to go out of plane and then be redirected again for in-plane guiding. Such a beam path is similar to one encountered in a classic periscope configuration in which, in the simplest case, two 45° mirrors are used to shift the optical axis by a fixed z offset. As it turns out, the same approach can be used on a microscale, where in place of mirrors, two identical deflectors are used, one in each communicating layer. A schematic diagram of such an arrangement is shown in Fig. 4.37.

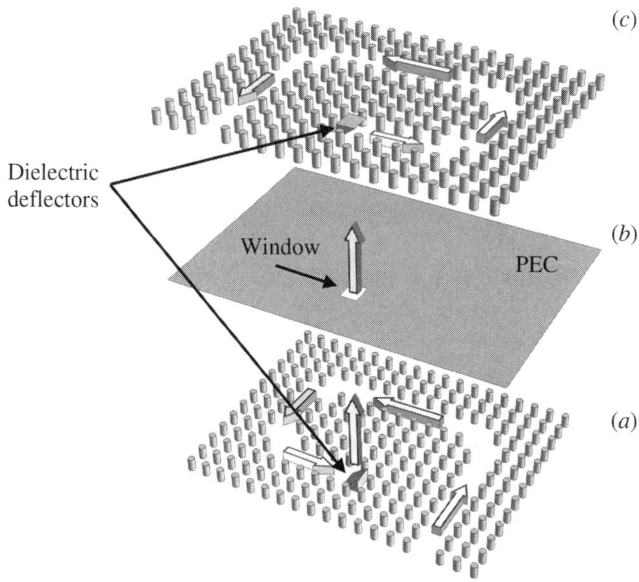

Figure 4.37 Dielectric deflector used as an optical via to couple light between adjacent layers. (a) The bottom layer contains a waveguide and a dielectric deflector designed to couple between the top and the bottom layers, such that an optical beam that propagates through the waveguide will be reflected upward toward the top layer. (b) The confining layer contains a window for vertical coupling. (c) The top layer contains another deflector, which reflects an optical beam toward the waveguide in the top layer.

The operation of the two-layer optical interconnection network is shown in FDTD simulation in Fig. 4.38, where four snapshots are taken of the electromagnetic wave as it traverses the circuit. Light is launched in the layer-2 waveguide. Figure 4.38b shows the progression of the beam along the single-defect photonic-crystal waveguide some time after the beam is turned on. As it reaches the dielectric deflector, the beam is redirected toward layer 4 where it emerges as shown in Fig. 4.38c, and follows the PhC waveguide patterned therein; see Fig. 4.38d.

4.4.10 Coupled Photonic-Crystal Waveguides

When two parallel dielectric waveguides are placed in close proximity in such a way that the evanescent tails of their respective modes overlap, energy of the electromagnetic wave that propagates within one of them can be partially or completely transferred to the other. Such an arrangement is referred to as coupled dielectric waveguides, or a directional coupler. Because the line defects in photonic crystals exhibit guiding properties similar to those of dielectric waveguides, when two such defects are placed parallel in close proximity, it is expected that light launched into one of them will couple to the other just as in the conventional case.

Thus, coupled photonic-crystal waveguides (CPhCWs), which consist of two closely spaced parallel waveguides, provide a means of transferring an electromagnetic wave from one waveguide to another. Such a system can be used for a variety

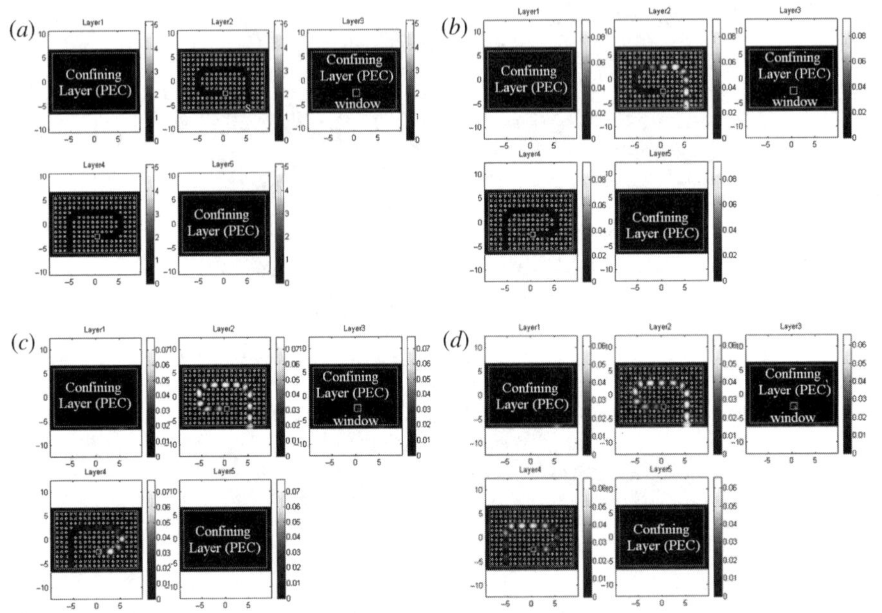

Figure 4.38 Four consecutive snapshots of FDTD simulation for the structure shown in Fig. 4.37. Layers (1, 3, and 5) are made of a perfect electric conductor (PEC) for vertical confinement, whereas layers 2 and 4 are finite-height photonic-crystal waveguides. Source (S) is located at the bottom right corner of layer 2.

of applications, which include switching, power division, and frequency/polarization filtering. Below, the concept of coupled photonic-crystal waveguides is examined in more detail, and several applications are discussed.

4.4.10.1 Theory of Coupled Photonic-Crystal Waveguides. Consider a pair of waveguides shown in Fig. 4.39. If an electromagnetic wave is launched into port 1, it propagates along waveguide 1 from left to right. Because the mode of the propagating wave extends beyond the physical boundaries of waveguide 1, it interacts with waveguide 2. As a result, optical power is gradually transferred from waveguide 1 to waveguide 2 as the wave traverses the length of their proximity L. If L is long enough, all power is transferred to waveguide 2. At this point, field configuration is exactly opposite to the launch configuration—all power is now in waveguide 2. Therefore, if the interaction region continues further, optical power begins to transfer from waveguide 2 back to waveguide 1. The back-and-forth transfer of the optical power between the waveguides continues as long as the waveguides remain in close enough proximity that coupling between them can occur.

In the situation described above, complete power transfer from one waveguide to another is possible only if the propagation constants corresponding to the respective modes of each waveguide taken in isolation are the same. This equality of propagation constants, which is also known as phase synchronization, occurs naturally when the two waveguides are identical. In that case, all the guided modes of the waveguides

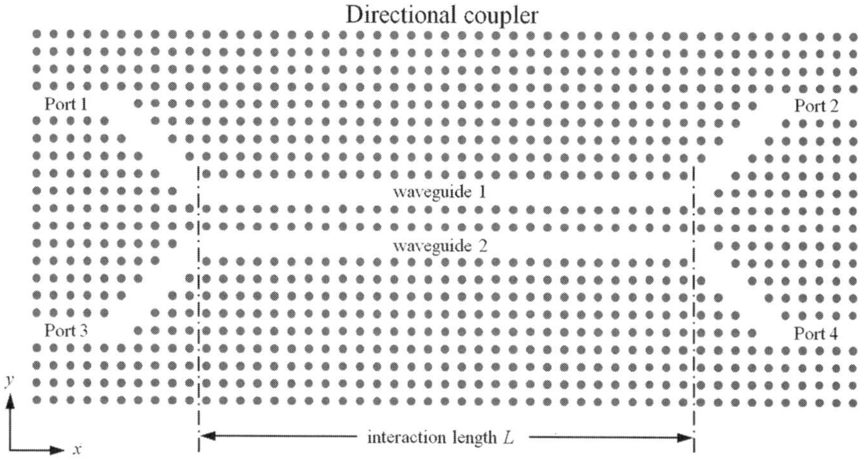

Figure 4.39 A CPhCW system that consists of two closely spaced PhC waveguides of length L, separated by two PBG layers.

are in phase synchronism and can couple to each other at all wavelengths. For the device of Fig. 4.39, the two waveguides are identical and are both single-mode as shown in Section 4.3.

Another way to look at such a device is to think of it as a single waveguide, but one that supports two modes, as opposed to two single-mode waveguides. Because the structure is symmetric, its eigenmodes must be either symmetric or antisymmetric with respect to the center-line mirror reflection; see Fig. 4.40a and b. In general, in the presence of nonzero coupling between the individual defect lines, the propagation constants of the two eigenmodes (symmetric and antisymmetric) will differ. This means that if both modes are launched into the waveguide, the phase between them will change as they propagate along; i.e., if they start in phase, after some length of the waveguide, they will be out of phase and then in phase again, and so on.

More formally, following Eq. (4.6), the electromagnetic field of an eigenmode in the waveguide is expressed as

$$F_k(x, y, z) = u_k(x, y, z)\exp(-jkx), \tag{4.11}$$

where the waveguide direction is along the x-axis, such that $u_k(x + a, y, z) = u_k(x, y, z)$, and a is the photonic-crystal pitch. Such a representation is possible because, as before, the waveguide structure is periodic along the propagation direction. Thus, for the two eigenmodes designated as odd and even, one can write

$$F_{ek_e}(x, y, z) = u_{ek_e}(x, y, z)\exp(-jk_e x), \tag{4.12}$$

$$F_{ok_o}(x, y, z) = u_{ok_o}(x, y, z)\exp(-jk_o x), \tag{4.13}$$

where k_e (k_o) is the propagation constant of the even (odd) mode.

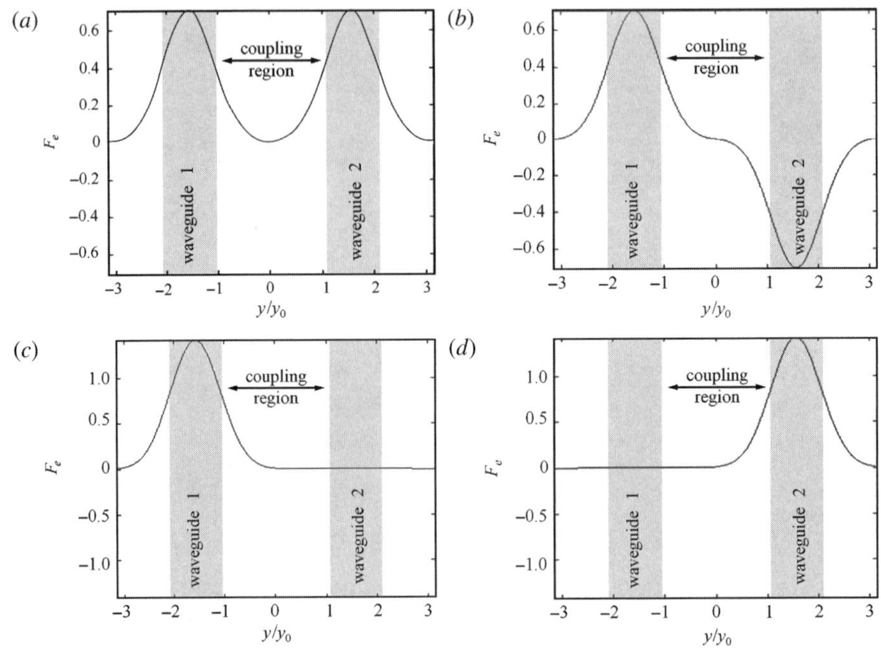

Figure 4.40 Schematic representation of even u_{ek} and odd u_{ok} eigenmodes of the CPhCW system shown in Fig. 4.39. The plot shows the overlap of each mode with the corresponding waveguide and with the coupling region in between. (a) Even mode, (b) odd mode, (c) even mode + odd mode, (d) even mode–odd mode. y_0 is the individual waveguide width.

Adding the even and odd modes results in a field distribution that is concentrated mostly in waveguide 1 (Fig. 4.40c). This corresponds to a field configuration at the beginning of the coupling region if light is launched into port 1 in Fig. 4.39. Conversely, the difference of the two eigenmodes has most of the field concentrated in waveguide 2, Fig. 4.40d, as it would at the beginning of the coupling region if light were launched into port 3 of the structure in Fig. 4.39. Formally, these statements can be written as

$$u_{1k}(x,y,z) = \frac{1}{\sqrt{2}}[u_{ek_e}(x,y,z) + u_{ok_o}(x,y,z)], \tag{4.14}$$

$$u_{2k}(x,y,z) = \frac{1}{\sqrt{2}}[u_{ek_e}(x,y,z) - u_{ok_o}(x,y,z)], \tag{4.15}$$

where the numerical factor $1/\sqrt{2}$ is introduced to preserve the normalization of the eigenmode, and k is the arithmetic average of k_e and k_o, $k = (k_e + k_o)/2$. Note that the only difference between a field being concentrated in waveguide 1, Eq. (4.14), and a field being concentrated in waveguide 2, Eq. (4.15), is in the sign between the two terms inside parentheses. Given that a minus sign is equivalent to a π-phase, the change in relative phase between the even and the odd modes of the two-waveguide structure is equivalent to the transfer of optical power between waveguide 1 and waveguide 2 that constitute it. Such a phase change occurs naturally as the wave propagates along the waveguide as long as the propagation constants of the even and odd modes are not identical; i.e., $k_e \neq k_o$.

Equations (4.14) and (4.15) are identical to those of a system of coupled dielectric waveguides. The difference is in the definitions of u_{ek} and u_{ok}, which are now dependent on x, the direction of propagation. However, the general behavior of light propagating in either of the structures is the same. Thus, the coupling length L_c is similarly defined as

$$L_c = \frac{\pi}{2|\kappa|}, \quad \text{where } \kappa = \frac{k_e - k_o}{2}. \tag{4.16}$$

L_c is the distance (length of the coupling region) it takes for the electromagnetic wave launched into one of the waveguides to transfer completely to the other. To see how this energy transfer happens in the PhC case, assume that light is launched into waveguide 1; i.e., at $x = 0$, the field distribution is concentrated therein,

$$F(x = 0, y, z) = u_{1k}(x = 0, y, z). \tag{4.17}$$

To obtain the field distribution at an arbitrary position along the waveguide, Eqs. (4.14) and (4.15) are used to express u_{1k} in terms of u_{ek} and u_{ok}, and then the latter are replaced with F_{ek} and F_{ok} per Eqs. (4.12) and (4.13) to take into account the propagation of the modes along the coupled-waveguides system. As a result, the field distribution at an arbitrary position x along the waveguides is

$$\begin{aligned} F(x) &= \frac{1}{\sqrt{2}}[F_{ek_e}(x) + F_{ok_o}(x)] \\ &= \frac{1}{\sqrt{2}}[u_{ek_e}(x)e^{-jk_e x} + u_{ok_o}(x)e^{-jk_o x}]. \end{aligned} \tag{4.18}$$

where the y, z arguments have been omitted to reduce clutter in the formula. Equation (4.18) can be rearranged to express the field distribution in terms of κ

$$F(x) = \frac{e^{-jkx}}{\sqrt{2}}[u_{ek_e}(x)e^{-j\kappa x} + u_{ok_o}(x)e^{j\kappa x}], \tag{4.19}$$

where $k = (k_e + k_o)/2$. The eigenmodes u_{ek} and u_{ok} are then expressed in terms of field distributions u_{1k} and u_{2k} defined in Eqs. (4.14) and (4.15),

$$F(x) = e^{-jkx}[u_{1k}(x)\cos(\kappa x) - u_{2k}(x)j\sin(\kappa x)]. \tag{4.20}$$

According to Eq. (4.20), for $x = 0$, the field distribution is u_{1k}, which is consistent with light being launched into waveguide 1, whereas at $x = L_c$, the field distribution becomes u_{2k}, up to a phase factor, and is concentrated in waveguide 2; in other words, the field transferred completely from waveguide 1 to waveguide 2 over a distance L_c.

By performing similar algebra for light launched into waveguide 2, the following field distribution is obtained at an arbitrary position x:

$$F(x) = e^{-jkx}[-u_{1k}(x)j\sin(\kappa x) + u_{2k}(x)\cos(\kappa x)]. \tag{4.21}$$

Equations (4.20) and (4.21) can be combined into a single matrix equation,

$$\begin{bmatrix} F_{1k}(x) \\ F_{2k}(x) \end{bmatrix} = e^{-jkx}\begin{bmatrix} \cos(\kappa x) & -j\sin(\kappa x) \\ -j\sin(\kappa x) & \cos(\kappa x) \end{bmatrix}\begin{bmatrix} u_{1k}(x) \\ u_{2k}(x) \end{bmatrix}, \tag{4.22}$$

which describes the transfer of energy between the coupled PhC waveguides as light propagates along. The way to understand Eq. (4.22) is by examining the equations that constitute it, Eqs. (4.20) and (4.21). Thus, $F_{1k}(x, y, z)$ on the LHS of Eq. (4.22) is the field distribution obtained when the wave is launched into waveguide 1 at $x = 0$; i.e., $F_{1k}(x = 0, y, z) = u_{1k}(x = 0, y, z)$. Similarly, $F_{2k}(x, y, z)$ is the field distribution obtained when the wave is launched into waveguide 2 at $x = 0$; i.e., $F_{2k}(x = 0, y, z) = u_{2k}(x = 0, y, z)$. (Note that we reintroduced variables y and z into the arguments of these functions.)

To see the system described above in action, consider the example shown in Fig. 4.41, in which the coupled waveguides are obtained by removing two rows of

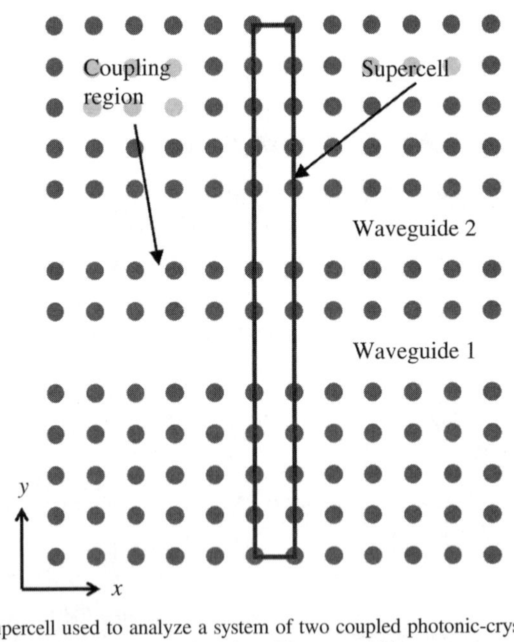

Figure 4.41 The supercell used to analyze a system of two coupled photonic-crystal waveguides. Each waveguide can support a single mode in isolation. The waveguides are separated by two photonic-crystal layers. The photonic-crystal structure used is an array of infinitely long dielectric rods ($\varepsilon_r = 11.56$, $r = 0.2a$) in air background.

dielectric rods from an otherwise perfect two-dimensional photonic crystal. The dielectric constant of the cylindrical rods is $\varepsilon_r = 11.56$, and their radius is $r = 0.2a$, where a is the lattice constant. The defect-free photonic crystal has a bandgap in the spectral range $0.23 \leq a/\lambda \leq 0.41$ for TM polarization.

Because $\kappa = (k_e - k_o)/2$, Eq. (4.22) implies that the operation of a coupled-waveguide system is highly sensitive to the propagation constants k_e and k_o. To obtain an accurate measure of these quantities, a supercell is defined, as shown with a rectangle in Fig. 4.41, for calculating the guided modes using PWEM—following the approach described in Section 4.3.1.1. The dispersion diagram obtained this way is shown in Fig. 4.42 using a solid line. The two curves in the photonic bandgap correspond to the two modes, even and odd, supported by this structure. To determine which curve corresponds to which mode type, the associated field distributions must be examined. Two such field distributions at the same frequency, obtained via the PWEM routine described in Section 4.3.1.1, are presented in Fig. 4.43. Based on these field distributions, we can conclude that the lower of the two curves in Fig. 4.42 corresponds to the even mode.

Similar results can be obtained using FDTD. To excite both even and odd modes for the calculation of the dispersion diagram, the source of the TM-polarized wave is located at the input of waveguide 1, whereas the classification of modes to even and odd is obtained by exciting the respective symmetric and antisymmetric waves in the coupled-waveguides system [88]. The resulting band diagram obtained this way, which is shown in Fig. 4.42 as solid circles, matches very well the diagram obtained using PWEM.

The coupling length L_c of the PhC coupled-waveguide system is calculated from Eq. (4.16), where the values of k_e and k_o are found in the dispersion diagram of

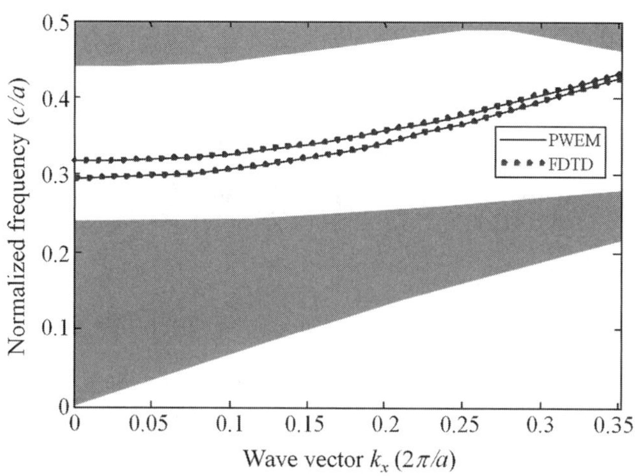

Figure 4.42 Dispersion diagram calculated for the coupled photonic-crystal waveguide supercell shown in Fig. 4.41. The dispersion diagram shown was calculated using both the PWEM (solid line) and the FDTD method (solid circles).

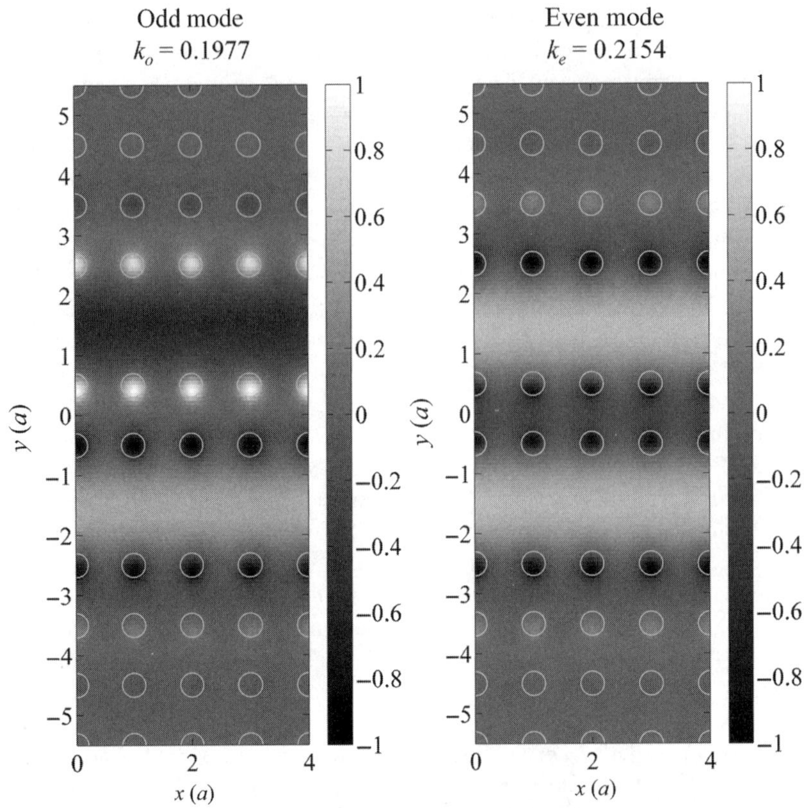

Figure 4.43 Modal field distribution for the odd mode with $k_o = 0.1977$ and for the even mode with $k_e = 0.2154$ for the CPhCW system of silicon rods in air background shown in Fig. 4.41. The frequency for both modes is the same $f = 0.35\ c/a$.

Fig. 4.42 for the frequency of the launched optical signal. To illustrate this procedure, assume that the frequency of the launched wave is $f = 0.35\ c/a$; a close-up of the dispersion curves for the guided modes around this frequency is shown in Fig. 4.44. The propagation constant of the even mode is $k_e = 0.2154\ (2\pi/a)$, whereas that of the odd mode is $k_o = 0.1977\ (2\pi/a)$. Using Eq. (4.16), the coupling length is calculated to be

$$L_c = \frac{\pi}{|k_e - k_e|} = \frac{\pi}{|0.2154 - 0.1977|(2\pi/a)} \approx 28.25a. \tag{4.23}$$

Thus, if light is launched into waveguide 1, in little over 28 periods of the photonic crystal, it will be completely transferred to waveguide 2. At $f = 0.35\ c/a$, this translates to less than 10 free-space wavelengths.

4.4.10.2 Applications of Coupled Photonic-Crystal Waveguides. The unique properties of coupled PhC waveguides analyzed above can be used in integrated

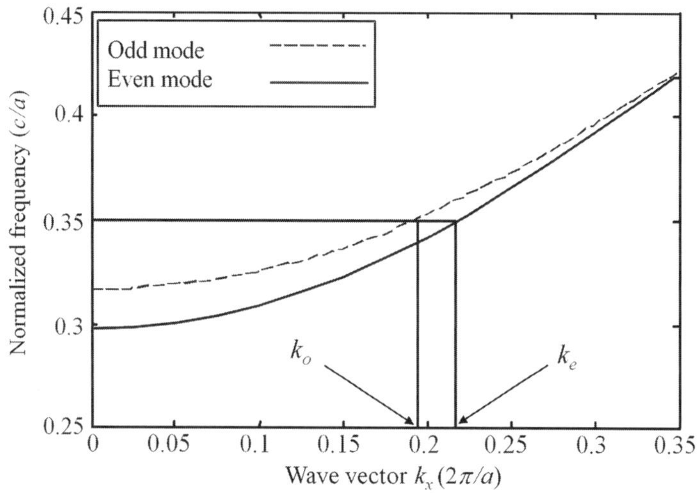

Figure 4.44 Modal dispersion curves of the even and the odd modes of the coupled photonic-crystal structure shown in Fig. 4.41.

photonic applications to process optical signals on the chip scale. Below, several such applications are discussed in more detail.

Frequency-Selective Filter. The frequency dependence of the coupling length, which is a result of the dispersive characteristics of PhC waveguides and the coupling-induced splitting between even and odd modes, can be used for wavelength-selective routing of optical signals. For the example discussed in Section 4.4.10.1, the coupling length changes by a factor of about five within a mere 15% change in frequency, as shown in Fig. 4.45. Therefore, it is expected that highly sensitive and compact filters can be realized within this framework.

To illustrate this possibility in the near-infrared part of the spectrum, the system of coupled PhC waveguides of Figs. 4.39 and 4.41 is scaled so that the period of the underlying photonic crystal is $a = 543$ nm. Assuming that light is launched into port 1, the fraction of the output power in ports 2 and 4 are calculated using Eq. (4.20),

$$P_2(x = L) = \cos^2\left[\frac{\pi L}{2L_c(f)}\right], \tag{4.24}$$

$$P_4(x = L) = \sin^2\left[\frac{\pi L}{2L_c(f)}\right], \tag{4.25}$$

where L is the length of the coupling region between the two waveguides, or the interaction length, and L_c is the coupling length calculated using Eq. (4.16). From Eqs. (4.24) and (4.25), it is clear that for a fixed interaction length L, the division of power between ports 2 and 4 is a strong function of the launched-signal frequency.

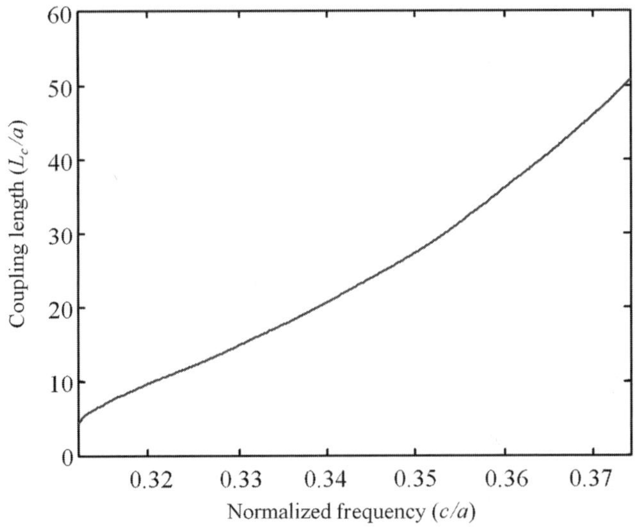

Figure 4.45 Coupling length versus frequency for the coupled photonic-crystal waveguide system shown in Fig. 4.41.

Furthermore, the wavelength sensitivity can be tuned in the design stage of the filter by varying the interaction length L, which is present in the numerator of the trigonometric-function arguments in Eqs. (4.24) and (4.25). For the interaction length $L = 28a$, the spectral response of the filter obtained using Eqs. (4.24) and (4.25) is shown in Fig. 4.46. This result has been verified with direct FDTD

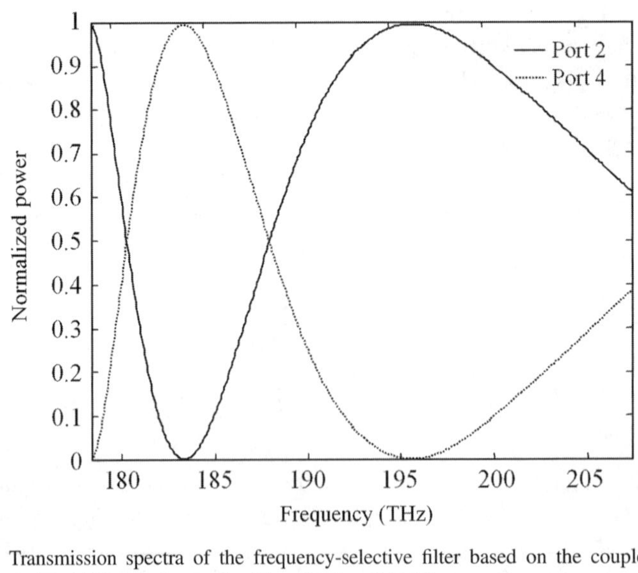

Figure 4.46 Transmission spectra of the frequency-selective filter based on the coupled-waveguides concept shown in Fig. 4.39 calculated using Eqs. (4.24) and (4.25).

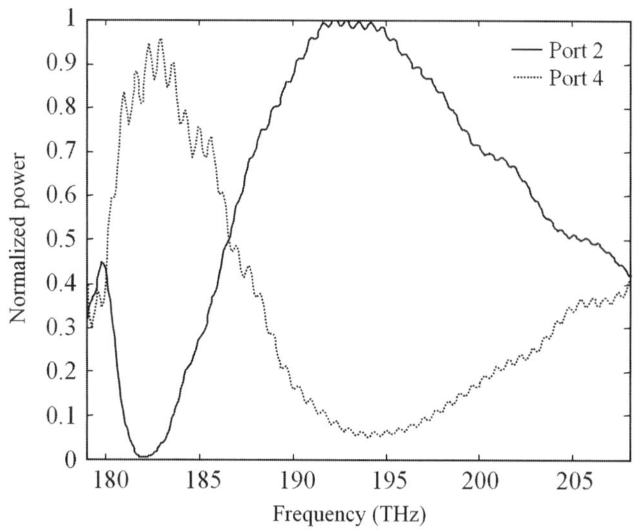

Figure 4.47 Transmission spectra of the system shown in Fig. 4.39 found using FDTD simulations with a broad-band input pulse launched into port 1.

simulations by launching a broadband pulse into port 1 and by observing the fields at ports 2 and 4. The spectral content of the output signals, which is shown in Fig. 4.47, agrees with the results obtained using coupled-waveguides theory of Fig. 4.46. The ripples observed in the graph are the result of imperfect transmission through bends in the waveguides, which are required to define the length of the interaction region, as well as reflections from the ends of the waveguides caused by imperfect absorption at PML boundary of the computational region. Based on these results, it is concluded that such device can be used as a broadband filter. In the case analyzed here, the optical signal at a frequency $f = 195$ THz ($\lambda = 1.54$ μm) is extracted at port 4, whereas the signal at a frequency $f = 184$ THz ($\lambda = 1.63$ μm) exits through port 2, with the beam launched into port 1.

Intuitively, the operation of the system of coupled waveguides as a frequency-selective filter can be understood by observing that a monochromatic wave launched into, say, port 1 will bounce back and forth between the waveguides before reaching the far end of the interaction region and exiting through either port 2 or 4. For a given interaction length, the number of bounces between the waveguides, and therefore the destination output port, is a function of frequency, because L_c is a function of frequency. As a result, an optical beam launched into port 1 will exit through either port 2 or 4 depending on its wavelength, which is a defining property of a frequency-selective filter.

By cascading such coupled waveguides, as shown in Fig. 4.48, the incoming signal can be directed into one of multiple channels depending on its frequency; hence, this device constitutes a multiple-wavelength-selective filter. Applying Eqs. (4.24) and (4.25) to the three interaction regions in Fig. 4.48 yields the following expressions

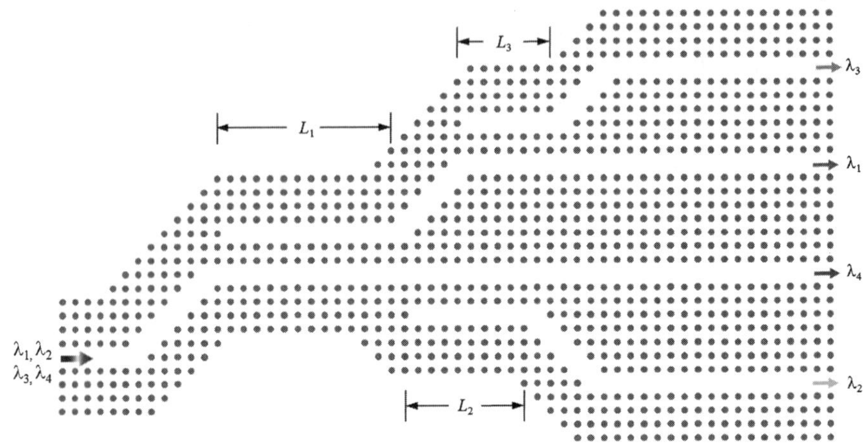

Figure 4.48 A system of coupled photonic-crystal waveguides engineered to perform as a four-channel frequency-selective filter.

for the (normalized) powers P_1, P_2, P_3, and P_4 in the four output ports:

$$P_1 = \sin^2\left(\frac{\pi L_1}{2L_c}\right)\cos^2\left(\frac{\pi L_3}{2L_c}\right), \tag{4.26}$$

$$P_2 = \cos^2\left(\frac{\pi L_1}{2L_c}\right)\sin^2\left(\frac{\pi L_2}{2L_c}\right), \tag{4.27}$$

$$P_3 = \sin^2\left(\frac{\pi L_1}{2L_c}\right)\sin^2\left(\frac{\pi L_3}{2L_c}\right), \tag{4.28}$$

$$P_4 = \cos^2\left(\frac{\pi L_1}{2L_c}\right)\cos^2\left(\frac{\pi L_2}{2L_c}\right), \tag{4.29}$$

The lengths of the coupling regions L_1, L_2, and L_3 are selected so as to extract desired frequencies at the respective output ports. An example of the spectral response of such a device is shown in Fig. 4.49, which was obtained using Eqs. (4.26) through (4.29). The wavelength selectivity of the four ports is evident and demonstrates the viability of using photonic-crystal coupled waveguides in filter design.

This approach to wavelength-selective filtering lends itself to photonic applications, which include wavelength-division multiplexing, such as wide wavelength division multiplexing (WWDM) and DWDM. In the former case, for 20-nm channel spacing, the required length of the coupling region is 182*a*, or 98.75 μm. For DWDM with 0.88-nm channel spacing, the length of the coupling region is 3780*a* or 2 mm, whereas 0.44-nm channel spacing requires 7560*a* or a 4-mm-long coupling region. For such small wavelength separations, the required interaction length is inversely proportional to the channel spacing.

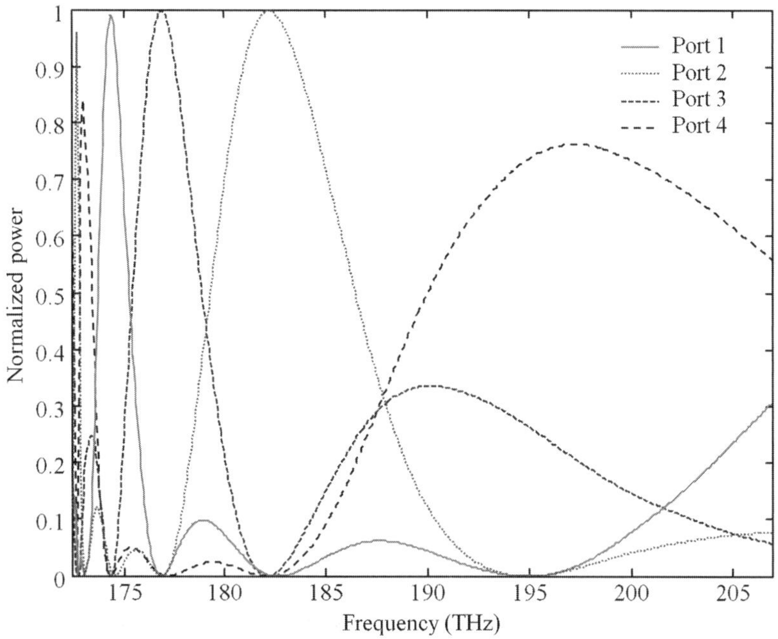

Figure 4.49 Transmission spectra of the system shown in Fig. 4.48.

Optical Switch. A crossbar switch, Fig. 4.50a, is a four-port device, with two input and two output ports that can exist in one of two states: the bar state, Fig. 4.50b, in which the signal from input ports 1 and 3 are sent to ports 2 and 4, respectively, and the cross state, Fig. 4.50c, in which the signal from the input port 1 is sent to output port 4, whereas the signal from the input port 3 is sent to output port 2.

As discussed earlier, the system of coupled waveguides shown in Fig. 4.39 has four optical ports, two input ports and two output ports, and the signals from the input ports can follow similar paths to those in the crossbar switch of Fig. 4.50. However, in the coupled-waveguides case, the signal destinations are determined by the geometry of the system, defined at the fabrication stage, and by the wavelength of the incoming signal. To make the device into a crossbar switch, it is necessary that external stimuli, not the input signal, control the destination ports. In other words, the parameters of the coupling regions must be modifiable while the device operates.

To determine what parameters are responsible for the state of the coupled-waveguides system, consider Eq. (4.22), which describes the distribution of the

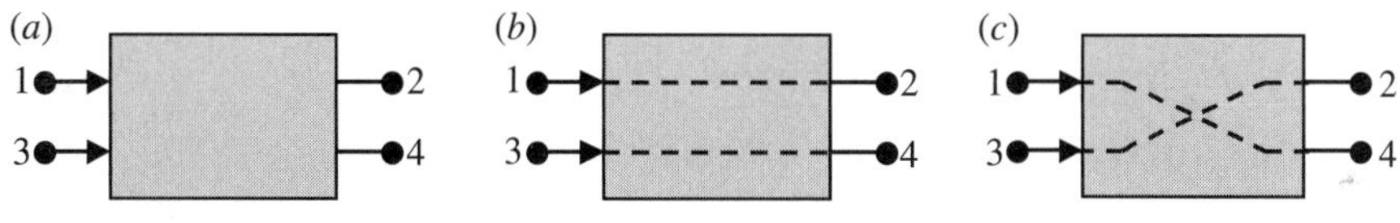

Figure 4.50 Crossbar switch with input ports 1 and 3 and output ports 2 and 4.

signal between the two waveguides of the coupled system. For the length of the coupling region L, and $\kappa = \kappa_{\text{bar}}$ such that $\sin(\kappa_{\text{bar}}L) = 0$, the off-diagonal elements of the matrix at the RHS of Eq. (4.22) vanish, and the equation becomes

$$\begin{bmatrix} F_{1k}(x = L) \\ F_{2k}(x = L) \end{bmatrix} = \pm e^{-jkx} \begin{bmatrix} 1 & 0 \\ 0 & 1 \end{bmatrix} \begin{bmatrix} u_{1k}(L) \\ u_{2k}(L) \end{bmatrix}, \tag{4.30}$$

which indicates that the signal launched into port 1 (waveguide 1) emerges at the end of the interaction region in waveguide 1 (port 2), and the signal launched into port 3 (waveguide 2) emerges at the end of the interaction region in the same waveguide 2 (port 4), just as in the bar state of the crossbar switch.

However, for the same interaction length L, but $\kappa = \kappa_X$ such that $\cos(\kappa_X L) = 0$, the diagonal elements of the matrix vanish, and the equation becomes

$$\begin{bmatrix} F_{1k}(x = L) \\ F_{2k}(x = L) \end{bmatrix} = \pm je^{-jkx} \begin{bmatrix} 0 & 1 \\ 1 & 0 \end{bmatrix} \begin{bmatrix} u_{1k}(L) \\ u_{2k}(L) \end{bmatrix}, \tag{4.31}$$

which indicates that the signals cross their paths; i.e., one that started in waveguide 1 at the beginning of the interaction region is in waveguide 2 at its end, and vice versa, which is similar to the cross state of the crossbar switch, Fig. 4.50c.

Thus, changing κ between κ_{bar} and κ_X switches between the bar and the cross states of the system. Such a change can be accomplished by modulating the dielectric properties of the materials, from which the PhC device is made. Recall that $\kappa = (k_e - k_o)/2$, and that the propagation constants of photonic-crystal waveguides can be very sensitive to changes in the refractive index of the constituent dielectric materials. Thus, a small change in the dielectric constant effected by, e.g., external electric field in electro-optic materials, can result in switching between the two states. Note also that $\sin(\kappa_{\text{bar}}L) = 0$ is equivalent to

$$\kappa_{\text{bar}}L = \pi\nu \Rightarrow \kappa_{\text{bar}} = \frac{\pi\nu}{L}, \tag{4.32}$$

and $\cos(\kappa_X L) = 0$ is equivalent to

$$\kappa_X L = \pi\left(\nu' + \frac{1}{2}\right) \Rightarrow \kappa_X = \frac{\pi(\nu' + 1/2)}{L}, \tag{4.33}$$

where ν and ν' are integers. Therefore, the change in κ required for switching

$$\Delta\kappa = \kappa_{\text{bar}} - \kappa_X = \frac{\pi}{L}(\nu - \nu' - 1/2) = \frac{\pi}{2L} \quad \text{for } \nu = \nu' \tag{4.34}$$

can be made arbitrarily small by increasing the interaction length L, at the cost of reducing the bandwidth of the device.

Another approach to switching in the coupled waveguides system is by suppressing the coupling between the waveguides. The concept is similar to conventional free-space optics where the reflectivity of a mirror is higher for increased conductivity of the mirror metal coating. Thus, it is expected that by sufficiently increasing the conductivity, or loss tangent, of the material in the coupling region between the waveguides, the region will act as a mirror to confine the wave and prevent its transition to the adjacent waveguide. In semiconductors, the change in conductivity can be controlled by (1) carrier injection from forward-biased p-n junctions on the rods, (2) depletion of doped rods with metal-oxide-semiconductor (MOS) gates, and (3) generation of electrons and holes by shining light upon the coupling region, where the energy of the photons used is higher than the bandgap of the semiconductor. To estimate how much the conductivity of a semiconductor can be varied, consider for example silicon. Because charge carriers produced using one of the methods listed above are no different than carriers produced by doping silicon with n-type or p-type materials, the graph shown in Fig. 4.51 provides a useful approximation of the range of values conductivity may assume.

To determine whether this variation of conductivity is sufficient for switching in the coupled-waveguides framework, consider the structure analyzed before, where the photonic crystal consists of silicon rods ($\varepsilon_r = 11.56$) arranged in a square lattice with pitch $a = 543$ nm, and the diameter of the rods is 217 nm. The coupled-waveguides system is that of Fig. 4.39, where the length of the interaction region $L = 28a$ is approximately equal to the coupling length L_c at the wavelength $\lambda = 1550$ nm.

Four snapshots of the coupled-waveguides system in operation, at different values of conductivity for the rods in the coupling region, are shown in Fig. 4.52. At low conductivity, Fig. 4.52a, the device behaves like a coupled-waveguides system discussed above, and the optical signal is completely transferred from one waveguide

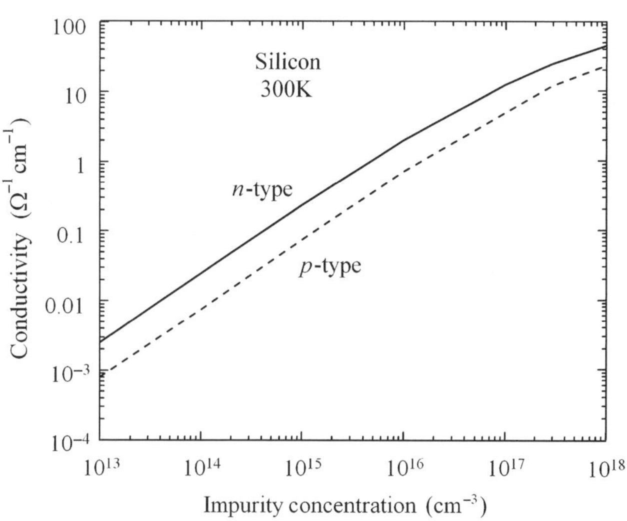

Figure 4.51 Dependence of conductivity σ on n- and p-doping concentrations.

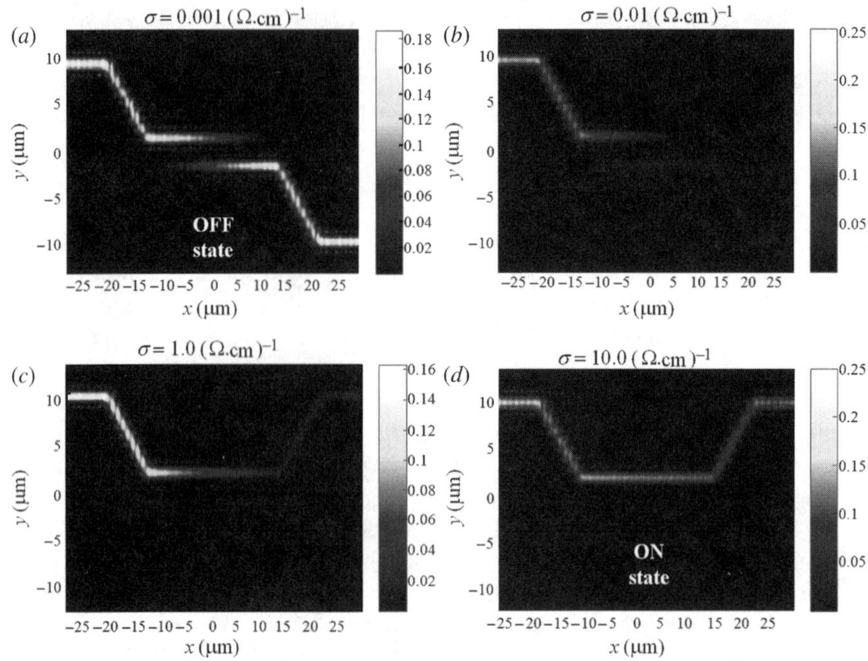

Figure 4.52 Four snapshots of FDTD simulations of the 2×2 electro-optical switch shown in Fig. 4.39. The change in conductivity within the practical range for silicon, as shown in Fig. 4.51, produces a transition between the cross state and the bar state of the switch.

to the other because the interaction length L is approximately equal to the coupling length L_c. At high conductivity, Fig. 4.52d, the signal is confined to the waveguide where it was launched. At intermediate values of conductivity, Fig. 4.52b and c, the signal is partially split and strongly attenuated by ohmic losses in the coupling region. More importantly, the change in conductivity σ between 0.001 $(\Omega\ \text{cm})^{-1}$ and 10.0 $(\Omega\ \text{cm})^{-1}$, which is within the limits shown in Fig. 4.51, results in a complete transition from the cross state to the bar state of the switch.

The operation of the switch is shown in more detail in Fig. 4.53, which presents the dependence of the power at the outputs as a function of conductivity of the rods in the interaction region. The switch is characterized by high transmission efficiency of nearly 100% in both cross and bar states, and by low cross talk, with the suppression in the range of 20 dB.

Splitter. Consider now the case of three coupled photonic-crystal waveguides as shown in Fig. 4.54a. The field distribution in this system can be expressed as a linear combination of three mutually orthogonal modes. The mirror symmetry of the structure with respect to the line running in the center of the middle waveguide allows for classifying the eigenmodes into symmetric, of which there are two, and antisymmetric, of which there is one. The eigenmodes are unchanged in shape as they propagate along the structure, although they pick up a relative phase determined

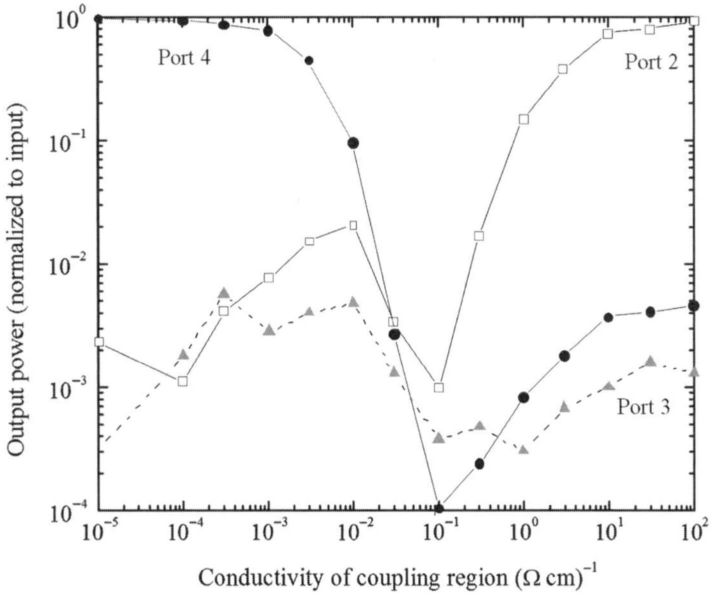

Figure 4.53 Switching behavior of the crossbar switch shown in Fig. 4.39. Light is launched into port 1.

by their respective propagation constants. As before, the hybrid modes with most of the field concentrated in any of the three coupled waveguides are obtained as linear combinations of the eigenmodes. These hybrid modes, in general, do not stay in the waveguide into which they are launched because of the difference in the propagation constants of the eigenmodes from which they are formed.

Consider, for instance, a monochromatic signal launched into the central waveguide. The resulting field configuration is symmetric; i.e., it has no antisymmetric component and therefore is a linear combination of the two symmetric eigenmodes. As in the case of two coupled waveguides where launching an unbalanced signal produced oscillations of power between the two constituent waveguides, the launching of light into the central waveguide produces oscillations that involve the two symmetric eigenmodes, with the coupling length inversely proportional to the difference in their propagation constants. In practical terms, it means that the signal originally in the center waveguide will transfer equally to the two side waveguides and then transfer back. These oscillations will repeat as long as the coupling between the waveguides continues.

Such behavior has been observed with the help of FDTD simulations. The structure and the snapshot of field distribution are shown in Fig. 4.54b, where a monochromatic wave was launched into the central waveguide.

Based on this behavior, the structure can be used for balanced splitting of optical signal into two. In this case, the side waveguides would need to separate from the central waveguide after an appropriate length, or the central waveguide could be simply terminated thereby eliminating the coupling.

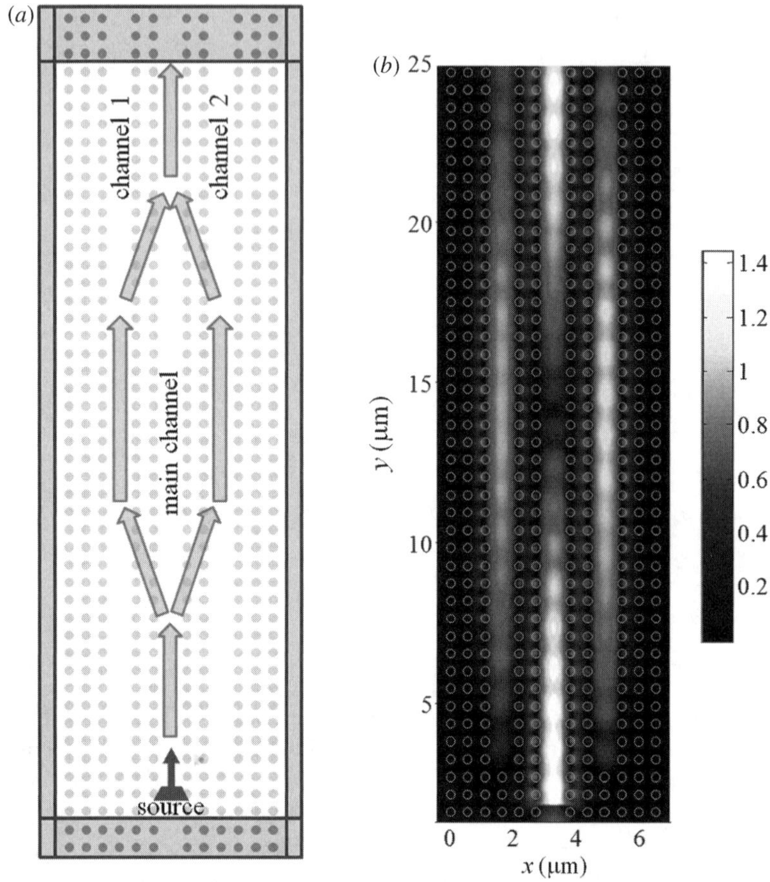

Figure 4.54 A coupled system of three PhC waveguides configured as a splitter and a combiner. (a) An overview of the device designed in a square lattice of silicon pillars ($\varepsilon_r = 11.56$, $r = 0.2a$) in air background. (b) An FDTD snapshot of the transient electromagnetic field propagating through the structure.

4.4.11 Other Applications of Photonic Bandgap

The designs that exploit the presence of photonic bandgap in photonic crystals discussed above constitute only a fraction of the number of applications contemplated for this phenomenon in the last two decades. Indeed, the field continues to expand, providing new and improved solutions for classic applications of conventional optics, as well as developing unique capabilities. Among such applications are:

- Control of spontaneous emission [89–91], which originally inspired photonic-bandgap materials and continues to be an area of vigorous research as new techniques to fabricate photonic crystals and probe them are developed [92–94].
- Enhancing the performance of antennas by mounting them on a photonic-bandgap substrate, thereby reducing surface waves and radiation into unwanted directions [95].

- Optical limiter, where a high-powered laser beam causes heat-induced change in the refractive index of one of the constituent materials. Such a change opens a bandgap at the laser wavelength and thereby attenuates the beam as it propagates through the photonic crystal [96].
- Coupled-cavity waveguides, where instead of a linear defect, a series of strongly coupled point defects are used to guide light along arbitrary trajectories in a photonic crystal [63, 97–99]. By properly designing the cavities and tuning the coupling between them, unique dispersive properties of such waveguides can be achieved.
- Optical isolator that takes advantage of nonreciprocity introduced by the application of an external magnetic field to route incident and reflected beams along different paths in the photonic crystal [100].

And there are many others. The authors hope that the unavoidably incomplete list provides the reader with a good starting point to explore the field, and that at the same time it gives a flavor of the wealth of applications in which the photonic-bandgap phenomenon can be applied.

REFERENCES

1. E. Yablonovitch, T. J. Gmitter, R. D. Meade, A. M. Rappe, K. D. Brommer, and J. D. Joannopoulos, "Donor and Acceptor Modes in Photonic Band-Structure," *Phys. Rev. Lett.*, vol. 67, pp. 3380–3383, 1991.
2. E. Yablonovitch, "Photonic Band-Gap Crystals," *J. Phys. Condens. Matter*, vol. 5, pp. 2443–2460, 1993.
3. R. P. Stanley, R. Houdre, U. Oesterle, and M. Ilegems, "Impurity Modes in One-Dimensional Periodic-Systems—the Transition from Photonic Band-Gaps to Microcavities," *Phys. Rev. A*, vol. 48, pp. 2246–2250, 1993.
4. V. Karathanos, A. Modinos, and N. Stefanou, "Planar Defects in Photonic Crystals," *J. Phys. Condens. Matter*, vol. 6, pp. 6257–6264, 1994.
5. P. R. Villeneuve and M. Piche, "Photonic Bandgaps in Periodic Dielectric Structures," *Prog. Quantum Electron.*, vol. 18, pp. 153–200, 1994.
6. D. Cassagne, A. Barra, and C. Jouanin, "Defects and Diffraction in Photonic Crystals," *Superlattice. Micros.*, vol. 25, pp. 343–346, 1999.
7. S. Y. Lin, J. G. Fleming, M. M. Sigalas, R. Biswas, and K. M. Ho, "Photonic Band-Gap Microcavities in Three Dimensions," *Phys. Rev. B*, vol. 59, pp. R15579–R15582, 1999.
8. R. D. Pradhan and G. H. Watson, "Impurity Effects in Coaxial-Connector Photonic Crystals: A Quasi-One-Dimensional Periodic System," *Phys. Rev. B*, vol. 60, pp. 2410–2415, 1999.
9. G. J. Schneider, S. Hanna, J. L. Davis, and G. H. Watson, "Defect Modes in Coaxial Photonic Crystals," *J. Appl. Phys.*, vol. 90, pp. 2642–2649, 2001.
10. K. Cho, T. Hirai, and T. Kawa, "Propagating Mode in the Photonic Gap of 1D Resonant Bragg Reflector," *J. Lumin.*, vol. 100, pp. 283–289, 2002.

11. T. Ueta, K. Ohtaka, N. Kawai, and K. Sakoda, "Limits on Quality Factors of Localized Defect Modes in Photonic Crystals due to Dielectric Loss," *J. Appl. Phys.*, vol. 84, pp. 6299–6304, 1998.
12. A. Yariv, *Optical Electronics in Modern Communications*, 5th ed. New York: Oxford University Press, 1997.
13. M. Agio, E. Lidorikis, and C. M. Soukoulis, "Impurity Modes in a Two-Dimensional Photonic Crystal: Coupling Efficiency and Q Factor," *J. Opt. Soc. Am. B*, vol. 17, pp. 2037–2042, 2000.
14. A. Mekis, J. C. Chen, I. Kurland, S. H. Fan, P. R. Villeneuve, and J. D. Joannopoulos, "High Transmission Through Sharp Bends in Photonic Crystal Waveguides," *Phys. Rev. Lett.*, vol. 77, pp. 3787–3790, 1996.
15. P. R. Villeneuve, S. H. Fan, J. D. Joannopoulos, K. Y. Lim, G. S. Petrich, L. A. Kolodziejski, and R. Reif, "Air-Bridge Microcavities," *Appl. Phys. Lett.*, vol. 67, pp. 167–169, 1995.
16. J. Rarity, C. Weisbuch, and North Atlantic Treaty Organization, Scientific Affairs Division, *Microcavities and Photonic Bandgaps: Physics and Applications*, Boston: Kluwer Academic Publishers, 1996.
17. B. D'Urso, O. Painter, J. O'Brien, T. Tombrello, A. Yariv, and A. Scherer, "Modal Reflectivity in Finite-Depth Two-Dimensional Photonic-Crystal Microcavities," *J. Opt. Soc. Am. B*, vol. 15, pp. 1155–1159, 1998.
18. J. Vučković, O. Painter, Y. Xu, A. Yariv, and A. Scherer, "Finite-Difference Time-Domain Calculation of the Spontaneous Emission Coupling Factor in Optical Microcavities," *IEEE J. Quantum Electron.*, vol. 35, pp. 1168–1175, 1999.
19. D. J. Ripin, K. Y. Lim, G. S. Petrich, P. R. Villeneuve, S. H. Fan, E. R. Thoen, J. D. Joannopoulos, E. P. Ippen, and L. A. Kolodziejski, "One-Dimensional Photonic Bandgap Microcavities for Strong Optical Confinement in GaAs and GaAs/AlxOy Semiconductor Waveguides," *J. Lightwave Technol.*, vol. 17, pp. 2152–2160, 1999.
20. P. Sabouroux, G. Tayeb, and D. Maystre, "Experimental and Theoretical Study of Resonant Microcavities in Two-Dimensional Photonic Crystals," *Opt. Commun.*, vol. 160, pp. 33–36, 1999.
21. R. D. Meade, A. Devenyi, J. D. Joannopoulos, O. L. Alerhand, D. A. Smith, and K. Kash, "Novel Applications of Photonic Band-Gap Materials—Low-Loss Bends and High Q-Cavities," *J. Appl. Phys.*, vol. 75, pp. 4753–4755, 1994.
22. G. S. Solomon, M. Pelton, and Y. Yamamoto, "Single-Mode Spontaneous Emission from a Single Quantum Dot in a Three-Dimensional Microcavity," *Phys. Rev. Lett.*, vol. 86, pp. 3903–3906, 2001.
23. C. J. M. Smith, H. Benisty, S. Olivier, M. Rattier, C. Weisbuch, T. F. Krauss, R. M. De La Rue, R. Houdre, and U. Oesterle, "Low-Loss Channel Waveguides with Two-Dimensional Photonic Crystal Boundaries," *Appl. Phys. Lett.*, vol. 77, pp. 2813–2815, 2000.
24. J. Arentoft, T. Sondergaard, M. Kristensen, A. Boltasseva, M. Thorhauge, and L. Frandsen, "Low-Loss Silicon-on-Insulator Photonic Crystal Waveguides," *Electron. Lett.*, vol. 38, pp. 274–275, 2002.
25. Y. Desieres, T. Benyattou, R. Orobtchouk, A. Morand, P. Benech, C. Grillet, C. Seassal, X. Letartre, P. Rojo-Romeo, and P. Viktorovitch, "Propagation Losses of the Fundamental Mode in a Single Line-Defect Photonic Crystal Waveguide on an InP Membrane," *J. Appl. Phys.*, vol. 92, pp. 2227–2234, 2002.

26. M. Lončar, D. Nedeljkovic, T. P. Pearsall, J. Vučković, A. Scherer, S. Kuchinsky, and D. C. Allan, "Experimental and Theoretical Confirmation of Bloch-Mode Light Propagation in Planar Photonic Crystal Waveguides," *Appl. Phys. Lett.*, vol. 80, pp. 1689–1691, 2002.
27. J. M. Lourtioz, H. Benisty, A. Chelnokov, S. David, and S. Olivier, "Photonic Crystals and the Real World of Optical Telecommunications," *Ann. Telecommun.*, vol. 58, pp. 1197–1237, 2003.
28. M. J. Adams, *An Introduction to Optical Waveguides*. New York: Wiley, 1981.
29. M. Koshiba, *Optical Waveguide Theory by the Finite Element Method*. Boston: Kluwer Academic, 1993.
30. D. W. Prather, A. Sharkawy, and S. Shi, "Design and Applications of Photonic Crystals," in W. A. Goddard III, D. W. Brenner, S. E. Lyshevski, and G. J. Iafrate, Eds. *Handbook of Nanoscience, Engineering, and Technology*, Boca Raton, FL: CRC Press, 2002, pp. 211–232.
31. S. H. Fan, J. N. Winn, A. Devenyi, J. C. Chen, R. D. Meade, and J. D. Joannopoulos, "Guided and Defect Modes in Periodic Dielectric Wave-Guides," *J. Opt. Soc. Am. B*, vol. 12, pp. 1267–1272, 1995.
32. M. D. B. Charlton, S. W. Roberts, and G. J. Parker, "Guided Mode Analysis, and Fabrication of a 2-Dimensional Visible Photonic Band Structure Confined within a Planar Semiconductor Waveguide," *Mat. Sci. Eng. B*, vol. 49, pp. 155–165, 1997.
33. J. S. Foresi, P. R. Villeneuve, J. Ferrera, E. R. Thoen, G. Steinmeyer, S. Fan, J. D. Joannopoulos, L. C. Kimerling, H. I. Smith, and E. P. Ippen, "Photonic-Bandgap Microcavities in Optical Waveguides," *Nature*, vol. 390, pp. 143–145, 1997.
34. M. D. B. Charlton and G. J. Parker, "Nanofabrication of Advanced Waveguide Structures Incorporating a Visible Photonic Band Gap," *J. Micromech. Microeng.*, vol. 8, pp. 172–176, 1998.
35. M. D. B. Charlton, G. J. Parker, and M. E. Zoorob, "Recent Developments in the Design and Fabrication of Visible Photonic Band Gap Waveguide Devices," *J. Mat. Sci. Mat. Electron.*, vol. 10, pp. 429–440, 1999.
36. I. El-Kady, M. M. Sigalas, R. Biswas, and K. M. Ho, "Dielectric Waveguides in Two-Dimensional Photonic Bandgap Materials," *J. Lightwave Technol.*, vol. 17, pp. 2042–2049, 1999.
37. K. Sakoda, *Optical Properties of Photonic Crystals*. New York: Springer, 2001.
38. K. Sakoda, "Optics of Photonic Crystals," *Opt. Rev.*, vol. 6, pp. 381–392, 1999.
39. A. R. McGurn, "Photonic Crystal Circuits," *Physica B*, vol. 296, pp. 201–209, 2001.
40. S. Y. Lin, E. Chow, V. Hietala, P. R. Villeneuve, and J. D. Joannopoulos, "Experimental Demonstration of Guiding and Bending of Electromagnetic Waves in a Photonic Crystal," *Science*, vol. 282, pp. 274–276, 1998.
41. A. Talneau, L. Le Gouezigou, and N. Bouadma, "Quantitative Measurement of Low Propagation Losses at 1.55 μm on Planar Photonic Crystal Waveguides," *Opt. Lett.*, vol. 26, pp. 1259–1261, 2001.
42. E. Chow, S. Y. Lin, J. R. Wendt, S. G. Johnson, and J. D. Joannopoulos, "Quantitative Analysis of Bending Efficiency in Photonic-Crystal Waveguide Bends at $\lambda = 1.55$ μm Wavelengths," *Opt. Lett.*, vol. 26, pp. 286–288, 2001.
43. A. Yariv and P. Yeh, *Optical Waves in Crystals: Propagation and Control of Laser Radiation*, Hoboken, N.J.: Wiley, 2003.

44. O. J. Painter, A. Husain, A. Scherer, J. D. O'Brien, I. Kim, and P. D. Dapkus, "Room Temperature Photonic Crystal Defect Lasers at Near-Infrared Wavelengths in InGaAsP," *J. Lightwave Technol.*, vol. 17, pp. 2082–2088, 1999.

45. A. Chutinan and S. Noda, "Waveguides and Waveguide Bends in Two-Dimensional Photonic Crystal Slabs," *Phys. Rev. B*, vol. 62, pp. 4488–4492, 2000.

46. P. Lalanne, "Effective Medium Theory Applied to Photonic Crystals Composed of Cubic or Square Cylinders," *Appl. Opt.*, vol. 35, pp. 5369–5380, 1996.

47. T. Baba, N. Fukaya, and J. Yonekura, "Observation of Light Propagation in Photonic Crystal Optical Waveguides with Bends," *Electron. Lett.*, vol. 35, pp. 654–655, 1999.

48. A. Chutinan and S. Noda, "Analysis of Waveguides and Waveguide Bends in Photonic Crystal Slabs with Triangular Lattice," *Jap. J. Appl. Phys. Part 2*, vol. 39, pp. L595–L596, 2000.

49. N. Fukaya, D. Ohsaki, and T. Baba, "Two-dimensional Photonic Crystal Waveguides with 60 Degrees Bends in a Thin Slab Structure," *Jap. J. Appl. Phys. 1*, vol. 39, pp. 2619–2623, 2000.

50. S. Y. Lin, E. Chow, S. G. Johnson, and J. D. Joannopoulos, "Demonstration of Highly Efficient Waveguiding in a Photonic Crystal Slab at the 1.5-μm Wavelength," *Opt. Lett.*, vol. 25, pp. 1297–1299, 2000.

51. M. Tokushima, H. Kosaka, A. Tomita, and H. Yamada, "Lightwave Propagation Through a 120 Degrees Sharply Bent Single-Line-Defect Photonic Crystal Waveguide," *Appl. Phys. Lett.*, vol. 76, pp. 952–954, 2000.

52. Y. S. Kivshar, P. G. Kevrekidis, and S. Takeno, "Nonlinear Localized Modes in Waveguide Bends," *Phys. Lett. A*, vol. 307, pp. 287–291, 2003.

53. K. Miura and Y. Ohtera, "Reduction of Propagation and Bending Losses of Heterostructured Photonic Crystal Waveguides by Use of a High-Delta Structure," *Opt. Lett.*, vol. 28, pp. 734–736, 2003.

54. S. Olivier, H. Benisty, M. Rattier, C. Weisbuch, M. Qiu, A. Karlsson, C. J. M. Smith, R. Houdre, and U. Oesterle, "Resonant and Nonresonant Transmission Through Waveguide Bends in a Planar Photonic Crystal," *Appl. Phys. Lett.*, vol. 79, pp. 2514–2516, 2001.

55. S. Olivier, H. Benisty, C. Weisbuch, C. J. M. Smith, T. F. Krauss, R. Houdre, and U. Oesterle, "Improved 60 Degrees Bend Transmission of Submicron-Width Waveguides Defined in Two-Dimensional Photonic Crystals," *J. Lightwave Technol.*, vol. 20, pp. 1198–1203, 2002.

56. J. Moosburger, M. Kamp, A. Forchel, S. Olivier, H. Benisty, C. Weisbuch, and U. Oesterle, "Enhanced Transmission Through Photonic-Crystal-Based Bent Waveguides by Bend Engineering," *Appl. Phys. Lett.*, vol. 79, pp. 3579–3581, 2001.

57. A. Chutinan, M. Okano, and S. Noda, "Wider Bandwidth with High Transmission Through Waveguide Bends in Two-Dimensional Photonic Crystal Slabs," *Appl. Phys. Lett.*, vol. 80, pp. 1698–1700, 2002.

58. A. Sharkawy, D. Pustai, S. Y. Shi, and D. W. Prather, "High Transmission Through Waveguide Bends by Use of Polycrystalline Photonic-Crystal Structures," *Opt. Lett.*, vol. 28, pp. 1197–1199, 2003.

59. B. L. Miao, C. H. Chen, S. Y. Shi, J. Murakowski, and D. W. Prather, "High-Efficiency Broad-Band Transmission Through a Double-60 Degrees Bend in a Planar Photonic

Crystal Single-Line Defect Waveguide," *IEEE Photon. Technol. Lett.*, vol. 16, pp. 2469–2471, 2004.

60. S. G. Johnson, C. Manolatou, S. H. Fan, P. R. Villeneuve, J. D. Joannopoulos, and H. A. Haus, "Elimination of Cross Talk in Waveguide Intersections," *Opt. Lett.*, vol. 23, pp. 1855–1857, 1998.
61. S. H. Fan, S. G. Johnson, J. D. Joannopoulos, C. Manolatou, and H. A. Haus, "Waveguide Branches in Photonic Crystals," *J. Opt. Soc. Am. B*, vol. 18, pp. 162–165, 2001.
62. T. Sondergaard and K. H. Dridi, "Energy Flow in Photonic Crystal Waveguides," *Phys. Rev. B*, vol. 61, pp. 15688–15696, 2000.
63. M. Bayindir, B. Temelkuran, and E. Özbay, "Photonic-Crystal-Based Beam Splitters," *Appl. Phys. Lett.*, vol. 77, pp. 3902–3904, 2000.
64. A. Taflove and S. C. Hagness, *Computational Electrodynamics: The Finite-Difference Time-Domain Method*, 2nd ed. Boston: Artech House, 2000.
65. S. H. Fan, P. R. Villeneuve, and J. D. Joannopoulos, "Channel Drop Tunneling Through Localized States," *Phys. Rev. Lett.*, vol. 80, pp. 960–963, 1998.
66. S. H. Fan, P. R. Villeneuve, J. D. Joannopoulos, and H. A. Haus, "Channel Drop Filters in Photonic Crystals," *Opt. Express*, vol. 3, pp. 4–11, 1998.
67. S. H. Fan, P. R. Villeneuve, J. D. Joannopoulos, M. J. Khan, C. Manolatou, and H. A. Haus, "Theoretical Analysis of Channel Drop Tunneling Processes," *Phys. Rev. B*, vol. 59, pp. 15882–15892, 1999.
68. J. Fu and S. L. He, "Analysis of Higher Order Channel Dropping Tunneling Processes in Photonic Crystals," *Chin. Phys. Lett.*, vol. 17, pp. 737–739, 2000.
69. E. Centeno, B. Guizal, and D. Felbacq, "Multiplexing and Demultiplexing with Photonic Crystals," *J. Opt. A*, vol. 1, pp. L10–L13, 1999.
70. M. Imada, S. Noda, A. Chutinan, M. Mochizuki, and T. Tanaka, "Channel Drop Filter Using a Single Defect in a 2-D Photonic Crystal Slab Waveguide," *J. Lightwave Technol.*, vol. 20, pp. 845–850, 2002.
71. H. Takahashi, S. Suzuki, and I. Nishi, "Wavelength Multiplexer Based on SiO_2-Ta_2O_5 Arrayed-Wave-Guide Grating," *J. Lightwave Technol.*, vol. 12, pp. 989–995, 1994.
72. C. Dragone, "Efficient $N \times N$ Star Couplers Using Fourier Optics," *J. Lightwave Technol.*, vol. 7, pp. 479–489, 1989.
73. K. O. Hill, Y. Fujii, D. C. Johnson, and B. S. Kawasaki, "Photosensitivity in Optical Fiber Waveguides—Application to Reflection Filter Fabrication," *Appl. Phys. Lett.*, vol. 32, pp. 647–649, 1978.
74. H. Kosaka, T. Kawashima, A. Tomita, M. Notomi, T. Tamamura, T. Sato, and S. Kawakami, "Photonic Crystals for Micro Lightwave Circuits Using Wavelength-Dependent Angular Beam Steering," *Appl. Phys. Lett.*, vol. 74, pp. 1370–1372, 1999.
75. M. Koshiba, "Wavelength Division Multiplexing and Demultiplexing with Photonic Crystal Waveguide Couplers," *J. Lightwave Technol.*, vol. 19, pp. 1970–1975, 2001.
76. A. Sharkawy, S. Y. Shi, and D. W. Prather, "Multichannel Wavelength Division Multiplexing with Photonic Crystals," *Appl. Opt.*, vol. 40, pp. 2247–2252, 2001.
77. H. Kosaka, T. Kawashima, A. Tomita, M. Notomi, T. Tamamura, T. Sato, and S. Kawakami, "Superprism Phenomena in Photonic Crystals," *Phys. Rev. B*, vol. 58, pp. R10096–R10099, 1998.

78. H. Kosaka, T. Kawashima, A. Tomita, M. Notomi, T. Tamamura, T. Sato, and S. Kawakami, "Superprism Phenomena in Photonic Crystals: Toward Microscale Lightwave Circuits," *J. Lightwave Technol.*, vol. 17, pp. 2032–2038, 1999.
79. K. Busch, "Photonic Band Structure Theory: Assessment and Perspectives," *Comptes Rendus Physique*, vol. 3, pp. 53–66, 2002.
80. T. Baba and M. Nakamura, "Photonic Crystal Light Deflection Devices Using the Superprism Effect," *IEEE J. Quantum Electron.*, vol. 38, pp. 909–914, 2002.
81. B. Gralak, S. Enoch, and G. Tayeb, "Anomalous Refractive Properties of Photonic Crystals," *J. Opt. Soc. Am. A*, vol. 17, pp. 1012–1020, 2000.
82. A. Sharkawy, D. W. Prather, and S. Shi, "Photonic Crystal Multi-Channel Drop Filter and Hetero-Structure Beam Splitter," *Proc. SPIE*, vol. 4290, pp. 62–71, 2001.
83. A. Sharkawy, S. Shi, and D. W. Prather, "Heterostructure Photonic Crystals: Theory and Applications," *Appl. Opt.*, vol. 41, pp. 7245–7253, 2002.
84. P. Halevi and E. Ramos-Mendieta, "Tunable Photonic Crystals with Semiconducting Constituents," *Phys. Rev. Lett.*, vol. 85, pp. 1875–1878, 2000.
85. S. W. Leonard, J. P. Mondia, H. M. van Driel, O. Toader, S. John, K. Busch, A. Birner, U. Gosele, and V. Lehmann, "Tunable Two-Dimensional Photonic Crystals Using Liquid-Crystal Infiltration," *Phys. Rev. B*, vol. 61, pp. R2389–R2392, 2000.
86. D. Kang, J. E. Maclennan, N. A. Clark, A. A. Zakhidov, and R. H. Baughman, "Electro-Optic Behavior of Liquid-Crystal-Filled Silica Opal Photonic Crystals: Effect of Liquid-Crystal Alignment," *Phys. Rev. Lett.*, vol. 86, pp. 4052–4055, 2001.
87. A. Sharkawy, S. Shi, and D. W. Prather, "Optical Networks on a Chip Using Photonic Bandgap Materials," *Proc. SPIE*, vol. 4436, pp. 179–189, 2001.
88. L. L. Liou and A. Crespo, "Dielectric Optical Waveguide Coupling Analysis Using Two-Dimensional Finite-Difference in Time-Domain Simulations," *Microwave Opt. Technol. Lett.*, vol. 26, pp. 234–237, 2000.
89. V. Weisskopf and E. Wigner, "Berechnung der Naturlichen Linienbreite auf Grund der Diracschen Lichttheorie," *Z. Phys.*, vol. 63, pp. 54–73, 1930.
90. D. Kleppner, "Inhibited Spontaneous Emission," *Phys. Rev. Lett.*, vol. 47, pp. 233–236, 1981.
91. E. M. Purcell, "Spontaneous Emission Probabilities at Radio Frequencies," *Phys. Rev.*, vol. 69, p. 681, 1946.
92. E. Yablonovitch, "Inhibited Spontaneous Emission in Solid-State Physics and Electronics," *Phys. Rev. Lett.*, vol. 58, pp. 2059–2062, 1987.
93. M. Boroditsky, T. F. Krauss, R. Coccioli, R. Vrijen, R. Bhat, and E. Yablonovitch, "Light Extraction from Optically Pumped Light-Emitting Diode by Thin-Slab Photonic Crystals," *Appl. Phys. Lett.*, vol. 75, pp. 1036–1038, 1999.
94. M. Boroditsky, R. Vrijen, T. F. Krauss, R. Coccioli, R. Bhat, and E. Yablonovitch, "Spontaneous Emission Extraction and Purcell Enhancement from Thin-Film 2-D Photonic Crystals," *J. Lightwave Technol.*, vol. 17, pp. 2096–2112, 1999.
95. R. Gonzalo, P. de Maagt, and M. Sorolla, "Enhanced Patch-Antenna Performance by Suppressing Surface Waves Using Photonic-Bandgap Substrates," *IEEE Trans. Microwave Theory Technol.*, vol. 47, pp. 2131–2138, 1999.
96. H. B. Lin, R. J. Tonucci, and A. J. Campillo, "Two-Dimensional Photonic Bandgap Optical Limiter in the Visible," *Opt. Lett.*, vol. 23, pp. 94–96, 1998.

97. A. M. Zheltikov, N. I. Koroteev, S. A. Magnitsky, and A. V. Tarasishin, "Self-Phase Modulation and Compression of Short Laser Pulses in a Nonlinear Photonic Crystal," *Izvestiya Akademii Nauk Seriya Fizicheskaya*, vol. 63, pp. 717–724, 1999.
98. M. Bayindir and E. Özbay, "Heavy Photons at Coupled-Cavity Waveguide Band Edges in a Three-Dimensional Photonic Crystal," *Phys. Rev. B*, vol. 62, pp. R2247–R2250, 2000.
99. M. Bayindir, B. Temelkuran, and E. Özbay, "Propagation of Photons by Hopping: A Waveguiding Mechanism Through Localized Coupled Cavities in Three-Dimensional Photonic Crystals," *Phys. Rev. B*, vol. 61, pp. 11855–11858, 2000.
100. Z. Wang and S. H. Fan, "Optical Circulators in Two-Dimensional Magneto-Optical Photonic Crystals," *Opt. Lett.*, vol. 30, pp. 1989–1991, 2005.

5

Engineering Photonic-Crystal Dispersion Properties

5.1 INTRODUCTION

Photonic bandgaps provide the means for strong confinement of light in photonic crystals (PhCs). Such confinement was used in the devices described in Chapter 4 to realize waveguides, resonators, and other defect-based devices at scales comparable with the wavelength of light. However, photonic crystals can affect the behavior of light in a nontrivial manner even in the absence of photonic bandgap, or defects, thanks to a wealth of phenomena that ensue from the coherent interaction of an electromagnetic wave with a periodic structure. To study and understand these interactions, a conceptual tool in the form of a dispersion diagram is often employed. Borrowed from the field of solid-state physics, which deals with periodic systems on the atomic scale, a dispersion diagram shows the functional dependence of the frequency/energy of a wave propagating in a crystal on its corresponding wave vector/(pseudo)momentum. Band diagrams and dispersion relations were discussed previously in Chapter 2, which outlines the theoretical foundations and conceptual utility of the dispersion diagram. These diagrams were also discussed extensively in Chapters 3 and 4 when characterizing numerical methods for analyzing PhCs and studying the operation of photonic-bandgap-based devices, respectively. Some examples of band diagrams were presented in Chapter 4 where bandgap-based resonators and waveguides were analyzed. Although in Chapter 4 the main focus was on the bandgap, i.e., the range of frequencies where propagating electromagnetic waves cannot exist (absent defects) in the crystal, in this chapter, the allowed spectral region is investigated. The importance of the dispersion diagram in such investigations stems from the notion of group velocity,

$$\mathbf{v}_g = \nabla_k \omega, \tag{5.1}$$

encountered already in Chapters 2 and 4, which describes how a wave packet and the energy it carries propagate in a medium. According to Eq. (5.1), group velocity is a vector pointing in the direction of the steepest ascent of the function $\omega = \omega(\mathbf{k})$. As

---

Photonic Crystals: Theory, Applications, and Fabrication. By D. W. Prather, S. Shi, A. Sharkawy, J. Murakowski, and G. J. Schneider

such, it is perpendicular to the equifrequency surface [or contour for a two-dimensional (2D) system] in k-space. It will be shown in this chapter that photonic crystals admit a variety of shapes for equifrequency contours, as well as dispersion surfaces and, therefore, correspondingly varied wave-propagation phenomena. This is in contrast to the dispersion surfaces in conventional materials, whose limited shapes are described in Chapter 2. The growing interest in the dispersion characteristics of photonic crystals bore fruit in the discovery of the superprism effect [1], self-collimation [2–4], negative refractive index [5–10], and left-handed materials [11–13].

The dispersion properties of photonic crystals can be exploited to control precisely the propagation of electromagnetic waves. As a result, an alternative exists to conventional defect-based PhC waveguides, in the form of devices that require no defects for confinement and yet can arbitrarily route light. As will be shown later in this chapter, such devices are capable of exceptionally low propagation loss when implemented in slab photonic crystals, mainly thanks to low out-of-plane scattering because the propagation mode resides below the light line. At the same time, wavelength discrimination, which in the photonic-bandgap–based devices is realized using point-defect cavities, can be obtained by means of wavelength-dependent angular beam steering [14] in a properly engineered photonic crystal.

In this chapter, phenomena related to the unique dispersion characteristics of photonic crystals are discussed, as well as ways these phenomena are exploited to realize microphotonic devices. Thus, in Section 5.2, the concept of the dispersion diagram is explored in more detail. The emergence of superprism and self-collimation effects based on the shapes of equifrequency contours in the 2D case, and equifrequency surfaces in the three-dimensional (3D) case, is discussed in Sections 5.3 and 5.4. Examples of photonic crystals that exhibit self-collimating behavior are provided, along with their experimental characterization and applications to beam routing, splitting, and reconfigurable optical switching. Section 5.5 deals with the left-handed behavior of light in photonic crystals, where the Poynting vector and phase velocity point in opposite directions. Negative refraction that results from such left-handedness is used in Section 5.5.1 for super-resolution imaging in three dimensions. The chapter is concluded by examining a PhC that exhibits both left-handedness and self-collimation, as well as the high sensitivity of the refraction angle to the angle of incidence—the hallmark of a superprism.

5.2 DISPERSION IN PHOTONIC CRYSTALS

In linear and homogeneous, but not necessarily isotropic, materials, the interaction of the medium with the electromagnetic wave is described by the dielectric tensor, which leads to the general anisotropic dispersion surface, as discussed in Chapter 2. As the frequency is varied, the shape of the dispersion surface is unaffected for ordinary, lossless, homogeneous media, as shown in Fig. 5.1. However, as mentioned in Section 5.1, the dispersion of photonic crystals can be considerably more varied than that of the materials found in nature. Figures 5.2 and 5.3 show examples of dispersion diagrams for several two-dimensional photonic crystals. Each diagram comprises several (in principle, an infinite number of) surfaces stacked vertically in the ω direction

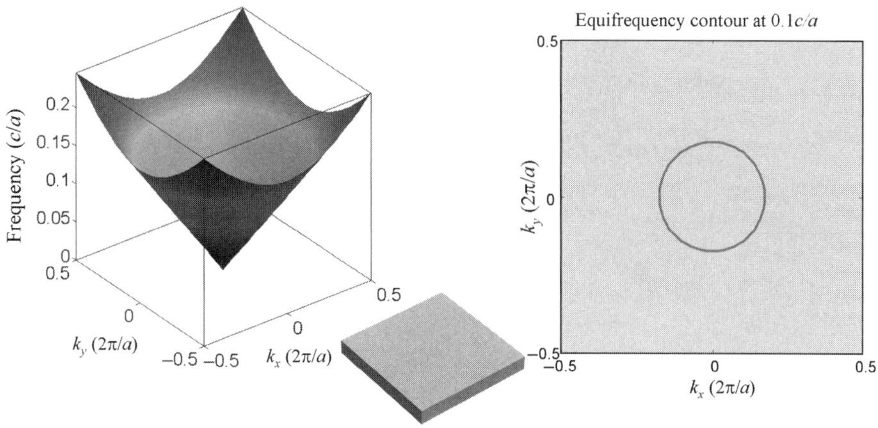

Figure 5.1 Dispersion of a homogeneous unpatterned slab. The dispersion surface is approximately conical, and the isotropy of the material manifests itself as a circular equifrequency contour.

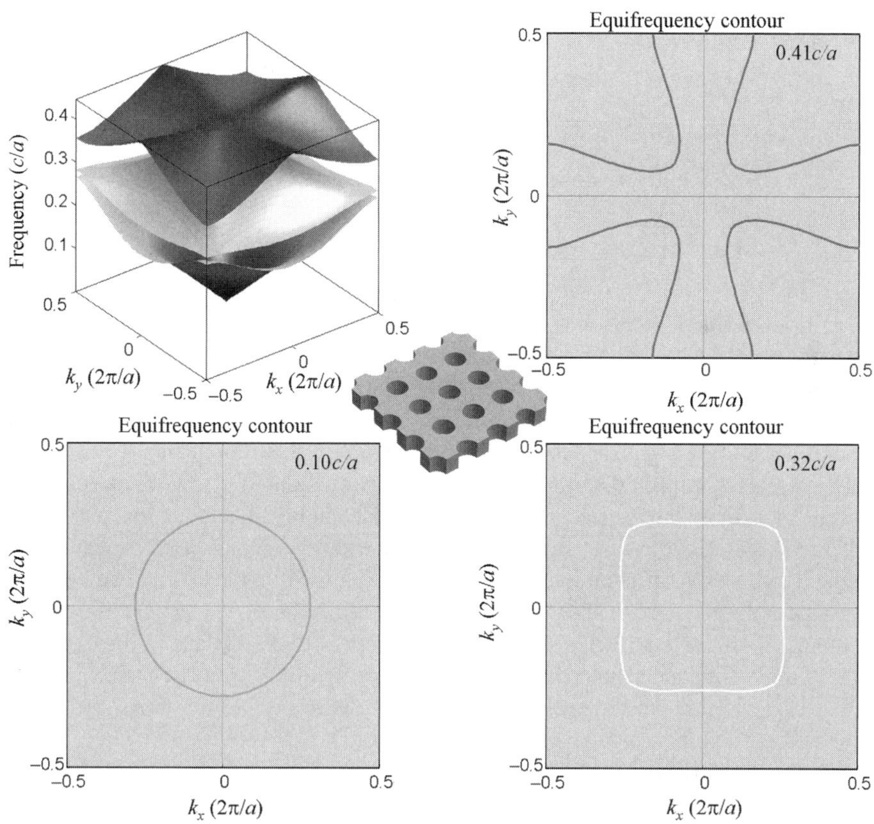

Figure 5.2 Dispersion of a homogenous square-lattice photonic-crystal slab. The dispersion surfaces depict the functional dependence of the frequency of modes that propagate in the corresponding PhC on the wave vectors. Equifrequency contours are obtained by intersecting a constant-frequency plane with the dispersion surface.

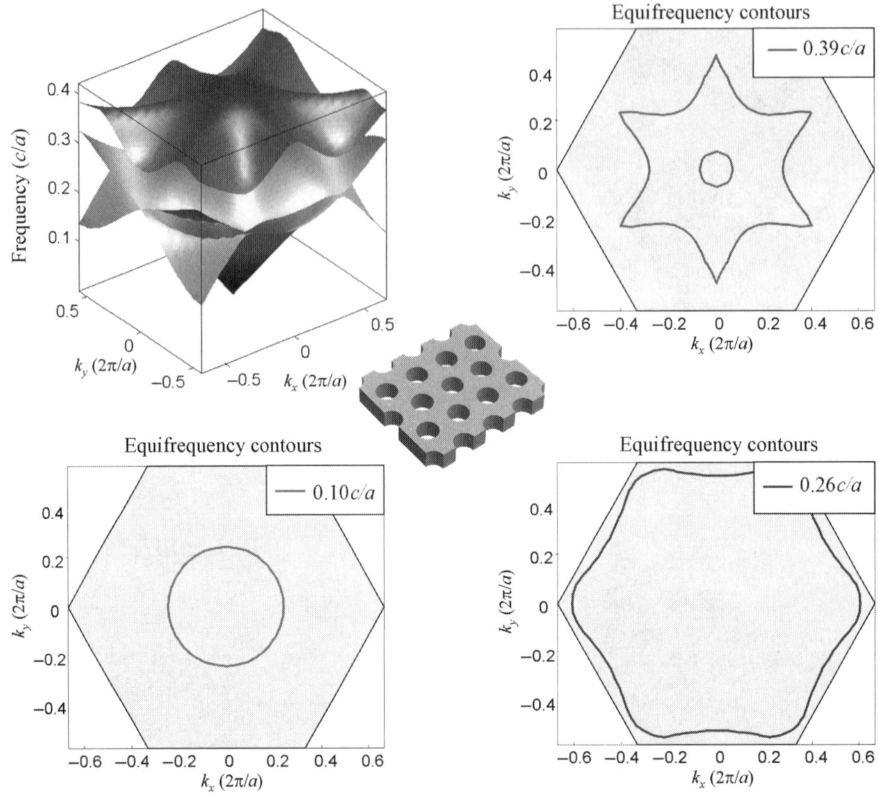

Figure 5.3 Dispersion of a homogenous triangular-lattice photonic-crystal slab. The dispersion surfaces depict the functional dependence of the frequency of modes that propagate in the corresponding PhC on the wave vectors. Equifrequency contours are obtained by intersecting a constant-frequency plane with the dispersion surface. Shaded gray is the first Brillouin zone.

because of folding of the dispersion relations into the first Brillouin zone (see Chapter 2). Thus, the dispersion diagram for a two-dimensional photonic crystal consists of a family of (dispersion) surfaces that represent the multivalued function $\omega = \omega(\mathbf{k})$, where the intersections of a constant-k line with the different surfaces correspond to particular modes at different frequencies that can exist within the lattice. The intersection of a constant-ω plane with a dispersion surface is referred to as an equifrequency contour (EFC). Several such contours are projected onto the k-plane in Figs. 5.2 and 5.3. In three dimensions, the contours become equifrequency surfaces (EFS), and in either case, they characterize the relationship between all allowed wave vectors in the structure and their corresponding frequencies.

To obtain such a rendering, it is sufficient to compute the eigenfrequencies for wave vectors at all k-points within, in addition to those on the boundary of, the *irreducible* Brillouin zone. By exploiting the lattice symmetries, the dispersion surface for the entire Brillouin zone is obtained.

It should be noted that the EFCs displayed in Figs. 5.2 and 5.3 may take a variety of shapes, including square-like with four sides almost flat as well as star-like, in which sharp corners are present. Furthermore, the slope of the dispersion surface can be either positive or negative with respect to the magnitude of the wave vector **k**. These observations will have important consequences in the applications discussed in the sections that follow where the straight sections of an EFC produce self-collimation and negative slope leads to left-handed behavior.

5.3 SUPERPRISM EFFECT

When a beam of light crosses the interface between two homogeneous dielectrics, it undergoes refraction governed by Snell's law

$$n_1 \sin(\alpha_1) = n_2 \sin(\alpha_2), \tag{5.2}$$

where α_1 and α_2 are the incident and refracted angles, and n_1 and n_2 are the refractive indices of the respective two materials. At optical frequencies, the refractive index of ordinary material is limited to about 4 or less. As a result, to obtain high variation of the refracted angle with small change of the incident angle, it is necessary to approach the interface from the high-index side and to work near the critical incident angle. Unfortunately, this last condition limits the utility of a device based on this principle because near the critical angle the transmission of light across the interface is diminished, and the beam energy is reflected back.

In contrast, thanks to the variety of dispersive properties exhibited by photonic crystals, as discussed briefly in the previous section and exemplified by the dispersion diagrams in Figs. 5.2 and 5.3, the limitation of Snell's law (5.2) no longer applies. The following example shows how in this context, high variation of the refracted angle is possible at near-normal incidence.

Consider a two-dimensional photonic crystal that consists of circular air holes drilled in dielectric with a permittivity of $\varepsilon_r = 8$ ($n = 2.828$) and arranged in a triangular lattice such that $r/a = 0.24$, where r is the hole radius and a is the lattice constant. The equifrequency contours for this structure at normalized frequency 0.4 are shown in Fig. 5.4. Assume that the photonic crystal occupies the lower half of the xy plane ($y < 0$), such that the interface between free space and this photonic crystal is along the x direction, and that the incident plane wave has wave vector $\mathbf{k}_i$ as indicated with an arrow in the same figure. As the wave enters the PhC from free space, the x-component of the incident k-vector is conserved, which means that the k-vector of the wave that propagates inside the PhC must lay on the dash-dotted vertical line in Fig. 5.4. This line intersects the EFC of the photonic crystal at four points: two for band 2 and two for band 3, so that potentially, four waves with four different propagation vectors inside the PhC can be created by a single incident wave. However, energy flow considerations eliminate two of these possibilities—see Section 5.6 for details. For the moment, we will also disregard band 2 and focus on band 3 only. Accordingly, the k-vector in band 3 that satisfies the conservation of the x-component is indicated with an arrow labeled as $\mathbf{k}_0$ in Fig. 5.4. The group velocity associated with this wave vector is $\mathbf{v}_0$ and is shown as a black arrow pointing in the direction of the steepest ascent of ω, perpendicular to the EFC, as required by Eq. (5.1). Because the group-velocity vector is coincident with the energy flow of the optical beam,

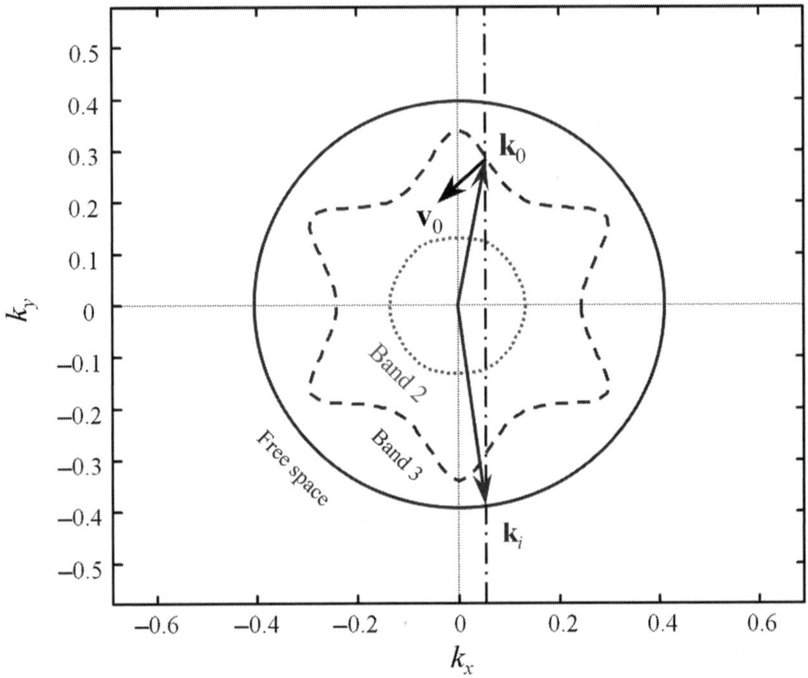

Figure 5.4 Equifrequency contours corresponding to a photonic-crystal structure of air holes in the dielectric described in the text, at a normalized frequency of 0.4. Contours from the second and third bands are displayed. The wave vector of the wave propagating inside the PhC is determined using the conservation of the wave vector component parallel to the free-space/PhC interface, k_x in this case. Group-velocity vector $\mathbf{v}_0$ points in the direction perpendicular to the EFC.

$\mathbf{v}_0$ also shows how the optical beam is refracted as it traverses the free-space/PhC interface. It should be clear now how the high sensitivity of the refracted beam to the incident angle is achieved in a photonic crystal characterized by sharp corners in its equifrequency contours. As the incident angle moves closer to normal, i.e., the x-component of $\mathbf{k}_i$ approaches zero, the wave vector $\mathbf{k}_0$ of the wave inside the photonic crystal traces the sharp corner of the EFC, which results in rapid change of the direction of $\mathbf{v}_0$ that is always perpendicular to the EFC. In the case described here, the variation of the incident angle in a mere $\pm 4^\circ$ range produces a whopping $\pm 50^\circ$-wide swing of the refracted beam at the normalized frequency 0.4. Such behavior justifies the name "superprism" given to a structure that exhibits it [1, 15].

The presence of another wave vector having a component parallel to the interface conserved, but lying in a different band, results in the splitting of the incident beam into two refracted beams. Thus, for the particular example discussed here, the energy of the incident beam will be divided into two refracted beams: the superprism beam described above and one that behaves more conventionally thanks to the nearly circular EFC of band 2. For photonic crystals with multiple bands that intersect the line of constant parallel k-vector component, the incident beam will be split into the corresponding number of refracted beams where the fraction of energy carried by each depends on the details of the PhC and of the interface with free space, as well as the polarization state of the incident wave.

In addition to high sensitivity to the incident angle, the PhC described above displays high sensitivity to the wavelength of the incident light as illustrated in Fig. 5.5. In this case, the change in frequency by 10% produces a 52° variation of the refracted angle, which further justifies the use of the name "superprism" for this structure.

For the example discussed above, the photonic crystal was made of material with a fairly high dielectric constant. However, the superprism phenomenon is also exhibited by PhCs made of low-dielectric materials such as polymers. This has potentially profound consequences because it allows functional PhC devices that work in the optical regime to be made from standard polymer photoresists used in photolithography. The difficulty lies in selecting the design parameters so as to remove the degeneracy

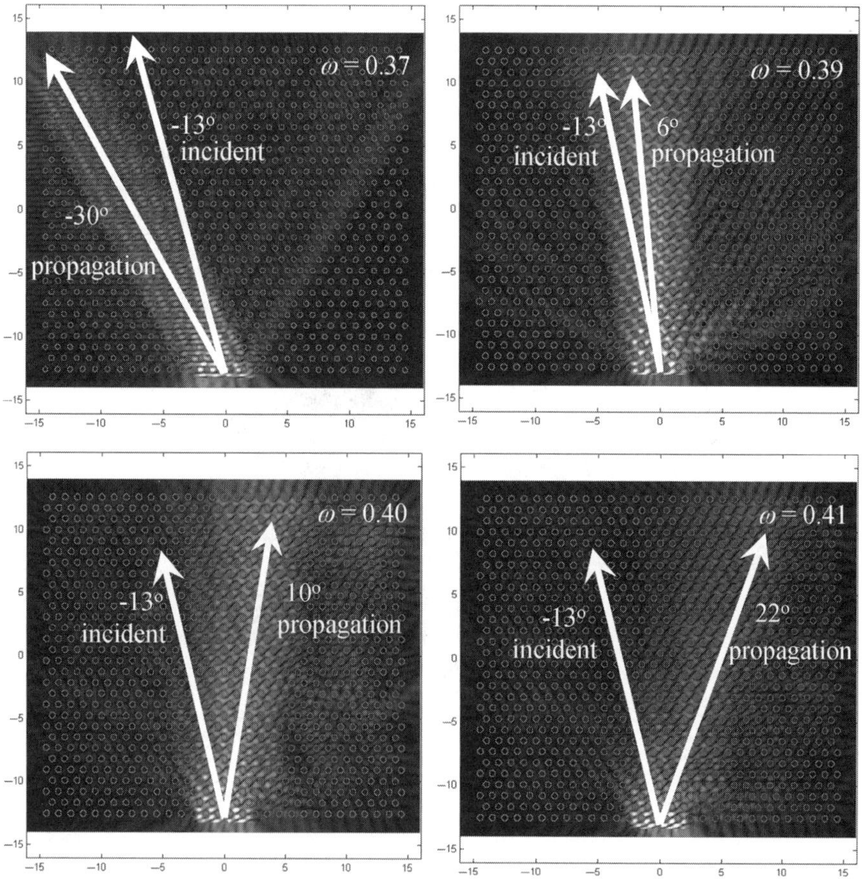

Figure 5.5 Effect of wavelength change on the direction of the beam propagation inside a photonic crystal. With the incident angle of 13°, the 10% change in normalized frequency $f_n = \omega a/2\pi c$ produces a corresponding change of the refracted angle by 52°.

(multiple bands corresponding to a single frequency) that often plagues low-contrast PhCs. Furthermore, the selection of the type of structure that is amenable to layer-by-layer fabrication is important, so that the capabilities of modern semiconductor fabrication equipment are used efficiently in realizing them. Fortunately, proper selection of the face- centered-cubic (FCC) woodpile-geometry parameters, which was described in this book as one of the first examples of 3D photonic crystals, endows it with superprism properties. For the "log" width $d = 0.205a$ and the refractive index of the constituent dielectric $n = 1.57$, the equifrequency surface at $f = 0.94c/a$ takes the shape shown in Fig. 5.6, where the presence of relatively sharp edges and corners, which are the hallmark of superprism, is evident.

To visualize and quantify the angular sensitivity of the three-dimensional photonic crystal more conviently, the angle ψ is defined as follows:

$$\cos\psi \equiv \hat{\mathbf{k}}_i \cdot \hat{\mathbf{v}}_0, \tag{5.3}$$

where $\hat{\mathbf{k}}_i$ and $\hat{\mathbf{v}}_0$ are unit vectors that point in the directions of the incident k-vector and the group velocity of the corresponding wave inside the photonic crystal, respectively. Thus, ψ is a measure of how far the refracted beam inside the photonic crystal deviates from the original propagation direction of the incoming wave; the more rapid the change of ψ as a function of the incident wave angle(s) and/or its frequency, the more pronounced the superprism effect.

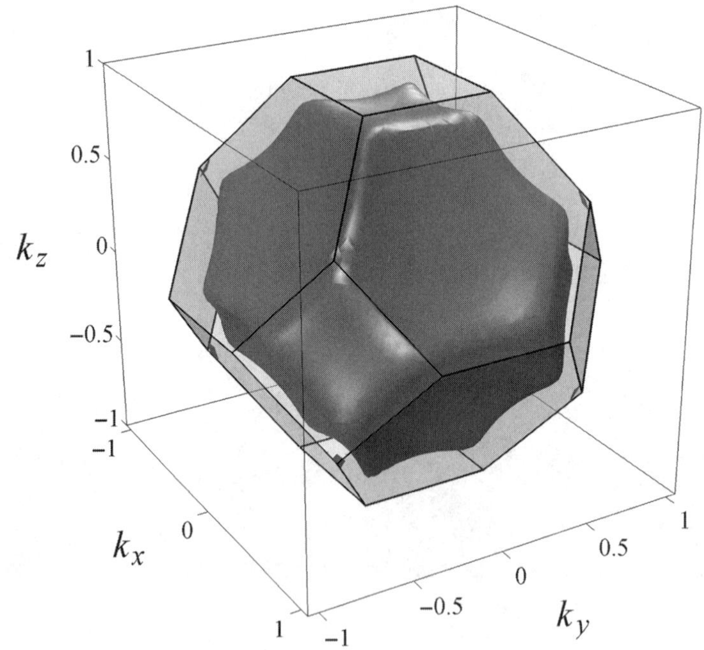

Figure 5.6 The EFS of the third band of a polymer FCC woodpile PhC at frequency $0.94c/a$. The refractive index of the polymer is $n = 1.57$, and the log width is $d = 0.205a$, where a is the FCC lattice constant.

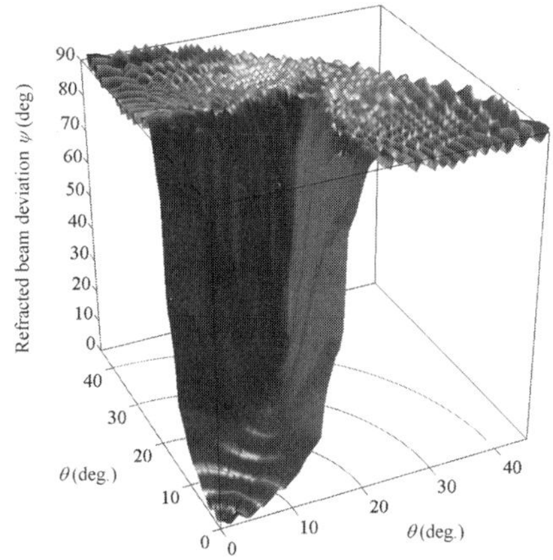

Figure 5.7 ψ as a function of the incident angle at frequency $0.94c/a$.

Figure 5.7 shows a three-dimensional plot of the variation of ψ as a function of the azimuth (φ) and elevation (θ) angles of the incident wave. According to this figure, the most rapid variation occurs for the incident angle θ in the range between 15° and 30° depending on the azimuth angle φ selected. Close examination of the data used to generate Fig. 5.7 indicates that a 50° change in the refracted angle for 5° change in the incident angle θ is easily achieved for $\varphi = 0^\circ$ or 45°. For a narrower range of incident angles, it is also possible to obtain refracted angle variation as high as 30° for a 1° change of θ.

A similar analysis performed for the frequency dependence of the refracted angle reveals that a 10° variation of the refracted beam propagation direction can be achieved for a 1% change in the wavelength of the incident beam. This conclusively demonstrates the superprism effect in a photonic crystal made of low-index dielectric.

It should be noted that an alternative mechanism for the superprism effect also exists and has been described in Ref. 16. In this case, the large variation of the refracted angle is achieved using a flat portion of the dispersion diagram rather than a sharp corner.

The high sensitivity of the angle of propagation on the incident angle and/or the wavelength offered by the superprism can be exploited in practical applications. Uses contemplated so far include passive wavelength demultiplexing [17–19] and beam steering [20–22], as well as rapid active beam steering [23], that relies additionally on the presence of the electro-optic effect in the constituent materials.

5.4 SELF-COLLIMATION

An optical beam propagating in a linear, homogeneous, isotropic material will necessarily diverge as it traverses the space occupied by said material. The divergence is the result of the finite beam width, which implies the presence of k-vector components perpendicular to the beam propagation direction. Over sufficient distance, these components are responsible for "pulling the beam apart." The wider the beam, the smaller the transverse components can be made, and the slower the divergence. Therefore, if we wish to have the beam propagate over long distances with low divergence, we must start with an appropriately wide beam. Unfortunately, for on-chip optical connection, space limitations dictate that we use narrow beams, of the order of a single wavelength, and have them propagate over long distances of hundreds or thousands of wavelengths. These two requirements seem mutually exclusive, if we try to implement them in homogeneous materials, therefore, waveguides, such as those discussed in Chapter 2, are typically employed. It would seem that we can do little to control the divergence of a narrow beam in the absence of waveguiding structures, because the k-vector content (i.e., its angular spectrum) of a beam is dictated by its width—no exceptions. However, it should be recognized that what is really desired is not a narrow range of k-vector directions, but rather a narrow range of *group velocity* directions, because it is the group (not the phase) velocity that defines energy transfer. Although in the case of isotropic materials the two angular ranges, k and v_g, coincide, it is not the case in general.

Consider the definition of group velocity provided by Eq. (5.1), which implies that the propagation direction of a wave packet is perpendicular to the EFC. As a result, if one were to invent material that exhibits square EFC for a certain frequency, then for wave vector directions spanning 90° ($\pm 45^\circ$ from the normal to the flat EFC edge), the group velocity would have the same direction—perpendicular to one side of the square. In other words, even though phase velocity, coincident with the wave vector, covers a wide range of angles, the wave packet travels in one particular direction: perpendicular to the side of the square equifrequency contour. Clearly then, for self-collimation, a square EFC is desirable. Although a perfect square seems to be beyond reach, one contour shown in Fig. 5.2 comes close.

To elucidate self-collimation further, consider Fig. 5.8b, where the EFC of an isotropic material (free space) is shown as a circle, and the EFC of the self-collimating photonic crystal at the same frequency is indicated with a dashed curve. At the boundary between the isotropic material and the self-collimating material, the component of the k-vector tangential to the boundary, designated in Fig. 5.8 as the x-direction, is preserved. As a result, a wave with a wave vector $\mathbf{k}_0$ in free space assumes $\mathbf{k}_p$ as the wave vector inside the photonic crystal. More importantly, at $\mathbf{k}_p$, group velocity obtained from Eq. (5.1) points in the direction perpendicular to the dashed contour, which is almost exclusively the y direction for a wide range of incident $\mathbf{k}_0$ angles. This range of angles is indicated in the figure by the range between $\mathbf{k}_0$ and $\mathbf{k}_0'$. In other words, the wave is only allowed to propagate along those directions that are normal to the sides of the square-like contour, and it is possible to vary the incident wave vector $\mathbf{k}_0$ over a wide range of angles and yet maintain a narrow range of propagating directions $\nabla_k \omega$ within the PhC.

Although the frequency-space analysis presented above provides useful intuition as to what types of structures exhibit self-collimation behavior, real-space simulations offer a more striking illustration of this phenomenon. Figure 5.9 shows a comparison of the electromagnetic field distribution produced by a point source placed in a

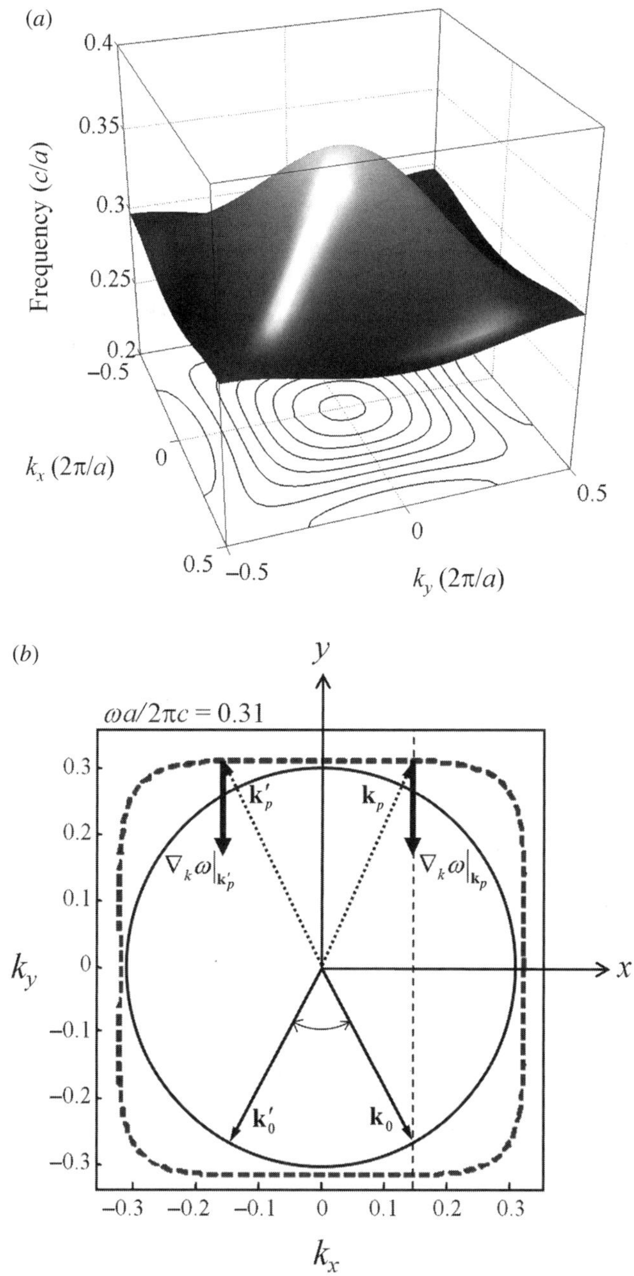

Figure 5.8 Dispersion diagram of a PhC designed to have a square EFC for certain frequencies. (a) A dispersion surface. (b) Illustration of self-collimation using a square EFC. $\mathbf{k}_0$, $\mathbf{k}'_0$ are incident wave vectors of waves that originate in homogeneous material with a circular EFC shown with a solid curve; $\mathbf{k}_p$, $\mathbf{k}'_p$ are the corresponding wave vectors in the PhC (dashed EFC), and $\nabla_k\omega|_{\mathbf{k}}$ is the group velocity in the PhC corresponding to wave vector $\mathbf{k}$.

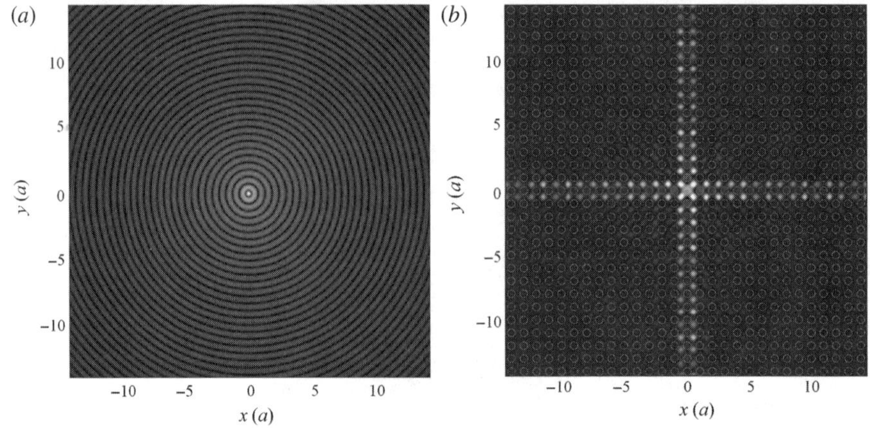

Figure 5.9 Propagation of light originating at a point source in (a) homogeneous silicon slab, and (b) PhC consisting of a rectangular lattice of air holes in silicon slab, wherein $r/a = 0.3$.

homogeneous slab with the field distribution produced by the same point source placed in a slab perforated by a square lattice of air holes. For the homogeneous material, the EFC is a circle, as shown in Fig. 5.1, and therefore light waves emanating from the source propagate isotropically within the plane as shown in Fig. 5.9a. On the other hand, if the EFC is nearly square, then waves originating at a point source located at the center of the PhC lattice, as in Fig. 5.9b, are limited to propagation along the x and y directions.

Although the verification of self-collimation in a photonic crystal by embedding a point source within the lattice is intuitive and convincing, experimental realization of such a scheme is not straightforward, especially for materials such as silicon, which is known to resist efforts to make it emit light. Therefore, in the next section, an alternative approach to experimentally demonstrating self-collimation is presented.

5.4.1 Experimental Demonstration of Self-Collimation

To demonstrate self-collimation convincingly using an external light source, it is necessary to have a wide angular distribution of k-vectors as the beam enters the PhC lattice. This happens naturally for a broad beam focused to a tight spot. In a slab waveguide, such focusing is readily achievable with a parabolic mirror similar to a J-coupler commonly used for coupling light into defect-based photonic-crystal waveguides [24].

Therefore, in this experiment, a photonic crystal consisting of circular holes, $r = 133$ nm, arranged in a square grid with a pitch $a = 442$ nm was patterned in a 260-nm-thick layer of silicon in a silicon-on-insulator (SOI) substrate, along with a J-coupler for focusing light from a wide ridge waveguide to the self-collimating lattice. Figure 5.10a shows the steady-state electromagnetic field distribution obtained using FDTD simulations where the J-coupler focuses a wide incoming beam, and the photonic crystal maintains the focused beam's narrow width as it propagates.

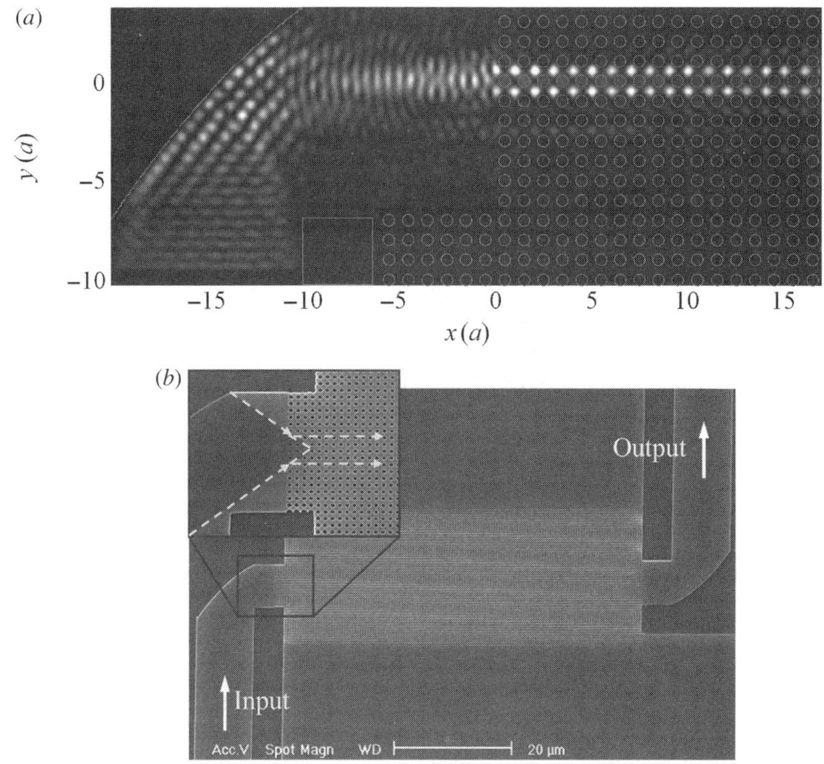

Figure 5.10 Design for experimental testing of self-collimation. (a) A steady-state field profile during device operation. (b) Scanning electron micrograph of a fabricated self-collimating PhC and J-coupler pair (for coupling into and out of the self-collimating lattice).

The data in Fig. 5.10a show that the range of angles in the focused beam produced by this J-coupler design is well matched to the self-collimation properties of this PhC lattice.

For experimental implementation, the focal spot of the J-coupler is located within the lattice, four lattice periods beyond the PhC lattice boundary, to avoid high scattering of a tightly focused spot from the first row of holes, and thereby to improve coupling efficiency [25]. Because the *k*-vector content of the focused beam is the same at the focal point and away from it, such a design does not affect the conclusions of the self-collimation test. However, such a design generates a beam that is wide enough within the PhC lattice to ensure that the high-index silicon regions between the holes are sufficiently illuminated, while at the same time it is narrow enough to allow the observation of lateral confinement.

To characterize propagation loss in such waveguides, infrared light with $\lambda = 1480\,\text{nm}$ is coupled into the PhC lattice via the J-coupler as described above, and out-of-plane scattering is observed at the silicon/PhC interfaces as shown in Fig. 5.11. The presence of a tightly focused spot at the far end of the PhC lattice

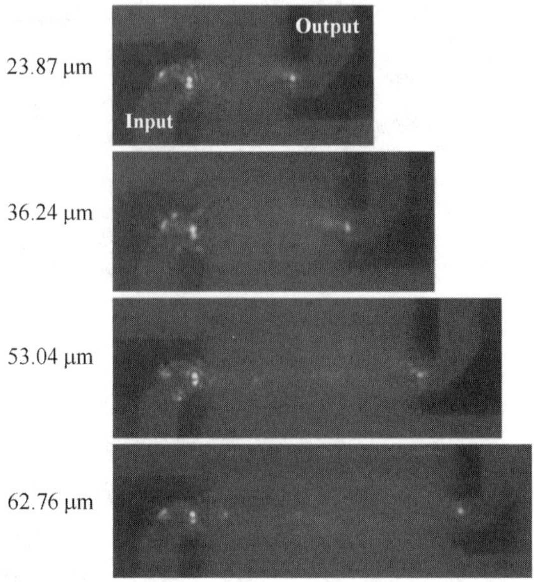

Figure 5.11 Image captured by a near-infrared (IR) camera of the scattered light, $\lambda = 1480\,\text{nm}$, at the PhC/silicon boundaries. The bright spot at the output proves lateral confinement of light within the PhC lattice.

demonstrates the lateral confinement and self-collimation of the initially divergent light as it propagates along the length of the PhC. The conspicuous absence of a light trail in Fig. 5.11 suggests low out-of-plane losses in this guiding structure. To characterize propagation loss quantitatively, several PhC dispersion waveguides were fabricated with lengths ranging from 10 μm to 80 μm, in SOI substrates and the cutback method [26] was employed. The loss was obtained from a linear fit of $\log(P_{\text{out}}/P_{\text{in}})$ versus waveguide length [27], where P_{in} (P_{out}) is the scattered light at the beginning (end) of the PhC dispersion structure. In using this measure, it is implicitly assumed that the amount of scattered light from each interface is proportional to the amount of light entering or exiting the PhC lattice. These measurements imply propagation loss as low as 1.1 dB/mm, which is a considerable improvement over the propagation losses measured experimentally for PhC line-defect waveguides [26, 28].

The improvement of guiding efficiency in slab photonic-crystal waveguides based on self-collimation as compared to bandgap-based waveguides can be explained by the relation of the guided modes to the light cone. In line-defect waveguides, the defect modes are often degenerate and a guided mode has its counterpart above the light cone at the same frequency. As a result of coupling between these modes induced by fabrication irregularities, the optical power initially in a guided mode can be transferred to the mode lying above the light cone, and to free-space radiation from there [29]. Such optical power transfer, which would lead to the loss of guided power, cannot take place in the case of self-collimation-based waveguides because no modes above the light cone exist at the same frequency as the self-collimated modes.

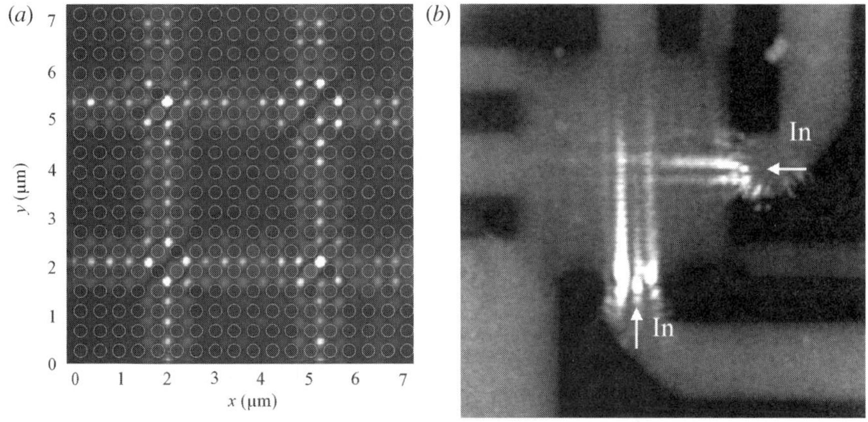

Figure 5.12 Simultaneous propagation and crossing of optical beams in a dispersion-guiding PhC. (a) A three-dimensional FDTD simulation. (b) Experimental results.

The lack of out-of-plane radiation in this case is clearly demonstrated in Fig. 5.11, which shows no light trail along the propagation path.

The discussion above makes it clear that such a structure is a viable candidate for efficient in-plane guiding of electromagnetic waves without the use of line defects or structural waveguides. Furthermore, the lack of waveguide structures allows multiple beams to be directed across one another in a high-density fashion without limitations arising from structural interactions or cross talk. This possibility is realized in a structure shown in Fig. 5.12. Therein, multiple beams cross one another without interaction in a manner similar to free-space propagation. The width of a light trail is of the order of 1 μm, which allows a high density of interconnection. Ultimately, this allows for very small, yet high-capacity, photonic circuits by alleviating the limitations of structural interactions that introduce additional loss at the crossing points. Note that for the purpose of experimental illustration, Fig. 5.12b, the underlying oxide layer is left intact, which lowers the light cone. As a result, the normally guided frequencies are now radiated out of plane, which thereby allows for qualitative observation of light propagation at the cost of propagation losses.

5.4.2 Self-Guiding Heterolattice

Self-guiding in the conduction band of a PhC is fairly intuitive, because light propagates along the cardinal directions of the lattice. Indeed, one could argue that the lines of material between the rows of holes of the photonic crystal constitute parallel straight waveguides that intersect at 90°, and that this is the reason why the optical beams do not spread as they propagate. However, the band diagrams of Fig. 5.2 and 5.3 indicate that self-collimated guiding should also be possible for the valence band, where the propagation would occur along the diagonals of the square lattice. To demonstrate such guiding and thereby illustrate the importance of the band-diagram analysis for the determination of self-collimation, a hybrid photonic-crystal lattice is employed [3], wherein light is first coupled to a conduction band and then transferred to a valence

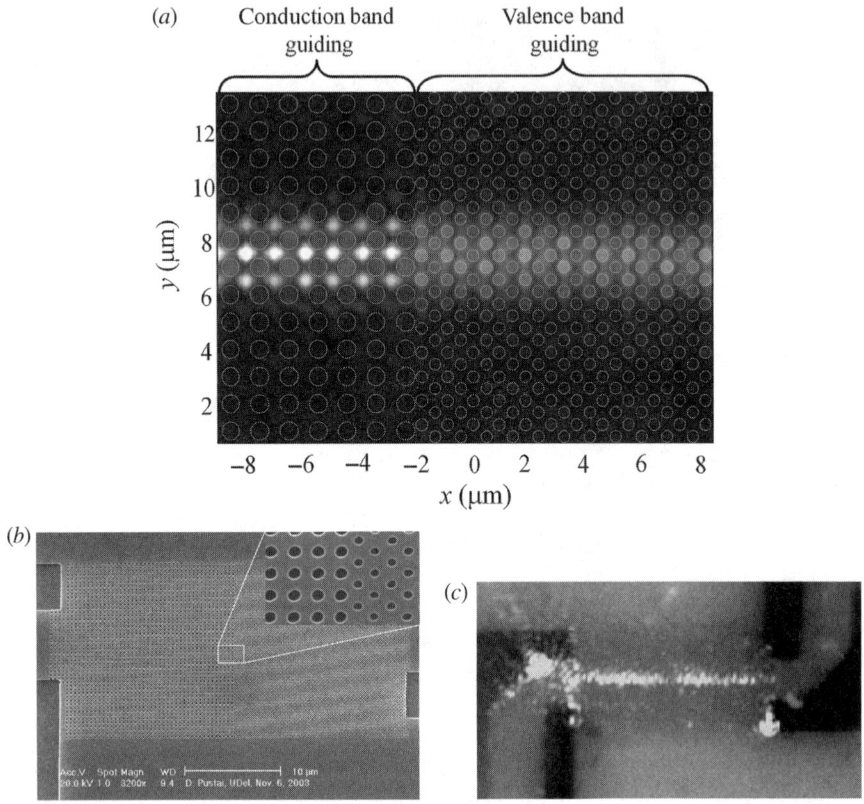

Figure 5.13 Self-collimation in a conduction band and a valence band. For the conduction band, the lattice parameters are $a = 442$ nm and $r = 133$ nm, whereas the valence band, with lattice rotated by 45°, has parameters $a = 252$ nm and $r = 75$ nm. (a) Steady-state H_z field for transverse-electric (TE) polarization obtained with FDTD simulations. (b) Scanning electron micrograph of the fabricated structure. (c) Top-down image of light guided through this structure at a wavelength $\lambda = 1480$ nm.

band of an adjacent PhC lattice, as shown in Fig. 5.13. To have self-collimation for the conduction band and for the valence band simultaneously at the same frequency, the dimensions of the respective photonic crystal are scaled accordingly. It is clear from Fig. 5.13a that light can propagate in the photonic crystal without divergence even when there is no obvious delineation of parallel waveguides. This result demonstrates that such behavior is best understood in the frequency domain with the help of band diagrams and equifrequency contours.

The phenomenon has been tested experimentally in the optical domain using structures fabricated in SOI. The conduction band self-collimation was achieved in a square-lattice photonic crystal with a pitch $a = 442$ nm and a hole radius $r = 133$ nm, where the equifrequency contour assumes a square-like shape as in Fig. 5.8. For valence-band self-collimation at the same wavelength $\lambda = 1480$ nm, the photonic-crystal lattice has a pitch $a = 252$ nm and a hole radius $r = 75$ nm. For proper coupling between the two photonic crystals, the lattice of the second is rotated by 45° with respect to the lattice of the first PhC. Figure 5.13c shows experimental validation of the concept where the tightly confined trail of light indicates efficient self-collimation of the optical beam as it propagates across the two photonic crystals.

Figure 5.14 shows the relevant equifrequency contours for the two photonic crystals. According to the EFCs shown in Fig. 5.14a and b, self-guiding occurs at

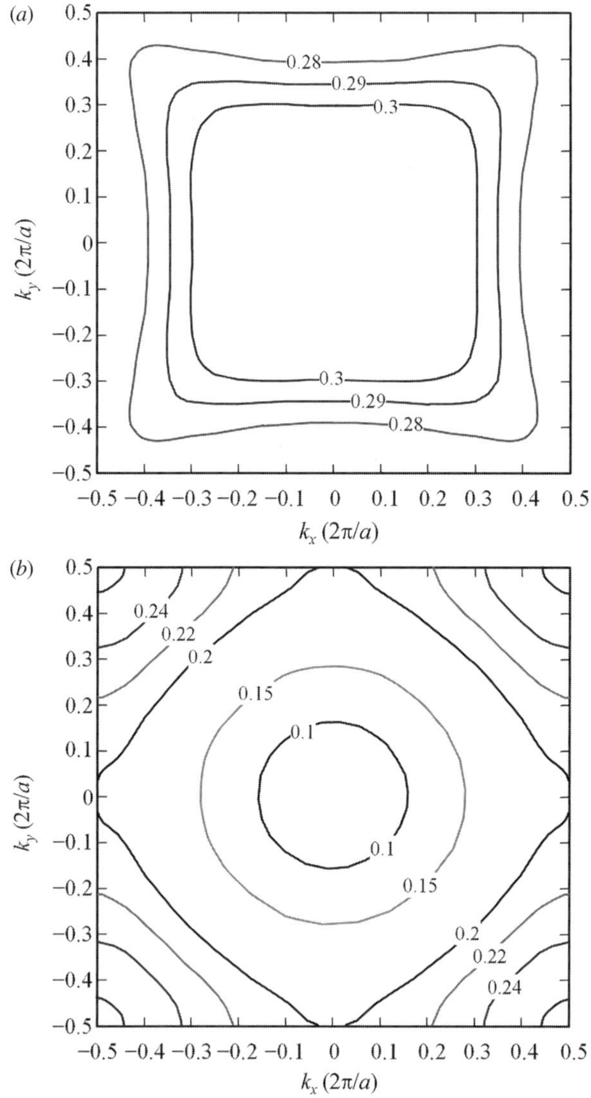

Figure 5.14 Equifrequency contours for the (a) conduction and (b) valence bands of the PhC described in the text. The labels in the contours indicate the normalized frequency a/λ.

normalized frequencies of 0.3 and 0.22 for the valence and the conduction bands, respectively.

These results demonstrate that by properly engineering the dispersion properties of photonic crystals, one can use them to guide light efficiently in a straight path on a chip scale with little beam divergence, low losses, and in the absence of engineered defects that perturb the regular lattice. However, in practical applications, it is also necessary to have the ability to redirect light. Therefore, a method of bending light paths in a self-collimating lattice is discussed next.

5.4.3 Redirecting Light in Self-Collimating PhCs

To make a functional integrated system, it is necessary that the signal-carrying connections between individual elements be routed along paths more general than straight lines. For example, electronic integrated circuits contain metal traces routed arbitrarily throughout the chip; similar capability is expected of any viable photonic alternative. Thus, in this section, a method for redirecting light in a self-collimating PhC is described.

Self-collimated propagation is unlike light guiding in which physical boundaries coerce light to follow a certain path. Rather, it is akin to free-space propagation, and therefore, one can use concepts familiar from the study of the latter to redirect light. Because in a square lattice, self-guiding takes place only along directions perpendicular to the four approximately flat sides of the square EFC, only 90° beam redirection is desired and indeed allowed. In other words, one needs a device integrated with the PhC that converts all wave vectors that constitute the initial optical signal into a group of wave vectors corresponding to the orthogonal direction.

In free-space optics, a collimated beam is redirected with a mirror [27], and one can use a similar approach in a self-collimating lattice. To this end, a small rectangular area, which behaves like a 45° mirror, is introduced along the Γ-M direction; see Fig. 5.15. To determine the interaction of the optical beam with such a "mirror," a full 3D FDTD method must be used instead of the simple geometrical optics employed in free-space optics, because the boundary conditions are more complex here due to the presence of the PhC structure. Furthermore, the PML boundary conditions at the top and bottom of the computational region must be placed at least 520 nm away from the slab to avoid evanescent modes, for which the PML is not entirely valid [30, 31].

Figure 5.16 shows the steady-state amplitude of the H_z component at the middle plane of a PhC slab for a 1-μm-wide Gaussian beam with a vertical extension of 260 nm launched along the x direction of the lattice. In this case, the distance of the mirror surface from the nearest hole is $d = 0.107a$, where a is the lattice constant; see Fig. 5.15c. The reflection efficiency of such a mirror is 82.9%, which is obtained by comparing the energy flow at the output with that of a reference device of the same length but without the mirror. To minimize the effect of local field fluctuations on the result, the energy flow is averaged over a region $5a$ long, $3a$ wide, and 260 nm high.

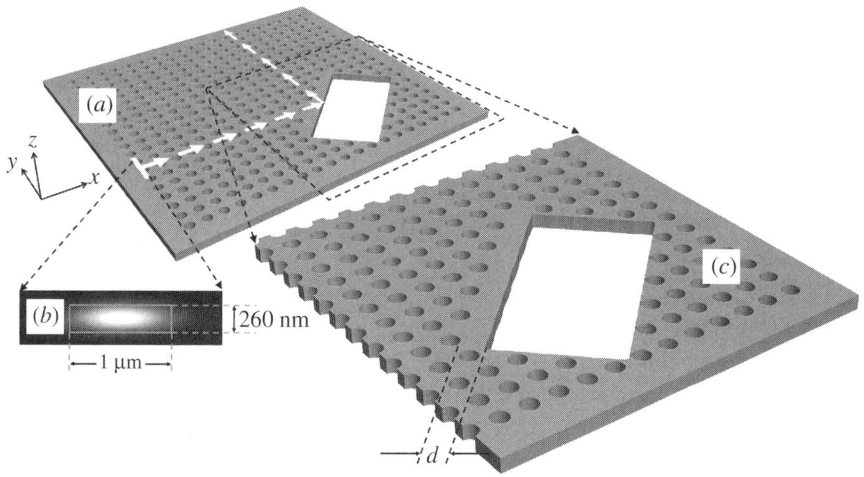

Figure 5.15 Design of a 45° mirror in a self-collimating PhC lattice. (a) Entire device. White arrows indicate the intended path of light. (b) Launched elliptical Gaussian beam profile. (c) Close-up of the mirror. d is the distance from the edge of the nearest hole to the mirror.

Although 82.9% is reasonably high, this figure is sensitive to the exact placement of the mirror with respect to the photonic-crystal lattice as illustrated in Fig. 5.17, which shows the efficiency as a function of d. The highest transmission is obtained for the mirror surface placed along the line indicated as A-B in the inset, which is about halfway between the holes of the photonic crystal. Another point of high efficiency is for the mirror cutting directly across the holes. Whereas the first

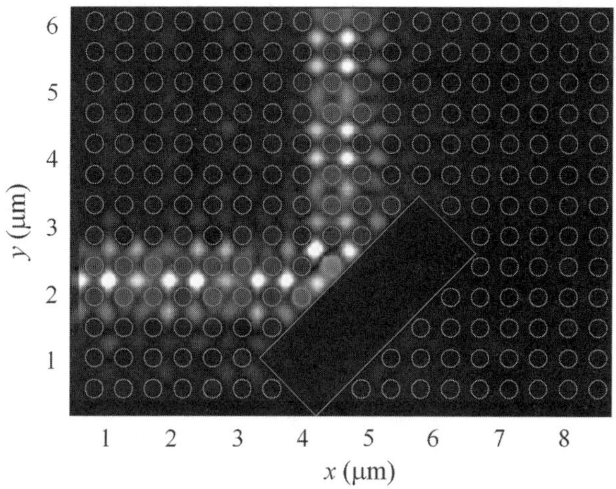

Figure 5.16 Steady-state amplitudes of the H_z component at the central plane of a photonic-crystal slab when $d = 0.107a$. The reflection efficiency, which is defined as the ratio of the bent-waveguide output energy to the output energy of a straight waveguide with the same length, is 82.9%.

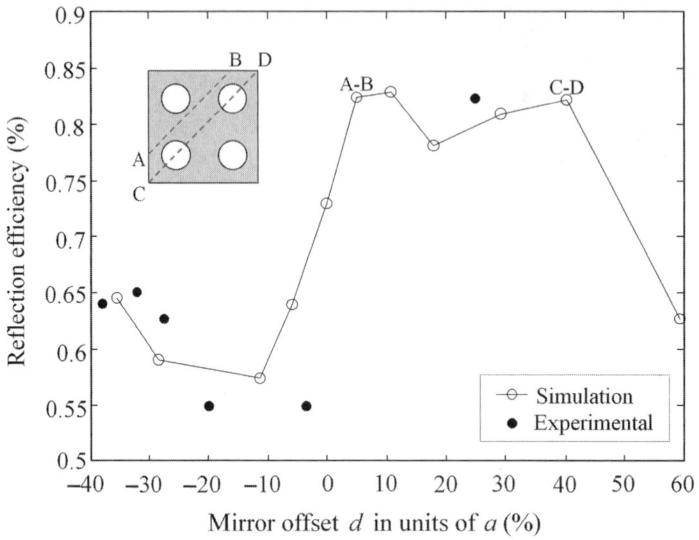

Figure 5.17 Reflection efficiency versus mirror offset d. The inset indicates the two mirror positions favored by the symmetry of the photonic crystal. Results obtained experimentally are shown as black dots.

high-transmission point can be intuitively understood as a result of the smoothness of the mirror surface, the second may need some explanation. From Figs. 5.13a and 5.16, it is clear that the electromagnetic field, in the case of conduction-band self-collimation, is concentrated in the high-index material between the rows of air holes of the PhC. At the same time, line C-D cuts directly across the holes—it is a symmetry line of the photonic crystal. As a result, the energy of the incoming mode, which is concentrated between the holes, finds its way to travel between the holes again after being reflected from the mirror [32].

The concept of using a dielectric mirror for redirecting light in a self-collimating photonic-crystal lattice has been validated experimentally. To this end, a device was fabricated in an SOI substrate with a 260-nm-thick device layer and was tested using a tunable infrared laser. The lattice parameters of the PhC are $a = 442$ nm and $r = 0.29a$. Defined lithographically, the mirror spanned three periods to ensure proper light confinement within the PhC region. Figure 5.18a shows scanning electron micrographs of the fabricated routing structures for the mirror at various positions, all fabricated in a single sample to compare the respective efficiencies quantitatively.

For optical characterization, infrared light at 1432 nm was launched into the PhC via a J-coupler as discussed in Section 5.4.1. Figure 5.18b, which was obtained using a near-IR camera, shows a top-down view of the device with light reflected by the mirror that was located at a distance $d = -0.251a$ from the nearest holes of the PhC lattice. To observe scattered light, the underlying oxide layer was intentionally left intact, which produced leaky modes above the light cone. The experimental efficiency, indicated in Fig. 5.17 by black dots, was measured after removing the oxide with hydrofluoric acid.

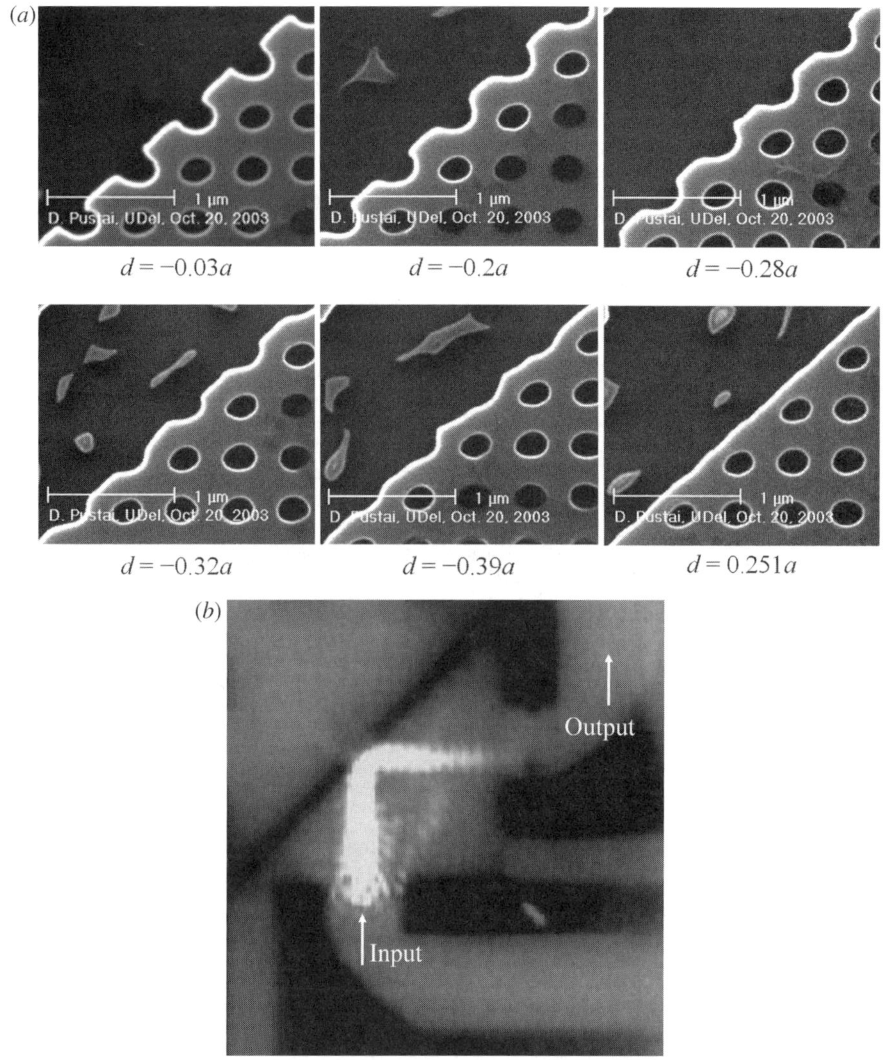

Figure 5.18 Experimental demonstration of routing in a self-collimating PhC. (a) Scanning electron micrographs of fabricated dispersion-based PhC waveguides with open areas that act as reflecting mirrors. (b) Near-IR camera image of a trail of light as it propagates through the self-collimating PhC lattice and is reflected by the etched mirror at $d = -0.251a$.

5.4.4 Beam Splitting in Self-Collimating PhC

In addition to arbitrary routing of signals throughout an integrated circuit, it is often necessary to split the signal and to deliver it to multiple destinations. In the case of electrical connections, this result is easily accomplished by ensuring the continuity of the properly patterned metal trace. For optical connections, the beam must be split, which requires more involved engineering. In Chapter 4, a single optical

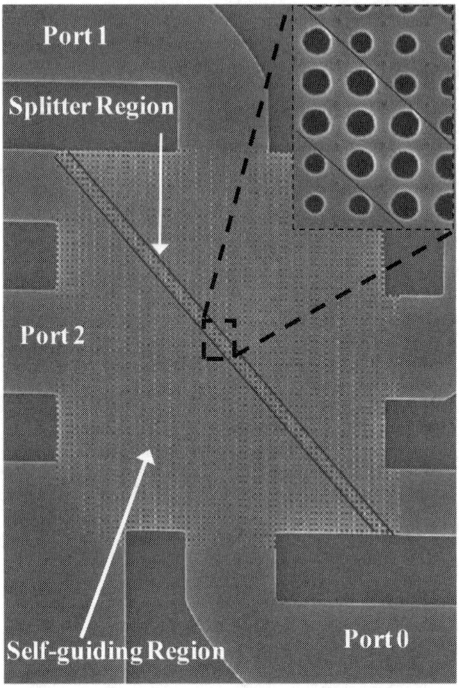

Figure 5.19 Scanning electron micrograph of a one-to-two beam splitter. The structure consists of a self-guiding PhC lattice and a PBG lattice. The latter is delineated by the diagonal solid lines. Light enters through Port 0 and is transmitted or reflected toward Ports 1 or 2.

beam was divided into multiple outputs with the help of directional couplers wherein two parallel waveguides were separated by several layers of photonic-bandgap (PBG) material that allowed transfer of energy between them. The imperfect confinement provided by a finite-size PBG lattice is also exploited here: A self-collimating lattice to guide light is combined with a PBG to partially reflect/transmit the beam. In other words, the beam splitter is a hybrid device that combines the two types of photonic crystals. The approach is similar in spirit to the 45° mirror of Section 5.4.3, which took cue from the study of free-space optics—the beam splitter is defined using a partially reflecting integrated mirror, which is made of PBG material.

Figure 5.19 shows a fabricated device based on this concept. The entire PhC region is filled with a self-collimating lattice—except for three rows of holes that run diagonally at 45°, which comprise a PhC lattice that exhibits a photonic bandgap. With the choice of the hole radius $r_g = 0.3a$, where $a = 442$ nm is the lattice constant, the photonic crystal is self-collimating for normalized frequencies between $0.28c/a$ and $0.31c/a$, as shown in Fig. 5.20. In this case, the wave is only allowed to propagate along two orthogonal directions Γ-X normal to the sides of the square [3]. For proper operation, the photonic bandgap of the beam-splitting region must overlap with the frequency band for self-guiding. This condition is achieved by increasing the hole size of the PhC to $r_g = 0.34a$ while maintaining the square geometry of the

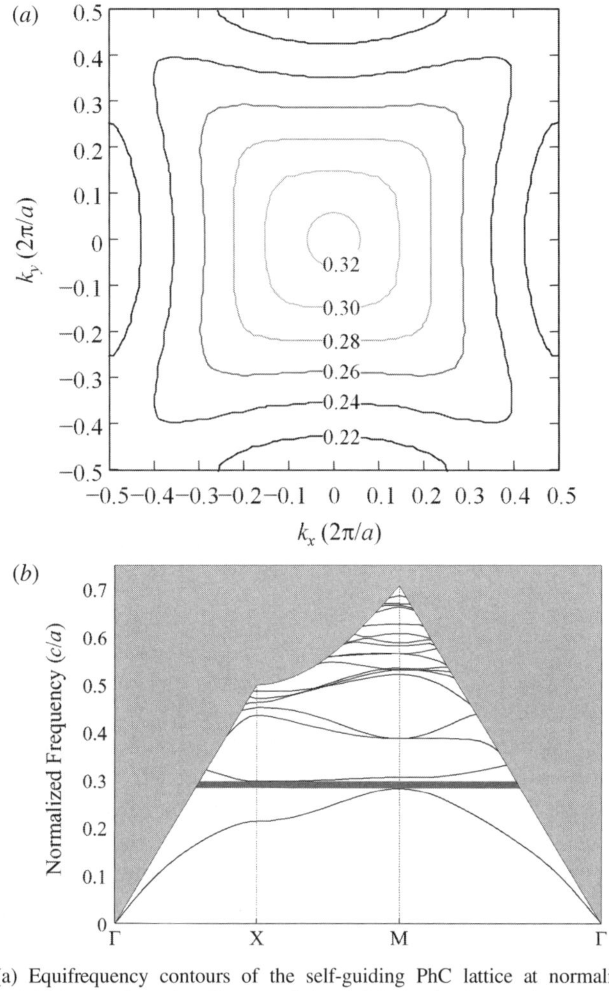

Figure 5.20 (a) Equifrequency contours of the self-guiding PhC lattice at normalized frequencies between 0.22 and 0.32. (b) Dispersion diagram of the PhC lattice in the splitting region. By increasing the diameter of the air holes in the splitting region, a bandgap is opened between normalized frequencies 0.2801 and 0.2977.

lattice with the same pitch $a = 442$ nm. As a result, photonic bandgap is created for normalized frequencies between $0.2801c/a$ and $0.2977c/a$, which is consistent with the self-guiding range listed above.

For experimental testing, tunable-laser light in the range of 1413–1503 nm ($a/\lambda = 0.3135$–0.2941) is end-fire coupled from a fiber to a ridge waveguide, and then from the ridge waveguide to the self-guiding lattice with a J-coupler [25], which is employed to observe lateral confinement. The transmission through and reflection from the splitter as a function of frequency are shown in Fig. 5.21, where the port numbers 1 and 2 refer to the designations of Fig. 5.19. The splitting ratio is frequency dependent because the operation of the device relies on coherent interaction

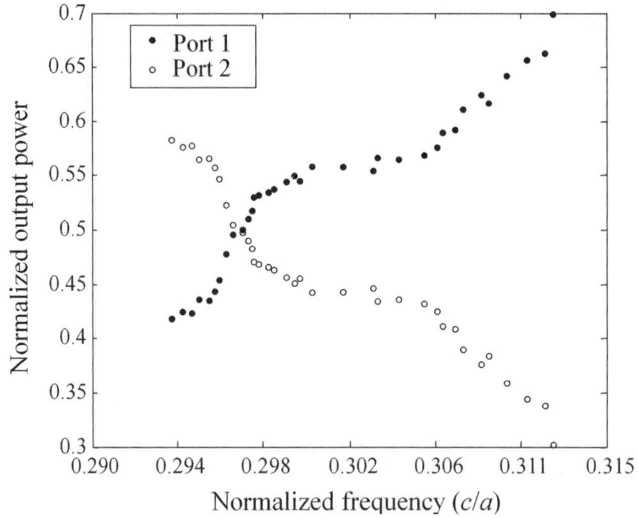

Figure 5.21 Fraction of output power in ports 1 and 2 versus normalized frequency for the air hole radius of the splitting region $r_s = 0.34a$. A 50–50 split is observed at a normalized frequency of 0.2982, which corresponds to a wavelength of 1482 nm.

of the electromagnetic wave with periodic dielectric structures. The 50–50 split is obtained for normalized frequency 0.2982 or wavelength 1482 nm.

Qualitatively, the frequency dependence of the beam splitting is shown in Fig. 5.22, in which three snapshots of the device are depicted for three different wavelengths launched into Port 0. In the language of photonic bandgaps, this behavior can be explained as follows. The wavelength $\lambda = 1453$ nm lies outside the bandgap of the beam-splitting region and inside its valence band. Therefore, light is allowed to propagate in such a photonic crystal. With only three layers of the PBG lattice, the self-guided beam has little chance to diverge and continues, well confined, past the splitting structure toward Port 1, with little power reflected into Port 2 (Fig. 5.22a).

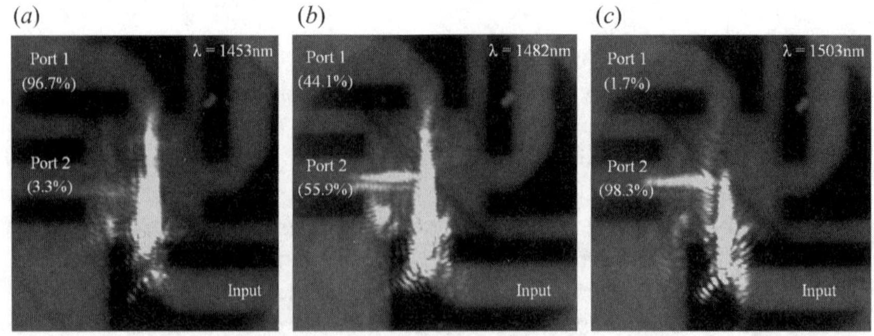

Figure 5.22 Top-down view of light propagation through the self-guiding and beam-splitting regions of the device shown in Fig. 5.19, at a wavelength of (a) 1453 nm, (b) 1482 nm, and (c) 1503 nm.

As the wavelength is increased to 1482 nm ($a/\lambda = 0.2982$), it approaches the band edge, $a/\lambda = 0.2977$, which manifests itself in increased reflection. This results in half of the power transmitted toward Port 1 and half reflected toward Port 2 as in Fig. 5.22b. Further increasing the wavelength places it inside the bandgap, which prevents light from entering the splitting region and produces high reflectivity. Consequently, most of the power is redirected toward Port 2 and little is transmitted toward Port 1, as in Fig. 5.22c.

As in previous experimental tests, the observation of light trail in Fig. 5.22 is made possible by leaving a silicon dioxide layer underneath the device silicon layer of SOI intact. Data presented in Fig. 5.21 are obtained after removing the underlying oxide to push the modes under the light cone and, thus, reduce out-of-plane loss.

In addition to frequency dependence, the splitting ratio also depends on the detailed geometry of the splitting region. For example, varying the size of the holes affects the position and the width of the stop-band, and as a result, the reflection efficiency. Furthermore, the photonic crystal that defines the splitting region need not have the same pitch or even the same lattice type as the self-collimating PhC. In fact, because the triangular lattice of air holes is known to exhibit a wider bandgap than a comparable square lattice, it may be desirable to use the former for the splitting mirror while retaining the square lattice for guiding. This geometry is presented in Fig. 5.23, where three layers of a triangular PBG material comprise the splitting region. However, as it turns out, the photonic bangap is so effective in this case that virtually all the energy of the incident beam is deflected, and little is able to propagate through the splitting region. In other words, the triangular lattice acts more like a mirror than a beam splitter required here. To reduce the reflection efficiency, the number of layers can be reduced, where the minimum is obviously one layer. Although it is difficult to justify calling a single layer of air holes a portion of a triangular lattice, the bottom line is really whether such a structure provides the desired functionality of beam splitting, and not how it is called or how it was obtained.

The fabricated device as well as its performance, both simulated and tested experimentally, is presented in Fig. 5.24. The goal of splitting the incident beam is clearly achieved—however, with an added bonus of splitting it into three rather than two outgoing beams. To understand this behavior, it should be noted that the single row of holes that comprises the splitter acts more as a grating than a semi-transparent mirror of free-space optics, as it consists of a periodic arrangement of identical holes.

For a free-space diffractive grating, the following condition must be satisfied:[1]

$$(\mathbf{k}_0 - \mathbf{k}_n) \cdot \mathbf{b} = 2\pi n, \tag{5.4}$$

where $\mathbf{k}_0$ is the wave vector of the incident beam, $\mathbf{k}_n$ is the wave vector of the nth order diffracted beam, $\mathbf{b}$ is the periodicity vector of the grating, and n is an integer. Furthermore, because diffractive grating is a linear device and assumed to be stationary, the frequency of the outgoing beams is the same as the frequency of the incoming

[1]Note: for simplicity, the discussion below is limited to two dimensions.

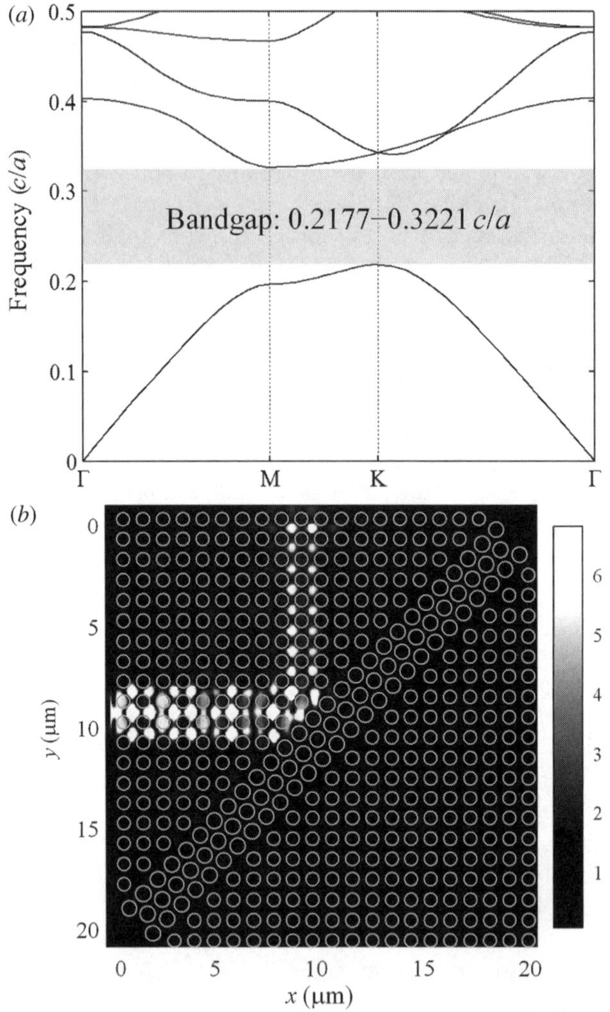

Figure 5.23 (a) Dispersion diagram of a triangular lattice with air hole of radius $r = 0.35a$. (b) Intensity distribution of light reflected by a splitting structure that consists of a three-layer triangular lattice of such air holes.

beam, which in free space translates to the dispersion relation

$$|\mathbf{k}_n| = |\mathbf{k}_0| = \omega/c \tag{5.5}$$

for all n. This means that in free-space optics, the wave vectors $\mathbf{k}_n$ of the diffracted beams lie at the intersections of lines defined by Eq. (5.4), which are perpendicular to the grating, with the circle defined by Eq. (5.5).

In self-collimating optics, on the other hand, there are two complications to this picture, both stemming from the periodic nature of the medium. First, the convenient

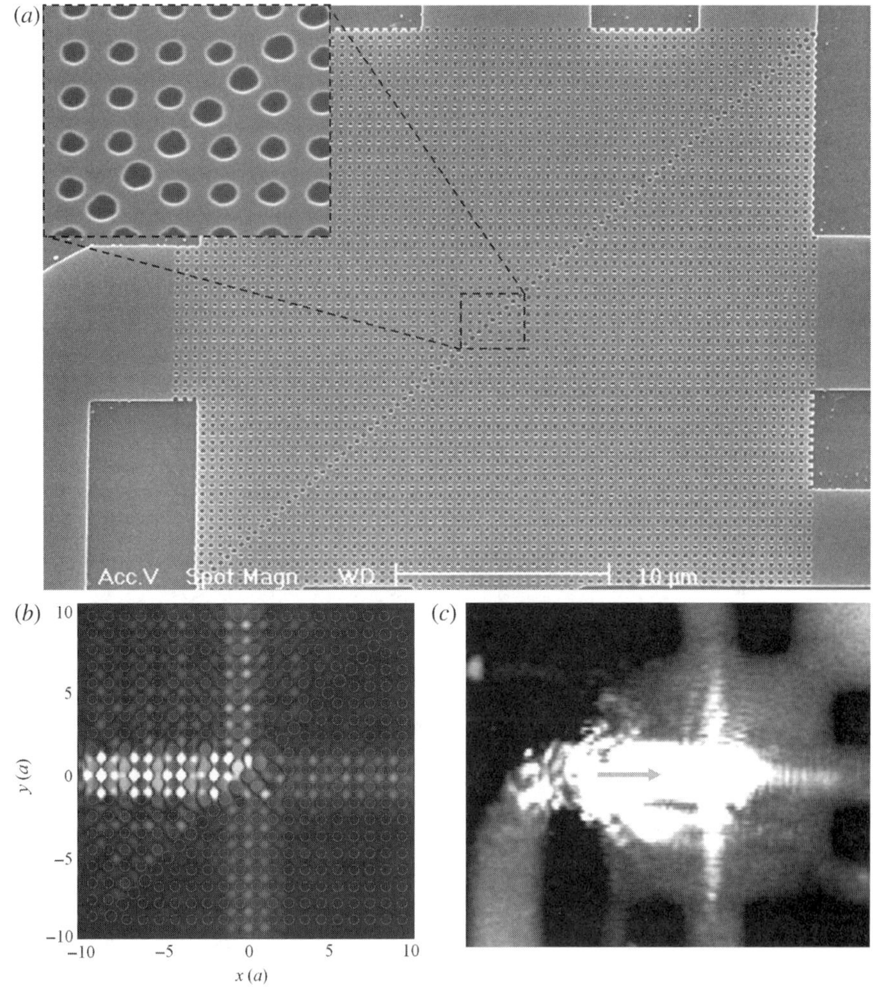

Figure 5.24 One-to-three beam splitter in a self-guiding PhC. (a) Scanning electron micrograph of the splitter fabricated in a silicon-on-insulator substrate. (b) Steady-state field at a frequency $f = 0.26c/a$; the incident beam is launched from the left along the x-axis. (c) Top-down view of the device during operation at a wavelength of $\lambda = 1600$ nm. Arrow indicates incident beam direction.

dispersion relation of Eq. (5.5) must be replaced by the suitable photonic-crystal dispersion relation obtained numerically. Thus, the possible diffracted orders are obtained from the intersections of lines defined by Eq. (5.4) with the appropriate equifrequency contour, e.g. from Fig. 5.20a rather than a circle of Eq. (5.5). Second, because the dispersion surfaces of conduction bands are folded into the first Brillouin zone where the length of k-vectors is limited, one may need to extend the graph to include the second Brillouin zone before using the analog of Eq. (5.4).

To illustrate the procedure, consider the specific example discussed above. At the frequency $f = 0.26c/a$, the equifrequency contour in this PhC is approximately square, as shown in Fig. 5.20a, and the self-collimated directions are parallel to the x- and y-axes. The center-point k-vectors for the self-collimated directions that corresponds to this frequency are approximately

$$\mathbf{k}_x = \pm 0.29\frac{2\pi}{a}\hat{\mathbf{x}}, \quad \text{and} \quad \mathbf{k}_y = \pm 0.29\frac{2\pi}{a}\hat{\mathbf{y}}, \tag{5.6}$$

where the minus sign in the first equation and the plus sign in the second equation are chosen to account for left-handedness of the material (see Section 5.5). After extending the diagram to the second Brillouin zone, these k-vectors become

$$\mathbf{k}_x = 0.71\frac{2\pi}{a}\hat{\mathbf{x}}, \quad \text{and} \quad \mathbf{k}_y = -0.71\frac{2\pi}{a}\hat{\mathbf{y}}. \tag{5.7}$$

$\mathbf{k}_x$ corresponds to the zeroth-order beam as its wave vector is unaffected by the presence of the grating, i.e., $\mathbf{k}_x = \mathbf{k}_0$. However, $\mathbf{k}_y$ is expected to correspond to a diffracted beam; therefore, it is written as $\mathbf{k}_y = \mathbf{k}_n$, where n is the order of diffraction. To identify n, Eq. (5.4) is used with the periodicity vector of the grating

$$\mathbf{b} = a\frac{\hat{\mathbf{x}} + \hat{\mathbf{y}}}{\sqrt{2}}, \tag{5.8}$$

so that

$$(\mathbf{k}_0 - \mathbf{k}_n) \cdot \mathbf{b} = 2\pi\frac{1.42}{\sqrt{2}} \approx 2\pi \times 1.004. \tag{5.9}$$

Given the uncertainty in reading the magnitudes of the k-vectors from equifrequency contours shown Fig. 5.20a, one may assert that $\mathbf{k}_y$ ($= \mathbf{k}_n$) in Eq. (5.7) corresponds to the first diffracted order ($n = 1$).

Thus, the emergence of the third beam in the splitter, in addition to the expected transmitted and reflected beams, can be attributed to the diffraction of the incident wave from the diagonally placed (periodic) row of holes. It should be noted that this effect is only possible if one of the diffractive orders coincides with the self-collimating direction perpendicular to the propagation of the incoming beam. In the example analyzed above, the periodicity of the grating equal to the periodicity of the lattice used for self-collimation provided the necessary condition for the first-order diffraction.

5.4.5 Optical Analog-to-Digital Converter

The emergence of high-speed parallel digital optical systems demands the use of an optical analog-to-digital converter (ADC). Optical or optoelectronic ADCs have received significant interest because of the expanding applications of optical technologies in areas such as telecommunication, sensors, and imaging.

Two processes are required to convert an analog, time-varying signal into the digital data domain, namely, *sampling* and *digitizing*. In principle, both processes are implemented by the analog-to-digital converter. In practice, however, considerable time, which ranges from microseconds to milliseconds, is required for the ADC to digitize an analog input. During this time, the analog input must remain stable if an accurate digital representation of the analog input is to be obtained.

Analog-to-digital conversion or digitizing is a two-step process that involves quantization and coding of the analog input. Quantization is the process of transforming a continuous analog signal into a set of discrete output states. These output states must then be coded, which involves assigning a binary bit pattern to each state. The two steps can be seen clearly in the operation of parallel or "flash" ADCs, where a set of discrete states is generated using a bank of comparators and the binary code is produced by encoding the outputs of the comparators. A conceptual diagram of a flash ADC is shown in Fig. 5.25.

The two design parameters of an ADC that are important from a system point of view are its resolution and its conversion time. The resolution R of the ADC is

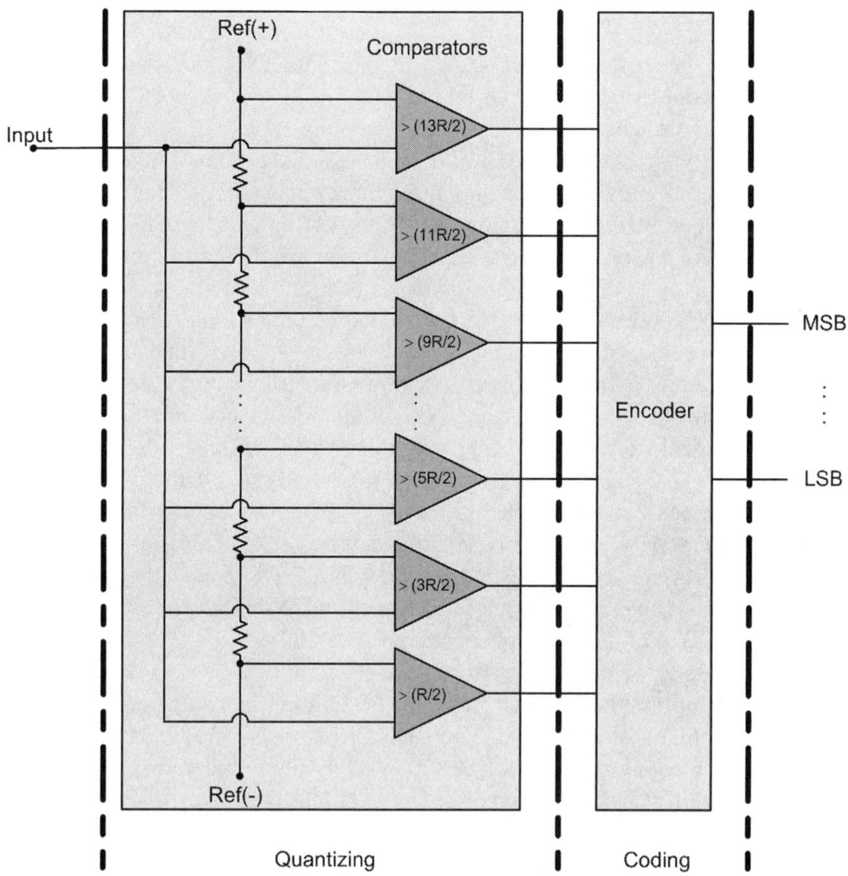

Figure 5.25 Block diagram of a flash analog-to-digital converter. MSB: most significant bit; LSB: least significant bit.

Table 5.1 Analog-to-Digital Conversion Resolution

ADC Word Length (bits)	% Resolution
8	0.39
10	0.098
12	0.024
16	0.0015

important because the system resolution is determined by this parameter. It is usually defined as the analog amount needed to produce a change in the least significant bit of the ADC. For an n-bit ADC, the resolution may be expressed as

$$R = \frac{\text{Full scale input signal amplitude}}{2^n}. \tag{5.10}$$

In many cases, the resolution of an ADC is quoted somewhat loosely as the with, n, of the binary word that is generated by the ADC. Table 5.1 shows the relationship between ADC word length and resolution.

The conversion time (t_c) of an ADC is the time required for the ADC to quantize the analog input and to provide, at its output, the binary equivalent of the analog input. The maximum permissible conversion time of an ADC in an analog input subsystem is determined, in general, by the requirements of the sampling theorem, because the conversion time must be less than the time between samples.

Optical ADCs offer better performance than conventional electronic ADCs because optical signals do not interact with electronic noise and radiation, and thus, they are immune to electromagnetic interference. Moreover, all-optical ADCs would eliminate the complexity and speed limitation of electrical-to-optical and optical-to-electrical conversions in photonic networks. Hitherto, several optical/ optoelectronic ADC approaches have been proposed and demonstrated, which included the Mach-Zender interferometer ADC [33, 34], optical fiber temporal-spectral mapping ADC [35], optoelectronic thyristor ADC [36], and the optical alternating layer ADC [37]. However, these designs are either optically or electrically complex or cannot be easily integrated. To address these limitations, an optical ADC was designed based on the concept shown in Fig. 5.26.

A cascaded array of beam splitters and detectors splits the power of an incoming optical signal into multiple channels by specific splitting ratios that correspond to a digital representation of the signal. The splitting ratios are chosen so that a strong signal will have enough energy for detection along the entire waveguide path. As the signal strength is reduced, the power in the output ports progressively becomes too weak for detection at the most distant detectors. Thus, the digital representation follows the characteristics shown on the right of Fig. 5.26 from the maximum to minimum power level. Based on these initial specifications, an optical ADC device was designed in a self-collimation photonic-crystal structure combined with an array of

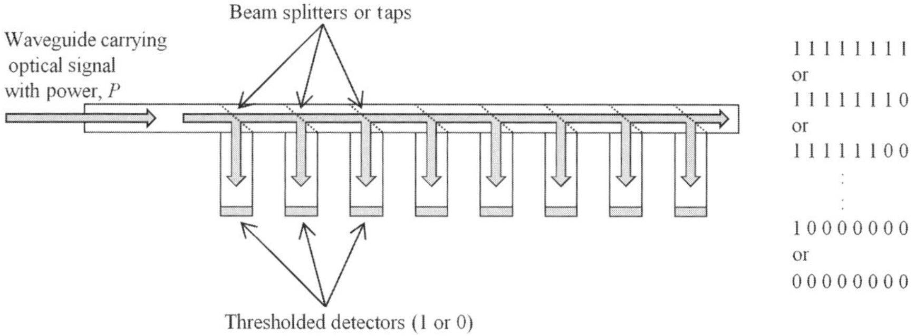

Figure 5.26 Analog-to-digital converter.

weighted splitters, as shown in Fig. 5.27. It is worth noting here that a threshold detector, which is used for event detection, is also a simple 1-bit digitizer. In this case, the quantization of a continuously variable analog signal is implemented using a single detection threshold to generate two discrete output states. These states are coded by the threshold detector at the moment of quantization as a single binary digit.

As a demonstration, a 2-bit ADC is realized using three splitters in a self-guiding photonic-crystal square lattice (see Fig. 5.27) to achieve four distinct states. Each splitter is designed by changing the size of air holes along a 45° line with respect to the propagation direction of the self-guided beam to break the lattice symmetry and thus perturb the beam. This results in an optical ADC that is compact and can be easily integrated with other photonic components.

To construct a 2-bit optical ADC, one needs four distinguishable states, which requires three splitters with splitting ratios of 50 : 50, 66 : 34, and 100 : 0, respectively. The quantization scheme of the 2-bit optical ADC is illustrated in Fig. 5.28. When the power of the incident wave is lower than twice the threshold value of the detectors, namely $P_{\text{in}} < 0.67P_{\text{th}}$, the output powers at three output ports are smaller than the threshold level ($P_{\text{th}} = 15\%$ full-scale input optical power), and the quantized states

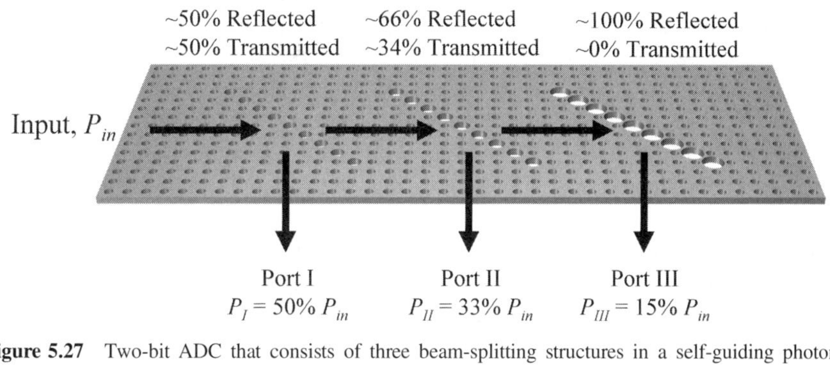

Figure 5.27 Two-bit ADC that consists of three beam-splitting structures in a self-guiding photonic crystal.

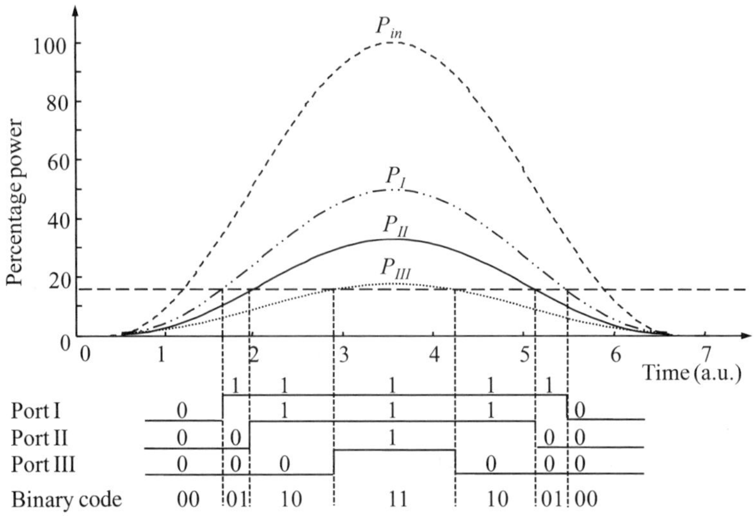

Figure 5.28 Quantization scheme for a 2-bit optical ADC excited with a time-variable optical signal.

of three output ports yield a state "000." When the input power is increased beyond $0.67P_{\text{th}}$ but is less than $0.88P_{\text{th}}$, the output power at port I reaches the threshold value of the detector associated with that port, whereas the outputs at the remaining two ports are still below threshold. Therefore, the state of the three-output ports in this case can be recognized as "100." Similarly, when the input power is increased beyond $0.88P_{\text{th}}$ but is less than $0.91P_{\text{th}}$, the output power at ports I and II reaches the threshold value of the detector associated with these ports, whereas the output at port III is still below threshold, and a "110" state can be extracted. Finally when the input power is greater than $0.91P_{\text{th}}$, the threshold value for detecting a high state at ports I, II, and III can be reached, and a "111" state is extracted. Consequently, we have four different states "000," "100," "110," and "111" at different levels of incident power. These four states can be coded into the binary representation as "00," "01," "10," and "11" for a 2-bit optical ADC.

To validate the design concepts of the 2-bit optical ADC experimentally a self-guiding photonic-crystal lattice was fabricated in the 260-nm-thick silicon device layer of a SOI substrate with a 1-μm-thick SiO_2 insulating layer. E-beam lithography and inductively coupled plasma (ICP) dry etching was employed to pattern and transfer the structure to the silicon device layer. The underlying SiO_2 layer was removed using buffered oxide etching (BOE) to produce a suspended photonic-crystal slab. Figure 5.29 shows scanning electron microscopy (SEM) micrographs of the fabricated splitters needed for the ADC design. The self-guiding region consists of a square array of air holes with a diameter of 220 nm and a lattice constant a of 450 nm. The three splitters were created by increasing the diameter of the air holes along the [11] direction to 286 nm, 434 nm, and 694 nm, respectively.

To test these fabricated splitters, the sample was cleaved and light from a tunable laser was coupled into the input facet of a tapered dielectric waveguide from a tapered

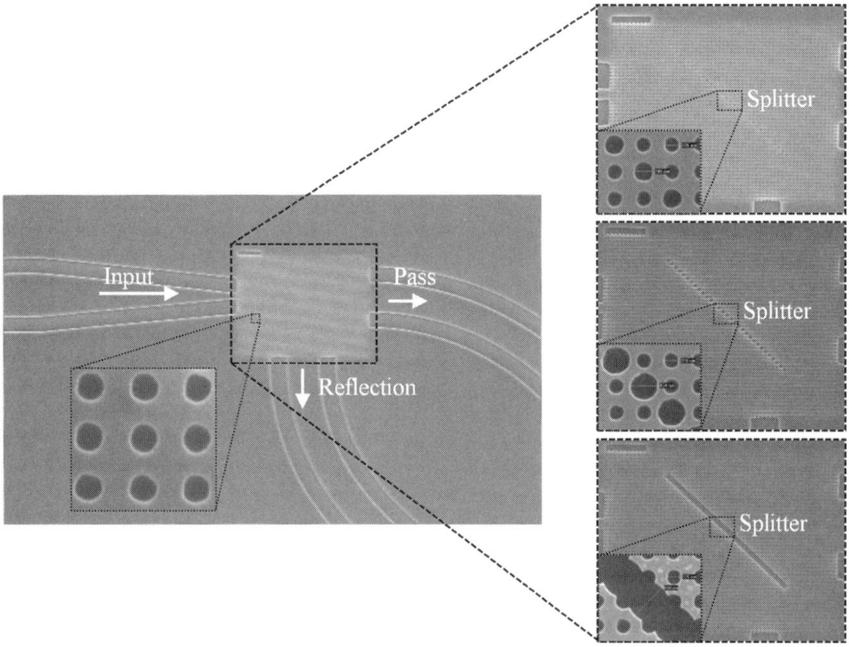

Figure 5.29 SEM images of a fabricated splitter units in a self-guiding photonic-crystal lattice. Shown in the figure are three different splitters with variable splitting ratios.

polarization-maintaining fiber. The tapered waveguide is interfaced with the self-guiding lattice so that after the wide beam from the fiber is narrowed down by the tapered waveguide, it enters the PhC lattice and is guided along the Γ-X direction.

When the self-guided beam propagates through the uniform lattice and arrives at the splitting structure, it is partially transmitted and partially reflected. The splitting ratio is determined by the radius of the perturbed air holes.

Figure 5.30 demonstrates this behavior, where (a) contains top-down images captured by an IR charge-coupled device (CCD) camera for the case with the radius of the splitting structure of 184 nm at the wavelength of 1560 nm. To characterize the relation between the splitting ratio and the size of splitting air holes, we varied the radius of splitting air holes and measured the output powers for each case. Figure 5.30b is a plot of the resulting data, which again shows that the output power at port I decreases as the radius of the air holes in the splitting region increases; the opposite is true for Port II. The output power was normalized at each port to the total output power to eliminate the influence of various coupling and scattering losses in the splitting region on the characterization of splitting ratios.

Based on these experimental results for the self-guiding splitters, a 2-bit ADC was fabricated and characterized on the same SOI wafer described above. Figure 5.31a shows an SEM image of the fabricated device. The self-guiding regions consist of air holes patterned periodically in a square lattice. Again, the radius of the air holes is 110 nm and the lattice constant is 450 nm. The radii of the perturbed air

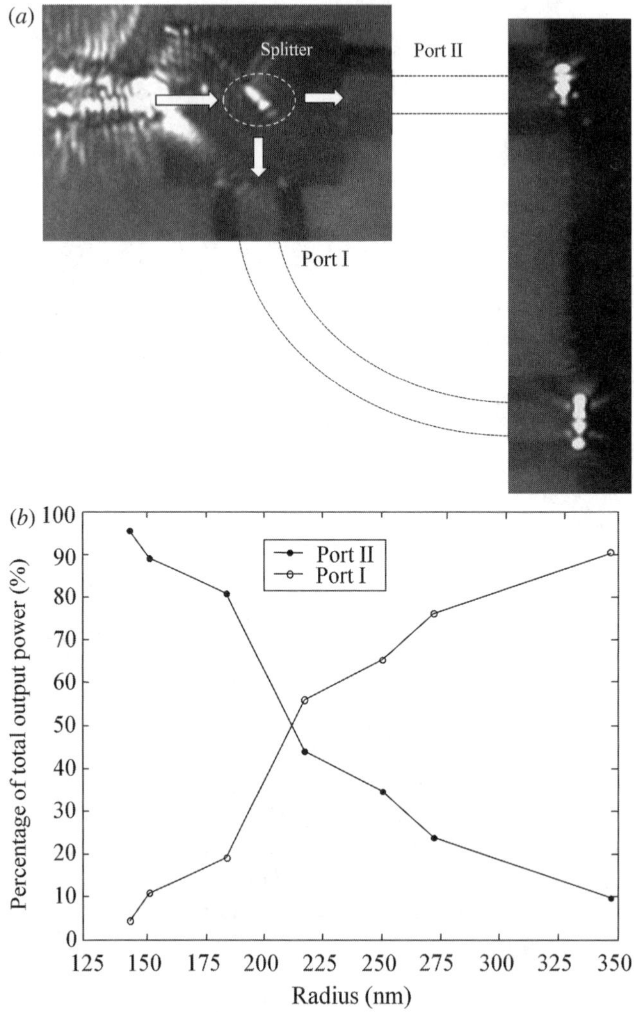

Figure 5.30 Experimental characterization of splitters shown in Fig. 5.29, for use in a 2-bit ADC. (a) Top-down view of the structure excited with input light at 1560 nm. (b) Percentage optical power at ports I and II measured for the range or $r = 125$ nm to $r = 350$ nm.

holes in three splitting regions are 225 nm, 245 nm, and 335 nm, respectively. Figure 5.31b shows the captured IR images of output ports at $\lambda = 1560$ nm at different incident powers. From Fig. 5.31, one can observe that the output powers at ports I, II, and III will reach a threshold level at different incident powers. To clarify this concept further, Fig. 5.32 plots the measured output powers at output ports I, II, and III versus the launched laser power when it varies from 100 μW to 1.5 mW linearly. The vertical axis has arbitrary unit to show the relative output power at each port. The fabricated device has four well-defined states, as required for a 2-bit

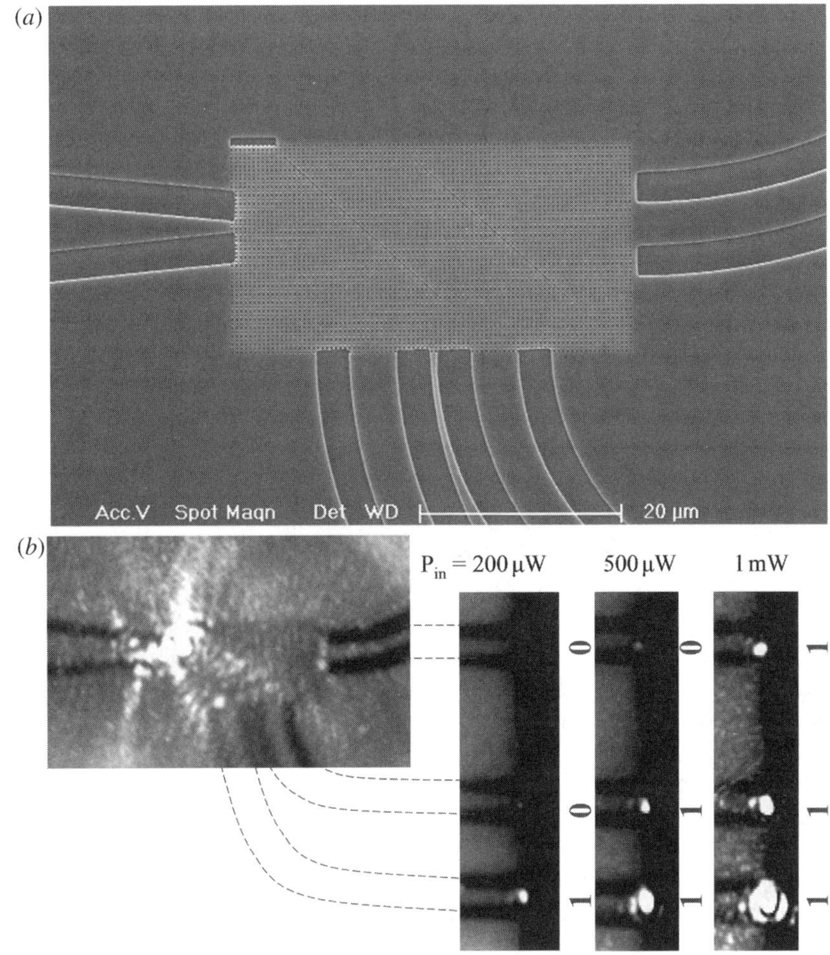

Figure 5.31 Fabrication and characterization of a 2-bit ADC: (a) the SEM picture of the device and (b) the captured IR images of output ports at $\lambda = 1560$ nm.

ADC. For instance, if the dashed line is set as the threshold level, when the incident power is 100 μW, the outputs at ports I, II, and III can be quantized as "001," which can be coded as the state "01" of the 2-bit ADC. Accordingly, the 2-bit ADC reaches the states "10" and "11" when the incident power is 180 μW and 900 μW, respectively.

5.4.6 Self-Collimation in Three-Dimensional Photonic Crystals

The concept of self-collimation discussed above for two dimensions can be readily extended to three dimensions. In this case, the equifrequency contours become

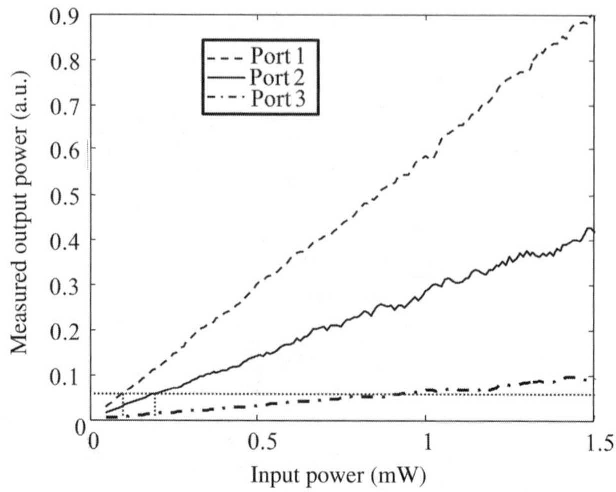

Figure 5.32 Plot of the measured output powers at output ports I, II, and III versus the incident power when it varies from 100 μW to 1.5 mW linearly.

EFSs, and the self-collimation is achieved for a frequency range where the EFS has approximately flat surfaces over a sufficiently wide solid angle in *k*-space. As before, the propagation of optical waves in such structures is best understood by examining their dispersive properties.

Self-collimation in two dimensions was found for a photonic crystal based on a square-lattice geometry. Because a cube is a 3D analog of a square, 3D self-collimation is naturally sought in a simple cubic lattice. When made of silicon and air, such photonic crystals exhibit only a small omnidirectional bandgap for certain filling fractions. However, their dispersion properties outside the bandgap are considerably more interesting for the present application.

As an illustration, consider a photonic crystal made of silicon with spherical voids with a radius $r = 0.55a$ placed at the nodes of a cubic grid with a lattice constant a, as shown in Fig. 5.33a. Because the radius is larger than half the distance between the centers of the spheres, the voids overlap. The interaction between the electromagnetic wave and this periodic structure is most conveniently described once again with a dispersion diagram that characterizes the relationship between the frequency ω of the propagating wave and its associated wave vector $\mathbf{k}$, as in Fig. 5.33b. It is obtained by casting Maxwell's equations into an eigenvalue problem with Bloch-wave boundary conditions, and then by solving it using the plane-wave expansion method (PWEM) [38], as described in Chapter 3. To reduce computational cost, only the dispersion diagram for *k*-vectors within an *irreducible* Brillouin zone are calculated, and the band diagram for the full Brillouin zone (BZ) is obtained by exploiting the symmetry of the lattice. The simple cubic lattice considered here has three high-symmetry points, which are indicated in the inset of Fig. 5.33b, and the volume of the irreducible Brillouin zone is 1/48th of the full Brillouin zone volume in this case. The region of interest for self-collimation lies around a normalized

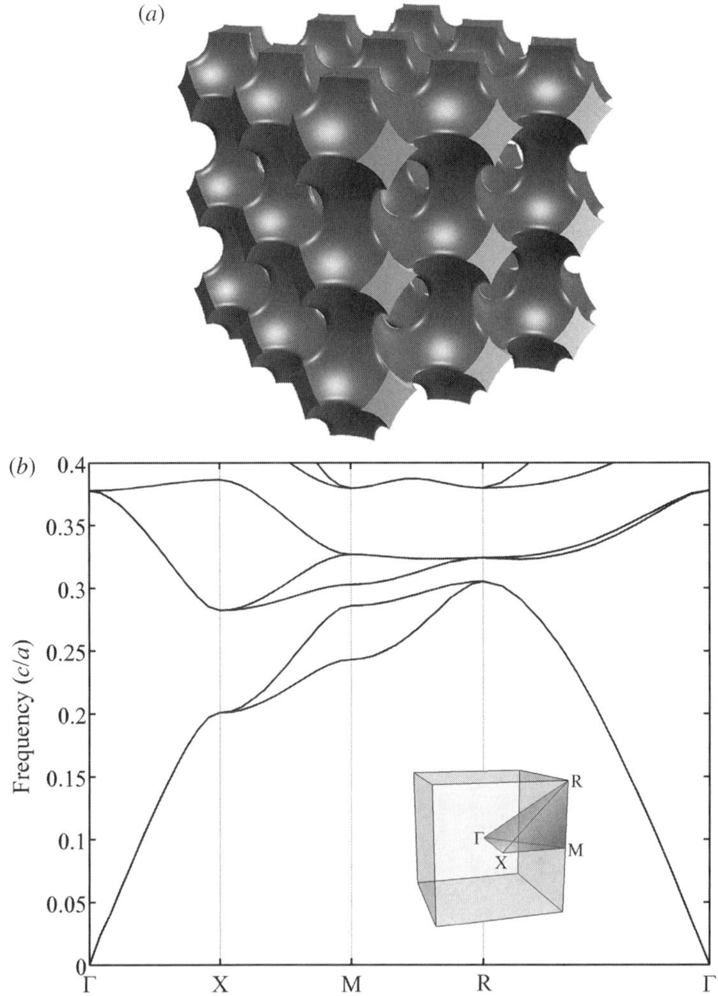

Figure 5.33 Simple cubic photonic crystal consisting of spherical voids on a cubic grid in a silicon matrix. (a) Perspective view of the lattice. (b) Band diagram—the *k*-vectors on the horizontal axis are traced along the edge of the Brillouin zone. The inset shows the positions of the high-symmetry points in the first irreducible Brillouin zone.

frequency $f = 0.34c/a$, which appears unremarkable in the dispersion diagram Fig. 5.33b, except for the degeneracy between bands three and four along the Γ-X direction. Similar to the 2D case, the self-collimation in 3D reveals itself through equifrequency surfaces rather than through a dispersion diagram, like the one presented in Fig. 5.33b, which merely depicts the variation of eigenfrequencies along the edge of the irreducible first Brillouin zone.

To determine the presence of self-collimation, eigenfrequencies must be calculated for all points in the Brillouin zone and not just along its edges, which makes the problem considerably more computationally intensive than obtaining the dispersion diagram of Fig. 5.33b. To this end, the first irreducible Brillouin zone is discretized, and a set of eigenfrequencies that correspond to each of the 3D k-vectors is found. Then, a single frequency is selected, and the k-vectors that show this frequency among its corresponding eigenfrequencies are connected to form a surface—the EFS. Figure 5.34 shows an EFS for $f = 0.34c/a$. Because of degeneracy, the EFS consists of several disjoint and overlapping surfaces. The section across the EFS at $k_z = 0$ shown in Fig. 5.34b reveals the shape of the resulting equifrequency contours, where

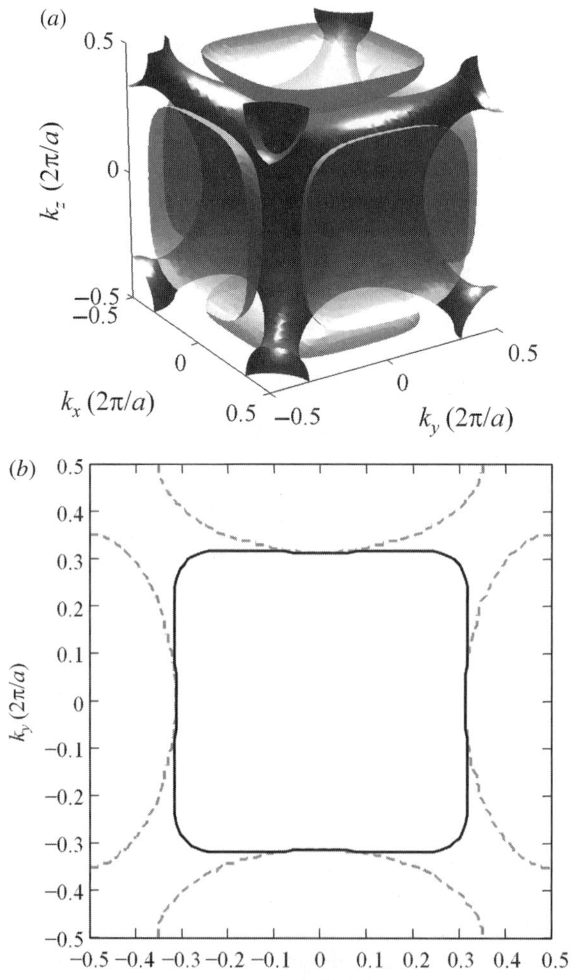

Figure 5.34 Equifrequency surface (a), and its section at $k_z = 0$ (b), for the three-dimensional photonic crystal of Fig. 5.33, at a normalized frequency $f = 0.34c/a$. Of the bands degenerate at the high-symmetry point, the third band has a square-shaped EFS, which is suitable for beam-routing applications.

the solid line corresponds to the third band and the dashed line represents the fourth band. From this figure, it is clear that band three has a square-like shape with flat sides, which makes it suitable for light guiding and routing applications, whereas band four is devoid of this feature.

As explained at the beginning of Section 5.4, the importance of the EFS shape stems from the relation $\mathbf{v}_g = \nabla_k \omega(\mathbf{k})$, which means that the group velocity coincides with the direction of the steepest ascent of the dispersion function $\omega(\mathbf{k})$ and is perpendicular to the EFS. As a result, the cubic shape of EFS, with significantly flat surfaces, allows propagation of light along the three orthogonal directions perpendicular to the sides of the cube with little divergence. This property has a practical application for waveguiding or beam routing in the absence structural features such as waveguides. However, to realize the 3D self-collimation-based optical routing, the EFC has to maintain the square shape over a range of k_z so as to allow the finite vertical extension of the beam. This is verified by taking multiple cross sections at different values of k_z such that $|k_z| \leq 0.25(2\pi/a)$, and by observing whether the respective EFCs have a square-like shape. The results, which are not shown here, indicate that indeed self-collimation is present in this structure for a wide range of incident k-vectors.

To illustrate explicitly the self-collimating behavior of this 3D photonic crystal, a dipole source is placed at the center of the silicon region, and FDTD simulations [39] are run over a cube $40a$ on a side bound by a perfectly matched layer (PML) [31]. The source can be, in principle, either a magnetic-current dipole or an electric-current dipole, each exciting a different mode. The FDTD mesh size is chosen at $a/20$, which produces 512 million grid points. However, the symmetry of the problem allows the reduction of the computational region to $1/8$ of its original size for simulations. In practical terms, this translates to using perfect electric conductor (PEC) boundary conditions for the xz and yz planes and perfect magnetic conductor (PMC) boundary conditions for the xy plane for a magnetic current dipole chosen as the source. The simulations were performed on a 30-node personal computer (PC) cluster, where the central processing unit (CPU) of each node runs at 3 GHz and is equipped with 1.5 GB of memory.

In an isotropic medium, a circular radiation pattern is generated in the xy plane by such a dipole source and a figure-eight-shape pattern in a plane passing through the z-axis. In contrast, when the medium is the photonic crystal discussed here at a frequency $f = 0.34c/a$, the electromagnetic field propagation is expected to be limited to x, y, and z directions because of the cubic shape of the EFS. Indeed, as illustrated in Fig. 5.35, which shows the H_z field component in the xy and xz planes, the wave is tightly confined within a few lattice periods. This confinement is most pronounced in the xy plane where the field barely extends beyond $3a$ and shows uncanny resemblance to the 2D self-collimation case.

The total field distribution $H_{\text{total}} = \sqrt{H_x^2 + H_y^2 + H_z^2}$ is shown in Fig. 5.36 in real space and in the spatial-frequency domain (k space). The real-space distribution, Fig. 5.36a, illustrates tight confinement of the propagating field as it emanates from the magnetic-current dipole source placed at the center of the computational region. Figure 5.36b is obtained by calculating the 3D Fourier transform of the H_z field component and by overlaying the result on the dispersion diagram, or

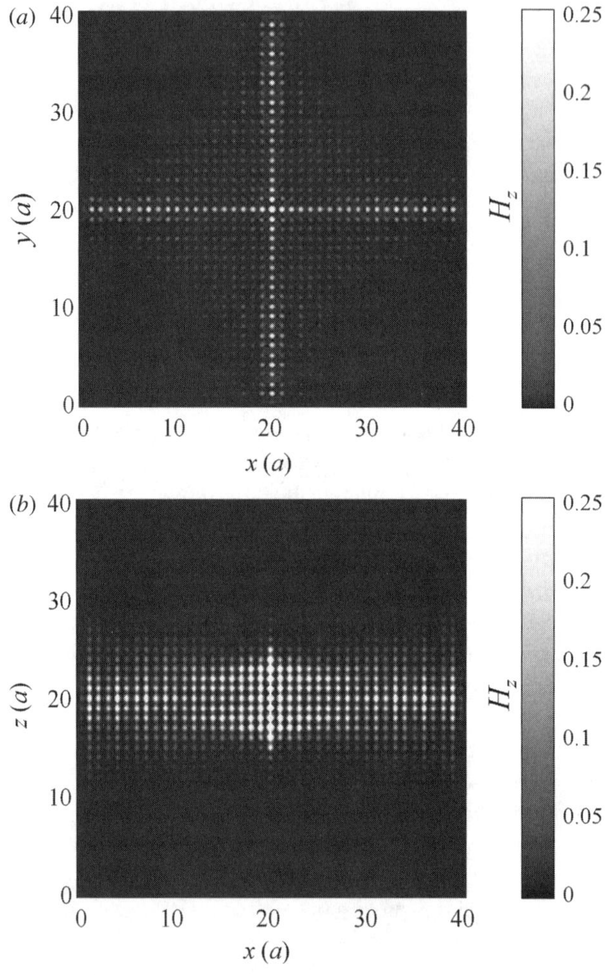

Figure 5.35 Propagation of an electromagnetic field in a self-collimating 3D photonic crystal. Distribution of the z-component of the magnetic field in the (a) xy plane and (b) xz plane. The source is a magnetic-current dipole placed at the center of the structure in silicon.

EFC plot, at $k_z = 0$. It shows that the magnetic-current dipole source excites predominantly the third band. This way, a direct link is established between the shape of the EFC of the excited band and the propagation properties of the wave in real space: The square-like EFC corresponds to self-collimated propagation. Furthermore, it is found that a magnetic-current dipole source excites this mode.

To verify the propagation properties of light in the other of the two degenerate bands, the respective mode needs to be excited using a different source. Thus, an electric-current dipole source is placed in the center of the silicon region of the photonic crystal, and the FDTD simulations are run until steady state is reached. As before, the symmetry of the structure is exploited by carving out $1/8$ of the computational

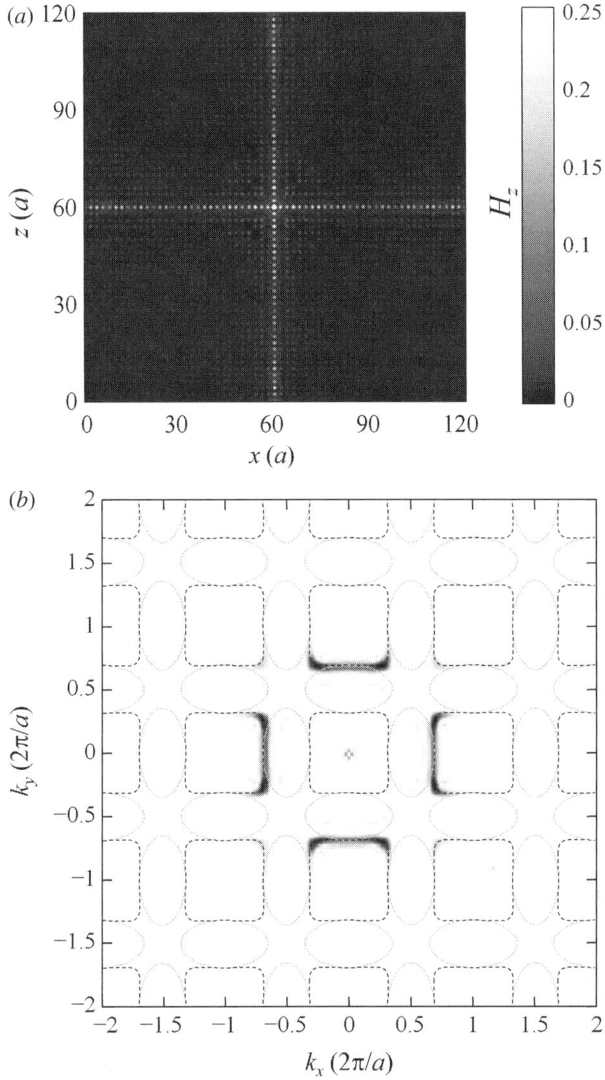

Figure 5.36 (a) Total field distribution produced by a magnetic-current dipole source in real space. (b) Field distribution in the spatial-frequency domain (indicated by gray levels in arbitrary units, darker is greater) overlaid on the dispersion diagram obtained using PWEM.

region and by selecting the proper boundary conditions. In this case, PMC is used at xz and yz planes, whereas PEC is used for the xy plane.

The electric field distribution of the steady state is shown in Fig. 5.37a. In contrast to Figs. 5.35 and 5.36a, no self-collimation is observed in this case. Rather, the field spreads evenly in all directions. The examination of the field distribution in the spatial frequency domain overlaid on the EFC, shown in Fig. 5.37b, confirms

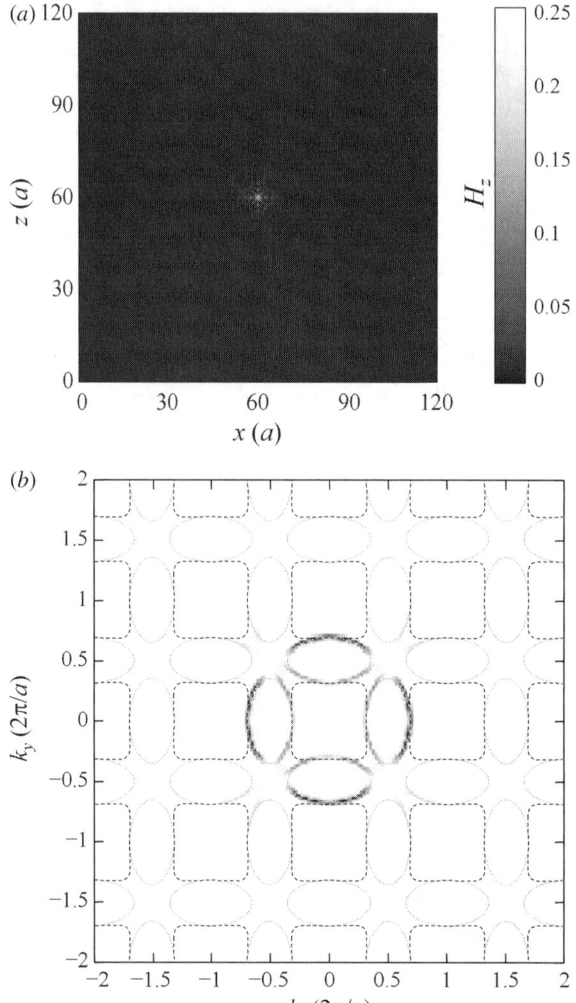

Figure 5.37 Field distribution produced by a dielectric-current dipole source placed in the center of the computational region occupied by a 3D photonic crystal. (a) Field distribution in real space shows no directional propagation. (b) Field distribution in spatial-frequency domain (indicated by gray levels in arbitrary units, darker is greater) overlaid on the dispersion diagram indicates that the energy is concentrated in the fourth band.

that the electric-current dipole source excites the fourth band, which displays no self-collimation properties.

The results presented above indicate that to take advantage of the self-collimation properties of a 3D photonic crystal, one must be careful in making sure that the band exhibiting the proper behavior is excited. In other words, in addition to selecting the right frequency, the choice of the type of source plays an essential role in demonstrating self-collimation computationally and, ultimately, in an experiment. This is beneficial for the realization of 3D self-collimation-based devices, because it indicates that little or no mixing occurs between the degenerate bands in this PhC when properly excited.

5.4.7 Experimental Verification of 3D Self-Collimation

Numerical investigations of self-collimation in three dimensions have been first reported in Refs. 40 and 41. The results presented there offered a tantalizing glimpse of the possibilities offered by structures that can confine and guide light in 3D in the absence of structural waveguides. However, experimental demonstration of the phenomenon proved elusive because of difficulties in fabricating the required intricate three-dimensional geometries. Although hitherto the effect has not been shown for optical frequencies, the use of high-dielectric-constant materials and computer-controlled precise machining allowed for testing the phenomena in the gigahertz range, where the scale of the features is more amenable to fabrication [42]. Below, we describe the procedure used for building and testing a self-collimating three-dimensional photonic crystal in the radio-frequency (RF) region of the electromagnetic spectrum, and we summarize the obtained results.

The photonic crystal fabricated for experimental testing of 3D self-collimation is obtained by drilling three sets of holes through blocks of high-index material, such as HiK500F from Emerson & Cumming, with the dielectric constant $\varepsilon_r = 30$. The cylindrical holes with a diameter $2r = 3.2$ mm intersect at right angles within a cubic unit cell as shown in the inset of Fig. 5.38a. The lattice constant of the photonic crystal is $a = 5.0$ mm, which results in the $\sim$10-GHz range as the spectral region of interest. The entire test structure shown in Fig. 5.38a, obtained by stacking six blocks, is 170 mm long, 85 mm wide, and 76 mm high.

According to the band diagram shown in Fig. 5.38b, a narrow bandgap opens between 14.4 GHz and 15 GHz. However, the region of interest for this chapter lies below—for a frequency range centered at 12.3 GHz, where self-collimation is expected. Figure 5.39a shows the EFSs for the third and fourth bands at 12.3 GHz. The third band is also depicted alone in Fig. 5.39b to illustrate its cube-like appearance with substantially flat sides. The horizontal cross sections of the EFS for several different values of the z-component of the propagation vector are shown as contours in 5.39c, which make it clear that the third band should exhibit self-collimation because of the square-like appearance of the contours with largely flat segments in the vicinity of the cardinal directions. In contrast, the EFCs of the fourth band are flat only when the value of k_z is increased, and they are highly curved at $k_z = 0$. This makes the fourth band unsuitable for self-collimation applications as the beam would be highly divergent if the distribution of propagation vectors of the incident beam were concentrated around the x or y directions.

As pointed out in Section 5.4.6, when testing for self-collimation, one must be careful to excite the proper band at the right frequency. In the case discussed here, it is desired to launch energy into the third band, because the PWEM simulations reveal that it exhibits self-collimating behavior at 12.3 GHz, whereas the fourth band does not. FDTD simulations show that, similar to the case discussed in Section 5.4.6, the third band is excited by a magnetic current source, whereas an electric current

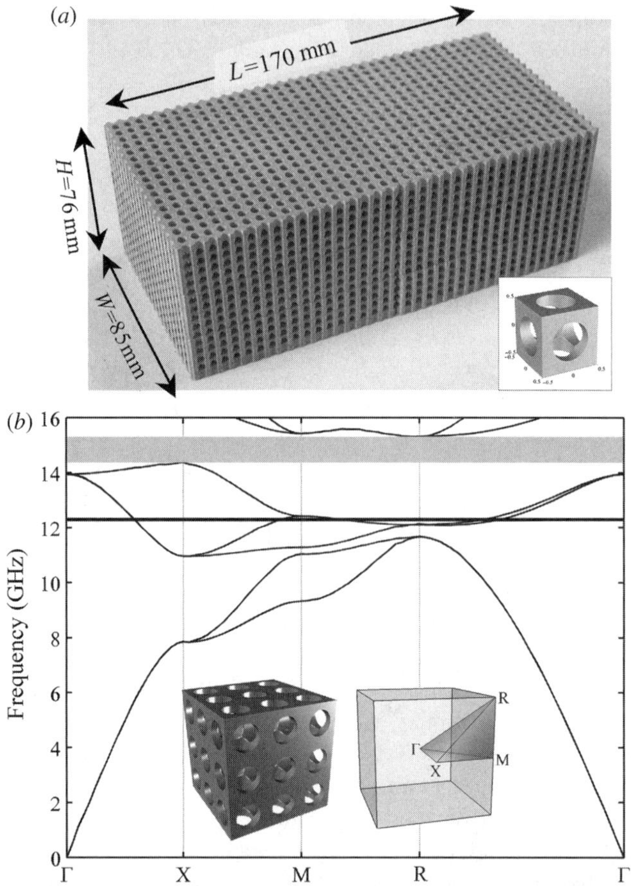

Figure 5.38 Photonic crystal used for testing self-collimation in three dimensions. (a) The fabricated device obtained by drilling three sets of holes using a high-precision, computer-controlled machine [43, 44] through blocks of high-index material consists of 18 × 15 × 36 unit cells. The overall dimensions are $W = 85$ mm, $H = 76$ mm, and $L = 170$ mm. The inset shows a single unit cell. (b) Band diagram of the photonic crystal with the inset indicating the positions of high symmetry points. The shaded region between 14.4 GHz and 15 GHz is a bandgap, and the horizontal line near 12.3 GHz is the range of self-collimation.

source excites the fourth band. Because magnetic charges, and therefore magnetic current, have not been found in nature, to observe self-collimation experimentally, the source must be configured in such a way as to make it *equivalent* to a magnetic current dipole. This can be accomplished by using a waveguide with a flat copper ground plane having a square aperture 1 cm by 1 cm at the end and by ensuring that the electric component of the incident field is polarized along the vertical axis [45]. In the experiment, the RF signal is provided by a vector network analyzer.

To map out the electromagnetic field throughout the volume of the photonic crystal, a monopole antenna was used, which was made of a 1.8-mm-diameter coaxial

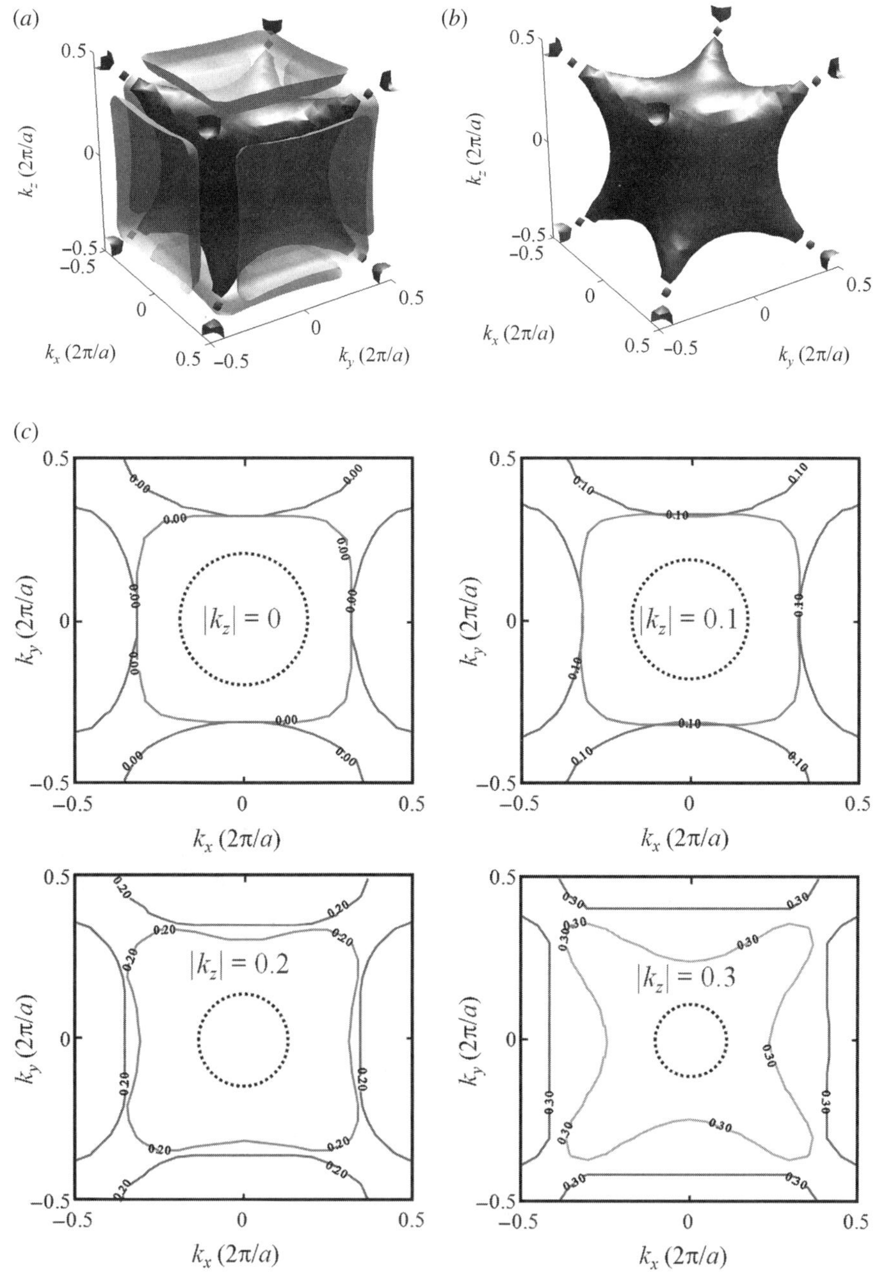

Figure 5.39 EFSs for the third and fourth bands at 12.3 GHz. (a) Third and fourth bands are degenerate. (b) EFS of the third band is shown separately. (c) Equifrequency contours obtained by sectioning the EFS of (a) with a horizontal plane at four different values of k_z. The dotted circle in the center of each diagram is the EFC of free space.

cable with 1.0 mm of the core exposed. The antenna was mounted on a motorized, computer-controlled *x-y-z* stage and connected to the input of a vector network analyzer to detect the transmission from the source to the detector, and thus the electromagnetic field produced by the source at the detector location. Because the diameter of the coaxial cable used for the detector antenna, 1.8 mm, is smaller than the diameter of the holes in the tested photonic crystal, 3.2 mm, the antenna can be inserted into the photonic crystal, and the electromagnetic field distribution can be mapped throughout its volume. Although the presence of the detector antenna perturbs the measured electromagnetic field to some extent, the effect is deemed minimal.

With this arrangement, the field inside the 3D photonic crystal was scanned in a point-by-point, and hole-by-hole, fashion, where the step size along the hole axis was 1.0 mm. At each location, the frequency was swept from 11.0 GHz to 17.0 GHz in 0.1-GHz increments. The result is a three-dimensional distribution of the electromagnetic field in the form of a matrix with $N_x \times N_y \times N_z = 75 \times 17 \times 34$ elements. To estimate the beam size and shape within the PhC, a standard definition of full width at half maximum (FWHM) was used, where the effect of random variation was reduced by averaging over the five highest values in each *xy* slice to find the reference maximum value in that slice. The shape of the beam at half maximum is shown in Fig. 5.40 for the frequency of the input signal $f = 12.2$ GHz. The figure

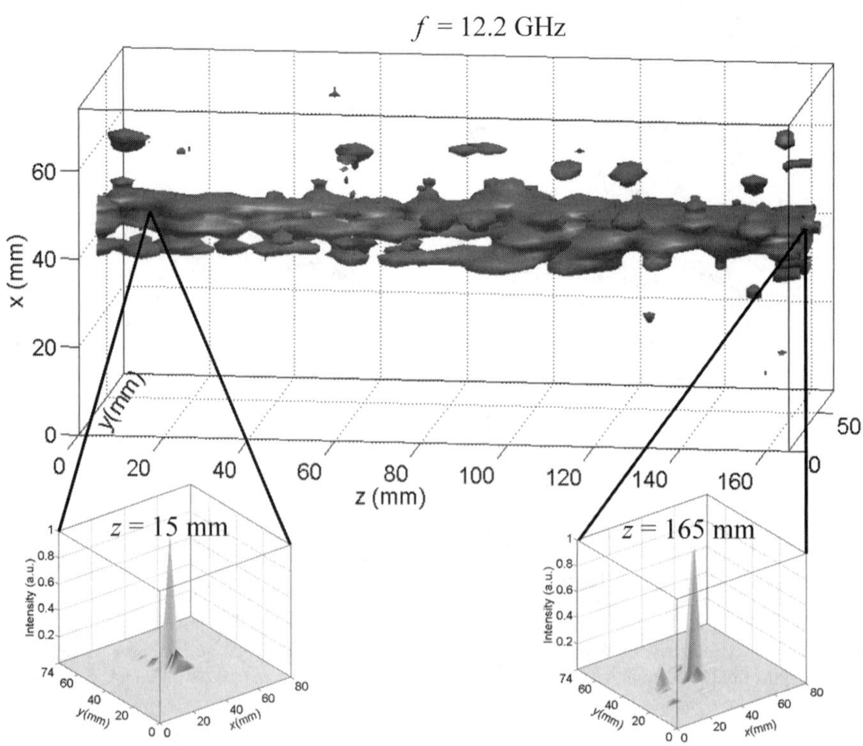

Figure 5.40 Profile of an RF beam propagating in a three-dimensional photonic crystal shown in Fig. 5.38a. The depicted surface corresponds to half of the maximum value at each *xy* slice.

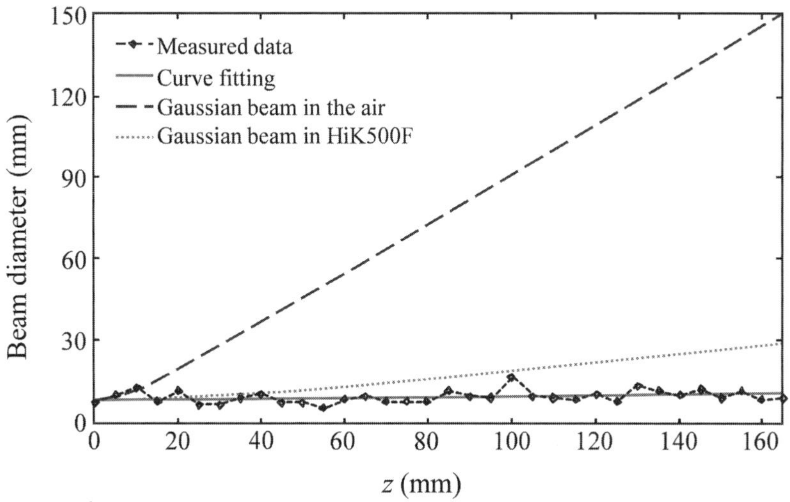

Figure 5.41 The evolution of the beam diameter in the tested photonic crystal compared with beam expansion in free space and uniform high-index material.

clearly shows a narrow, well-collimated beam with little divergence along the propagation direction. The isolated hot spots are thought to be the result of field disturbance by the coaxial cable that leads to the monopole probe used for field detection, which is inserted into the photonic crystal from the bottom.

The minimal expansion of the beam shown in Fig. 5.40 as it propagates within the photonic crystal is quantified in Fig. 5.41, where the beam diameter is tracked at the consecutive *xy* slices. It shows a very slow increase in size from 8.5 mm to 10.5 mm over a distance of 165 mm. For comparison, a Gaussian beam with the initial waist of 8.5 mm propagating in free space expands to 150 mm over this distance, whereas the same beam propagating in uniform material with $\varepsilon_r = 30$ (the same as the material used to build the photonic crystal) expands to 29 mm. This result shows unequivocally that it is indeed the presence of the photonic-crystal structure that enables beam propagation with little or no divergence—or self-collimation.

Because in practical applications the collimated beam would be used to transfer electromagnetic signal or energy between different locations, it is essential to examine losses along the propagation direction. To this end, the power at the first *xy* slice is obtained by integrating the beam intensity and used as a reference input power, with respect to which the transmission is measured. Figure 5.42a shows the decay of the signal as it propagates along the photonic crystal. For a self-collimated frequency, the total attenuation over the 165-mm length of the tested photonic crystal is 15 dB, whereas the minimum transmission of −79 dB is observed when signal with a frequency inside the photonic bandgap is launched. For the self-collimated signal, most of the power loss is observed within the first few periods of the photonic crystal. This insertion loss can be attributed to the impedance mismatch between the source and the photonic crystal. A similar problem arises in 2D self-collimation

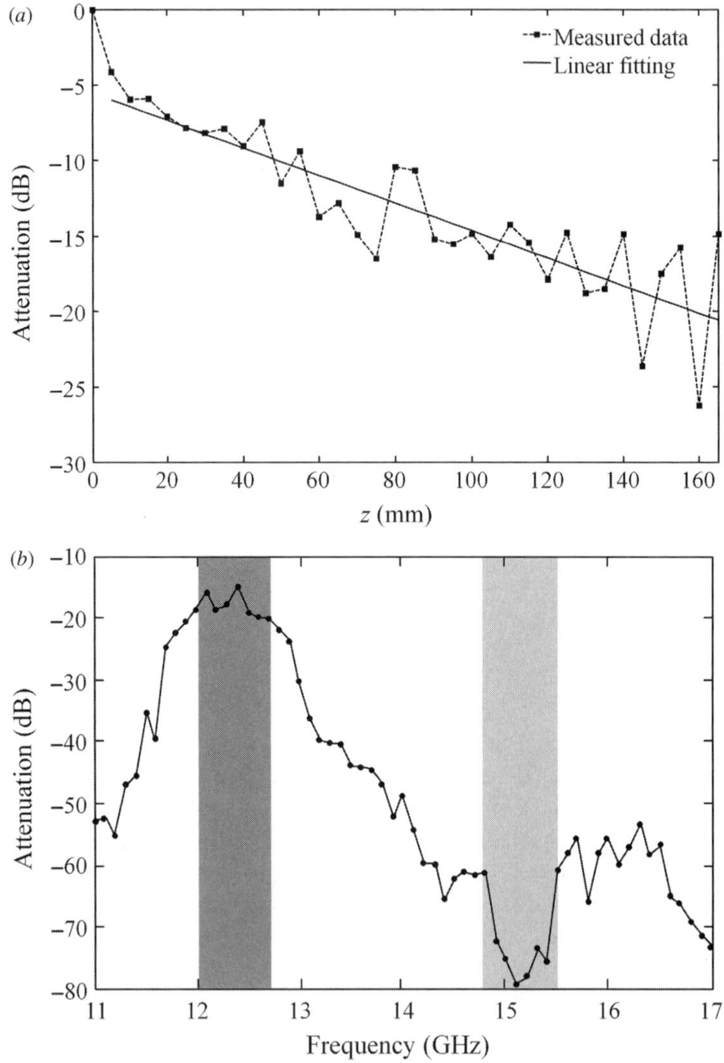

Figure 5.42 Energy transport in a self-collimating photonic crystal. (a) Attenuation of the input signal at the self-collimated frequency $f = 12.4\,\text{GHz}$ along the propagation direction. (b) Attenuation versus frequency. Indicated in dark gray is a frequency band where self-collimation takes place. The photonic bandgap is shaded light gray.

and was reported in Ref. 46. Beyond that entry region, the decay is linear on the dB scale.

It is worth noting that a structure similar to the one used for the simulations, namely a simple cubic lattice of overlapping air holes, seems to be amenable to fabrication at IR length scales using standard microelectronics processing methods; see Chapter 6 for details.

5.5 LEFT-HANDED BEHAVIOR AND NEGATIVE REFRACTION

In the absence of charges and currents, Maxwell's curl equations take the following form:

$$\begin{aligned} \nabla \times \mathbf{E} &= -\frac{1}{c}\frac{\partial \mathbf{B}}{\partial t} \\ \nabla \times \mathbf{H} &= \frac{1}{c}\frac{\partial \mathbf{D}}{\partial t} \end{aligned}. \tag{5.11}$$

For a plane wave traveling through isotropic and homogeneous material, such that the space-time dependence of the field is expressed as $\exp[j(\mathbf{k}\cdot\mathbf{r}-\omega t)]$, Eqs. (5.11) take the form

$$\begin{aligned} \mathbf{k} \times \mathbf{E} &= \mu\frac{\omega}{c}\mathbf{H} \\ \mathbf{k} \times \mathbf{H} &= -\varepsilon\frac{\omega}{c}\mathbf{E} \end{aligned}, \tag{5.12}$$

where μ and ε are, respectively, the permeability and the permittivity of the material. In ordinary dielectrics found in nature, both μ and ε are positive, which leads to the right-handed relation among the triplet of vectors **E**, **H**, and **k**, where the thumb points in the direction of **E**, whereas index and middle fingers point in the directions of **H** and **k**, respectively. However, if μ and ε are both negative, then the vector triplet **E**, **H**, and **k** satisfies the left-handed relation. Hence, materials where both the permittivity and the permeability are negative are referred to as left-handed.

The significance of this distinction becomes apparent when one considers the flow of energy associated with the electromagnetic wave, which is determined by the Poynting vector $\mathbf{S} = \mathbf{E} \times \mathbf{H}$. It follows that for right-handed materials, the direction of the energy flow coincides with the wave propagation vector, i.e. **S** is parallel to **k**. In contrast, in left-handed materials, **S** and **k** are antiparallel—in other words, the energy flow is opposite to the phase velocity of the wave. This property has profound consequences for the behavior of electromagnetic waves traversing such materials, as discovered by Veselago [12, 47] and by Pendry [13]. For example, in such materials, the Doppler shift is reversed; i.e., the frequency of light emitted by a source moving toward the observer is reduced, and Cherenkov radiation points the other way with respect to the moving particle that generates it. Furthermore, such materials enable a flat lens with no optical axis that faithfully reconstructs images containing subwavelength features thanks to the amplification of evanescent waves [13].

One can use Eqs. (5.11) to derive a wave equation for the electric or magnetic field and to obtain the propagation speed of the corresponding wave equal to $c/\sqrt{\varepsilon\mu}$, where $\sqrt{\varepsilon\mu}$ is referred to as the refractive index of the material and designated with a letter n. Note that in principle, n can be either positive or negative

$$n = \pm\sqrt{\varepsilon\mu}, \tag{5.13}$$

because the square root does not determine the sign; conventionally in right-handed materials, the + sign is chosen. If both μ and ε are negative, then it would seem that we have the same freedom in selecting the sign in Eq. (5.14), but careful consideration reveals that the opposite sign to that assumed for right-handed materials must be used [13]. In other words, the refractive index of left-handed materials is negative and so is the refraction of light at the boundary between left-handed and right-handed material.

Such a negative refraction makes an axis-free lens possible as illustrated in Fig. 5.43. In this case, light rays that originate at a point to the left of a negative-index slab undergo negative refraction at the boundary between the two materials. They follow on to converge to a focus inside the negative-index material, and after the second boundary, they produce a focal point on the other side of the slab. Because such a lens is translationally invariant, no optical axis is present, and the same focusing and imaging takes place regardless of the position of the object along the surface.

The phenomena described above are a direct consequence of the fact that the electromagnetic energy moves in the opposite direction to the wave vector and only indirectly follow from the negative sign of μ and ε. In highly structured materials, such as photonic crystals, using the Poynting vector to determine the energy flow can be awkward. However, because group velocity also points in the direction of energy flow and is well defined in photonic crystals, it can be used instead of the Poynting vector. Thus, left-handedness in this case can be defined as having group and phase velocities that point in opposite directions. Furthermore, if an EFC is circular with a radius k, the effective refractive index for the frequency ω corresponding to such contour can be defined following Snell's law as

$$\frac{k/\omega}{c}, \tag{5.14}$$

where c-is the speed of light in vacuum. These definitions will be used in subsequent sections to analyze the properties of certain photonic crystals.

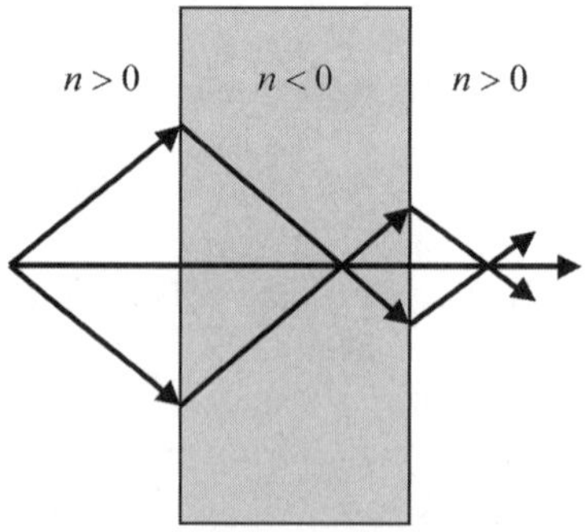

Figure 5.43 Optical-axis-free lens made from negative-index material. Rays of light that originate at a point undergo negative refraction at the boundary between the positive-index medium and the negative-index medium. As a result, the rays are focused first inside the negative-index material and then outside.

5.5.1 3D Subwavelength Imaging by a Photonic-Crystal Flat Lens

A practical three-dimensional photonic crystal that exhibits negative refraction over a relatively wide bandwidth has been first described by Luo et al. [48]. It consists of air cubes placed in the nodes of a body-centered-cubic (BCC) lattice and embedded in high-index dielectric. The sides of the air cubes are parallel to the sides of the conventional BCC unit cell and are three quarters of their length. Because the side of the air cube is more than half of the BCC-cell side, the air cubes overlap to produce a three-dimensional mesh shown in the right inset of Fig. 5.44a. The spherical shape of the EFS at $f = 0.36c/a$, as shown in Fig. 5.44b, allows the use of Eq. (5.14) for determining the effective refractive index of such material. Furthermore, according to the band diagram of Fig. 5.44a, for the third band, between points Γ and N as well as between points Γ and H, the slope of the dispersion curve is negative (frequency decreases with increasing $|k|$), which indicates that the group velocity is opposite to phase velocity for this band. Combined with the spherical shape of the EFS, it makes this photonic crystal suitable for the demonstration of negative refraction and its use as a flat lens. The frequency preferred for such application is $0.36c/a$, where the dispersion curve intersects the light line, shown in Fig. 5.44a as diagonal lines originating at the Γ point, thereby producing the effective refractive index of -1, which is important for minimizing aberrations.

The layered appearance of this photonic crystal makes it amenable to fabrication using planar machining and stacking. However, the elementary cell shown in Fig. 5.45a, although highly symmetric, is difficult to realize because of the presence of undercut bridges. To facilitate the fabrication process, a different unit cell is chosen, Fig. 5.45b, which is obtained by shifting the original one in the z direction by a fraction of the lattice period. An entire layer of such unit cells can be fabricated by machining both sides of a flat sheet of the dielectric material using a high-precision computer numerical control (CNC) machine [43]. Stacking such prefabricated sheets produces a three-dimensional photonic crystal as shown in Fig. 5.45c, where 20 layers were used. Note that the front surface of the slab corresponds to a (101) plane of the photonic crystal, because the BCC unit-cell cubes are rotated around the vertical axis by 45° with respect to the orientation of the lens facets; this rotation is necessary to couple from free space to the intended PhC mode that exhibits negative refraction [48]. The dimensions of the photonic crystal are as follows: thickness, 25 mm; width, 140 mm; and height, 127 mm. The material is a low-loss dielectric called HiK 500F, manufactured by Emerson & Cuming, with $\varepsilon_r = 25$ at microwave frequencies. For $a = 6.35$ mm, the region of interest for negative refraction is in the microwave part of the spectrum around 17 GHz.

The performance of thus realized photonic-crystal flat lens has been tested using two types of sources, each with subwavelength spatial extent: a monopole and a pinhole. In the first case, a 1.8-mm-thick coaxial cable with 1 mm of the core exposed at the end was connected to the output of a vector network analyzer and placed 3 mm away from the photonic crystal, close to its center, on what will be referred to as the object side. A similar monopole configuration was used for a detector, which was placed on an x-y-z computer-controlled scanning stage, on the image side of the

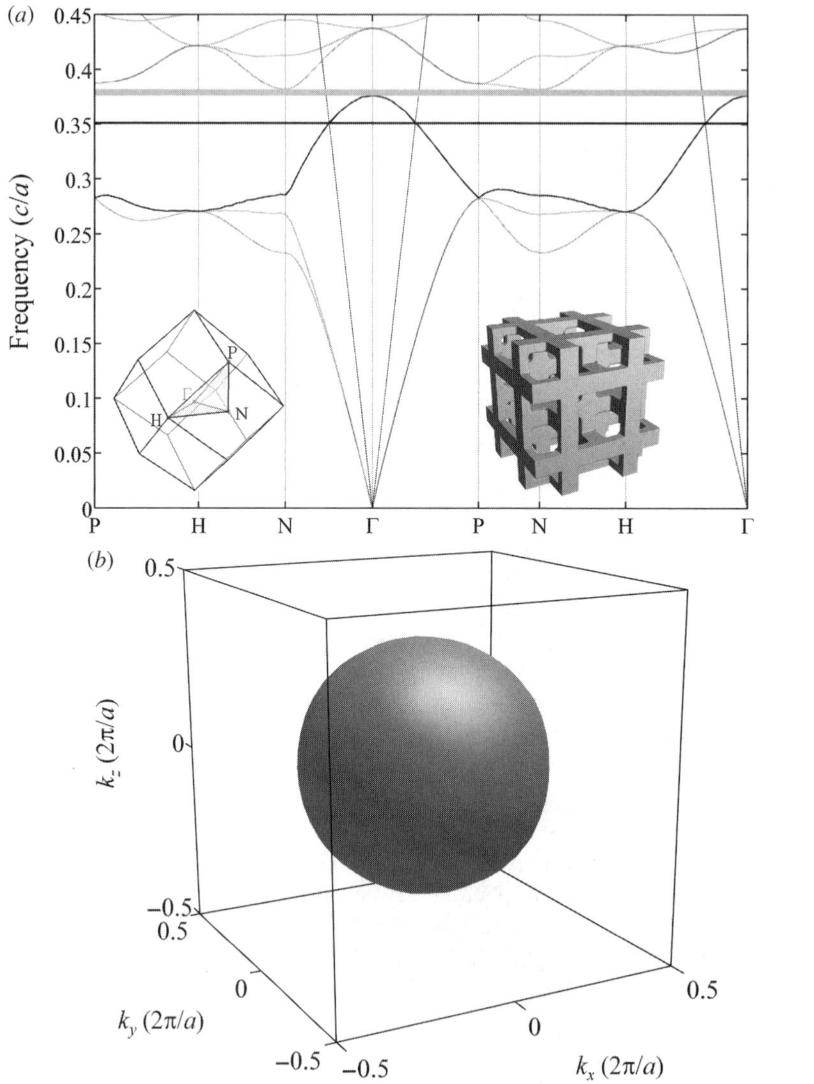

Figure 5.44 Band diagram (a) and equifrequency surface at $f = 0.36c/a$ (b) of a BCC photonic crystal shown in the inset of (a). Dielectric constant of the material is $\varepsilon_r = 25$. The straight diagonal lines that originate at the Γ point correspond to the dispersion curves of free space, i.e., light lines.

lens, and connected to the input of the same network analyzer. The orientation of the source and the detector monopoles were the same to produce and detect x-polarized waves, where the z direction is perpendicular to the photonic-crystal-lens surface. The lens is mounted on a Teflon bar, which has a low dielectric constant, and therefore, its presence can be neglected in analyzing the results.

The spatial field distribution is obtained by scanning the detector across all three directions, while at each position sweeping the frequency across the range of

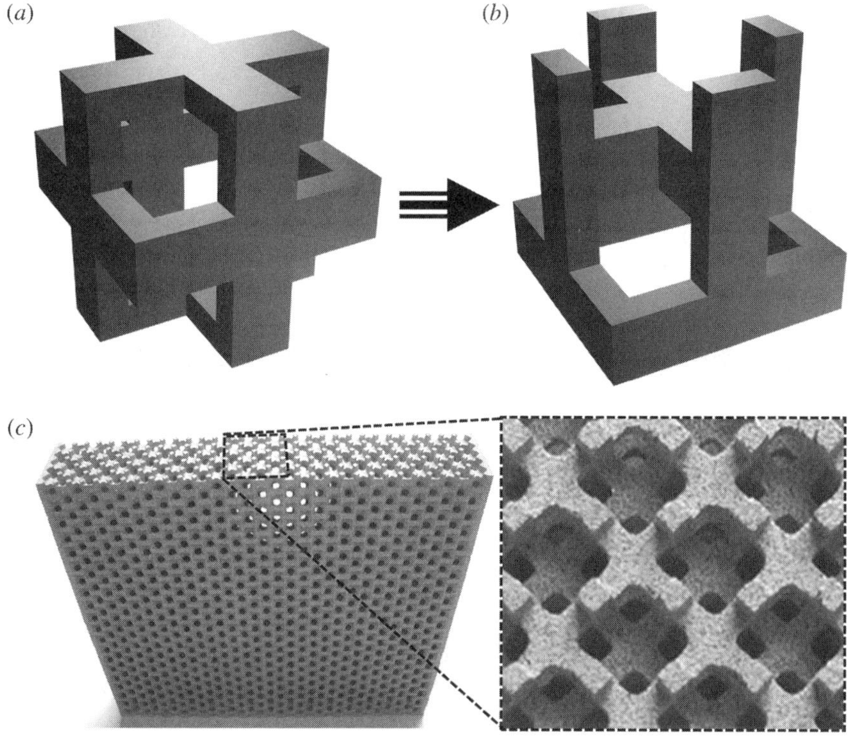

Figure 5.45 (a) Conventional cubic unit cell of the BCC photonic crystal shown in the inset of Fig. 5.44a. (b) The modified unit cell after shifting by $(3/8)a$ along the vertical direction. (c) Three-dimensional PhC fabricated following a layer-by-layer process (20 layers in total) with a lattice constant $a = 6.35$ mm.

interest. Figure 5.46 shows the resulting image of the monopole source produced by the photonic-crystal lens at a frequency $f = 16.4$ GHz, in an xy plane 12 mm away from the lens surface. The spot size, defined using the FWHM criterion, is about 9 mm by 7 mm, which is considerably smaller than the wavelength of the radiation $\lambda = 18.3$ mm; i.e., the lens produced a subwavelength image.

The distribution of the field in the yz plane is shown in Fig. 5.47. It demonstrates how the image of the radiating monopole antenna is formed on the far side of the 3D photonic-crystal lens and away from its surface. The amplitude distribution of Fig. 5.47a shows a bright spot on the image side of the lens, while the phase distribution of Fig. 5.47b shows how the diverging spherical waves are centered at the focal point.

The evolution of the focusing is additionally demonstrated in Fig. 5.48, which shows scans obtained in different xy planes (different separations from the lens surface) at the same frequency. Therein, one can see that the field distribution close to the surface of the lens is dispersed, becomes tight in the image plane, and then spreads beyond it. In other words, the electromagnetic waves converge first and then diverge into far field, which again proves that the subwavelength image is not simply the result of the proximity of the detector to the source, but rather that it is a genuine effect produced by the PhC lens.

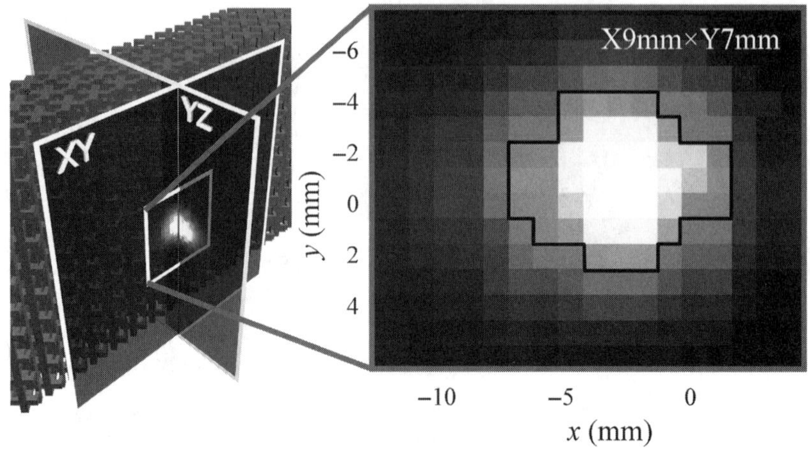

Figure 5.46 Image of a microwave monopole emitting at $f = 16.4$ GHz produced by a photonic-crystal flat lens 12 mm from its surface.

It should be noted that the asymmetry of the field distribution observed in Fig. 5.48 is the result of the corresponding asymmetry of the source and of the detector. The monopole antennas are at the respective ends of coaxial cables, which are parallel to the x-axis. Thus, the system, which includes the source and the detector, exhibits mirror symmetry with respect to y reflection but not with respect to x reflection. This asymmetry manifests itself in the appearance of the images.

To alleviate the asymmetry problem, the PhC lens was also tested using a symmetric source. To this end, a 3-mm pinhole was produced in aluminum foil, which was then secured to a piece of 5-mm-thick, low-density foam board, whose thickness determines the distance of the source from the lens surface. Because the dielectric constant of the foam-board material is around 1, it has a negligible effect on the propagation of the electromagnetic field. The foam board was then placed flush with the photonic-crystal lens with the foil side facing away and was illuminated by a horn antenna connected to the output of a vector network analyzer. Thus, a source with 3-mm-wide aperture placed 5 mm away from the lens is obtained. Because the aperture is considerably smaller than half the wavelength used for illumination, the field emanating from it is largely evanescent. The detector portion of the system remained unchanged from the experiment described above.

The resulting field distribution is shown in Fig. 5.49. The size of the obtained spot is even smaller than with the monopole source, and measures 7 mm by 6 mm FWHM, which averages 6.5 mm or 36% of the wavelength used in this experiment. This confirms the capabilities of the PhC lens to image beyond the diffraction limit of conventional optics, which is equal to 0.51λ. The improvement is the result of evanescent-field amplification as discussed in the seminal paper by Pendry [13]. The degradation of the feature size in the image as compared to the object (6.5 mm vs. 3.0 mm) can be understood if the periodicity of the photonic crystal is taken into consideration. According to Ref. 8, the PhC imposes an upper limit on the spatial

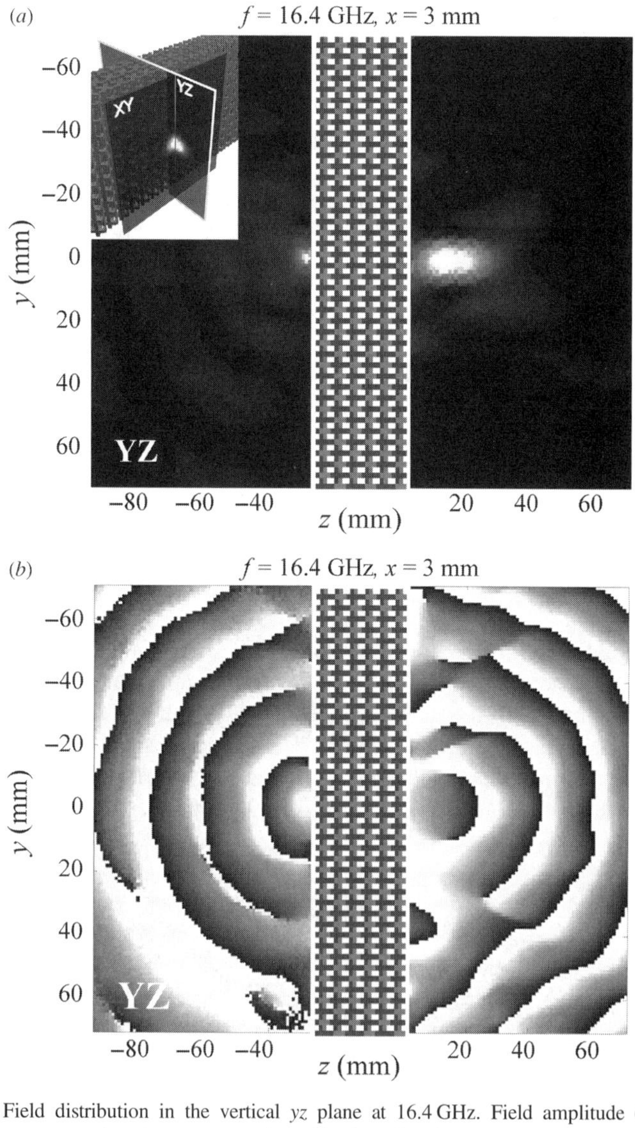

Figure 5.47 Field distribution in the vertical yz plane at 16.4 GHz. Field amplitude (a) and phase (b) produced by a monopole antenna source oriented in the x direction.

frequency of the evanescent waves that can be amplified. This cutoff provides the lower bound on the image size, which in this case approaches the periodicity of the PhC used for the lens just as predicted.

The distance of the image from the lens is different than expected from a simple effective-index theory where the distance between the source and its image is equal to twice the lens thickness for $n_{\text{eff}} = -1$. The discrepancy is most likely the result of the structure of the lens material and its boundary. In particular, it seems that the physical thickness of the lens, 25 mm, is larger than its effective thickness, i.e.,

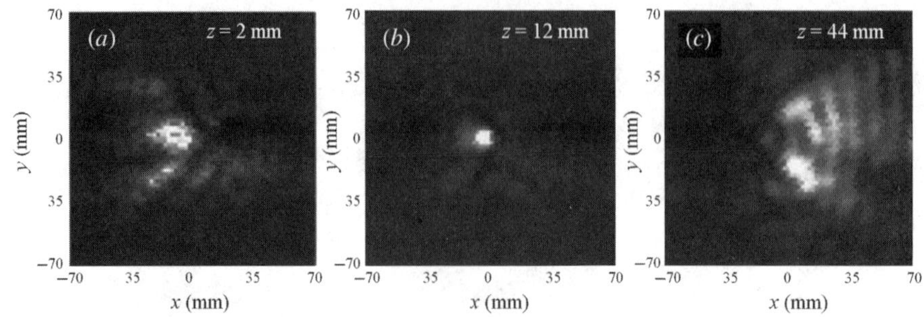

Figure 5.48 Amplitude distribution in three xy planes for three different separations from the lens surface. (a) $z = 2$ mm, (b) $z = 12$ mm, (c) $z = 44$ mm. The electromagnetic wave first converges and then diverges into far field. The asymmetry observed in the field distribution is the result of the asymmetric detector and source, where the coaxial cable that leads to the monopole antenna is parallel to the x-axis.

the thickness seen by the electromagnetic wave. According to this interpretation, the experimental results indicate that the effective lens thickness is about 20.5 mm or in other words, the lens effective boundary is shifted by about 2.25 mm inside the material.

The image formation is shown in Fig. 5.50, where the intensity and phase distributions are obtained at a frequency $f = 16.4$ GHz for yz and xz planes. The high-intensity spot along with the closed concentric phase circles in the scanned planes indicate how the image is formed. The FWHM in the vertical yz plane is measured to be 7 mm × 9 mm, whereas that in the horizontal xz plane it is 8 mm × 9 mm. So, the overall size of the spot is $(8 \times 7 \times 9)\,\text{mm}^3$ or $(0.38 \times 0.34 \times 0.49)\,\lambda^3$,

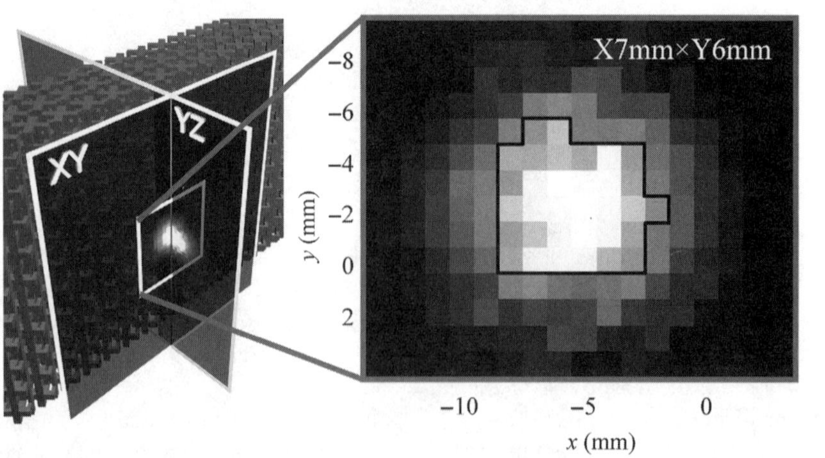

Figure 5.49 The image of a point source produced by a 3D PhC flat lens in xy plane 11 mm away from the lens surface. The point source is in the form of a 3-mm-wide pinhole in aluminum foil placed 5 mm away from the lens on the object side and illuminated with an electromagnetic field at a frequency of 16.4 GHz.

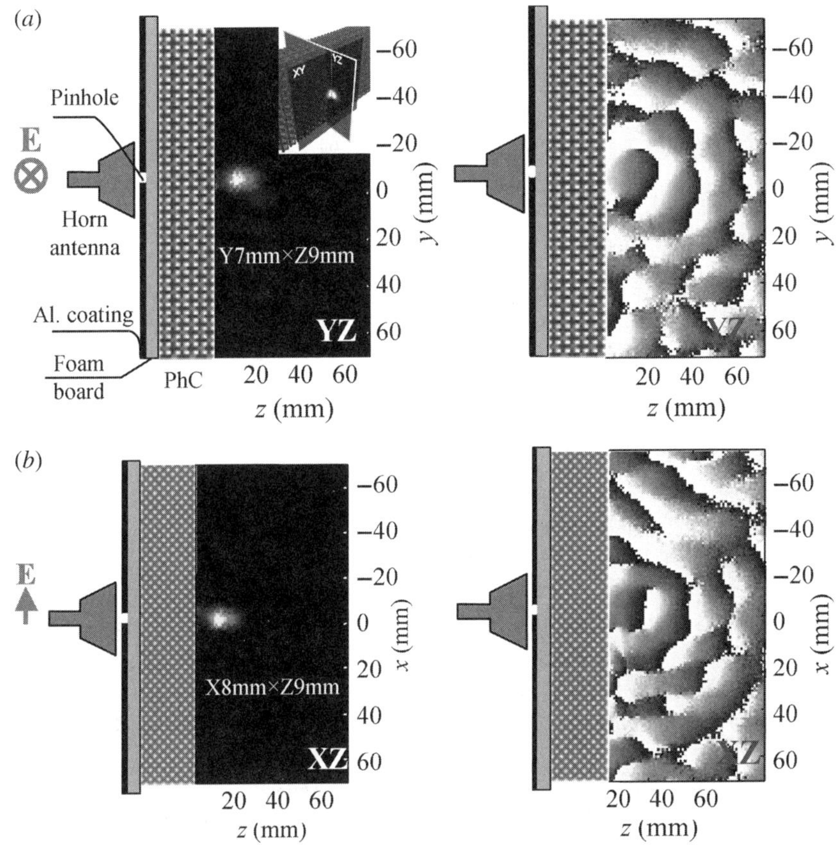

Figure 5.50 Intensity and phase distribution in (a) vertical yz plane, and (b) horizontal xz plane. The frequency of the incident wave is $f = 16.4\,\text{GHz}$, which corresponds to a free-space wavelength $\lambda = 18.3\,\text{mm}$.

which demonstrates subwavelength imaging resolution in all three dimensions. This implies the ability to image subwavelength features three-dimensionally using such negative-refraction flat lenses. In other words, a super-resolution 3D real image of an object that is nearly identical to the original in every aspect, including orientation, can be reconstructed on the other side of the lens.

Additional validation of the superresolution is provided by introducing a second pinhole with the same diameter, and distance 10 mm away from the first pinhole, and repeating the scan. Figure 5.51 shows the intensity distribution obtained at a distance $z = 11$ mm away from the lens surface, where two spots 10 mm apart, which corresponds to 0.55λ, can be clearly discerned. The magnification of the system is 1, which is expected because the lens is flat and has translation symmetry in the xy plane. Closer inspection of the linear intensity scan shown in Fig. 5.51b indicates that the potential resolving power of the lens is even better, and that coherent point sources 8 mm apart should still be resolvable.

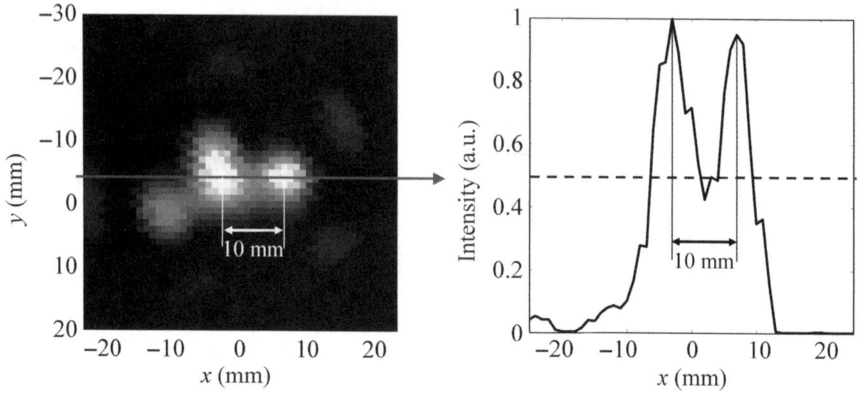

Figure 5.51 (a) Image of two closely spaced pinholes produced by a left-handed PhC lens. (b) Intensity profile along the horizontal line of (a).

5.6 SUPERPRISM, NEGATIVE REFRACTION AND SELF-COLLIMATION

The wealth of dispersive phenomena available in photonic crystals can be illustrated with a single device that exhibits all properties analyzed in this chapter. It should be pointed out that such a device is not only a theoretical curiosity but also could find practical applications in optical signal processing for enhancing optical switching and in scanning applications.

The structure is a square lattice of circular air holes in silicon with the ratio of the hole radius to the lattice period equal to 0.3. The equifrequency contour at normalized frequency $a/\lambda = 0.26$ is shown in Fig. 5.52a where the arrows along the contour point in the direction of the frequency gradient. The k-vector components span the range from $-2\pi/a$ to $2\pi/a$, which means that more than the first Brillouin zone is covered; therefore, the diagram contains several disjoint contours that appear periodically in the reciprocal space. Because the EFC is square-like with substantially flat sides, self-collimation is expected as explained in Section 5.4. Furthermore, the frequency gradient points inward—in the direction of diminishing k, which means that the photonic crystal is left-handed, as explained in Section 5.5. Therefore, negative refraction is expected at the boundary between the PhC and a right-handed medium such as free space. A detailed explanation of how these two phenomena emerge in this photonic crystal is presented below with the help of a series of diagrams shown in Fig. 5.52.

Since the electromagnetic wave crosses the boundary between the photonic crystal and free space, media with different dispersion properties, an EFC for free space is also depicted in Fig. 5.52b, using a dashed curve, in addition to the PhC EFC; the EFC of free space is a simple circle. Arrows along the free-space EFC in Fig. 5.52b point in the direction of frequency gradient, outward in this case, which is consistent with nondispersive nature of free space.

The photonic-crystal boundary is assumed to be parallel to the [11] direction of the crystal—along the square diagonal. The direction of this boundary is depicted

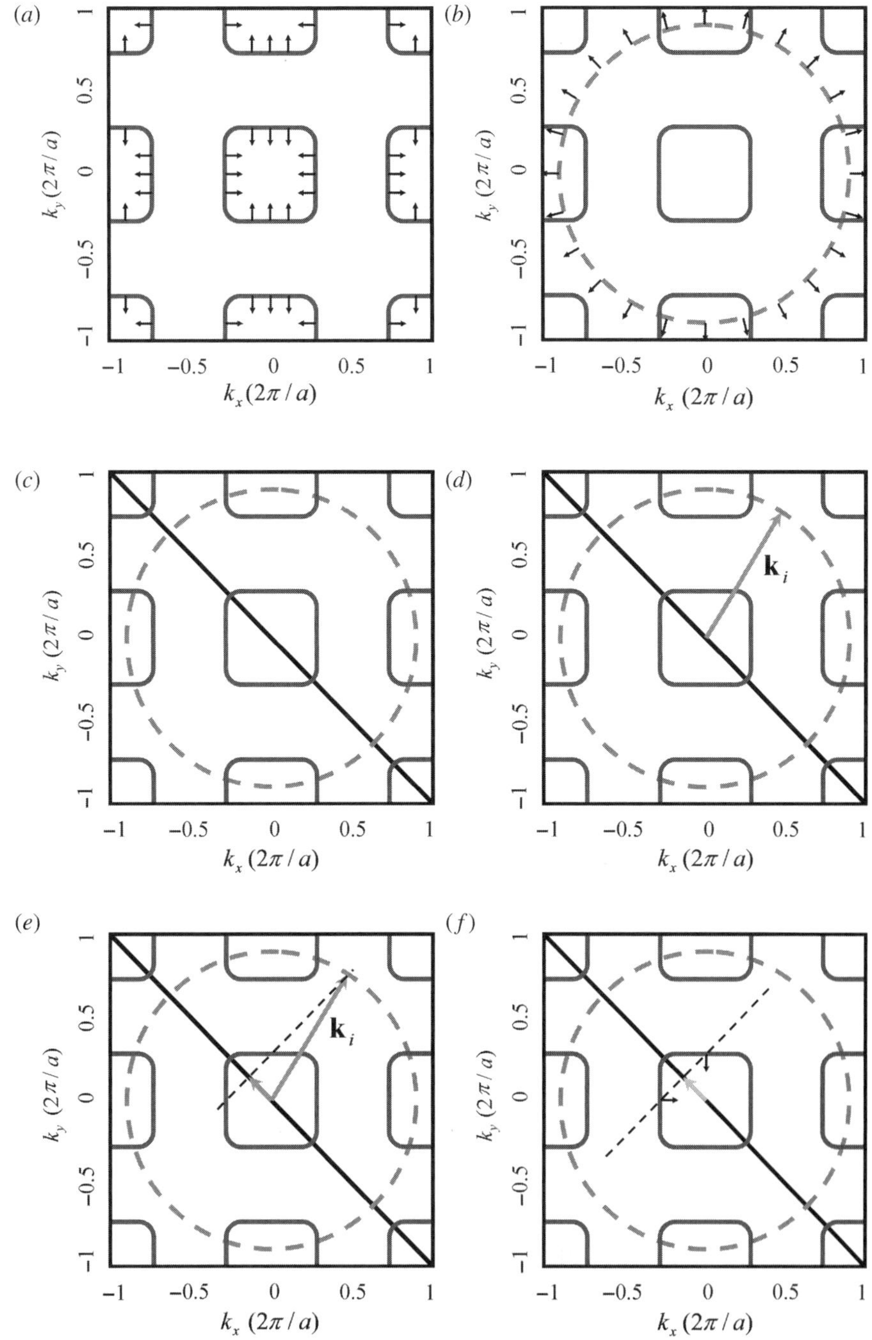

Figure 5.52 Refraction of a wave at the boundary between free space and photonic crystal. See text for details.

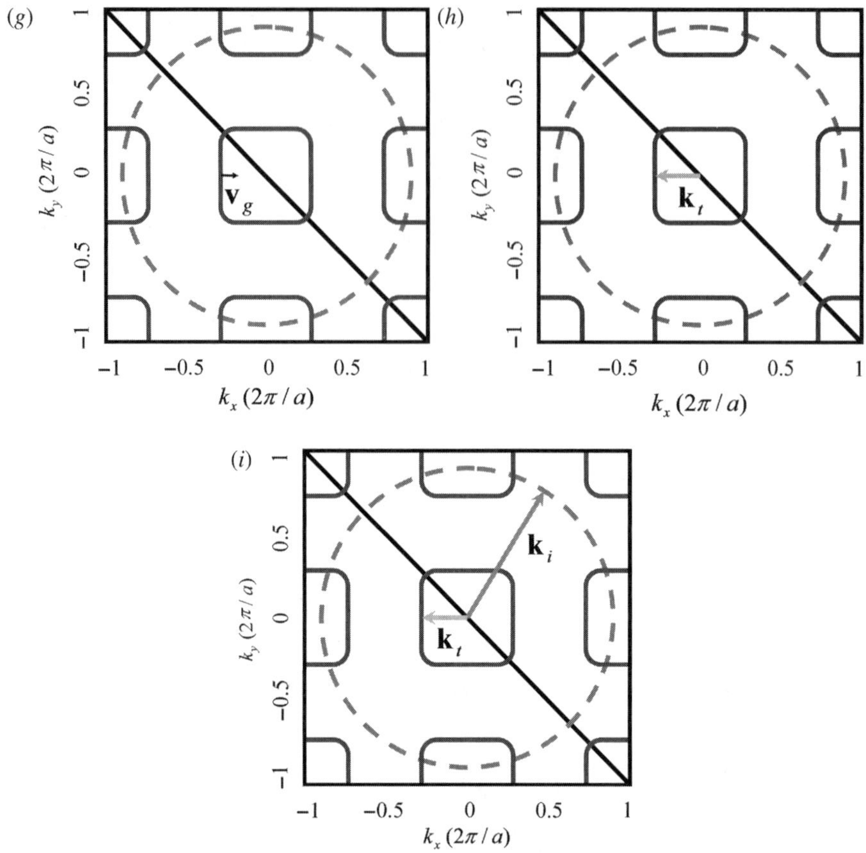

Figure 5.52 (*Continued*)

schematically in Fig. 5.52c, which shows a diagonal Γ-M line in addition to the EFCs of the photonic crystal and free space. It should be noted that the line in Fig. 5.52c is *not* the boundary of the photonic crystal itself—that boundary is in the real space, not in the frequency space shown in the diagram. Rather, it indicates the *direction* of the PhC cut by singling out all *k*-vectors parallel to the boundary.

Assume now that a plane wave propagating in free space is incident on the PhC boundary. The *k*-vector of the plane wave is shown as a gray arrow in Fig. 5.52d that starts at the origin of the *k*-space and ends at the free-space EFC. The component of the *k*-vector parallel to the boundary between the two media must be conserved. This component is indicated in Fig. 5.52e, which shows the projection of the incident wave *k*-vector $\mathbf{k}_i$ on the diagonal, black ΓM line. It is identical to the tangential component of the propagation vector inside the photonic crystal shown as a light gray arrow in Fig. 5.52f. For the tangential component inside the PhC, two different PhC propagation vectors exist; they are determined by the intersections of a line perpendicular to the PhC cut direction, shown as a dashed line in Fig. 5.52e and f, with the

EFC of the photonic crystal. Which of the two possibilities will be realized is decided based on the energy flow. By definition, in the case analyzed here, the energy flows from free space and into the photonic crystal. Therefore, the group velocity vector associated with the propagating wave, which shows the direction of energy flow, must point toward the photonic crystal. Of the two possibilities in Fig. 5.52f, only one satisfies this condition, and it is shown separately in Fig. 5.52g. As a result, the propagation vector inside the photonic crystal, $\mathbf{k}_t$, points toward free space as indicated with a gray arrow in Fig. 5.52h. The k-vectors of the incident and refracted waves are shown again together in Fig. 5.52i, and it is seen that the phase velocities in free space and in photonic crystal both point toward the free-space/PhC boundary. In addition, it should be noted that the group velocities point opposite ways with respect to the boundary normal, which can be ascertained by comparing the direction of $\mathbf{v}_g$ in Fig. 5.52g and the direction of the incident free-space k-vector $\mathbf{k}_i$, which is parallel to the free-space group velocity, Fig. 5.52d. Therefore, negative refraction in this photonic crystal is established.

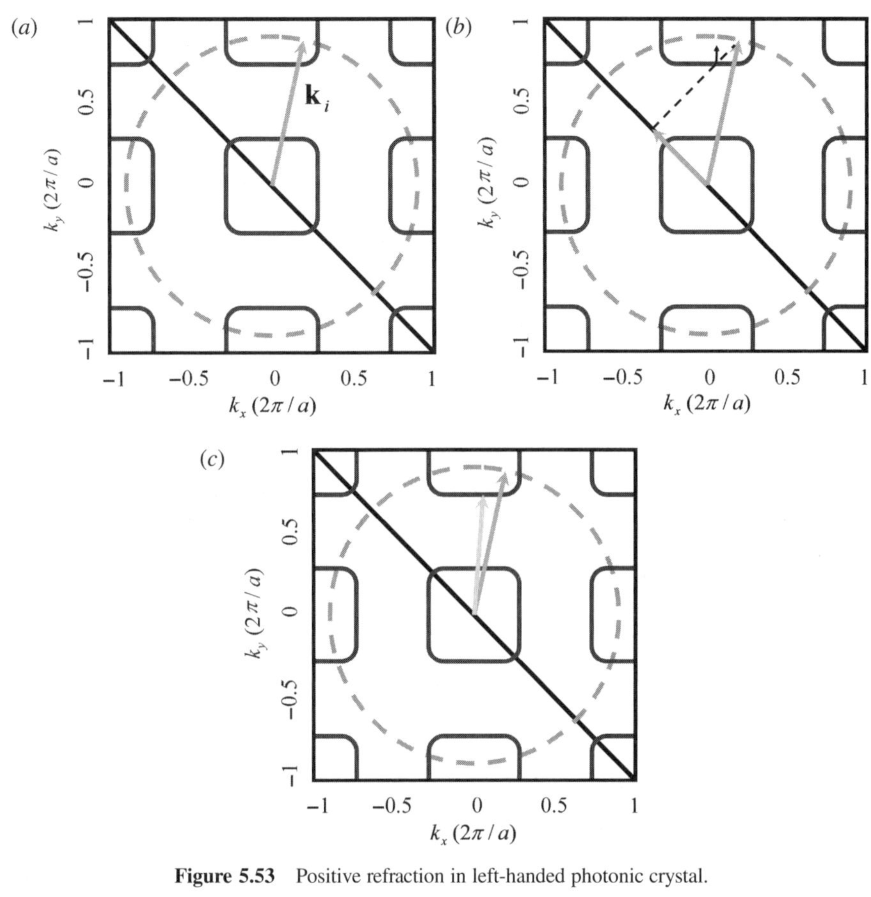

Figure 5.53 Positive refraction in left-handed photonic crystal.

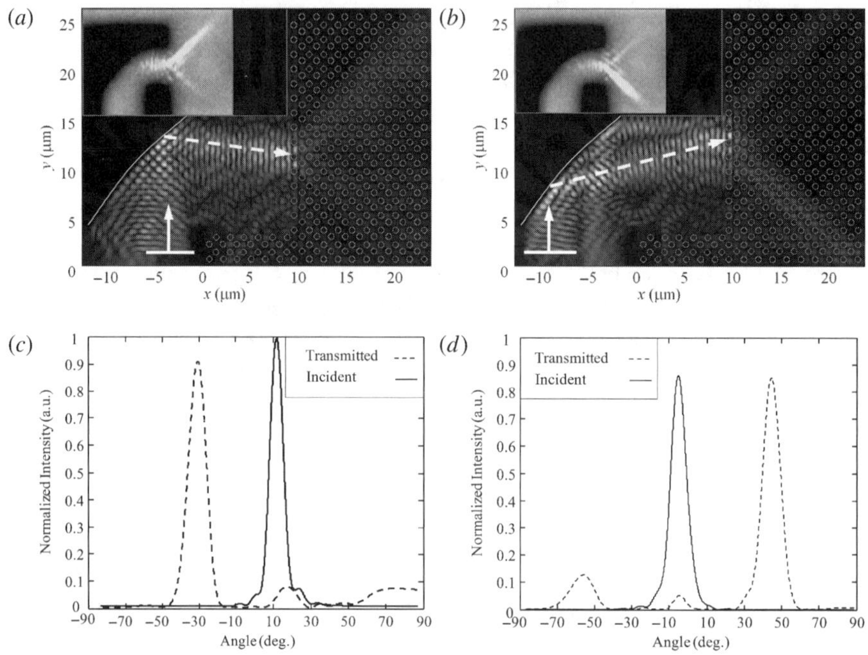

Figure 5.54 Negative refraction, self-collimation, and superprism effects observed in a single photonic crystal at near infrared wavelength. (a) and (b) show FDTD steady-state electromagnetic field and the corresponding experimental results in the insets. The solid line in (c) and (d) shows the distribution of the incidence angle of the incoming wave, whereas the dashed line shows the corresponding distribution of the refracted beam angle for the situations depicted in (a) and (b), respectively.

Interestingly, this material also admits positive refraction at the free-space/PhC boundary. To understand how this happens, consider Fig. 5.53 where a construction similar to that presented above is shown. In this case, the incident k-vector $\mathbf{k}_i$, Fig. 5.53a, is such that its projection line on the direction parallel to the free-space/PhC boundary does not intersect the EFC in the first Brillouin zone. Instead, the EFC from the second Brillouin zone must be taken into account, Fig. 5.53b. As a result, the incident and refracted k-vectors are nearly parallel, as shown in Fig. 5.53c, and so are the vectors of group velocities in the two media. Because both lie on the same side of the boundary normal, the usual positive refraction takes place.

The negative refraction described above in the spatial frequency domain has also been observed in real-space simulations and experimentally as shown in Fig 5.54. To control the incidence angle, different portions of the J-coupler are illuminated. In Fig. 5.54a, the top portion of the J-coupler is illuminated by placing the Gaussian source to the right of the input waveguide axis. As a result of this asymmetric illumination, the beam reflected and focused by the J-coupler is also asymmetric, and the incidence angle on the slab/PhC boundary is about 12°. This angle can be ascertained by following the bright light trail in the steady-state results in Fig. 5.54a and from Fig. 5.54c, which shows the angular distribution of the incident and refracted *k*-vectors. At the boundary, the beam undergoes negative refraction—it points in the direction with an angle opposite to the incidence angle with respect to the boundary normal. Indeed, according to Fig. 5.54c, the refracted angle is about -33°. Furthermore, inside the photonic crystal the beam maintains its lateral extension, Fig. 5.54a, which is consistent with self-collimation anticipated from the *k*-space discussion above. The negative refraction and self-collimation have also been observed experimentally, as shown in the inset of Fig. 5.54a, where a silicon slab PhC was tested using infrared laser radiation.

The incidence angle is changed by moving the source to the other side of the input waveguide axis, which results in the illumination of the bottom portion of the J-coupler and the corresponding shift in the asymmetry of the reflected and focused beam. At the incidence angle of -6°, the beam is refracted at about 42° and is self-collimated in the photonic crystal medium, as seen in Fig. 5.54b and d.

Thus, this single photonic-crystal structure exhibits all dispersive phenomena discussed in this chapter.

5.7 SUMMARY

Photonic crystals exhibit a wealth of dispersive phenomena independent of the presence or absence of a photonic bandgap, which is a region in the spectrum where the propagation of electromagnetic waves is forbidden. They include superprism effect, negative refraction, and self-collimation, which are all discussed in detail in this chapter. Understanding of their origin provides tools for custom engineering photonic-crystal properties to fit a wide range of photonic applications, all at scales comparable with the wavelength of light. As further progress is made in the field, the range of interesting and useful phenomena will only grow and enhance the ability to mold the behavior of light to fit requirements of the nascent field of microphotonics.

REFERENCES

1. H. Kosaka, T. Kawashima, A. Tomita, M. Notomi, T. Tamamura, T. Sato, and S. Kawakami, "Superprism Phenomena in Photonic Crystals," *Phys. Rev. B*, vol. 58, pp. 10096–10099, 1998.
2. H. Kosaka, T. Kawashima, A. Tomita, M. Notomi, T. Tamamura, T. Sato, and S. Kawakami, "Self-Collimating Phenomena in Photonic Crystals," *Appl. Phys. Lett.*, vol. 74, pp. 1212–1214, 1999.
3. J. Witzens, M. Lončar, and A. Scherer, "Self-Collimation in Planar Photonic Crystals," *IEEE J. Sel. Topics Quantum Electron.*, vol. 8, pp. 1246–1257, 2002.
4. D. N. Chigrin, S. Enoch, C. M. S. Torres, and G. Tayeb, "Self-Guiding in Two-Dimensional Photonic Crystals," *Opt. Express*, vol. 11, pp. 1203–1211, 2003.
5. B. Gralak, S. Enoch, and G. Tayeb, "Anomalous Refractive Properties of Photonic Crystals," *J. Opt. Soc. Am. A*, vol. 17, pp. 1012–1020, 2000.

6. E. Cubukcu, K. Aydin, E. Özbay, S. Foteinopoulou, and C. M. Soukoulis, "Negative Refraction by Photonic Crystals," *Nature*, vol. 423, pp. 604–605, 2003.
7. P. V. Parimi, W. T. T. Lu, P. Vodo, and S. Sridhar, "Photonic Crystals—Imaging by Flat Lens Using Negative Refraction," *Nature*, vol. 426, pp. 404, 2003.
8. C. Y. Luo, S. G. Johnson, J. D. Joannopoulos, and J. B. Pendry, "Subwavelength Imaging in Photonic Crystals," *Phys. Rev. B*, vol. 68, pp. 045115, 2003.
9. C. M. Krowne, "Negative-Refractive Bicrystal with Broken Symmetry Produces Asymmetric Electromagnetic Fields in Guided-Wave Heterostructures," *Phys. Rev. Lett.*, vol. 93, pp. 053902, 2004.
10. A. Berrier, M. Mulot, M. Swillo, M. Qiu, L. Thylen, A. Talneau, and S. Anand, "Negative Refraction at Infrared Wavelengths in a Two-Dimensional Photonic Crystal," *Phys. Rev. Lett.*, vol. 93, pp. 073902, 2004.
11. E. J. Reed, M. Soljačić, and J. D. Joannopoulos, "Reversed Doppler Effect in Photonic Crystals," *Phys. Rev. Lett.*, vol. 91, pp. 133901, 2003.
12. V. G. Veselago, "Properties of Materials Having Simultaneously Negative Values of Dielectric (Xi) and Magnetic (Mu) Susceptibilities," *Soviet Phys. Solid State, USSR*, vol. 8, pp. 3571–3573, 1966.
13. J. B. Pendry, "Negative Refraction Makes a Perfect Lens," *Phys. Rev. Lett.*, vol. 85, pp. 3966–3969, 2000.
14. H. Kosaka, T. Kawashima, A. Tomita, M. Notomi, T. Tamamura, T. Sato, and S. Kawakami, "Photonic Crystals for Micro Lightwave Circuits Using Wavelength-Dependent Angular Beam Steering," *Appl. Phys. Lett.*, vol. 74, pp. 1370–1372, 1999.
15. H. Kosaka, T. Kawashima, A. Tomita, M. Notomi, T. Tamamura, T. Sato, and S. Kawakami, "Superprism Phenomena in Photonic Crystals: Toward Microscale Lightwave Circuits," *J. Lightwave Technol.*, vol. 17, pp. 2032–2034, 1999.
16. C. Y. Luo, M. Soljačić, and J. D. Joannopoulos, "Superprism Effect Based on Phase Velocities," *Opt. Lett.*, vol. 29, pp. 745–747, 2004.
17. K. B. Chung and S. W. Hong, "Wavelength Demultiplexers Based on the Superprism Phenomena in Photonic Crystals," *Appl. Phys. Lett.*, vol. 81, pp. 1549–1551, 2002.
18. B. Momeni and A. Adibi, "Optimization of Photonic Crystal Demultiplexers Based on the Superprism Effect," *Appl. Phys. B*, vol. 77, pp. 555–560, 2003.
19. T. Matsumoto, S. Fujita, and T. Baba, "Wavelength Demultiplexer Consisting of Photonic Crystal Superprism and Superlens," *Opt. Express*, vol. 13, pp. 10768–10776, 2005.
20. L. J. Wu, M. Mazilu, and T. F. Krauss, "Beam Steering in Planar-Photonic Crystals: From Superprism to Supercollimator," *J. Lightwave Technol.*, vol. 21, pp. 561–566, 2003.
21. L. J. Wu, M. Mazilu, T. Karle, and T. F. Krauss, "Superprism Phenomena in Planar Photonic Crystals," *IEEE J. Quantum Electron.*, vol. 38, pp. 915–918, 2002.
22. T. Baba and M. Nakamura, "Photonic Crystal Light Deflection Devices Using the Superprism Effect," *IEEE J. Quantum Electron.*, vol. 38, pp. 909–914, 2002.
23. D. Scrymgeour, N. Malkova, S. Kim, and V. Gopalan, "Electro-Optic Control of the Superprism Effect in Photonic Crystals," *Appl. Phys. Lett.*, vol. 82, pp. 3176–3178, 2003.
24. D. W. Prather, J. Murakowski, S. Y. Shi, S. Venkataraman, A. Sharkawy, C. H. Chen, and D. Pustai, "High-Efficiency Coupling Structure for a Single-Line-Defect Photonic-Crystal Waveguide," *Opt. Lett.*, vol. 27, pp. 1601–1603, 2002.

25. D. W. Prather, S. Y. Shi, D. M. Pustai, C. H. Chen, S. Venkataraman, A. Sharkawy, G. J. Schneider, and J. Murakowski, "Dispersion-Based Optical Routing in Photonic Crystals," *Opt. Lett.*, vol. 29, pp. 50–52, 2004.

26. M. Notomi, A. Shinya, K. Yamada, J. Takahashi, C. Takahashi, and I. Yokohama, "Structural Tuning of Guiding Modes of Line-Defect Waveguides of Silicon-on-Insulator Photonic Crystal Slabs," *IEEE J. Quantum Electron.*, vol. 38, pp. 736–742, 2002.

27. R. G. Hunsperger, *Integrated Optics: Theory and Technology*, 5th ed. New York: Springer Verlag, 2002.

28. T. Baba, A. Motegi, T. Iwai, N. Fukaya, Y. Watanabe, and A. Sakai, "Light Propagation Characteristics of Straight Single-Line-Defect Waveguides in Photonic Crystal Slabs Fabricated into a Silicon-on-Insulator Substrate," *IEEE J. Quantum Electron.*, vol. 38, pp. 743–752, 2002.

29. A. Adibi, Y. Xu, R. K. Lee, A. Yariv, and A. Scherer, "Properties of the Slab Modes in Photonic Crystal Optical Waveguides," *J. Lightwave Technol.*, vol. 18, pp. 1554–1564, 2000.

30. J. P. Berenger, "An Effective PML for the Absorption of Evanescent Waves in Waveguides," *IEEE Microwave Guided Wave Lett.*, vol. 8, pp. 188–190, 1998.

31. J. P. Berenger, "A Perfectly Matched Layer for the Absorption of Electromagnetic Waves," *J. Comp. Physics*, vol. 114, pp. 185–200, 1994.

32. C. H. Chen, A. Sharkawy, D. M. Pustai, S. Y. Shi, and D. W. Prather, "Optimizing Bending Efficiency of Self-Collimated Beams in Non-Channel Planar Photonic Crystal Waveguides," *Opt. Express*, vol. 11, pp. 3153–3159, 2003.

33. H. F. Taylor, "Optical Analog-to-Digital Converter - Design and Analysis," *IEEE J. Quantum Electron.*, vol. 15, pp. 210–216, 1979.

34. B. Jalali and Y. M. Xie, "Optical Folding-Flash Analog-to-Digital Converter with Analog Encoding," *Opt. Lett.*, vol. 20, pp. 1901–1903, 1995.

35. M. Y. Frankel, J. Kang, and R. D. Esman, "High-Performance Photonic Analogue-Digital Converter," *Electron. Lett.*, vol. 33, pp. 2096–2097, 1997.

36. J. Cai and G. W. Taylor, "Demonstration of an Optoelectronic 4-Bit Analog-to-Digital Converter Using a Thyristor Smart Comparator," *Opt. Commun.*, vol. 184, pp. 79–88, 2000.

37. L. Brzozowski and E. H. Sargent, "All-Optical Analog-to-Digital Converters, Hardlimiters, and Logic Gates," *J. Lightwave Technol.*, vol. 19, pp. 114, 2001.

38. S. G. Johnson and J. D. Joannopoulos, "Block-Iterative Frequency-Domain Methods for Maxwell's Equations in a Planewave Basis," *Opt. Express*, vol. 8, pp. 173–180, 2001.

39. A. Taflove and S. C. Hagness, *Computational Electrodynamics: The Finite-Difference Time-Domain Method*, 2nd ed. Boston, Artech House, 2000.

40. D. Prather, S. Shi, S. Venkataraman, Z. Lu, J. Murakowski, and G. Schneider, "Self-Collimation in 3D Photonic Crystals," *Proc. SPIE*, vol. 5733, pp. 84–93, 2005.

41. J. W. Shin and S. H. Fan, "Conditions for Self-Collimation in Three-Dimensional Photonic Crystals," *Opt. Lett.*, vol. 30, pp. 2397–2399, 2005.

42. Z. L. Lu, S. Y. Shi, J. A. Murakowski, G. J. Schneider, C. A. Schuetz, and D. W. Prather, "Experimental Demonstration of Self-Collimation Inside a Three-Dimensional Photonic Crystal," *Phys. Rev. Lett.*, vol. 96, pp. 173902, 2006.

43. Z. L. Lu, J. A. Murakowski, C. A. Schuetz, S. Y. Shi, G. J. Schneider, and D. W. Prather, "Three-Dimensional Subwavelength Imaging by a Photonic-Crystal Flat Lens Using Negative Refraction at Microwave Frequencies," *Phys. Rev. Lett.*, vol. 95, pp. 153901, 2005.
44. Z. L. Lu, C. H. Chen, C. A. Schuetz, S. Y. Shi, J. A. Murakowski, G. J. Schneider, and D. W. Prather, "Subwavelength Imaging by a Flat Cylindrical Lens Using Optimized Negative Refraction," *Appl. Phys. Lett.*, vol. 87, pp. 091907, 2005.
45. C. A. Balanis, *Advanced Engineering Electromagnetics*, New York: Wiley, 1989.
46. P. T. Rakich, M. S. Dahlem, S. Tandon, M. Ibanescu, M. Soljacic, G. S. Petrich, J. D. Joannopoulos, L. A. Kolodziejski, and E. P. Ippen, "Achieving Centimetre-Scale Supercollimation in a Large-Area Two-Dimensional Photonic Crystal," *Nature Mat.*, vol. 5, pp. 93–96, 2006.
47. V. G. Veselago, "Electrodynamics of Substances with Simultaneously Negative Values of Sigma and Mu," *Soviet Phys. Uspekhi, USSR*, vol. 10, pp. 509, 1968.
48. C. Y. Luo, S. G. Johnson, and J. D. Joannopoulos, "All-Angle Negative Refraction in a Three-Dimensionally Periodic Photonic Crystal," *Appl. Phys. Lett.*, vol. 81, pp. 2352–2354, 2002.

6

Fabrication

A photonic crystal (PhC) is a structure with a periodic dielectric constant. As such, it extends by definition throughout the entire three-dimensional (3D) space. Certainly, it is not possible to realize such a device in practice as it would require filling the entire universe with periodically arranged matter. Clearly then, what can be built even in principle is not the mathematical ideal but its finite version. Other practical considerations entail even more departure from the conceptual perfection as it is translated into the physical world. Although the granularity of matter, i.e., the fact that it consists of finite-size atoms, limits in principle the possible periodicities, the use of finite-precision instruments limits the practically realizable fidelity of obtained patterns even more. Finally, the maturity of available technologies that can pattern matter at scales required for PhC operation at optical wavelengths, or lack thereof, forces compromises even at the fundamental level of finite periodicity. Thus, the term "photonic crystal" is also applied to situations where the periodicity is reduced from three to two dimensions or even more to one dimension, where such a structure has originally been termed a dielectric stack.

In the sections below, we describe several methods that have been conceived and successfully applied to the realization of photonic crystals that fall under such an extended definition detailed above. Although we attempt to cover a wide range of approaches, the selection is dictated mostly by our own interests, and thus, we write about what got us excited as the most promising technologies. Because the methods to fabricate one-dimensional photonic crystals, i.e., dielectric stacks, have been well researched and can be found elsewhere (e.g., Ref. 1, Part II), we focus here on two-dimensional (2D) and three-dimensional photonic crystals. Furthermore, we omit the entire area of photonic-crystal fibers as the extensive research in that field warrants a book of its own (for example, see Refs. 2 and 3).

6.1 TWO-DIMENSIONAL PHOTONIC CRYSTALS

To control the behavior of light in the context of a photonic crystal, we need periodicity of the dielectric constant in all three spatial dimensions. At visible and near-infrared wavelengths, this poses a technological challenge, becuase the required

Photonic Crystals: Theory, Applications, and Fabrication. By D. W. Prather, S. Shi, A. Sharkawy, J. Murakowski, and G. J. Schneider

feature size to be attained reaches the sub-100-nm regime, and the material must be structured volumetrically. In Section 6.2, we detail several approaches to solving the 3D fabrication problem. The difficulty of realizing regular and well-controlled 3D structures at the nanometer scale has prompted the researchers to determine whether the essential features of photonic crystals can be obtained at lower than three-dimensional geometries. Theoretical studies revealed that two-dimensional photonic crystals exhibit many properties of their three-dimensional analogs, which makes them an attractive subject of research. Indeed, some researchers [4] believe that the best way to use three-dimensional photonic crystals in practical applications is by using them essentially as mere "cladding material" to isolate optically planar/ 2D photonic-crystal layers where all the action takes place. Such an approach adds to the importance of studying the behavior of light and possibility of controlling it in two-dimensional photonic crystals.

The term *2D photonic crystal* can be understood in (at least) two ways. In the "classical" sense, referring to 2D geometry implies that all variation happens in two dimensions, whereas everything is constant along the third dimension. An example, in the context of photonic crystals, would be an array of infinitely long parallel cylinders as schematically shown in Fig. 6.4a. When analyzing such a structure, the electromagnetic field is assumed to be uniform along the cylinder axes, which thereby reduces the dimensionality of the problem from three to two. Another way to interpret the term 2D photonic crystal is to assume that the field is confined to a two-dimensional plane, or to a thin slab, in such a way that only one mode (or two, if we take polarization into account) exists in the direction normal to the plane. An example of such a structure is shown in Fig. 6.1, which depicts a perforated thin dielectric slab suspended in free space. As before, in this case only the in-plane variability affects the propagation light confined to the slab, which essentially reduces the dimensionality of the problem to two from three. It should be pointed out, however, that in the latter interpretation, the structure is truly three-dimensional; therefore, to simulate the behavior of light confined to the slab, three-dimensional computational methods must be used to account for all phenomena that can occur therein. In particular, out-of-plane radiation cannot be explained using a two-dimensional model. Also, the *effective* dielectric constant of the slab is different (smaller) than the dielectric constant of the bulk material used for its realization because of the presence of evanescent tails that extend beyond the physical boundaries of the structure and into low dielectric claddings. Nonetheless, such geometries have proved extremely useful mainly because of relatively easy fabrication as well as the wealth of phenomena they offer for investigation.

Given the two possible interpretations of 2D photonic crystals, we will discus the methods to realize either. To distinguish between them, we will refer to the slab or z-confinement-based photonic crystals as planar or slab photonic crystals. We reserve the term 2D photonic crystal for structures that are considerably extended in the z direction so as to make them "infinite" for practical purposes; i.e., they are used in situations where the interaction of the electromagnetic wave with the z-boundaries can be neglected.

As pointed out above, the realization of photonic crystals for the optical and near-infrared wavelengths requires the ability to pattern features 100 nm and smaller

(a)

(b)

Figure 6.1 Planar photonic crystal that consists of a perforated dielectric slab suspended in low dielectric cladding: (a) artist's rendition and (b) physical realization and of the structure using silicon-on-insulator (SOI).

in size. Fortunately, semiconductor fabrication technology has reached a point where such a feat is achieved routinely: For example, commercially available microprocessors are fabricated using 45-nm technology (2009), which means that the critical dimension in the process is 45 nm. Furthermore, semiconductor devices rely on planar pattern transfer, making the fabrication technology ideally suited to the realization of planar and 2D photonic crystals, which require only two-dimensional patterning. Indeed, it is the advances in semiconductor fabrication technology that enabled the realization of 2D and planar photonic crystals for optical wavelengths, thus feeding the explosion of research in this area. For more information about the semiconductor fabrication tools, the reader is referred to Ref. 5; lithography and etching are extensively covered in Refs. 6 and 7, respectively. In the next section, we discuss the fabrication of planar photonic crystals and follow it with the discussion of 2D photonic crystals in Section 6.1.2.

6.1.1 Fabrication of Planar Photonic Crystals

Planar photonic crystals rely on the confinement of light to a thin, structured layer of material. Although at radio frequencies such confinement can be achieved by placing two flat conducting plates parallel to one another and launching the wave between them, high losses experienced by optical-frequency waves when interacting with metals preclude the use of such an approach there. Therefore, for optical waves, we typically rely on total internal reflection in high-dielectric-constant material, or index guiding, to confine the electromagnetic field in plane. This approach leads to the slab geometry shown in Fig. 6.1a, where a dielectric slab with a high refractive index is suspended in low-index cladding material. To achieve the highest index contrast, the cladding is typically air. The coherent scattering characteristic of a photonic crystal is achieved by drilling a regular array of holes in the slab. The slab has to be thin enough to support only one transverse mode, which for near-infrared wavelengths propagating in silicon means a few hundreds of nanometers. A slab, or rather a membrane, this thin would be difficult to handle on its own, so a thicker substrate is provided and the membrane is suspended above it, supported around the perimeter, outside of the optical device area. An example of such a structure is shown in Fig. 6.1b, where the perforated membrane that constitutes the photonic crystal is made of silicon, and the handle substrate is also silicon.

The fabrication begins with an SOI wafer; Fig. 6.2a. Such substrates can be obtained commercially at a cost of few hundred dollars per 4″ wafer and consist of a several-hundred-micron-thick silicon handle with a layer of silicon dioxide followed by a device layer of crystalline silicon. The thickness of the oxide layer is between a couple of hundred nanometers and a few microns. The device-layer thickness can range from a few hundred nanometers to tens of microns. Typical values for near-infrared wavelengths would be ~500 nm of oxide and ~250 nm of silicon.

For electron-beam lithography, the sample is then coated with poly-methyl-methacrylate (PMMA)—electron-beam (e-beam) sensitive resist; Fig. 6.2b. The resist has to be thin to allow for the fine features of the photonic crystal. On the other hand, it has to be sufficiently thick to withstand the harsh processing that follows. A 200-nm layer is in the range of the happy medium that allows both. It is obtained by spinning a solution of 4%-solid-content PMMA at 3000 rpm for 45 seconds. The sample is then baked at 180°C for 3 minutes to evaporate the solvent.

To pattern the photonic crystal, shown in Fig. 6.2c, direct-write electron-beam lithography is employed in research environments for its ability to define small features and low-cost overhead in rapid prototyping, as no lithographic masks are required. The pattern is then transferred to the underlying silicon device layer of the SOI substrate by dry reactive-ion plasma etching that allows for vertical sidewalls for highly anisotropic processes; Fig. 6.2d. In some cases, an intermediate hard mask is used where the pattern in PMMA is first transferred to silicon dioxide or silicon nitride, which then forms a mask for the silicon etch. Skipping this intermediate step allows better fidelity of the pattern as long as the etch recipe provides sufficient selectivity over the soft PMMA etch mask. The combination of high etch selectivity over a soft mask with high anisotropy for obtaining vertical sidewalls is afforded

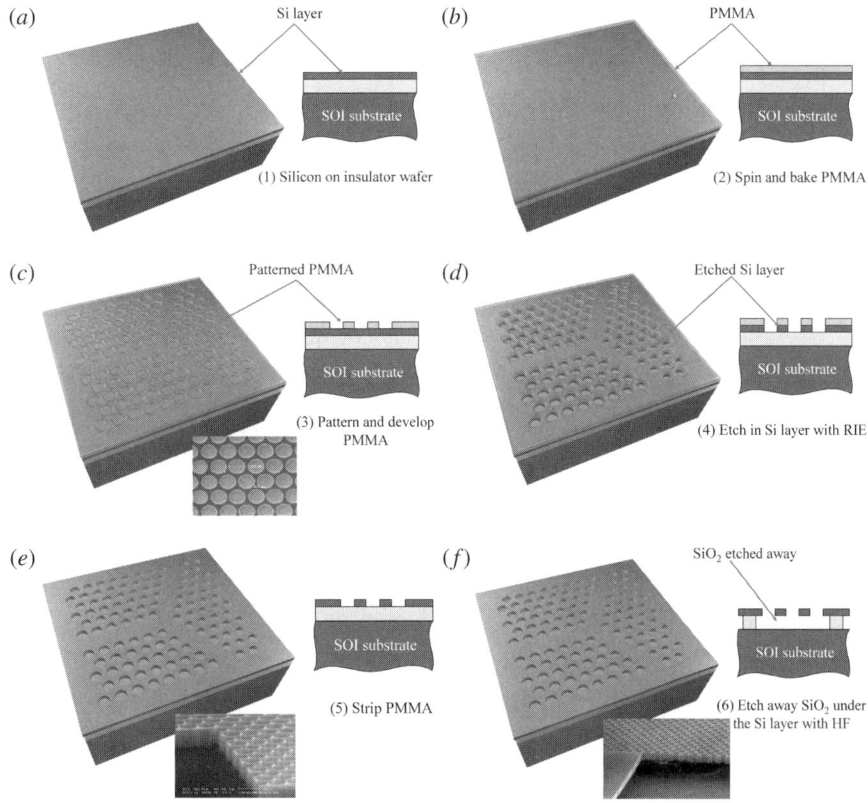

Figure 6.2 Fabrication of a planar photonic crystal using SOI.

by the use of time-multiplexed, a.k.a. Bosch, processes, where the etching steps alternate with passivation-layer deposition steps—see Section 6.2.7 for more information. The corrugation of the sidewall that is the result of this time-multiplexed process typically has little effect on the propagation of light in the structure because of its small, deep-subwavelength dimensions, and can be minimized by reducing the duration of individual cycles, which reduces its period. Note that obtaining vertical sidewalls is essential for this type of photonic crystals because conical, as opposed to cylindrical, openings have been proven to result in considerable radiation losses for the propagation of light in the structure [8]. The same is true for any asymmetry along the z direction, which means that the same cladding material must be used above and below the slab.

Following the silicon etch, the PMMA is stripped; Fig. 6.2e. This step can be done by rinsing the sample in acetone or by oxygen plasma etch. If the plasma reactor used for pattern transfer has an oxygen line, the latter method is preferred as the PMMA strip can be performed as part of the same process. Finally, silicon dioxide underlying the photonic-crystal structure is removed by a wet etch in aqueous solution of hydrofluoric acid; Fig. 6.2f. The wet etch has to be timed properly—it must be long enough so as to remove the oxide all the way to the silicon substrate through the openings in the silicon device layer, but not too long so as to leave sufficient support around the structure to prevent its collapse.

The result of this process is a planar photonic crystal made of a perforated silicon membrane suspended above the substrate. The separation of the membrane from the substrate is determined by the thickness of the oxide layer and needs to be large enough to prevent the leakage of light to the substrate through the evanescent tail of the slab mode. The cladding is symmetric on both sides of the device layer, and maximum index contrast is achieved by leaving the membrane suspended in free space.

The use of silicon in the process described above prevents (at the time of writing this book) monolithic integration of light sources in the structure. To build in a laser, compound semiconductor materials are required. Because of the absence of a gallium-arsenide analog of SOI, a somewhat different approach is used to realize perforated GaAs membranes suspended above the bulk substrate. In this case, a pure GaAs layer is grown on top of AlGaAs; Fig. 6.3a. After patterning and etching a two-dimensional array of cylindrical holes that constitute a photonic crystal in the top GaAs layer (Fig. 6.3b), the sample is oxidized in a wet atmosphere of nitrogen and water vapor at ~400°C. The oxidation process is highly selective and affects the underlying AlGaAs but leaves pure GaAs intact. This allows for a thorough oxidation of the AlGaAs layer while preserving the 2D photonic-crystal structure pattern in the top GaAs layer; Fig. 6.3c. The oxide is subsequently removed by a

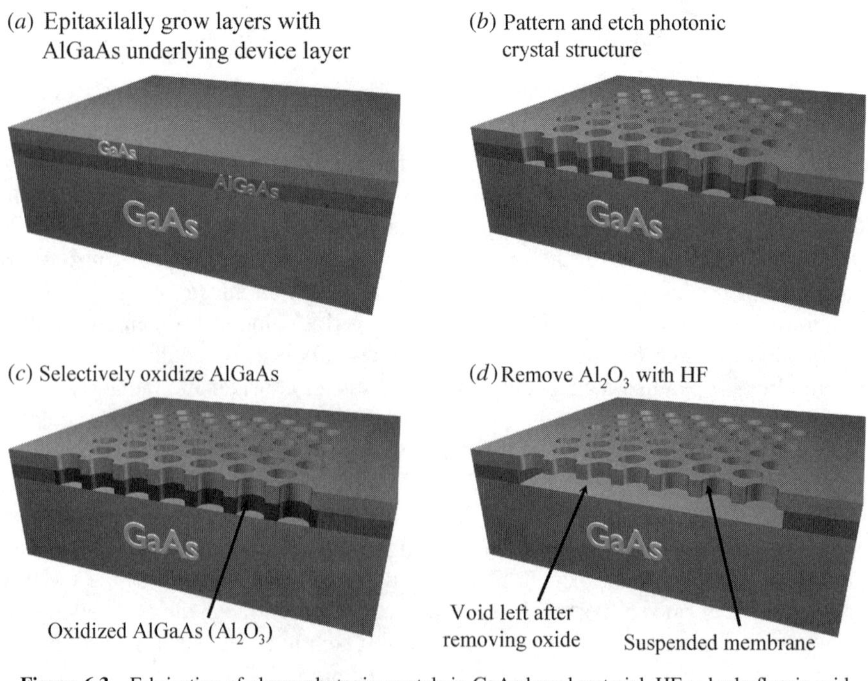

Figure 6.3 Fabrication of planar photonic crystals in GaAs-based material. HF = hydrofluoric acid.

buffered oxide etch of hydrofluoric acid diluted in water; Fig. 6.3d. The buffered oxide etch is also highly selective and stops precisely at the boundary between the oxide and the unaffected GaAs or unoxidized AlGaAs. Therefore, by controlling the depth, or profile, of the oxidation, it is possible to control the final geometry of the three-dimensional structure after etch.

For the patterning of the photonic crystal, the technique used is similar to that described above for the SOI case, except that a hard intermediate mask is typically used because organic resists usually present insufficient etch resistance to the harsh conditions of a chlorine-based III–V etch.

From the examples described above, it should be clear that to realize planar photonic crystals for optical frequencies, a layered substrate is used as a starting point. One of the buried layers is sacrificial; i.e., it is selectively removed at the end of the process in order to create a suspended perforated membrane. As noted above, having a suspended structure allows for the maximum index contrast between the device material and the cladding layer, which is generally desired for photonic crystals. At the same time, the photonic crystal is symmetric (top and bottom cladding have the same refractive index), which improves its in-plane guiding efficiency.

6.1.2 Fabrication of 2D Photonic Crystals

According to the naming convention expounded at the beginning of this section, 2D photonic crystals take the form of patterns extended in one direction, such as long rods or deep holes; see Fig. 6.4. Both the rods and the holes presented in Fig. 6.4 are cylindrical to emphasize the typical shape that is assumed when realizing photonic crystals for the optical wavelengths. For maximum dielectric-index contrast, one material that constitutes the photonic crystal should be free space with the refractive index of unity. Such a requirement makes the physical realization of the rod structure (Fig. 6.4a) problematic, as no structural support exists to maintain the position of the rods in space. This is true even if the ends of the rods are anchored, as

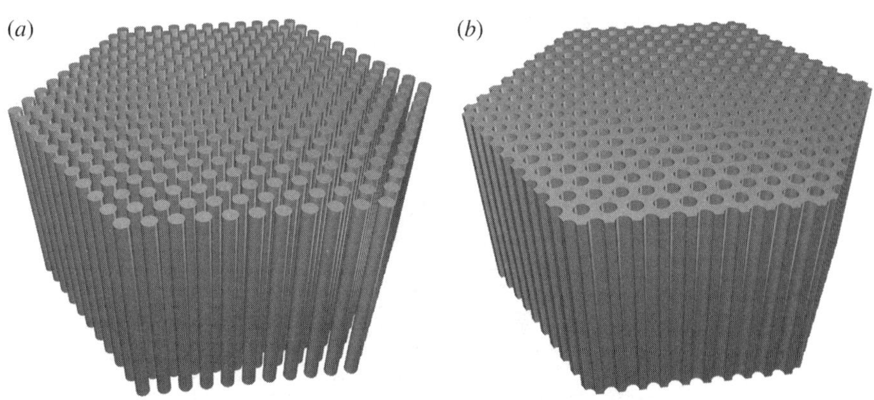

Figure 6.4 Two-dimensional photonic crystals. (a) Array of long (ideally infinite) parallel rods. (b) Thick (ideally infinitely thick) slab of dielectric material with deep cylindrical holes drilled throughout.

we would still have spaghetti-like structure between the anchoring planes with little structural stability. In contrast, a thick slab with holes drilled across remains rigid and can be handled with relative ease. Therefore, we will limit our discussion of 2D photonic crystals to the latter configuration; Fig. 6.4b.

To realize an array of deep cylindrical trenches in high-index material, at scales required for the operation at near infrared wavelengths, we again resort to the use of semiconductor technology for tools and inspiration. Thus, the two-dimensional pattern is produced on the surface of a flat substrate using standard lithography techniques. Then, this pattern is transferred into the substrate by highly anisotropic etching that can produce vertical sidewalls and high-aspect-ratio holes. To this end, either a dry or wet etch technique is used. High-anisotropy wet etching is possible in n-type silicon when the sample is additionally subject to electrical bias and illumination. In Section 6.2.5, we describe in detail the mechanism that leads to arrays of high-aspect-ratio cylindrical holes drilled into bulk silicon substrate. Whereas modulating the current and the illumination leads to a three-dimensional photonic crystal, maintaining the parameters constant throughout the etch allows for highly uniform (both in length and in diameter) cylinders to be formed in the substrate. Such an approach has been successfully applied to realize both regular 2D photonic crystals and photonic crystals with engineered defects; Fig. 6.5, Refs. 9–19.

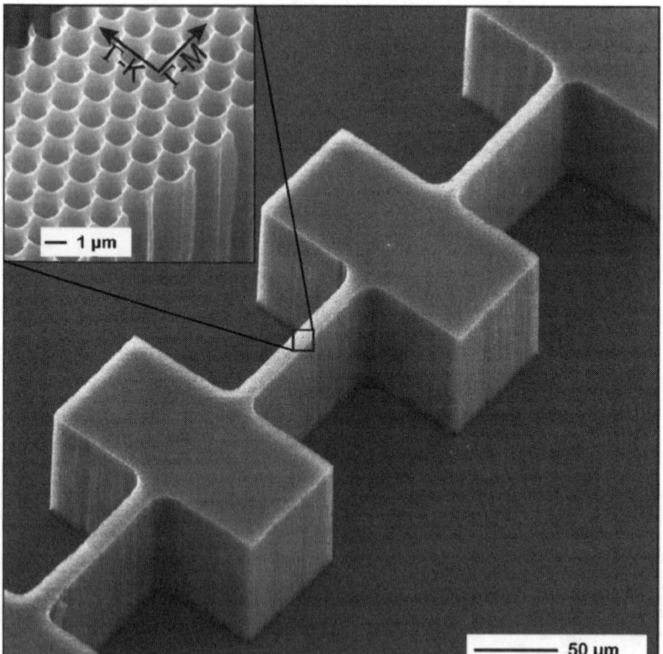

Figure 6.5 Two-dimensional photonic crystal realized by electro-photo-chemical etching of silicon. (Reprinted with permission from Ref. 13. © 1999, American Institute of Physics.)

Because, as pointed out above, silicon does not readily emit light, compound semiconductors have been employed to realize 2D photonic crystals with an embedded light source. The porous etching of compound semiconductors has been less investigated than that of silicon; instead, the deep high-aspect-ratio openings are obtained by dry etching. Although excellent results have been obtained this way, the achievable aspect ratio cannot compete with that of macroporous silicon, and the etch depth is still of the order of the wavelength of light for which the device is intended, and as such cannot really be considered a true 2D photonic crystal. To remedy this situation, confinement is provided by introducing refractive index variation along the z direction. Thus, as in planar photonic crystals, the in-plane guiding results from a higher refractive index of a core layer versus the cladding layers. The difference is that the index contrast between the core and the cladding is small, and hence, the evanescent tails of the mode extend far into the low index material. The thickness of the core is chosen such that single-mode propagation of light is ensured. Because a large fraction of the energy propagates in the cladding, it is necessary that both the core and the cladding be perforated to form a photonic crystal. Ideally, the structure would look similar to that presented in Fig. 6.6, where a higher index core is indicated with a darker color. With this approach, the structure can still be analyzed using two-dimensional modeling as long as effective index, as opposed to the index of the core, is used for the mode propagating in it.

This type of photonic crystal has been realized in III–V materials, including GaAs-based and InP-based substrates, beginning in mid-1990s; see Refs. 20–23. The ability

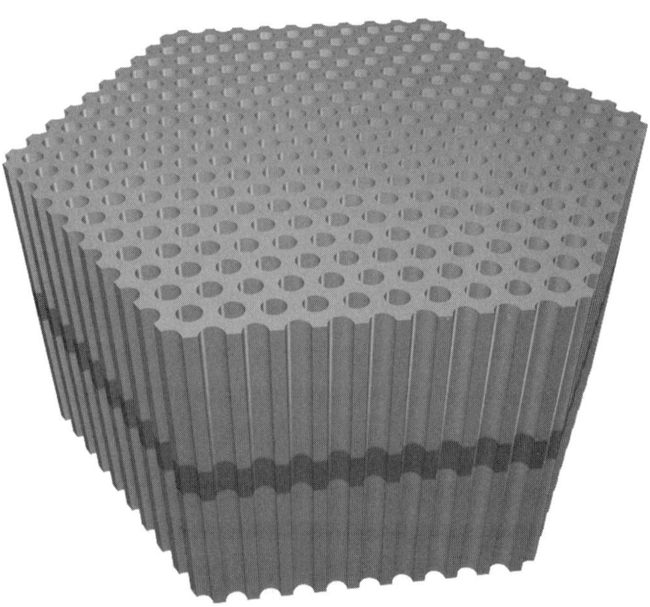

Figure 6.6 Two-dimensional photonic crystal with a higher index core layer, which is indicated in darker color, to provide in-plane confinement.

to introduce a source of light into the substrate itself offered the advantage of side-stepping a major hurdle in the experimental investigation of photonic crystals, that is, coupling light into the structure, which at the time was problematic because of large mode mismatch between a single-mode optical fiber and a photonic-crystal waveguide. The light source took the form of a quantum-well or a quantum-dot layer buried within the core and excited optically by photoluminescence [20, 24–29].

To realize these structures requires patterning features of the order of 100 nm and transferring them to the underlying substrate to the depth of several microns, while maintaining the size and shape of the original pattern. The high resolution is obtained by employing electron-beam lithography with PMMA as the e-beam resist. However, whereas PMMA provides the requisite resolution, it falls short of the necessary etch resistance for direct pattern transfer to the underlying semiconductor substrate. Therefore, the pattern is first transferred to an intermediate etch mask, typically consisting of a silicon dioxide (SiO_2) or silicon nitride (SiN_x) layer 100 to 500 nm thick, which offers high resistance to etching in the chemistries used for GaAs or InP. The etch depth must exceed the length of the evanescent tail of the guided mode to prevent light leakage into the substrate, but it is limited by the etch selectivity of the substrate versus the mask and by the redeposition of etch byproducts at the bottom of the trenches [30]. Over the past decade, improvements to the process allowed the realization of etch depths in excess of 4 μm for holes only a few hundred nanometers wide [28, 29, 31, 32]. This improvement was made possible by the development of highly efficient etching tools, such as chemically assisted ion-beam etchers (CAIBE) and inductively coupled plasma (ICP) etching systems, wherein the generation of plasma is decoupled from its extraction for etching. This decoupling allows independent control of plasma generation and etching conditions, which thereby provides the necessary control of the etch process critical to the realization of high-aspect-ratio features [31]. Furthermore, the discovery of SiN_x hardening under oxygen plasma [33] allowed even more improvement of the etch selectivity over the mask material. Similar SiO_2 mask hardening has been observed at a high voltage-to-current ratio as a result of the impact of high-energy ions in CAIBE [31]. The durability of the mask can also be improved by transferring it to nickel, which provides three times the selectivity afforded by SiO_2 when etching GaAs using chlorine-based chemistry [34].

In a typical fabrication process, first the substrate is prepared by epitaxially growing the required layers. To this end, metallo-organic vapor-phase epitaxy (MOVPE) [27, 29, 35–37] or molecular-beam epitaxy (MBE) [38] is used. The layers include lower cladding, core with embedded quantum wells or quantum dots, and upper cladding, plus a capping layer to protect the underlying structure from oxidation. The thickness and composition of each layer varies depending on the material system and on the purpose of the device and totals several hundred nanometers. For example, the structure studied by Ferrini et al. [27] consisted of an n-type InP substrate with 600-nm InP bottom cladding, a 434-nm GaInAsP waveguide core, and 200-nm InP top cladding. The waveguide layer had an embedded light source in the form of two strain-compensated quantum wells separated by a 30-nm spacer and located approximately in the middle of the guiding structure.

Following the epitaxy, an etch mask layer is deposited. This is usually silicon dioxide or silicon nitride, either sputtered or deposited by chemical vapor deposition (CVD). An alternative method has been used by O'Brien et al. [23] as well as Labilloy et al. [24], who first patterned and etched an epitaxially deposited AlAs, which they later oxidized in water-vapor atmosphere at 340°C to form a very robust aluminum-oxide etch mask. In either case, the pattern is first defined in a layer of PMMA using electron-beam lithography. It is desirable to have the PMMA layer as thick as possible to ensure reliable pattern transfer to the underlying hard mask. However, the scattering of electrons in the material, which results in proximity exposure and compromises the pattern resolution, limits this thickness. Using high-energy electrons alleviates this problem, and some researchers go as high as 100 keV to expose resist as thick as 500 nm [36–39] or more [28]. Otherwise, a 20- to 50-keV electron beam and 70- to 300-nm-thick resist prove adequate to realize desired structures.

Using the pattern developed in PMMA, the underlying hard mask is etched in a reactive ion etcher (RIE) with fluorinated gasses such as CHF_3/Ar [29, 36, 37, 39–42] or C_2F_6 [21, 23], with the former preferred as it offers better selectivity over e-beam resist thanks to the presence of hydrogen in the plasma [43]. Here, the goal is to etch through a hard mask layer thick enough to withstand the subsequent processing of the semiconductor substrate. The selectivity of the silicon nitride or silicon oxide etch over PMMA is of the order of one [28], which means that PMMA etches as fast as the hard mask, and can be somewhat improved by performing the RIE etch in short steps interspaced with N_2 flushing of the reaction chamber to cool down the polymer [44]. The resulting thickness of the hard mask is typically between 100 and 500 nm [21, 36].

After stripping the polymer, the hard mask is used to etch the underlying semiconductor substrate. The etch has to be deeper than the spatial extent of the mode to minimize the leakage of light into the substrate and thereby ensure low-loss propagation. Furthermore, the etch has to preserve the size and shape of the patterned features, and provide smooth sidewalls because any roughness acts as a scattering center for the propagating light. Maintaining vertical sidewalls throughout the etch is essential as conically tapered holes cause light leakage into the substrate thus contributing to guiding loss [8]. The depth is limited mainly by the etch selectivity and the redeposition of etch byproducts. The latter can be minimized by using chemically assisted etching wherein the etch products are volatile and are carried away by the vacuum system of the instrument. On the other hand, vertical sidewalls are achieved by highly directional ion bombardment. Thus, the right balance between chemical and physical components of the etch is essential for achieving good results [30, 35]. At the same time, the temperature of the substrate must be sufficiently high as to ensure volatility of the reaction byproducts. For example, $InCl_x$ is not very volatile, with vapor pressure of only 2 mTorr at 160°C. Therefore, if the InP substrate temperature is too low, an excess of indium will remain, which results in high-roughness etch. The typical temperature range used for InP is 185–250°C. The etch is performed using chlorine-based CAIBE, wherein pure Cl_2 is typically used for chemical etching, and Ar^+ ion-beam is used to activate the etch and provide its physical component [29, 31, 32]. In ICP, gasses such as $SiCl_4$ [40, 41, 45] or Cl_2 are often used.

The verticality of the sidewalls can be also improved by adding O_2 to the etch chemistry to passivate the sidewalls [42, 43, 46]; however, care should be taken not to exceed the necessary proportion of oxygen to avoid micromasking and grass formation [42].

The result of the process is a 2D photonic crystal that consists of an array of cylinders a few hundred nanometers wide that can extend more than 4 μm into the semiconductor substrate.

6.2 THREE-DIMENSIONAL PHOTONIC CRYSTALS: MICROMACHINING

In this section, we describe several methods used for the realization of structures periodic in all three spatial dimensions. The field of three-dimensional fabrication is far from mature, as can be ascertained by the presence of a plethora of approaches. Each approach seems to have a certain set of advantages and disadvantages as compared with others. Which one will emerge a winner in the race to commercialization depends equally on their intrinsic merits and on the specific applications for which it is intended. Likely, several methods will find their niches and some will be combined to take advantage of their respective strengths and overcome weaknesses. It is therefore essential for a researcher working in the field as well as for a potential entrepreneur to be aware of what technologies are available. The purpose of this section is to provide an overview what is out there.

6.2.1 Layer-by-Layer Fabrication

Microfabrication has been developed as a direct response to the needs of the microelectronics industry. Modern electronic circuits pack millions of interconnected devices on a single semiconductor chip. Typically, the devices occupy one layer, and the interconnection fabric is spread across a volume in the form of a multilayer net. The layers are built sequentially with tight tolerances to ensure that the connections properly line up between them. This method of creating complex circuits has been perfected over the decades of research and development in the industry.

Because any volumetric structure, such as a three-dimensional photonic crystal, can always be thought of as consisting of layers, it is natural to try to extend the methods of microelectronic industry for the purpose of creating photonic crystals. This approach became possible recently as the minimum feature size patterned for the creation of the latest generations of microprocessors shrunk to the size required for patterning photonic crystals that operate in the near-infrared regime, and at the same time, the overlay tolerances grew correspondingly smaller.

Although, as pointed out above, any structure can be thought of as consisting of layers, the trick is to conceive geometries that require a minimum number of layers to be patterned and aligned. For example, an array of spherical voids would require tens of carefully aligned layers to reproduce the spherical shape accurately. In contrast, simple cubes or cylinders with their axes perpendicular to the substrate plane need only a single pattering step. Accordingly, the photonic crystals most favored for

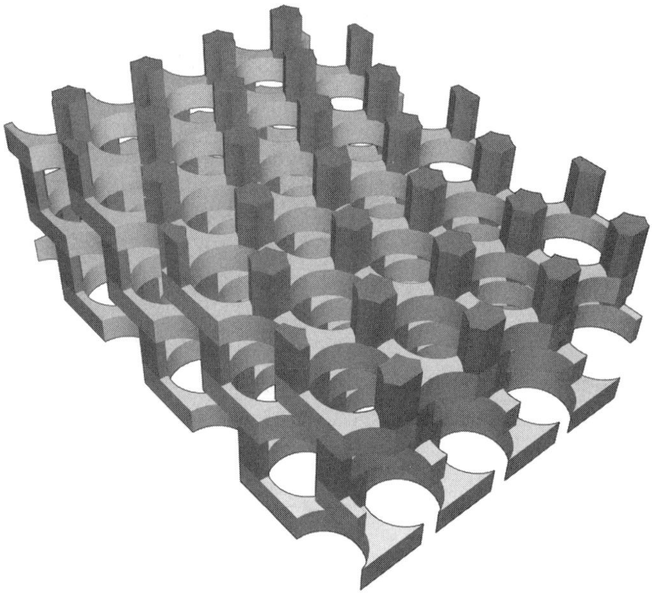

Figure 6.7 One vertical period and several horizontal periods of an FCC structure that consists of an array of cylindrical holes patterned in dielectric. Hole layers are light gray, whereas rod layers are dark gray.

layer-by-layer fabrication consist of blocks, cylinders, cylindrical voids, and other shapes that have piece-wise constant cross sections. An example of such a photonic crystal shown in Fig. 6.7 was first proposed by Johnson and Joannopoulos in Ref. 47, and extensively simulated in Ref. 48. It consists of stacked layers that alternate between an array of dielectric pillars, or rods, and cylindrical holes in a dielectric slab. They will be referred to as the rod layer and the hole layer, respectively; see Fig. 6.7. The hexagonal shape of the cross section of the rods is the result of the proposed fabrication process [47] that is designed to minimize the number of lithography/etching steps. The consecutive layers are offset so as to form a face-centered-cubic (FCC) lattice with the [111] direction perpendicular to the substrate surface and parallel to the cylinder axes. A single period in the z direction contains three pairs of hole and rod layers. The consecutive layer pairs can be thought of as shifted in the y direction so that the center of the hole falls directly above the junction of three holes below. Figure 6.8 schematically shows a top-down view of the structure with the centers of holes of three consecutive hole layers labeled A, B, and C. As can be observed from that figure, the centers of cylinders in a given layer are shifted by $a/(2\sqrt{3})$, where a is the lattice period, with respect to the layer below. Clearly, in this case, the layers do have piece-wise constant cross sections, as desired. Moreover, although it may appear that six lithography/etch steps would be required do define a single period of a photonic crystal (one for each rod and one for each hole layer), it will be shown below that only three are needed.

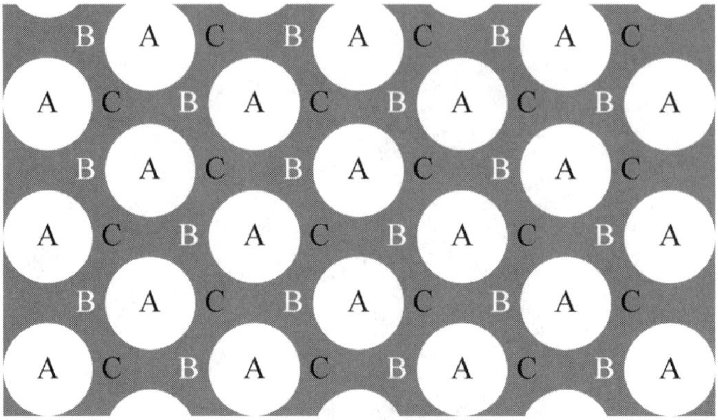

Figure 6.8 Cross-sectional top-down view of an array of cylindrical voids indicating their arrangement in consecutive layers. Letter positions correspond to lateral placement of voids in layers A, B, and C, respectively.

Another way to describe the structure is to consider an FCC arrangement of closely packed spherical voids and to replace the spheres with overlapping air cylinders. Figure 6.9 shows different perspectives of the structure to aid in the visualization of its three-dimensional nature. The way of looking at it as an arrangement of overlapping cylindrical voids suggests an efficient way of fabricating it: For each layer, one would etch an array of cylinders that extend across one hole layer and the two adjacent rod layers. This way, only three lithography/etch steps are required to pattern a single period of the photonic-crystal lattice. As a result, an increase in fabrication efficiency is achieved at the cost of reducing the flexibility of design because the shape and size of the rods are determined by the size and the shape of the holes and cannot be independently modified.

The fabrication of this photonic crystal would proceed as follows. An array of holes is patterned using standard lithography tools, such as ultraviolet, e-beam lithography, or X-ray lithography. The pattern is then transferred to a suitable etch-mask material coating the substrate, or the resist itself is used as the etch mask. Then, the pattern is etched into the substrate. For etching, any number of techniques can be used, such as RIE, ICP, and so on, [7]. The only requirement is that it yields reasonably vertical side-walls, low roughness, and sufficient etch depth for the size of the hole, or an aspect ratio, measured as the ratio of the depth of a hole to its diameter of at least 1.6. Following the etch, the mask is removed, and the holes are filled with material other than the substrate. For example, if the substrate is silicon, then the holes could be filled with silicon dioxide. The idea here is either to provide sufficient dielectric contrast between the two materials that the photonic crystal made out of them has sufficiently wide bandgap for the planned application, or to provide sufficient etch selectivity between them so that one of the materials can be removed in the last step and produce air voids as a result. As with etching, many back-filling techniques are available depending on the material to be used, which include electron-beam evaporation, thermal evaporation, sputtering, chemical-vapor deposition, and so on.

(a)

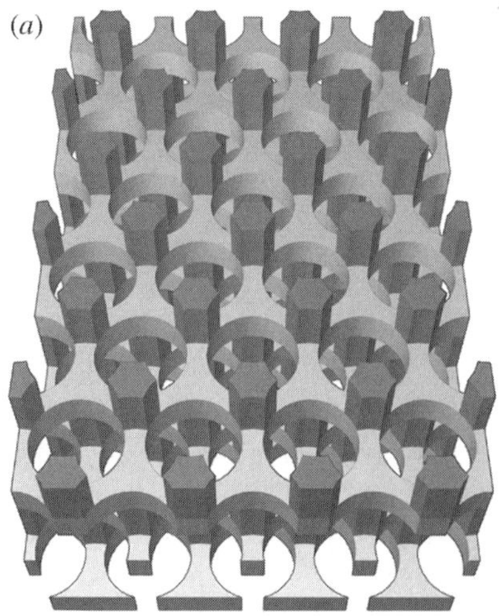

(b)

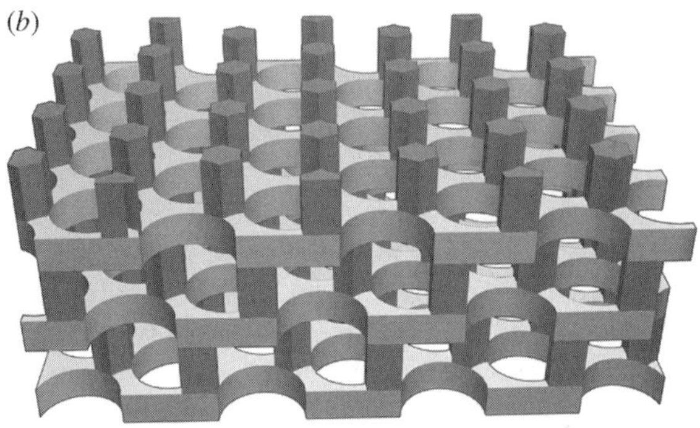

Figure 6.9 Array of cylindrical voids in dielectric forming an FCC structure.

The requirement is that the sidewalls of the structures are coated without leaving voids or shadowing. With directional deposition methods, such as sputtering or electron-beam evaporation, this can be achieved by mounting the substrate at an angle and rotating it during the deposition.

After the deposition, the sample is planarized to remove bumps and pits left by the deposition on a patterned surface and to prepare it for the next-level lithography/etching step. For planarization, chemical-mechanical polishing can be employed, which relies on wet slurries and pads to produce a high-quality, smooth surface. Alternatively, bias-sputter-etching can be used such as described in Ref. 49 and discussed in more detail later in this chapter. The advantage of the latter is that it is a dry process and can be combined with the preceding deposition in the same bias-sputtering reactor, thereby reducing the required processing. The planarization step yields a smooth surface with an array of cylindrical pits filled with the second type of material. Now, a new layer of material, which is identical with that of the substrate, is deposited. The deposition can take the form of sputtering, evaporation, or chemical vapor deposition, but if the substrate is crystalline, the best results are obtained by MBE, which amounts to the growth of crystal structure identical to the substrate. The difficulty with MBE is that it is typically a slow process, so growing hundreds of nanometers, as required for photonic crystals to work in the near-infrared regime, may test the patience of researchers and even more so that of profit-oriented industrial entities. The thickness of this layer should be equal to or slightly more than (to leave room for subsequent planarization) the combined thickness of the hole and rod layers.

Then, the next layer of holes is patterned, and aligned so that the centers of the holes in the new layer line up with the three-way junction of holes underneath, as indicated in Fig. 6.8 by the letter B. The overlay accuracy is much more critical in fabricating a photonic crystal than in the case of electronic circuits. For electronic circuits, it is only important that electrical contact is made between the layers. Shift as large as half of the diameter of the via can easily be tolerated. This is not the case for photonic crystals. A shift of a quarter of the photonic-crystal period would ruin its periodicity and produce an unusable device. A smaller shift can accumulate as new layers are built and similarly destroy the photonic crystal. In general, the more layers of photonic crystal are required for a given application, the tighter the overlay tolerances that have to be maintained to render a useful structure. We will come back to the issue of overlay accuracy and show how nanometer-level precision can be achieved later in this section.

The patterned holes are etched to form cylindrical voids, where the height of the cylinders is equal to the thickness of the hole layer plus two times the thickness of the rod layer. These cylindrical voids overlap the cylinders formed in the first step of the process. Therefore, the etch has to proceed across two different materials: the material of the substrate and the material that fills the cylindrical voids of the first layer. To define a high-quality photonic crystal, the etch recipe to be used here has to be such that the etch rates of the two materials are identical. The central portion of the cylindrical void produces the circular openings in the hole layer, whereas the top and the bottom portions sculpt the rods of the rod layers above and below it. Thus, every rod is produced by six cylinders: three from the layer below and three from the layer above.

After stripping the etch mask, the cylindrical openings are filled and the surface is planarized as before. A new film of material identical with the substrate is deposited or grown to the thickness of a combined hole and rod layers. Then, another array of holes is patterned. As before, the centers of the holes are lined up with the three-way junctions of the holes beneath. In this case, there are two possibilities: The holes could be lined up with the holes of the first layer, or they could fall above the three-way junction of the first and of the second layer. To produce an FCC structure, the latter is required, as indicated by "C" in Fig. 6.8. Filling the holes with the second material and planarizing the surface completes one period of the photonic crystal. The process can be repeated with the next layer aligned with the layer three layers below. After the desired number of periods is built, the material filing the holes can be selectively etched to increase the dielectric contrast and widen the photonic bandgap, or left in place if the contrast and the bandgap are sufficient for a given application.

From the fabrication process presented above, it is clear that the photonic crystal can be modified locally to introduce defects such as waveguides or resonators. These defects can be placed arbitrarily throughout the photonic crystal without disturbing the periodicity of the remainder of the lattice. This is a big advantage of this layer-by-layer fabrication. The price to pay for it is the extensive labor necessary to fabricate the structure as every period requires three separate layers, and each layer is a multistep process that involves deposition, planarization, lithography with tight alignment, and etching. Although all the steps belong to the set of processes used routinely in the semiconductor industry, one poses a particular challenge. It is the high-overlay accuracy mentioned above that is essential to maintain the regular periodicity of the photonic crystal. For the photonic crystal to work in the near-infrared regime, where the periodicity of the lattice is of the order of 500 nm, the overlay accuracy should be a few tens of nanometers or better. This task is difficult for conventional lithography. However, Qi and Smith [50] describe a method that can potentially achieve overlay accuracy in the 1-nm range.

In general, when building a structure in a layer-by-layer fashion, the features placed in a given layer have to be lined up with the features a layer below. For mask-based lithography, if we assume that masks themselves are undistorted, and the substrate surface is perfectly flat and parallel to the mask, there are three degrees of freedom for the relative mask-substrate movement: Translations in the x and y directions, and rotation in the xy plane. To achieve the high overlay accuracy, high precision is required in (1) the measurement of the displacement from the desired position and (2) the correction of the displacement. To measure and correct for rotational displacement, simple alignment marks can be used. High precision in this case is achieved by using a sufficiently long arm, or large separation between the marks, so that angles can be defined accurately. For translation, the situation is more difficult because the amount of measured and applied translation is directly related to the placement accuracy; that is, if we desire the placement accuracy of 10 nm, we have to be able to measure shifts as small as 10 nm or better. Given that the resolution of optical systems is limited to about onewavelength of the light used for imaging, or about half a micron for visible light, the use of visual inspection for this purpose may at first seem hopeless. However, the ability to measure a 10-nm displacement is not equivalent to the ability to resolve 10-nm features. An example that supports this statement is provided by a Vernier patterned on calipers used for fine measurements. Such instruments have been successfully used since the dawn of the industrial age for measurements with accuracy down to 50 μm with naked-eye readout. Another technique involves moiré patterns for alignment. A particular incarnation of this technique that is suitable for the discussed purpose has been described in detail in Ref. 51. It allows for the measurement of translational displacement with accuracy of the order of 1 nm using only an inexpensive microscope. The method is based on the observation that when two gratings with periods close to one another are superimposed, a moiré fringe pattern is created with a spatial beat frequency equal to the difference of the spatial frequencies of the two gratings. The latter can be easily observed using optical means. At the same time, small movement of the gratings with respect to one another results in large movement of the beat frequency pattern. For example [51], if gratings with periods $p_1 = 3.95$ μm and $p_2 = 3.60$ μm are chosen, then a moiré pattern is formed with a fundamental period of $P = (p_1 p_2)/(p_1 - p_2) = 40$ μm. In this case, a relative shift between the gratings of less than 4 μm produces a shift of the moiré pattern equal to the entire period P, that is, 40 μm. With the detection sensitivity of $P/1000$ and additional tricks, such as using a harmonic moiré, which improves the sensitivity by a factor of two, and counter-moving fringes, which also improves the sensitivity by a factor of two, it is possible to achieve shift sensitivity down to 1 nm [51]. This is more than sufficient for the purpose of creating high-quality, three-dimensional photonic crystals, for the near-infrared and visible range, in a layer-by-layer fashion such as the method described above.

Having identified a technique for measuring displacement using optical means with an accuracy of 1 nm, we turn our attention to the second item: correcting the detected misalignment with a comparable precision. In short, we seek a way to shift a mask with respect to a substrate by a predetermined vector in the *xy* plane with an accuracy of a few nanometers. Actually, the above statement is too strong. What we really seek is to shift the *image* patterned in the mask and projected onto the sample. If phrased this way, the problem is open to a wider range of solutions. One possibility is to mount the mask on a moving platform controlled by piezoelectric actuators, which are known to allow for extremely precise positioning. However, Povinelli et al. [48] proposed a different solution where the mask is rigidly attached to the sample. In this case, a controlled gap of 2 to 5 μm is left between the mask and the sample by introducing aluminum studs around the perimeter of the mask. For exposure, a Cu_L X-ray line, $\lambda = 1.32$ nm, is used. Radiation of this wavelength suffers little diffraction when used to expose features several hundred nanometers across even if the separation between the mask and the sensitive layer is a few microns. The shift of the image is achieved by tilting the entire mask-sample assembly as shown in Fig. 6.10 so that the shadow of the features in the mask appear at the surface of the sample displaced by $V = \theta \cdot G$, where θ is the tilt angle in radians and G is the gap between the mask and the sample. With a gap $G = 5$ μm and the tilt-angle precision 2×10^{-4} rad, a positioning accuracy of 1 nm is expected [50].

Having resolved, at least in principle, the fabrication issues for the realization of structure shown in Fig. 6.7, one can consider other geometries that would be practical with this approach but offer unique advantages compared with the FCC lattice of Fig. 6.7. One such structure has been proposed by Roundy and Joannopoulos [52] and is shown in Fig. 6.11. It also consists of alternating rod and hole layers. However, in contrast to the original multilayer FCC structure, this one exhibits square symmetry within each layer. This property makes it uniquely suited for on-chip routing applications because of the deep historic entrenchment of Manhattan geometry for this purpose in the microelectronic industry where electric wires are oriented at 90° angles. In addition, the photonic crystal of Ref. 52 needs only two sets of hole-rod layers per one photonic-crystal period, as opposed to three sets for the structure of Fig. 6.7, which may make it less laborious to fabricate if one uses the technique of etching through two rod layers and one hole layer in a single etch step similar to the one explained above.

Although the bandgap of this body-centered tetragonal lattice is somewhat narrower than the 21% of the FCC structure described in Ref. 47, its 18% is still

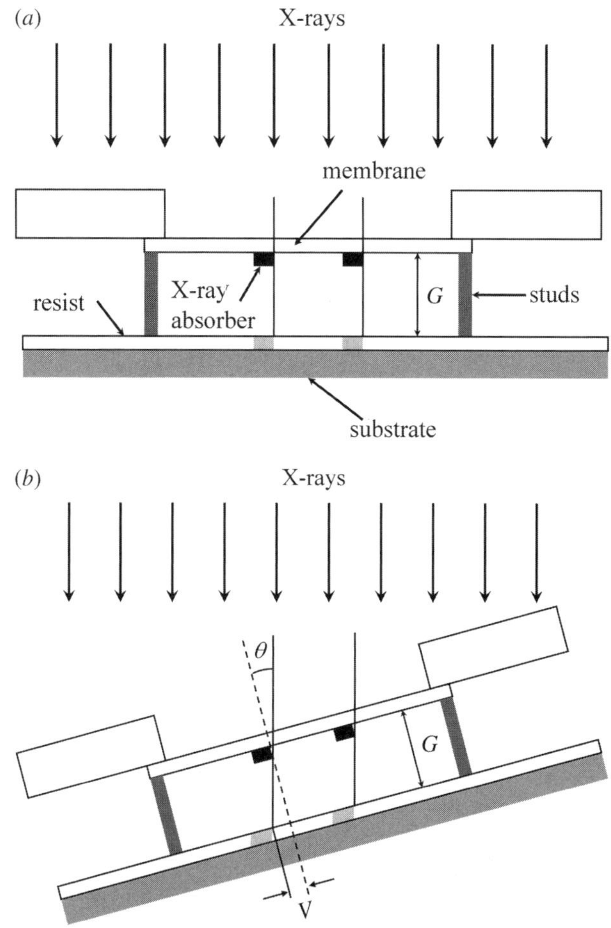

Figure 6.10 Shifting the mask image in the resist by tilting mask-sample assembly. The amount of shift is $V = \theta \cdot G$. The features in the mask are projected onto the sample in a parallel fashion thereby preserving their relative placement; i.e., the period of a photonic crystal is preserved. Because of the finite thickness of the X-ray absorber and of the layer of resist, the features may be distorted. However, for small angles, this distortion is minimal.

respectable and sufficient for many applications. Moreover, the presence of square symmetry within each layer suggests that it should be possible to design intersecting line defects with zero cross talk as described in Ref. 53 that would allow for arbitrary high-density routing of optical signals.

6.2.2 Woodpile Photonic Crystals

Historically, the first 3D photonic crystal that was proposed to be fabricated in a layer-by-layer fashion was the so-called woodpile structure. It consists of stacked, horizontally arranged dielectric rods that resemble a pile of logs—hence the name. Depending

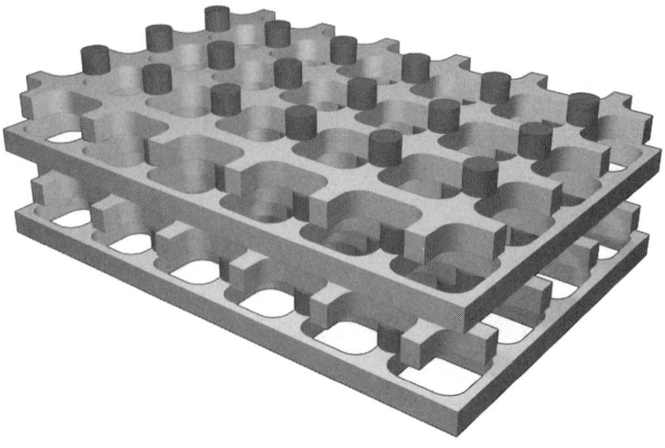

Figure 6.11 A photonic-crystal structure consisting of alternating hole and rod layers with square symmetry in each layer.

on the particular stacking geometry, many distinct symmetries are possible. One geometry that has attracted considerable attention is an arrangement where adjacent rod layers are rotated 90° with respect to one another, and the consecutive parallel layers are offset horizontally by half of the spacing between the logs. Such a structure is schematically depicted in Fig. 6.12. It is clearly periodic in all three dimensions. However, whereas the periodicity in the horizontal (x, y) directions is equal to the

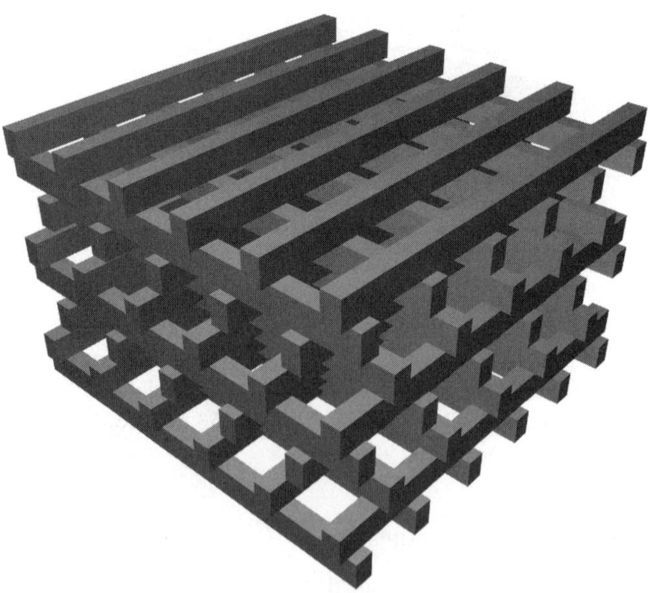

Figure 6.12 Three-dimensional arrangement of dielectric rods that form a woodpile structure.

distance between consecutive logs, the periodicity in the stacking, or z, direction is equal to four times the thickness of the rods, i.e., it repeats itself after four layers of the stack. This four-layer period is the result of the horizontal offset between the consecutive parallel layers equal to half of the horizontal period. In general, such a woodpile structure can be described as a face-centered-tetragonal (FCT) lattice. In the special case where the distance between the logs in a layer is a factor of $\sqrt{2}$ smaller than the vertical period of the lattice, it becomes a familiar FCC lattice. In this case, the dielectric rods define the horizontal $\langle 110 \rangle$ directions, whereas [001] points in the stacking direction. A cubic unit cell of such an FCC lattice cut out from the larger structure of Fig. 6.12 is presented in Fig. 6.13. It should be emphasized that in this case, the horizontal sides of the unit cell cube are at 45° with respect to the direction of the rods.

The woodpile photonic crystal as shown in Fig. 6.12 was first proposed, fabricated, and tested in 1994 [54]. It was made of alumina rods 15.24 cm long and 0.318 cm in diameter manually glued together at 1.123-cm separation between adjacent rods within a layer. The dimensions of the original structure make it suitable for the microwave region of the electromagnetic spectrum. For example, the stop band along the x axis, which corresponds to the [110] direction, has been measured to extend in this case between 11.9 GHz and 14.0 GHz [54].

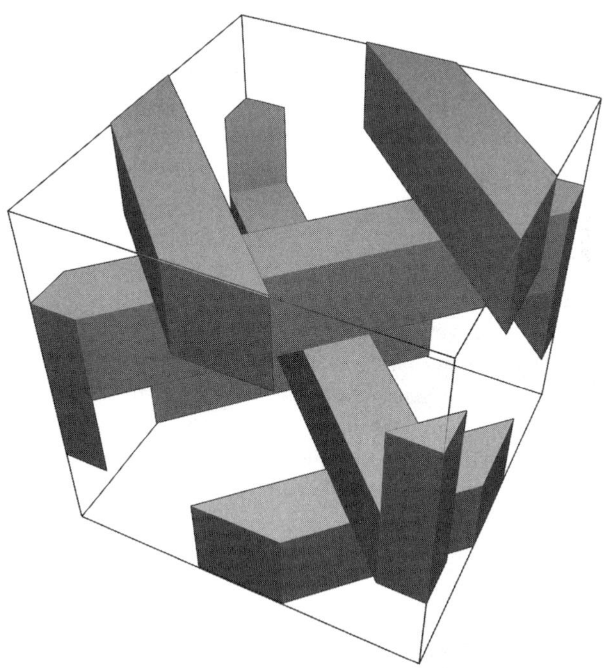

Figure 6.13 Unit cell of an FCC woodpile structure. It should be pointed out that the dielectric rods are at 45° with respect to the horizontal periodicity of the lattice. The unit cell has been chosen to emphasize the high symmetry of the structure and at the same time the lack of mirror symmetry with respect to the xy plane.

Although the fabrication method of Ref. 54 is not scalable to dimensions required for the photonic bandgap in the near-infrared region, the early work established the woodpile geometry as a viable structure for the use as a photonic crystal. Subsequent efforts of the same group resulted in reducing the rod cross-sectional dimensions first to 340 μm by 390 μm that allowed the photonic bandgap to open at 94 GHz [55] and then to 50 μm by 100 μm that yielded the photonic bandgap with a range of 370–520 GHz [56].[1] This has been achieved by transitioning, at least partially, to fabrication methods used in the microelectronics industry. Thus, the rods were defined photolithographically in silicon wafers and then etched in an aqueous solution of potassium hydroxide (KOH), which preferentially etches the {110} planes of the silicon crystal while leaving the {111} planes with their densely packed atoms relatively intact. In this case, the scale was dictated by the thickness of available silicon wafers: 390 μm for the former and 100 μm for the latter attempt, which were etched through to produce sets of parallel silicon rods anchored at each end to the remaining substrate. To assist in the alignment, four holes were made in the substrate at the same time as the rods were etched. The assembly of the photonic crystal then proceeds by manually stacking the patterned silicon wafers using a holder with four pins that mate with the alignment holes and guide the stacked wafers.

Although manual stacking of individual layers to build three-dimensional photonic crystals may seem far-fetched for fabricating them at scales suitable for near-infrared wavelengths, this approach has been followed by Aoki et al. [57]. To handle elements with submicron features, the researchers used micromanipulators under a scanning electron microscope (SEM) for the real-time observation of the assembly process; this technique was pioneered in the context of photonic crystals by Miyazaki et al. [58]. However, in contrast to the method presented in Ref. 58, where individual spheres forming a photonic crystal were manipulated, Aoki et al. [57] moved entire sheets of material with patterned rods, similarly to the macroscopic manual assembly of Refs. 55 and 56.

The fabrication process proceeds as follows. First, plates with parallel lines are patterned using standard lithography techniques and ICP etching that transfers the pattern into the InP layer of the substrate. The plates are then released by wet etch in sulfuric acid that selectively attacks the underlying InGaAs layer and leaves the plates suspended by narrow bridges that connect them to unpatterned areas. Subsequently, the plates are separated or broken off at the bridge junctions, from the substrate using micromanipulators, and stacked in an SEM specimen chamber. The micromanipulators consist of a single rod with a well-defined thickness at the tip. The plates stick to the rod and to the substrate because of electrostatic and van der Waals forces that act between the molecules of the tip and the material of the plate. At scales of several microns, these forces are much stronger than the gravitational force. Manipulating elements with this method is akin to handling sticky objects with a single finger [58]. The thickness of the tip is chosen depending on the size of the object to be moved, and it has to be conductive to dissipate the electrons of the scanning electron microscope probe beam and to prevent charging. They are typically

[1] These numbers refer to the stop band in the z or [001] direction.

made out of glass by pulling a heated fiber and snapping it once the desired thickness is reached. For electrical conductivity, the tips are coated with metal, such as gold, with an underlying layer of chromium for adhesion. The mechanical control of the micromanipulator is achieved by mounting it on precision piezoelectric actuators that allow three-dimensional movement with an accuracy of a few nanometers. The operator can then move the micromanipulator using a joystick linked to the piezo-actuators while observing his/her actions with the real-time video provided by the SEM.

To attain the high alignment accuracy required for quality photonic crystals, the plates are patterned with a set of alignment holes. Prior to placing a plate on a stack, two polystyrene microspheres with a diameter equal to the diameter of the holes are inserted into two alignment holes. The next plate is then placed on top of the stack and pushed down so that the holes patterned in it are guided to receive the spheres sticking out from the previous layer. Then, another sphere is inserted to fix the plate in place and two more for the next layer. With tight tolerances of the hole and sphere sizes, the alignment accuracy is only limited by the lithography and dry etching, and it has been found to be better than 50 nm [57]. The entire fabrication process is schematically shown in Fig. 6.14.

Whereas Aoki et al. [57] have been successful in applying the method to build demonstration devices 25 μm across, it is difficult to imagine extending the procedure to wafer-scale fabrication while maintaining the feature size required for infrared operation of the resulting photonic crystals. A particular difficulty lies in handling patterned plates several millimeters across and only a few hundreds of nanometers thick. At this aspect ratio, the plate would really be a thin, floppy membrane with spaghetti-like lines running across. Another problem has to do with embedded defects. For structural stability even at small scales, it is desirable that the rods that eventually form the woodpile be anchored at both ends. However, the simplest defects require that a section of a rod be missing. This results in a rod anchored on one end only. Although switching to different photonic-crystal geometry can alleviate this problem, the available topologies will always be limited to cases where every layer consists of a connected dielectric structure; i.e., no island of dielectric can exist that is separated from the rest of the layer.

On the other hand, the relative simplicity of the process as well as the flexibility in obtainable geometries and material used makes the method ideally suited for realizing small structures that can be built rapidly and tested experimentally to verify hypotheses that otherwise may take much longer to analyze computationally.

To create large-area woodpile photonic crystals, one might try to leverage the experience gained in the fabrication of microelectronic circuits. This approach was adopted by researchers at Sandia National Laboratories [59–64]. They have exploited the progress made in the microelectronics industry, in terms of minimum feature size and alignment accuracy, to benefit the field of photonic-crystals fabrication. The process is similar to that described above for the realization of the FCC structure with cylinders. It starts with depositing a layer of silicon dioxide on the substrate. The oxide is then patterned with a series of lines and etched to the desired depth. The trenches are then filled with polycrystalline silicon, and the surface is planarized using chemical mechanical polishing [65]. This step is important as it prevents the

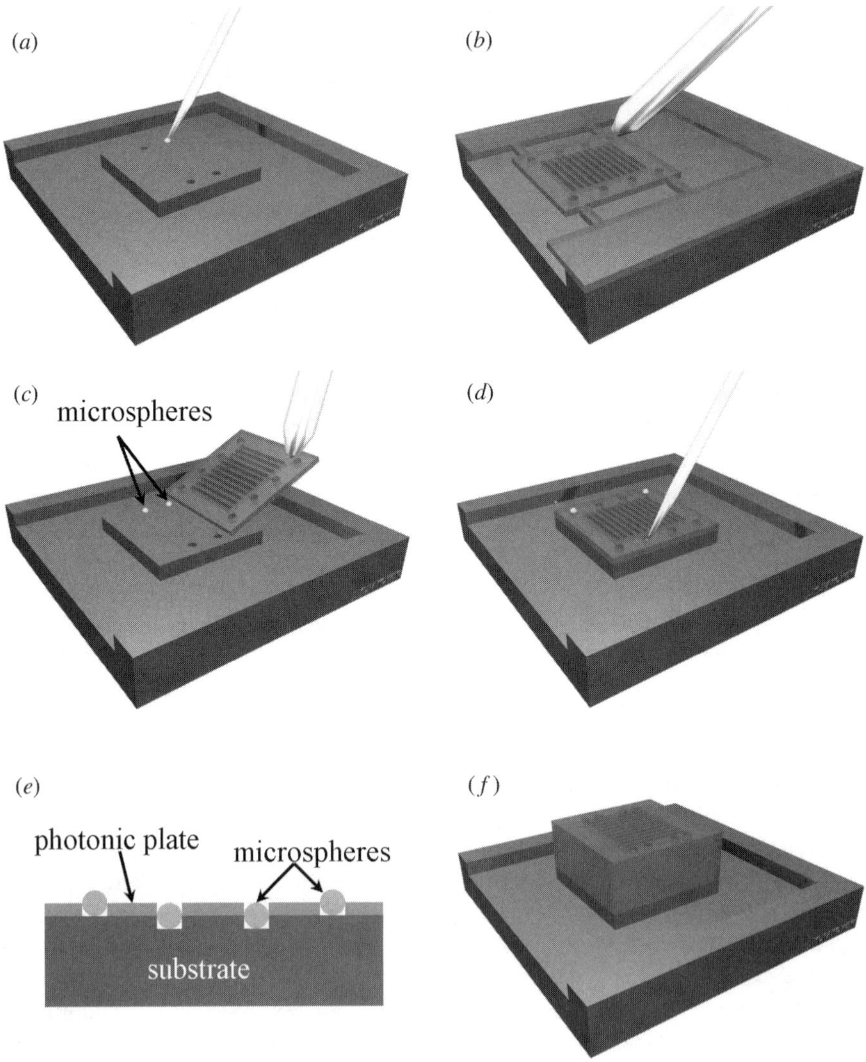

Figure 6.14 Procedure for building a woodpile photonic-crystal structure by stacking individual layers using micromanipulators. (a) Microspheres are placed in alignment holes of the substrate. (b) A plate that constitutes a single layer of a photonic crystal is separated from the substrate. (c) The plate is transferred to the prepatterned substrate. (d) An additional sphere is inserted into a remaining alignment hole to fix the plate on the substrate, and two more spheres are inserted for the alignment with the next layer. (e) Cross section of the alignment holes of the structure shown in (d). (f) Eight layers forming two complete periods of the woodpile photonic crystal.

propagation of surface topography created in one layer to subsequent layers. The process is then repeated for subsequent layers with proper alignment to form the woodpile structure. After the process is completed, the oxide is removed by aqueous solution of hydrofluoric acid. In this way, Lin et al. [59] fabricated high-quality woodpile photonic crystals with a horizontal dielectric rod separation of 4.2 μm, which makes it suitable for applications in the 10-μm wavelength regime. To scale the dimensions down and make the structure suitable for near-infrared wavelengths, which would be of particular interest for telecommunication applications that operate in the range of 1.55 μm, the group took advantage of a particular aspect of the woodpile structure that allows a higher density packing of features without upgrading the available lithography tools. The method, which is known as filleting, relies on the fact that a single layer of a woodpile structure consists of separate lines. Now, each line, as drawn in material, has two edges, or steps. Each of these two edges can be turned into a separate line all by itself by conformal coating and then anisotropic plasma etch that preferentially removes material from horizontal surfaces of the substrate. The material coating the vertical portion of the pattern remains unetched and forms a new, narrow line along the edge of the original line. The width of the new narrow lines is determined by the thickness of the deposited film, which is routinely controlled with nanometer-level accuracy and thus allows for precise definition of narrow, high-density lines.

The details of the process as first reported in Ref. 60 are explained below and summarized in Fig. 6.15. First, a multilayer stack is prepared on a carrier single-crystal silicon wafer; Fig. 6.15a. It consists of a 100-nm layer of SiN, 220-nm poly-Si, which is the desired final thickness of the rod layer, 50-nm SiN, and a 500-nm layer of SiO_2 deposited by plasma-enhanced CVD, which is the sacrificial material for the filleting. The oxide is then photolithographically patterned and anisotropically etched, stopping just short of the nitride layer underneath. Because the pattern in the oxide is used for filleting, the minimum feature size is equal to the periodicity of the resulting line array in the woodpile layer. The nitride between the oxide lines is exposed to a subsequent 30-s wet etch in a 6 : 1 mixture of ammonium fluoride and hydrofluoric acid at room temperature; Fig. 6.15c. This step also thins down the remaining oxide lines and thus allows for relaxing slightly the lithography requirements. To form the fillets, polysilicon is deposited conformally to coat both horizontal and, more importantly, vertical surfaces with a 180-nm layer, as depicted in Fig. 6.15d. Then the polysilicon is plasma etched using a highly anisotropic process to remove it from horizontal surfaces at both the top of the oxide lines and the bottom trenches between them, and to leave the sidewall coating intact. This forms polysilicon fillets clinging to the sacrificial oxide lines, as in Fig. 6.15e, which have now fulfilled their role and are removed in the next step; Fig. 6.15f. The remaining polysilicon fillets are used as an etch mask for transferring the pattern to the underlying nitride layer using wet etch (Fig. 6.15g) and then for anisotropically dry etching the device polysilicon underneath; Fig. 6.15h. Certainly, during the etching of the polysilicon layer, the fillets get etched as well because they are of the same material. However, because their height exceeds the thickness of that silicon layer, the latter will etch all the way to the first nitride layer before the fillets are gone. In addition, an extra layer of nitride between the fillets and the polysilicon layer will protect the underlying polysilicon in a linewise pattern in case the fillets do not hold. Finally, the gaps between the lines of polysilicon are filled with 300 nm of SiO_2, and then the surface is planarized back to the silicon nitride using chemical mechanical polishing; Fig. 6.15i, j.

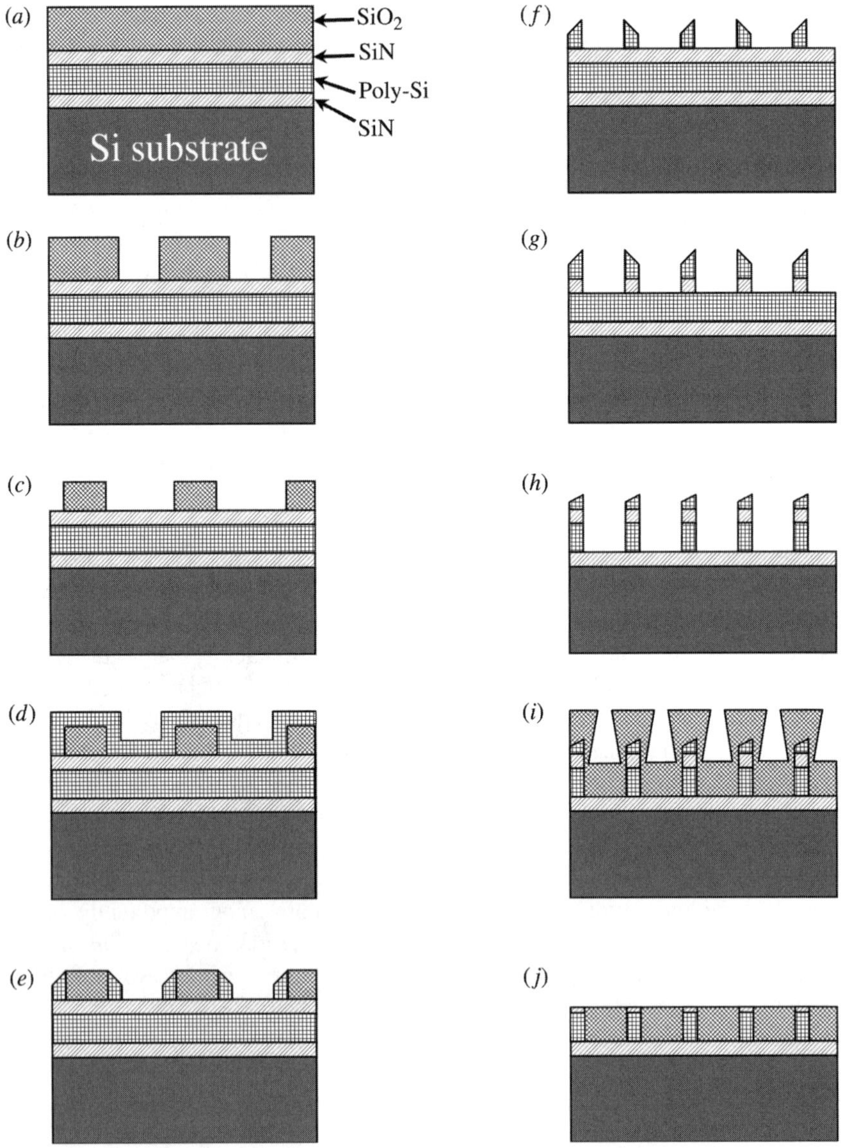

Figure 6.15 Process flow for the fabrication of a woodpile photonic crystal for near-infrared wavelengths by filleting. (a) Deposit 100 nm of SiN, 220 nm of poly-Si, 50 nm of SiN, and 500 nm of SiO_2. (b) Pattern and etch SiO_2 to form a pattern of lines; stop dry etch 50 nm before reaching the nitride layer. (c) Wet etch the oxide to remove approximately 90 nm. (d) Deposit 180 nm of polysilicon. (e) Dry etch anisotropically 180 nm of silicon to form fillets along the sidewalls of the oxide lines. (f) Remove the sacrificial oxide lines to form a set of free-standing fillets. (g) Wet etch the nitride layer using polysilicon fillets as mask. (h) Transfer the pattern with dry etching using the fillets and the nitride lines as an etch mask. (i) Deposit SiO_2 to fill-in the gaps between the lines. (j) Planarize the surface with chemical mechanical polishing to prevent the propagation of topography to subsequent layers, which are processed identically, except for the first nitride layer that acts as an etch stop in step (h).

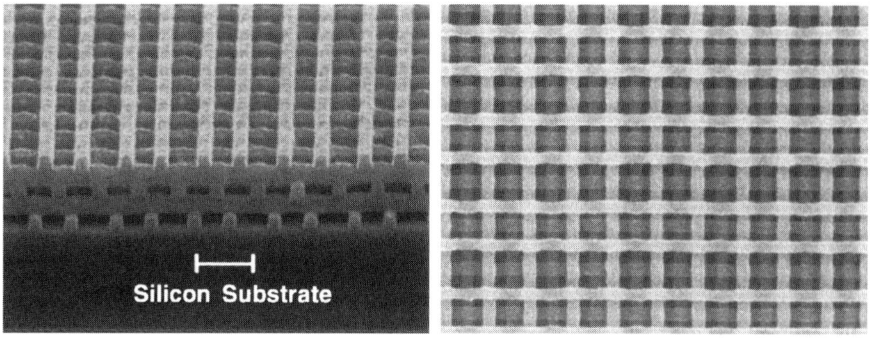

Figure 6.16 Woodpile photonic crystal fabricated using the filleting method of Ref. 61. Horizontal dimension bar indicates 1 μm. (Reprinted with permission from Ref. 61, © 1999, IEEE.)

The last step is critical as it ensures flatness of the surface for the next layer and therefore prevents the propagation of surface topography as new layers are fabricated. Thus, the first layer is completed, and the wafer is ready for the next layer, which is prepared in the same manner, except for the deposition of the first SiN nitride layer, which only serves as an etch stop for the first polysilicon etch. The process is then repeated with proper alignment between alternate layers until the desired number of layers is achieved. Finally, to increase the dielectric contrast between the two dielectrics that constitute the photonic crystal, silicon dioxide between the polysilicon lines is removed with a concentrated solution of hydrofluoric acid in water that selectively attacks the oxide while leaving polysilicon intact.

The results of this process as reported in Ref. 61 are presented in Fig. 6.16a, b. The woodpile photonic crystals show excellent periodicity and uniformity, with the exception of having pairs of lines more closely spaced than the distance between the pairs. The reason for this artifact can be traced back to the method of fabrication where the lines are created pairwise on either side of silicon dioxide lines. It stands to reason then that unless the process is perfectly controlled, the oxide lines may come out slightly too narrow, or too wide, to produce exact periodicity in the resulting fillet array. However, this artifact does not seem to have a significant effect on the optical performance of the resulting woodpile photonic crystal [60, 61].

The successful fabrication of the woodpile photonic crystal at scales suitable for its application in the near-infrared regions has been an important step on the road to widespread use of photonic crystals. However, the use of silicon and, in general, non-monocrystalline materials limits structures obtained therefrom to passive devices only. Although the potential field for passive photonic crystals is broad,[2] it is desirable to have the option to introduce sources and detectors as well as optical modulators into the structure. Given that heretofore attempts to coerce silicon to lase have been unsuccessful, one is compelled to investigate other materials that have been demonstrated to both produce and actively interact with light. Such a class of materials has been found in III–V compound semiconductors, which include GaAs, InP, and so on.

[2]See Chapters 4 and 5 for application of photonic crystals.

In addition to employing suitable active materials, we need fabrication methods to synthesize photonic crystals that preserve the intricate crystalline layers (i.e., quantum wells) on which functional electro-optic devices rely. One such method was proposed in 1996 by Noda et al. [66]; it uses wafer bonding to produce a stack of semiconductor rods that, when properly aligned, form a woodpile structure. The rationale behind the method is as follows: Because the handling of individual rods or sheets of them on scales required for infrared operation is at best difficult,[3] we keep them attached to a handle substrate hundreds of microns thick, which is removed only after they are placed on and bonded to another layer of the photonic crystal. This way, the fragile logs of the woodpile are mechanically supported at all times during and after the fabrication process.

The details of this approach, as reported in Refs. 66–76, are presented below. Referring to Fig. 6.17, the process can be summarized briefly as follows. A pattern of rods is defined in a suitable substrate, such as GaAs, using standard lithography and etching techniques; Fig. 6.17b. An identically patterned substrate is then bonded to it in such a way that the rods of the two are perpendicular to one another; Fig. 6.17c. The top handle substrate is then removed (Fig. 6.17d through f) to yield half a period of a woodpile supported by the other substrate; Fig. 6.17f. The process is then repeated with two structures identical to that of Fig. 6.17f, to form a full period of a woodpile shown in Fig. 6.17g. To produce a photonic crystal with the desired number of periods in vertical (normal to the substrate) direction, the process is then repeated a sufficient number of times. If a defect embedded in the photonic crystal is required, then one or more of the layers is accordingly modified prior to bonding.

Although conceptually the process described above is simple enough, several issues must be addressed for it to work and be practical. First, the bonding of the two patterned substrates has to be carried out in such a way as to result in a robust unitary structure that holds together under a reasonable amount of stress. Clearly, using simple glue, which works at macroscopic scales and was successfully used to realize woodpile PhC for microwaves [54], is out of the question when working at submicron scales. However, atomic-level flatness of the substrates allows for employing direct wafer fusion, wherein the substrates are brought into contact and allowed to create a bond under elevated temperature and pressure. In this case, the choice of the environmental conditions under which the fusion is carried out is of critical importance. In particular, the temperature must not be too high so as to destroy heterostructures, such as quantum wells, that may be built into the substrate(s); yet it has to be high enough to yield reliable bonds.

Another problem is the removal of the handle portion of the substrate after the bonding is realized. The difficulty here lies in the wide disparity of scales that are encountered. On the one hand, hundreds of microns of thickness have to be removed; on the other hand, the removal process has to stop within a few nanometers to maintain the quality of the periodic structure. This feat can be accomplished by judicious use and combination of a variety of material-removal techniques. While the bulk of the

[3]However, see pages 284–285 for a method where a woodpile is obtained by stacking patterned sheets on the microscale.

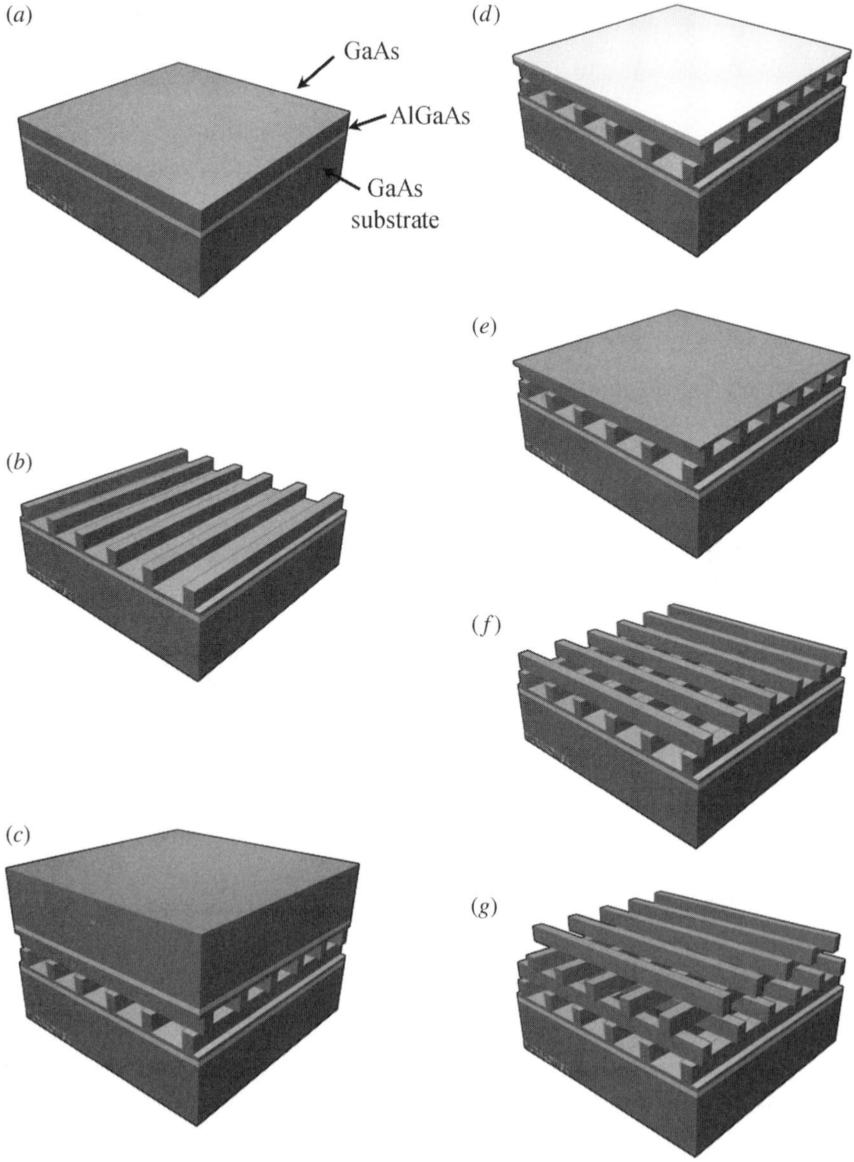

Figure 6.17 Fabrication process for realizing woodpile structures in near-infrared region based on wafer bonding of patterned substrates. (a) AlGaAs followed by GaAs layers are grown on GaAs substrate. The thickness of a GaAs layer is slightly more than the desired thickness of the resulting PhC layer. (b) A series of parallel rods are patterned and etched using standard lithography techniques. The etch depth is equal to the desired thickness of a single PhC layer and slightly less than the thickness of the top GaAs layer. (c) Two identically patterned wafers are bonded with rods of the two perpendicular to one another. (d) The handle portion of the top substrate is removed by mechanical polishing followed by wet chemical etch stopping at the AlGaAs boundary. (e) The AlGaAs layer is chemically etched stopping at the GaAs boundary. (f) The thin layer of GaAs is dry etched using RIE to expose the patterned rods. (g) The procedure (a)–(f) is repeated with the stacked PhC layers while ensuring proper alignment at step (c).

material is removed by mechanical polishing, the closer we get to the fine structure, the more delicate and refined the technique used.

Finally, when a full period is to be realized as shown in Fig. 6.17g, the two substrates have to be precisely aligned so that the rods of every other layer are mutually interspersed to form the woodpile structure. Here, the issue is that if optical means are to be used for alignment, then the substrate has to be transparent to the wavelength used. For GaAs or Si, this rules out visible light and near infrared is preferred in these cases. However, if the photonic crystal is designed for near infrared, then the periodicity of the linear structures falls beyond the resolution limit of this infrared wavelength, so the direct observation of the structures to be aligned is not possible. The way out of this quandary relies on decimating the linear structure in the alignment region and using an interferometric, rather than a direct-view, alignment technique, which in a natural way exploits the periodic nature of the pattern.

Below, we describe in more detail the critical steps of this fabrication method. The starting point is a GaAs wafer with a buffer layer of GaAs, and then $Al_{0.8}Ga_{0.2}As$, followed by a layer of GaAs grown using MBE or metallo-organic vapor-phase epitaxy (MOVPE) at 560°C to 600°C; see Fig. 6.17a. The thickness of the top GaAs layer is slightly more than the final thickness of a single woodpile layer. For a woodpile with a line periodicity of 4 μm, the thickness of the top GaAs layer was 1.5 μm to 2.0 μm, whereas the etch depth was 1.2 μm; see Refs. 68–71. The reason for this excess will be apparent later. After growth, a series of lines is patterned and etched in the top GaAs layer using RIE in $CH_4 + H_2$ plasma. The choice of etching species is dictated by the desire of a relatively slow etch to maintain a sufficient level of control over the etch depth, which is equal to the desired thickness of the woodpile layer. As a result, a thin layer of GaAs is left remaining on top of AlGaAs as shown in Fig. 6.17b. Two substrates thus prepared are almost ready for bonding.

Wafer Bonding Fig. 6.17c. Prior to bonding, the substrates need to be pretreated to remove contaminants and to make the surfaces more amenable to forming a bond. For this purpose, a 30-s wet etch in HF or a solution $HF : NH_4F$ in proportion of 1 : 6 is used [73]. Then, the two substrates are brought in contact so that the patterned lines are at right angles with respect to one another, and are pressed together at $1.7\ N/cm^2$ or more. The assembly including the two substrates and the holder is then loaded in a liquid-phase epitaxy furnace, where it stays in hydrogen atmosphere for 30 minutes. The choice of temperature depends on both the material(s) used and the size of the pattern. Originally, the Kyoto researchers used 650–700°C [66–70], with excellent results when the periodicity of the linear pattern was 4 μm or larger and the line width was at least 1 μm. However, as the feature size was reduced to make the structure suitable for near-infrared applications, where the required period is 0.7 μm and the line width is less than 200 nm, it turned out that this high temperature yields excessive mass transport, which distorts the final woodpile and compromises its optical performance. By testing temperatures within the range of 400–700°C, it was found that 500°C was the most appropriate temperature in terms of providing a strong bond while avoiding lattice distortion induced by mass transport and interdiffusion [74].

The resulting bonds were strong enough to withstand reasonably high pulling forces and, more importantly, to withstand the subsequent processing that included mechanical polishing of one of the handle substrates, even though the surface area that participates in the bond for the stripped pattern is significantly reduced as compared with flat, unpatterned wafers. An added advantage of using patterned samples was the disappearance of a bubble-formation problem that is a serious issue in a general wafer-bonding process. In the case of a woodpile structure, there is simply no possibility of having a bubble trapped between the two bonded surfaces because pathways are always available for the gas to escape between the patterned lines.

Although 500°C for wafer bonding works very well for identical GaAs substrates, there is a concern over whether this temperature is appropriate to bond dissimilar materials. The reason for using different materials stems from the desire to introduce light emitters into the photonic-crystal structure that would emit at a 1.3–1.55 μm wavelength, which is the most interesting spectral region from a telecommunication point of view, where GaAs does not emit. Such a material has been found in InP-based semiconductors. The problem is that GaAs and InP have different thermal expansion coefficients. Therefore, if two such wafers are bonded at an elevated temperature, then stress at the bond interface emerges during the cooling process. If the bonding temperature is too high, then the stress may be sufficient to cause delamination and thus destruction of the woodpile. In particular, it has been found that the temperature of 525°C is already too high and leads to delamination caused by stress [77]. Indeed, even temperatures as low as 300°C result in debonding. However, as it turns out, the stress can be relieved by thinning one of the substrates prior to bonding. The difficulty here lies with handling very thin substrates, which are inherently prone to break. This problem has been solved by Ogawa et al. [77] by introducing a multistep bonding process. Accordingly, after patterning the lines for the woodpile, the wafers are first fused at a low 220°C temperature. Then, the InP handle wafer is thinned down to less than ~1 μm. Finally, heating the sample to 525°C completes the bonding process. Using this method, an InP layer that includes an InGaAsP multiple quantum well (MQW) heterostructure designed to emit in the 1.5-μm range has been embedded in the middle of a woodpile photonic-crystal lattice. The optical performance of the structure has been verified by the observation of photoluminescence [76] where the MQW heterostructure was excited by a Ti-sapphire laser emitting at $\lambda = 900–940$ nm, which falls in the pass-band of the photonic crystal. This experimental result validates the process of bonding dissimilar patterned materials while preserving their intricate crystalline structures required for active interaction with photons.

Removing Handle Substrate [Fig. 6.17d–f]. After the bonding of the two patterned substrates [Fig. 6.17c], one of the wafers is thinned down to expose the top layer of rods in preparation for the possible next round of wafer bonding, as required for building the three-dimensional photonic crystal. This thinning down proceeds in three separate steps to accommodate the disparate scale requirements. Thus, the bulk of the substrate is mechanically polished down by mechanical means. With about 50 μm of GaAs remaining, the sample is transferred to a wet chemical etch bath that consists of $NH_4OH + H_2O_2$ in proportions of 1 : 20. This etchant attacks GaAs and dissolves it at a rate of about 1 μm/minute; therefore, it takes about 50 minutes to remove the remaining GaAs substrate. Also, it attacks GaAs selectively while leaving AlGaAs intact. Therefore, the thin layer of AlGaAs forms a barrier, or etch stop, which prevents removing the fine line structures forming the woodpile that are also made of GaAs. The selectivity between the GaAs and the AlGaAs is sufficiently high for this solution that even leaving the sample in it for extended period of time poses no danger of overetching. This is important if the procedure were to be adapted for commercial realization of photonic crystals as it offers a wide latitude of process parameters and, therefore, makes the fabrication robust. Subsequently, the AlGaAs layer is removed by applying a buffered HF solution to the sample; HF selectively attacks AlGaAs while leaving GaAs intact. At this point, the reason for retaining a thin layer of GaAs is clear. If the pattern of lines were to be etched all the way to the AlGaAs layer, then as soon as it went through the top AlGaAs layer, HF would start etching the supporting layer of AlGaAs and lift the bonded logs off the remaining bottom substrate. With a thin layer of GaAs present to separate the HF bath from the woodpile, the etchant has no access to the bottom layer of AlGaAs and the structure remains supported; see Fig. 6.17e. To expose the logs, the excess of GaAs is etched by RIE using a tightly controlled etch rate and stopping it exactly at the point when the flat GaAs area vanishes.

This way, a stack of two layers of logs positioned at right angles with respect to one another is formed. Extending the structure beyond the two layers can be done by repeating the procedure using a pair of two-layer stacks. However, to produce a woodpile, it is necessary that the layers be aligned properly. In particular, the logs of the third layer of the stack, although parallel to the logs of the first layer, must fall between them. The same requirement applies to layers two and four; see Fig. 6.17g. The alignment procedure is described next.

Alignment. Because the photonic crystal is a periodic structure just as each layer that forms it, it is natural to try and exploit this periodicity for aligning the layers. For a woodpile, the situation is particularly simple because each layer is a grating, the simplest nontrivial diffraction structure with a well-established principle of operation. However, whereas the effect of single diffractive grating on the propagation of light is widely taught in undergraduate courses, the interaction of light with a set of stacked gratings is not. Therefore, we provide here a simplified analysis suitable for the application to woodpile stacking.

Consider a situation presented schematically in Fig. 6.18a, where the periodicity of the gratings is T and the separation between grating planes is d. The lines of the gratings run parallel to one another and are perpendicular to the figure plane. The lateral offset between the two gratings is r. Light enters from the bottom of the figure and is diffracted at an angle θ. The amplitude of light diffracted at a particular angle depends on the periodicity of the gratings, the wavelength of light, their

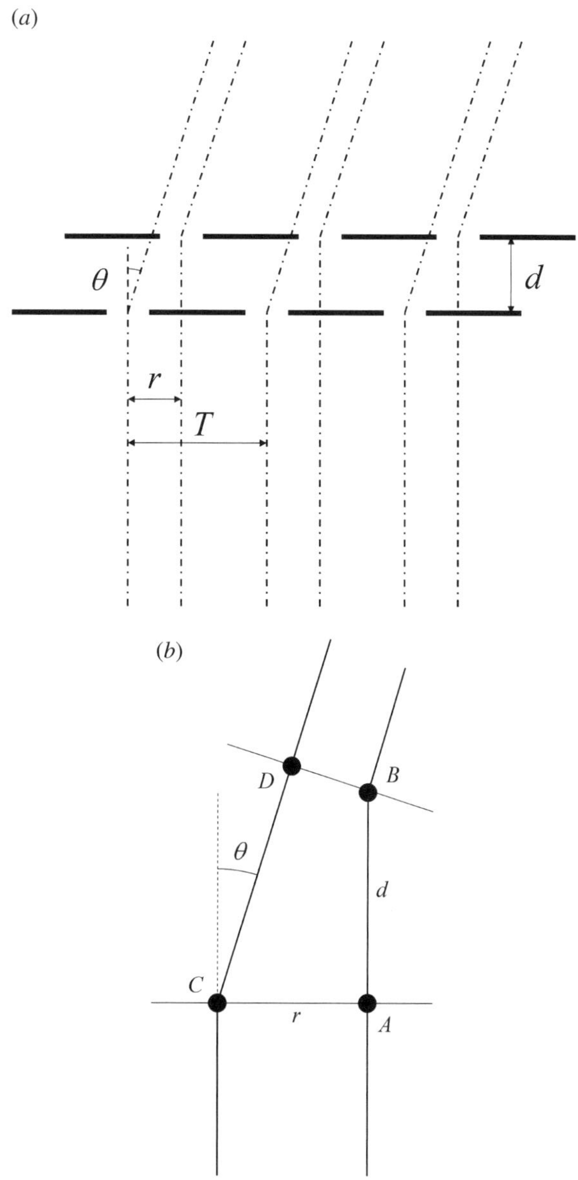

Figure 6.18 (a) Schematic representation of stacked gratings. Light enters from the bottom of the figure and is diffracted at an angle θ. (b) Portion of the figure (a) enlarged for the calculation of optical path-length difference.

separation, and the offset. This dependence can be found analytically for the simple situation of Fig. 6.18a by calculating optical path lengths and adding the phases coherently to find the ensuing diffraction pattern. To calculate the optical path-length difference between a pair of rays that pass through corresponding slots of the two gratings, we enlarged a portion of Fig. 6.18a in Fig. 6.18b; here, AC is a section of the front plane of the incoming wave and is perpendicular to ray AB, whereas BD corresponds to the front plane of the diffracted wave and is perpendicular to ray CD. Hence, what we are interested in is the difference between the lengths of intervals AB and CD. $|AB|$ is simply the separation between the grating planes, which is equal to d. $|CD|$ can be calculated using elementary geometry and is equal to $d\cos\theta + r\sin\theta$. Therefore, our optical path-length difference is

$$d\cos\theta + r\sin\theta - d = r\sin\theta - d(1-\cos\theta). \tag{6.1}$$

The entire diffracted beam is a sum of such pairs of rays that differ in phase by $k[r\sin\theta - d(1-\cos\theta)]$, where $k = 2\pi/\lambda$ and λ is the wavelength of light. Because the phase difference between consecutive pairs is equal to $kT\sin\theta$, the amplitude of the full beam can be expressed as

$$\Psi = \Psi_0 \sum_m \exp(jmkT\sin\theta)\{1 + \exp(jk[r\sin\theta - d(1-\cos\theta)])\}, \tag{6.2}$$

where Ψ_0 is the amplitude of a single component. The summation extends over the number of illuminated lines in the grating and technically should span a range from $-\infty$ to $+\infty$. In this case, the sum is straightforward to evaluate and is equal to zero unless the argument of the first exponential function in Eq. (6.2) is an integer multiple of $2\pi j$ for all m where it becomes infinite. It can be proven that

$$\sum_{m=-\infty}^{\infty} \exp(jmkT\sin\theta) = 2\pi \sum_{m=-\infty}^{\infty} \delta(kT\sin\theta - 2\pi m), \tag{6.3}$$

where δ is the Dirac delta function. The last expression in Eq. (6.3) is simply the manifestation of the presence of discrete diffraction orders that appear only at specific angles $\theta = \theta_m$, where the condition $\sin\theta_m = 2\pi m/kT$ is satisfied. With the help of Eq. (6.3), the amplitude of the full diffracted beam can be expressed as

$$\Psi = \left[2\pi\Psi_0 \sum_{m=-\infty}^{\infty} \delta(kT\sin\theta - 2\pi m)\right]\{1 + \exp(jk[r\sin\theta - d(1-\cos\theta)])\}. \tag{6.4}$$

In Eq. (6.4), the diffractive orders can clearly be identified. To calculate the intensity of the mth diffractive order, we simply multiply the amplitude of that order by its complex conjugate and obtain

$$\begin{aligned} I_m(r) &= I_{m0}|1 + \exp(jk[r\sin\theta_m - d(1-\cos\theta_m)])|^2 \\ &= 2I_{m0}[1 + \cos(k[r\sin\theta_m - d(1-\cos\theta_m)])], \end{aligned} \tag{6.5}$$

where I_{m0} is a constant. With the help of the condition for diffractive order angle $\sin\theta_m = 2\pi m/kT$; the intensity of the mth order can be rewritten as

$$I_m(r) = 2I_{m0}\left\{1 + \cos\left[\frac{2\pi m}{T}r - d\left(k - \sqrt{k^2 - (2\pi m/T)^2}\right)\right]\right\}, \tag{6.6}$$

which explicitly shows its dependence on the gratings' parameters and their relative position. Note that the lateral offset r, which is the only parameter adjusted during the alignment process other than the angle between the gratings, enters only in one position in Eq. (6.6), in the term $2\pi mr/T$, which is in the argument of the cosine function. The other term is a constant phase, which is fixed by the geometry of the logs patterned lithographically. In the absence of the constant phase term, the alignment would consist of shifting laterally one grating with respect to the other; i.e., adjusting r until the first diffractive orders simultaneously reach minimum intensity. At this point, $1 + \cos(2\pi r/T)$ becomes minimum, or equivalently, r is an odd-integer multiple of $T/2$, which places the logs of one layer exactly halfway between the logs of the other. The presence of the constant phase term complicates the situation somewhat. However, its direct inspection indicates that for the situation of interest here, it is relatively small, and considerably less than 2π. In this case, the intensities of the ±1st orders are shifted symmetrically in the opposite directions. Therefore, to adjust the relative positions of the two sets of logs, one would first find the positions corresponding to the minima of the $+1$st and -1st orders separately, and then move the gratings halfway between these two positions.

This procedure has been successfully used by Yamamoto and colleagues [68–70] to fabricate a full period of photonic crystals designed to work in the 10-μm region, where an alignment accuracy better than 100 nm has been achieved. Although in this case the logs that form the woodpile can be used directly for alignment, as the grating period is a comfortable 4 μm, scaling the procedure for use in near-infrared, $\lambda \sim 1.5$ μm where the periodicity is 0.7 μm, requires further consideration. The reason is that this small period does not allow for the observation of the diffracted beams because they become evanescent for the observation wavelength, which has to be longer than ~900 nm for GaAs to reach the transparency window. In this case, alignment regions outside of the device region are created by extending some of the logs. Extending just every third of the logs does the trick of increasing the period to 2.1 μm, where the method can be applied again [72].

6.2.3 Autocloning Technique

The autocloning technique is based on the observation that during plasma-assisted etching or deposition of material, certain shapes can be preserved without getting smoothed out even as thick layers are built up. This phenomenon has been first exploited in the context of multilayer microelectronic fabrication where each functional layer requires high-quality planarization before the next layer is patterned. Thus, a need arose for a clean process that would accomplish the planarization in a relatively short time and with the use of standard equipment. To address this need,

Mogami et al. [49] proposed planarization by a two-step radio-frequency (RF) bias-sputtering, where in the second step, the slope of a ridge is preserved while the angled surface moves in the process, which makes the ridge narrower until it disappears completely and leaves a smooth, flat surface.

The preservation of the slope, or angled surface, is best understood by considering the bias sputtering to be a combination of deposition and sputter etching, and then by analyzing the dependence of each on the angle of incidence of the ions and neutrals on the surface. In bias sputtering, a potential that is negative with respect to the plasma is applied to the sample. Therefore, positive ions, typically argon atoms stripped of the outer electrons, are accelerated toward the substrate and collide with it. The other electrode is floating and, in an equilibrium state, naturally develops a negative potential, and becomes a cathode, as the light, fast-moving electrons of the plasma leave the plasma region and land on it. The cathode holds a target made of material to be deposited on the substrate. Accordingly, the target is predominantly etched, whereas the substrate is both etched and deposited on with the target material.

The dependence of the deposition and etch rates on the incident angle is qualitatively depicted in Fig. 6.19. The curves reflect the physical basis of the processes. The deposition happens when a neutral particle ejected from the target by the bombardment of ions lands on the surface of the sample. The movement of the neutral particles in the plasma is largely random, and so the curve that represents the deposition rate is flat with a wide plateau at normal incidence. In contrast, the sputter etch is a strong function of the incident angle. This is because in the sputtering process, material is removed due to the transfer of kinetic energy from an ion striking the surface to the atoms forming it. This energy transfer is highly dependent on the incident angle. In addition, the velocity of ions is highly anisotropic, mostly aligned with the electric field. For incident angles approaching 90° or grazing incidence, the ions only skim the surface, not unlike a pebble cast across a lake that skips without penetrating water. Hence, little energy transfer occurs from the ion to the bound atoms, and little material removal is observed. At normal incidence, the ions penetrate deep into the material, and their energy is absorbed mostly by atoms far from the

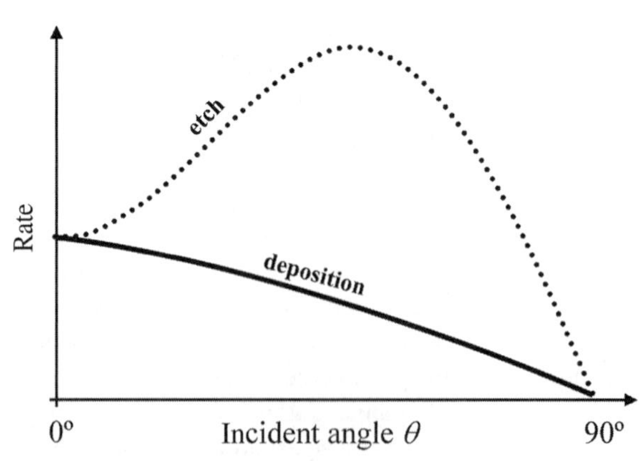

Figure 6.19 Qualitative dependence of the deposition and etch rate on the ion incident angle.

surface, which in turn transfer them to surface atoms. The indirect energy transfer results in its dissipation and moderate etch rate at normal incidence. For incident angles between approximately 40° and 70° the angle is steep enough for the ion to penetrate the surface and, thus, to yield high-energy transfer; yet it is shallow enough to result in the direct transfer of energy mostly to the atomic layers closest to the surface. Having acquired this energy, the atoms overcome the binding energy and escape. Consequently, the etch rate is highest in this range of incident angles as depicted in Fig. 6.19.

The rate at which the substrate surface moves during the bias-sputter process is equal to the difference of the deposition and etch rates of Fig. 6.19. For substrate bias voltage and the target power selected so that the deposition and the etch curves coincide at the incident angle equal to zero, as illustrated in Fig. 6.19, the horizontal surface of the substrate experiences no deposition or etch—it remains flat and its height does not change. However, any protrusion or a bump will be etched from all sides, and the angle of its sidewalls will stabilize at the maximum value of the deposition-minus-etch curve. This effect has indeed been observed by Mogami et al. [49], and it is shown in Fig. 6.20.

The preservation of slope during a bias-sputtering process inspired its use for the deposition of multiple alternating layers of low- and high-index materials that would preserve a corrugation patterned in the first layer. Such a structure would exhibit three-dimensional periodicity and thus fall in the category of photonic crystals. This method of fabricating 3D PhCs has been first proposed in 1997 by Kawakami [78]. Although at the time a satisfactory explanation of the precise mechanism of the process was lacking, the author demonstrated the possibility of building a three-dimensional photonic crystal by alternating bias-sputter deposition of silicon and silicon dioxide on a substrate prepatterned with a triangular lattice of pits several hundred nanometers deep, 200 nm wide, and 500 nm center-to-center separation. After the first few layers, the pattern stabilizes, and the slope, the depth of the pits, and certainly, the periodicity is preserved as new layers are deposited. Thus, in principle, an arbitrary number of layers that are exact replicas of one another can be deposited without compromising the quality of the periodic structure. This is good news for their use as photonic crystals, which require exact periodicity with few fabrication defects to exhibit the interesting properties discussed in other chapters of this book.

The entire process of creating a three-dimensional photonic crystal can be summarized as follows. First, the substrate is patterned using high-resolution lithography, such as electron-beam lithography, and etched in a reactive-ion reactor to define the two-dimensional geometry of the photonic crystal. It could be, for example, a regular, triangular lattice of holes, with a periodicity of 500 nm for a photonic crystal designed to work in the near-infrared regime. The sample is then cleaned to remove the etch mask and loaded into a bias-sputtering reactor where alternate layers of silicon and silicon dioxide are deposited under bias conditions that cause simultaneous etch and deposition, and promote the self-replication of the contour of the underlying layer. As a result, a structure similar to one shown in Fig. 6.21 is obtained.

As reported by Kawakami et al. [79, 80], the corrugated pattern quickly stabilizes, and after few layers, it remains essentially unchanged as new layers are built up.

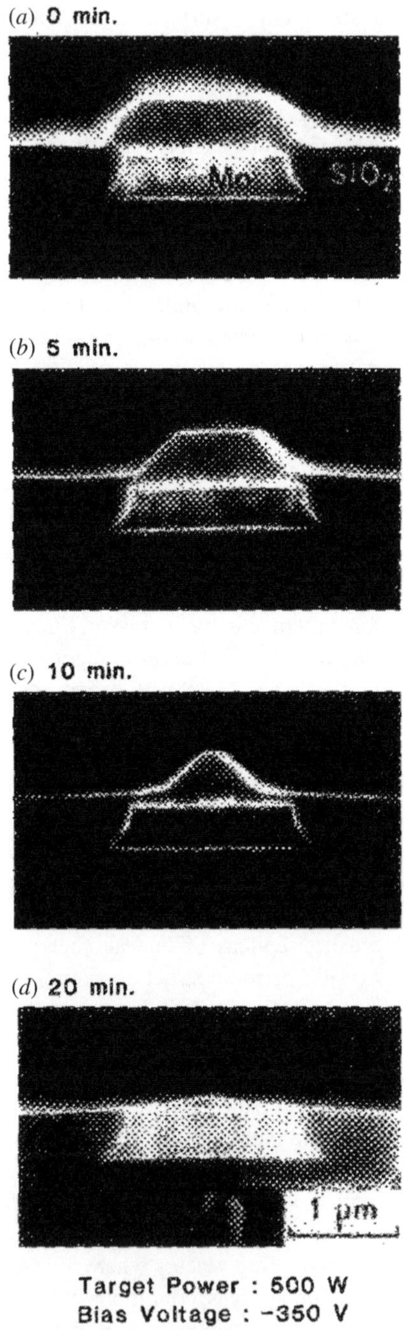

Figure 6.20 Movement of sidewalls during bias sputtering. Note that the slope of sidewalls is preserved and that the horizontal surface is neither etched nor deposited on and remains flat until the sidewalls connect. (Reprinted with permission from Ref. 49, © 1985, American Institute of Physics.)

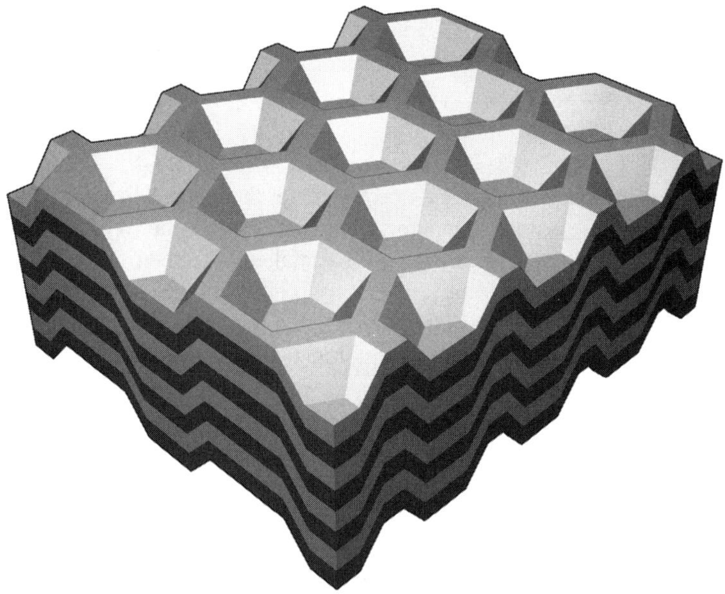

Figure 6.21 Schematic illustration of a three-dimensional photonic crystal obtained using the autocloning technique.

To understand how it is possible that angled facets are preserved, we need an accurate model of the processes that occur at the surface of the substrate as the etching and deposition are taking place. To simplify the analysis, we consider a two-dimensional situation; i.e., we assume that a series of long parallel grooves have been prepared in the substrate. In bias sputtering, essentially three processes are simultaneously taking place. One is sputter deposition of the target material and another is sputter etch. In addition, redeposition of sputter-etched material is taking place. As pointed out above, these processes are characterized by different dependences of the etch/deposition rate on the incident angle.

The deposition of particles from the target substrate can be modeled using a simple $\cos^n \phi$ function as the angular distribution of velocity, where n is the parameter of diffusion and ϕ is the angle from the vertical axis. For $n = 0$, there is no dependence on the angle; i.e., the flow of particles is isotropic. As n is increased, the distribution becomes narrower with the peak at $\phi = 0$. For n that is very large, the peak grows sharp, and as $n \to \infty$, the angular distribution becomes a Dirac delta function—the influx of particles is unidirectional. Then, if a point P is exposed to particles with the incident angle ranging between ϕ_1 and ϕ_2, the deposition rate can be expressed as

$$v_d(\mathrm{P}) = \frac{d}{I(n+1)} \int_{\phi_1}^{\phi_2} \cos^n \phi \cos(\phi - \theta)\, d\phi, \tag{6.7}$$

where θ is the angle between the normal to the surface at point P and the vertical axis,

$$I(n+1) = \int_{-\pi/2}^{\pi/2} \cos^{n+1}\phi \, d\phi$$

is the normalization factor, and d is a free parameter that characterizes the total influx of incident particles, so that when the integration ranges between $\phi_1 = \pi/2$ and $\phi_2 = \pi/2$ and the surface at point P is horizontal, $\theta = 0$, then the deposition rate $v_d(\mathrm{P}) = d$.

The angular dependence of sputter etching that is consistent with the qualitative curve of Fig. 6.19 can be modeled with the functional dependence

$$v_{b,c}(\mathrm{P}) = b\cos\theta + c\sin^2\theta\cos\theta, \tag{6.8}$$

where b and c are free parameters, both negative as $v_{b,c}$ represents the removal of material in this process, to be determined experimentally; they depend on the details of the bias-sputtering process.

Finally, redeposition is modeled in the same way as the primary deposition given by Eq. (6.7), with the exception that for the sake of simplicity the distribution of velocities is assumed isotropic in the range of angles available for redeposition

$$v_e(\mathrm{P}) = \frac{d}{I(n+1)} \int_{\phi_2}^{\phi+\pi/2} \cos(\phi - \theta)\, d\phi, \tag{6.9}$$

where the integration is carried over the range of angles that the redeposition particles can come from. The angles that appear in Eqs. (6.7), (6.8), and (6.9) are defined in Fig. 6.22.

The model described above is to a large extent a simplification of the numerous processes that take place simultaneously during the bias sputtering. In particular, only first-order redeposition is considered; that is, only the molecules ejected by the direct flux of ions are taken into account. However, in reality, also ions that are deflected by the tilted surface, which in turn strike the opposite side of the groove, can remove material from that opposite side. In addition, the molecules removed by sputtering that have high enough kinetic energy can, on impact at the surface, remove more material, rather than settle on it. Also, migration of molecules on the surface is ignored. Nevertheless, using the model of Eqs. (6.7), (6.8), and (6.9), and Fig. 6.22, Kawakami et al. [79, 80] demonstrated, by applying a 2D computer-simulation technique [81], that after a few layers, the shape of the corrugation stabilizes and the cross section exhibits an uncanny resemblance to the experimental results they obtained, see Fig. 6.23. Moreover, they showed that the stable shape depends little on the original pattern, but it is a strong function of the free parameters of the bias-sputter deposition model, and thus of the particular choice of experimental parameters during the process. This excellent agreement of the simulation results with the experiment validates the model used and, thus, provides a sound theoretical, albeit only numerical, foundation for the autocloning phenomenon.

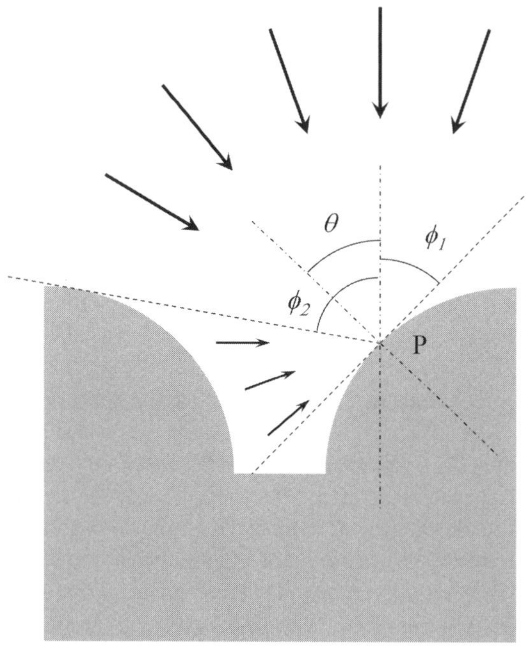

Figure 6.22 Bias-sputter deposition. Point P on the surface is considered. Long arrows represent primary deposition of target material. Short arrows represent redeposition of material sputtered from other regions of the substrate.

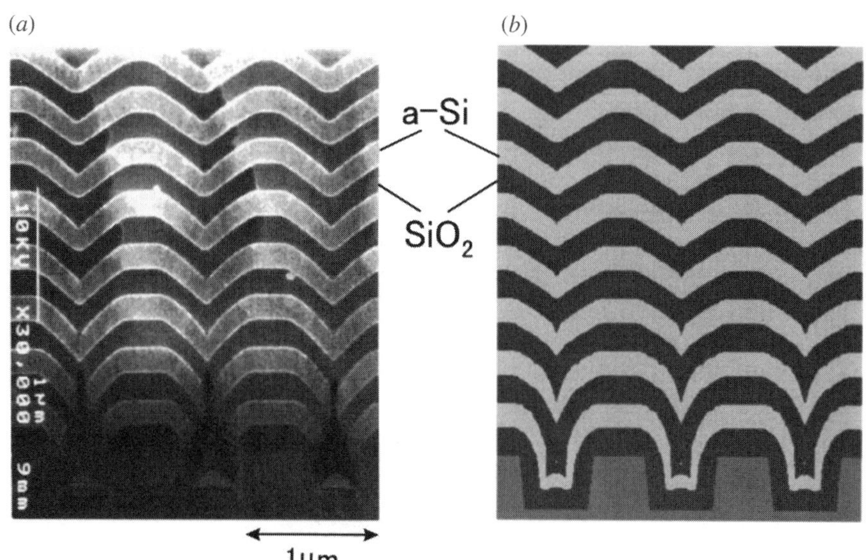

Figure 6.23 The comparison of an (a) experimentally obtained structure using an autocloning method; and (b) 2D computer simulations of the process. (Reprinted with permission from Ref. 79, © 1999, American Institute of Physics.)

The demonstration of repeatability of the autocloning method for the purpose of fabricating photonic crystals permits the extension of the process to include defects engineered in the regular lattice in order to build in functionality. However, it should be noted that although the structure presented in Fig. 6.21 is technically a photonic crystal, it does not exhibit a photonic bandgap. In particular, no bandgap exists in the horizontal direction for any polarization of the propagating light, which limits the applicability of the structure as presented in Fig. 6.21. On the other hand, even without a full bandgap present, the dispersive properties of such photonic crystals make it an interesting subject of study and may allow practical applications. The range of applications will broaden if defects can be introduced in a controlled fashion. First attempts at building in such defects and analyzing their properties have been shown in Refs. 80 and 82.

Accordingly, if a region in the substrate remains unpatterned, the flat surface will propagate as alternating layers are deposited. As in the case of a perfectly periodic structure, a stable pattern is achieved after several layers of deposition, which does not change as new layers are deposited. This produces a linear defect that runs through the photonic crystal in the vertical direction such as one shown in Fig. 6.24. Although the alternating dielectric layers in this region still produce a stop band since they form a *one*-dimensional (1D) photonic crystal, its characteristics are different than those of the *three*-dimensional photonic crystal. Consequently, it can be expected that transmission properties of this region will be modified. Indeed, this is what Hanaizumi et al. [82] and Kawakami et al. [80] observed. A judiciously chosen wavelength of light at $\lambda = 1.15\ \mu m$ appears to be transmitted in so-prepared waveguides, but it is blocked by the background region of the three-dimensional photonic crystal [80, 82] when the sample is illuminated in the direction normal to the surface. The pass band for the defect (a 1D photonic-crystal region) extends from $1.05\ \mu m$ to $1.2\ \mu m$, which is covered by a stop band in the 3D region.

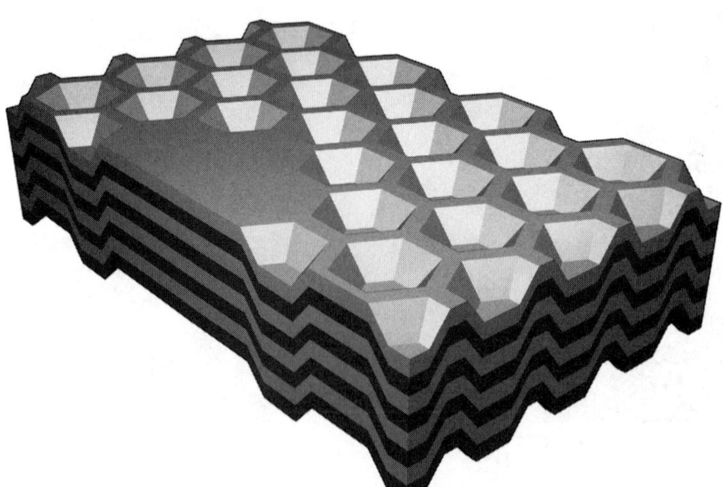

Figure 6.24 Three-dimensional photonic crystal fabricated using the autocloning technique with a linear defect running in the vertical direction.

The lack of photonic bandgap in the *xy* plane in the photonic crystal of Fig. 6.21 is the result of the relatively low effective modulation of the dielectric constant. This is the result of low-index contrast between the oxide and the amorphous silicon sputtered alternately as well as the particular structure where continuous sheets of high/low-index material are available for the electromagnetic wave to propagate in. In addition, a lack of connectivity in the vertical direction can be observed as the dielectric sheets form disjoint corrugated planes. Increasing the index contrast while maintaining the mechanical integrity of the structure is limited to replacing the materials used and can only bring marginal improvement. However, the continuity of the dielectric sheets can be disrupted by additional processing after the corrugated multilayer stack is fabricated. One of the simplest ways to accomplish the disruption of the dielectric sheets and thus to increase the effective modulation of the in-plane dielectric constant is to drill, or etch, cylindrical holes perpendicular to the corrugated surface. This approach would also introduce connectivity in the vertical direction, which has been pointed out to be an important feature for the full photonic bandgap to appear [83].

Such an approach to obtain a full three-dimensional bandgap in an autocloned structure was first suggested by Notomi et al. [84]. In this case, the starting point is a series of grooves rather than an array of pits as in Ref. 78. Bias sputtering alternate layers of silicon and silicon dioxide then yields a two-dimensional photonic crystal, where the photonic-crystal plane is perpendicular to the substrate plane. The holes are defined in the top surface, on the slopes of the grooves in a staggered pattern, to form a regular two-dimensional lattice; and then, they are drilled across the multilayer structure. The result is shown schematically in Fig. 6.25. Drilling vertical holes

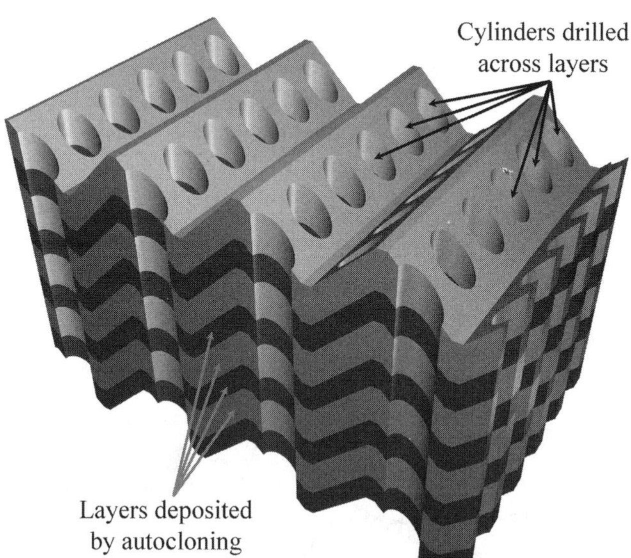

Figure 6.25 Three-dimensional photonic crystal obtained by growing corrugated alternating layers of low- and high-dielectric-constant material followed by anisotropic etch of an array of cylinders.

increases the in-plane modulation of the dielectric constant and opens a complete three-dimensional bandgap as wide as 11.8% of the mid-gap frequency [84].

The feasibility of fabricating such a structure with the use of standard microlithography tools has been subsequently demonstrated in Refs. 85–87. After the autocloning bias-sputter deposition of the corrugated multilayer stack, the array of holes has been patterned using an electron beam, then transferred to a nickel etch-mask, and finally etched using electron cyclotron resonance plasma etching with a CF_4—SF_6 gas mixture. Despite the patterning on sloped surfaces, the obtained etch results were encouraging; Fig. 6.26. The placement of holes requires sub-100-nm overlay accuracy to achieve the complete bandgap. This accuracy is possible with the standard lithography tools and has been demonstrated experimentally [85–87]. Thus, a method for fabricating a photonic crystal that exhibits a full three-dimensional photonic bandgap, in the form of a drilled alternating-layer photonic crystal, with the use of standard microfabrication tools, has been demonstrated.

Demonstrating the fabrication of a regular dielectric periodic structure that exhibits a photonic bandgap at near-infrared frequencies is only the first step in the development of a method to realize devices based on the photonic-bandgap phenomenon. The all-important issue is whether it is feasible to place, or engineer, defects in such a structure at will, defects that would serve as waveguides, resonators, mirrors, splitters, and so on. Without them, the practical use of photonic crystals is significantly limited. The first type of such a defect has been shown in Fig. 6.24 in an autocloned multilayer. It has been achieved by modifying the pattern in the substrate. Another degree of freedom in this fabrication method lies in the bias-sputtering parameters during the deposition of new layers. In particular, it is possible to control the thickness of individual layers. The combination of the modification of the regular pattern in the substrate and the control of the thickness of the individual layers can produce defects in the middle of the crystal that lie at their intersections. This approach has been exploited in Refs. 88–91, in which a variety of devices, mostly waveguide-based, has been fabricated and tested. Although the results presented therein are promising, they still do not realize the promise of photonic crystals such as small mode size or sharp bends in the waveguides, and the use of complete bandgap for guiding. However, the size of the mode, 3–5 μm in diameter, makes it particularly easy to couple to and from single-mode optical fibers characterized by a similar mode size.

Figure 6.26 Cross section of three-dimensional photonic crystal obtained by autocloning deposition of alternating Si and SiO_2, (a), followed by fluorine-based anisotropic etch of cylinders across the deposited layers, (b). (Reprinted with permission from Ref. 85, © 2000, American Institute of Physics.)

The disadvantage of this approach to creating defects in photonic crystals is that it requires the modification of large portions of the three-dimensional lattice to create localized variations for the defects. Indeed, the confinement in vertical direction is accomplished by changing the thickness of several entire layers of the lattice. This certainly limits the flexibility of the designs available with this approach. For example, if one were to wish to have a multilayer, in-plane photonic circuit with out-of-plane vias, the crossing of vertical light paths with the modified photonic-crystal layers may result in undesired effects, such as backreflection or resonances. To achieve the level of flexibility required for the fabrication of complex circuits embedded in autocloned photonic crystals, a new approach to creating defects in the middle of the crystal is required that would produce only local perturbation of the lattice, only where the defect is desired. Such a result remains elusive with this approach.

6.2.4 Glancing Angle Deposition (GLAD)

As we have just described, autocloning uses the specific behavior of matter at nanometer scales for the fabrication of three-dimensional photonic crystals. Another method that shares this trait exploits the observation that when thin solid films are deposited by evaporation on tilted substrates, they tend to form columnar and highly porous structures. This happens because of the shadowing affected by the newly deposited atoms that preclude the incoming atoms from taking arbitrary positions on the substrate. For the phenomenon to take place, it is necessary that the surface atom mobility be low; otherwise, the atoms tend to rearrange themselves so as to fill the voids. Such a condition is satisfied when, for example, chromium is evaporated on a sample held at room temperature. Examples of structures formed by evaporating chromium on unpatterned silicon substrates are shown in Fig. 6.27. The columnar porous structure of the deposited material can clearly be ascertained from these images. The tilt angle of the columns is a function of the incidence angle of the incoming atomic flux, but the two angles are not identical—the latter is considerably higher than the former. For example, the incidence angle of the incoming chromium flux for the structures of Fig. 6.27 was between 70° and 80°, whereas the angle of the column growth is considerably lower than that. This observation may seem counterintuitive if one has in mind a picture of a continuum flux being deposited on the surface. The problem is that such a model does not take into account the finite

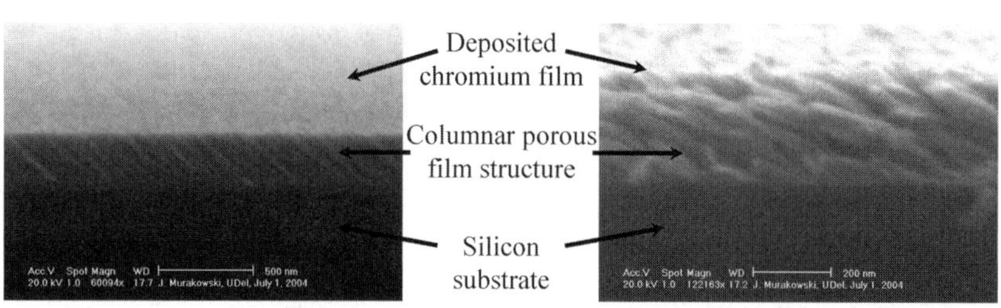

Figure 6.27 Cross-sectional SEM micrograph of a thin chromium film evaporated on a silicon substrate at an oblique angle. The columnar structure of the film is clearly visible. The tilt angle of the columns is around 45°, which is considerably smaller than the incidence angle of the atom flux.

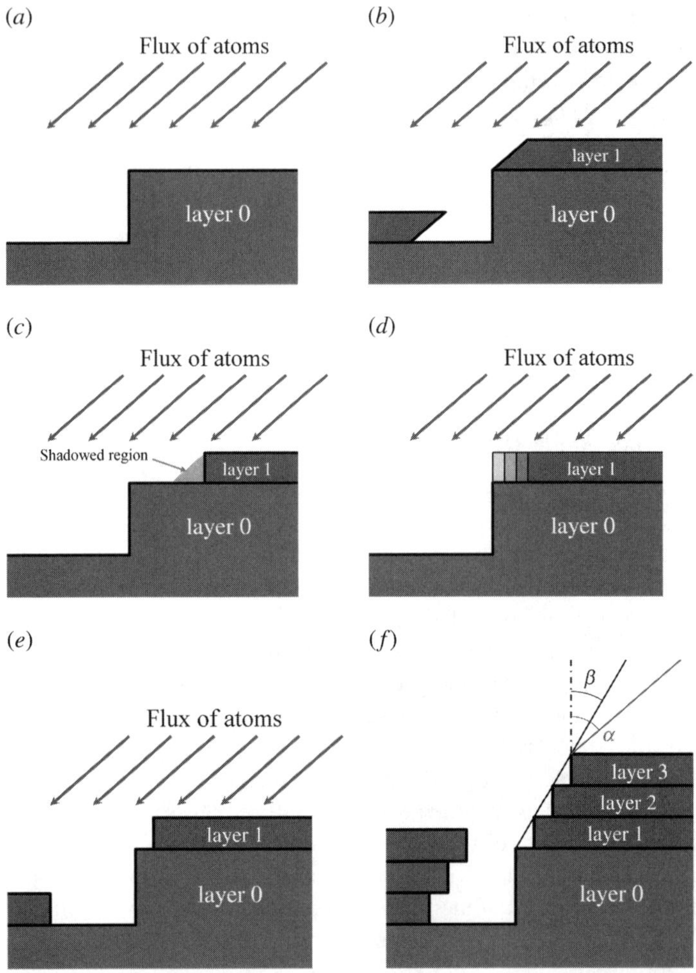

Figure 6.28 Coating a step with deposition at an angle. The continuum flux model, (b), fails to explain the disparity between the flux incidence angle and the angle of deposited material. Accounting for the finite size of deposited atoms, and therefore, the finite thickness of consecutive layers (c)–(f), provides such an explanation. See text for details.

size of atoms and, consequently, the finite thickness of each deposited layer. To better understand the relationship between the angle of flux incidence and the angle of column growth, let us consider a step being coated at an oblique incidence.

In Fig. 6.28a, we show a substrate with a step and a flux of atoms impinging on it.[4] Under a continuum flux model, one would expect that the additional material would

[4]The step does not have to be patterned on the substarte—the initial random distribution of the first atomic layer provides the necessary structure for subsequent deposition. We use it here merely to simplify the explanation.

simply follow the flux direction, as in Fig. 6.28b. However, if we think of new material being added in finite-thickness layers placed randomly on the surface, a different picture emerges. If the new layer, which is designated as "layer 1" in Fig. 6.28, lands on the substrate surface so that it ends some distance away from the step, as in Fig. 6.28c, then it fails to provide a shadow for its additional coating, and new material can still be deposited closer to the edge of the step. However, if the new layer terminates anywhere within the region indicated in Fig. 6.28d by varying grayscale shades, it does provide the shadow, which effectively prevents any new material from being deposited directly on layer 0. The probability of layer 1 terminating anywhere within this region is evenly distributed, so on average, we can assume that it ends in its center; Fig. 6.28e. The next layer sees layer 1 as the new step, and it is deposited accordingly. The structure thus grows, and the sidewall angle of the deposited material β is smaller than the incidence angle of the flux of incoming atoms α. From this description, it should also be clear that the two angles are related to each other according to the formula

$$\tan\alpha = 2\tan\beta, \tag{6.10}$$

which is often referred to as the tangent rule [92, 93].

Although Eq. (6.10) works for a wide range of materials, it is generally limited to $\alpha < 60^\circ$. This should not be surprising, as only the shadowing effect of finite-thickness layers has been considered, whereas the parallel momentum of incoming atoms has been ignored completely. For larger angles, the following formula provides a better description of columnar growth [94]:

$$\beta = \alpha - \sin^{-1}[(1-\cos\alpha)/2]. \tag{6.11}$$

In fact, the theory of columnar growth according to the description above [93] has been withdrawn by its authors. Nonetheless, the simplicity of the tangent rule makes it appealing [95] even though large deviations are observed experimentally, whereas the shadowing description provides useful intuition of the mechanism of layer formation, albeit lacking consideration of the parallel momentum of deposited atoms. For a review of models of oblique-incidence solid-film growth, the reader is referred to Ref. 95; for the purpose of the discussion here, it is sufficient to observe that (1) at oblique incidence, the deposited material tends to form columnar porous structure and (2) the angle of columnar growth is smaller than, and uniquely determined by, the angle of incidence of the incoming atomic flux, all other deposition parameters being equal.

The presence of a three-dimensional structure in films thus deposited offers a promise of using such a process to fabricate three-dimensional photonic crystals. However, before this happens, we need to extend the process to allow for the growth of structures periodic in three dimensions. To ensure in-plane periodicity, we need to provide a seed pattern on the substrate [96, 97]; otherwise the columns tend to be arranged randomly. This is easily accomplished using standard planar lithography methods used in the semiconductor industry, such as high-resolution ultraviolet (UV) lithography or electron-beam lithography followed by

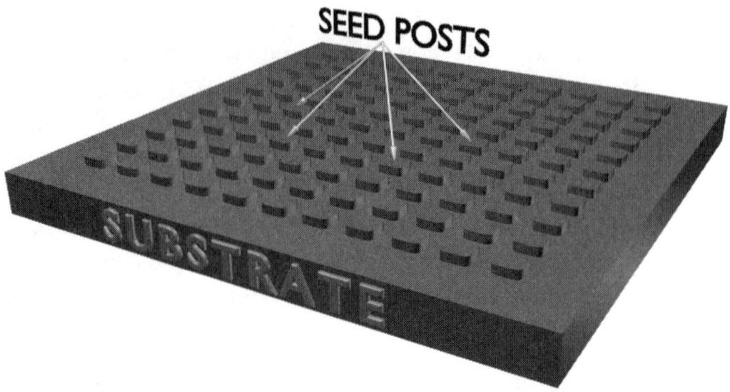

Figure 6.29 Substrate with a patterned seed layer for oblique deposition of tilted columns at predetermined locations.

dry or wet etching to transfer the pattern to the underlying structure. Alternatively, lithography followed by material deposition and lift off [98] can be used to arrive at a planar pattern similar to one shown in Fig. 6.29. When new material is deposited at an oblique angle on thus prepared substrate, the tilted columns tend to grow preferentially on top of the raised seed posts, which provide a shadow against the deposition on the flat portion of the surface.

To obtain periodicity in the third direction, perpendicular to the substrate surface, we can change the angle of atomic flux impinging on the sample, as proposed by Azzam [99]. This is most easily accomplished by placing the sample on a rotating platform above the source, with the angle adjusted externally, preferably using an automated controller. Figure 6.30 shows an experimental set up that allows for physical vapor deposition with a controlled incidence angle of the atomic flux. It includes a sample holder that can be rotated independently around two axes, with the sample facing the source. α rotation controls the incidence angle with respect to the surface normal, whereas φ rotation determines the direction, or azimuth angle, of the incoming flux. Such an arrangement allows for depositing arrays of spiral-shaped columns, where the lateral periodicity is determined by the pattern on the seed layer of the substrate (Fig. 6.29), and the geometry of individual spirals depend on the deposition parameters, such as angle of incidence, the rate of φ-rotation, and the deposition rate. The fill factor, or the fraction of the volume occupied by the material versus the total volume of the structure, is largely a function of the incidence angle α and adatom diffusion, but it can be varied to some extent by adjusting the size of the seed posts in the template. Fine control of the fill factor, or material porosity, which is largely independent of the growth angle of the columns, is achieved using more advanced GLAD techniques [100–102]. For example, to achieve material porosity that would normally correspond to an angle of deposition equal to 85°, but with the growth angle considerably smaller than that provided by Eq. (6.11), the sample is manipulated during the deposition using a "spin versus pause" technique [102]: The sample is spun around the φ axis during the deposition of a certain material thickness d_1, and then the spinning is paused for the deposition of material thickness d_2. For a given incidence

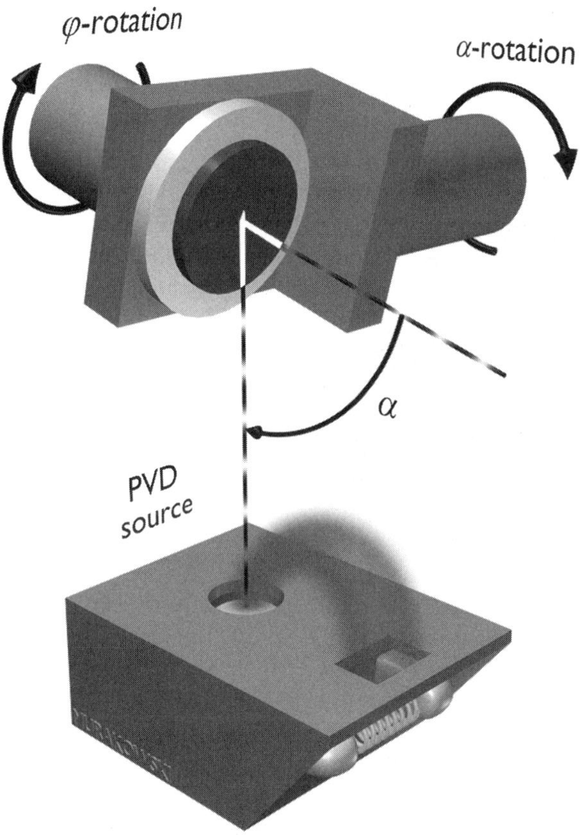

Figure 6.30 Oblique deposition of material by electron-beam evaporation. The heated element in the physical vapor deposition (PVD) source provides electrons, which are accelerated by direct current (DC) bias of several kV. The electron trajectories are bent by a strong magnetic field and focused inside a crucible filled with material, to be deposited. The high energy of impinging electrons heats up the material, which causes it to evaporate. The vapor condenses on all surfaces that are in the line of sight with the source, including the substrate placed directly above it. α and φ rotation adjustments allow arbitrary orientation of the sample surface with respect to the incoming atomic flux.

angle α, the ratio d_1/d_2 in conjunction with the rule of Eq. (6.11) determines the growth angle of the columnar structures, whereas the incidence angle alone determines material porosity. By advancing the "pause" position of the sample by a fixed amount at every "spin" rotation, helical structures are realized [102]. Furthermore, in GLAD, the helical structure of individual spirals need not have strictly circular symmetry. Indeed, as shown theoretically [96, 97, 103], square spirals similar to those of Fig. 6.31 offer the best combination of wide bandgap and robustness against fabrication variations.

The GLAD method offers the possibility of producing a three-dimensional photonic crystal with spiral-shaped rods of dielectric material. Such structures can also be inverted to produce three-dimensional photonic crystals that consist of air spirals embedded in high-index material. The process could proceed as follows [103].

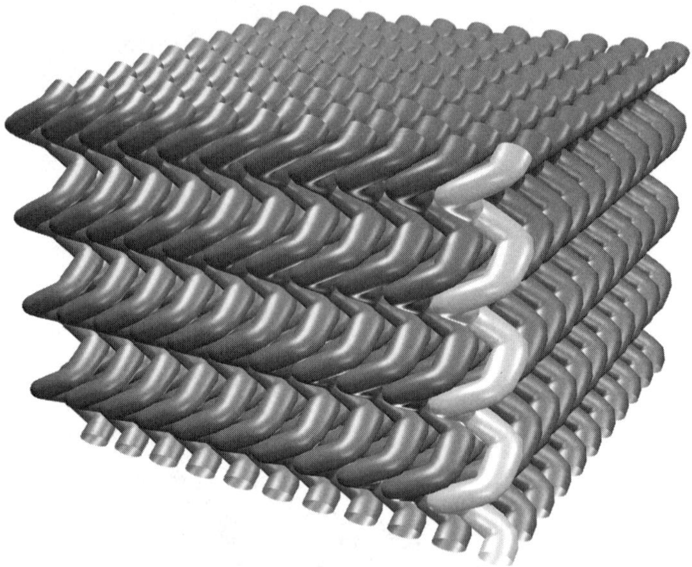

Figure 6.31 Array of spirals constituting a three-dimensional photonic crystal as would be obtained using the glancing angle deposition method. One column is shown in a different color to aid in the visualization of its geometry.

First, silicon dioxide spirals are deposited using the GLAD technique. The proper fill factor is achieved in postprocessing by additional conformal silica growth [104], which allows the realization of overlapping spirals [103]. Then, the structure is infiltrated with silicon by chemical vapor deposition—a process that conformally coats all silica surfaces and fills voids in the three-dimensional structures. Finally, silicon dioxide is removed selectively using an aqueous solution of hydrofluoric acid to reveal an array of spiral voids in silicon matrix.

Double inversion has also been achieved where the final structure is identical with the original array of spiral columns but is made of a different material [105]. This method can be useful if the final material is not suitable for GLAD process because of high mobility of adatoms as in metals with a low melting point, such as copper. In this case, the original GLAD-deposited porous film is infiltrated with a low-viscosity polymer, which is subsequently cured. The original spirals are then dissolved using an isotropic selective etch. For example, if the spirals are deposited using SiO_2, hydrofluoric acid can be used to dissolve them and reveal spiral pores in polymer. To obtain metallic spirals, the sample is electroplated by applying an electrical bias while it is immersed in a suitable electrolyte [105]. Finally, the organic mold is dissolved in acetone to reveal an array of spirals that are identical with the originally deposited one but are made of different material.

Arbitrary defects in the photonic crystal obtained using GLAD can be realized by interrupting the growth process, depositing a capping layer to seal the pores [101], patterning the layer according to the desired functionality, and then resuming the growth

[4]. Although the designed defects are confined to a plane, several of such defect planes can be introduced to achieve multilayer photonic-crystal architectures. The method is amenable to large-scale fabrication, wherein entire wafers would be patterned and subjected to square-spiral photonic-crystal growth as long as a sufficiently large deposition chamber is used to ensure negligible variation of incidence angle α across a wafer.

Although tremendous progress has been made in perfecting the GLAD process, the quality of structures realized thus far appears to be inadequate for photonic applications and the fine control of light promised by employing photonic crystals, despite the relatively robust photonic bandgap in this architecture [97]. It is not clear whether the disorder in columnar growth is inherent in the process, in that it relies on the low mobility of adatoms and on the stochastic nature of deposition, which tend to amplify irregularities, or if more research will provide the means to achieve structures closer to the ideal.

6.2.5 Macroporous Silicon

The methods described above rely on adding material to the substrate to produce a three-dimensional lattice of a photonic crystal. A complementary approach is based on removing material from the substrate. In this section, we focus on silicon wet etching to produce a lattice of voids at scales commensurate with near-infrared wavelengths. We begin our discussion by describing the mechanism of the reaction that leads to narrow, high-aspect-ratio openings in the silicon substrate, which is commonly referred to as porous silicon, first observed in the 1950s by Uhlir [106] and Turner [107]. For recent reviews of porous silicon formation, see Refs. 108 and 109.

6.2.5.1 Electro-Photo-Chemical Etching of Silicon. The mechanism of dissolution of a silicon electrode in aqueous solution of HF, according to Ref. 110, is presented in Fig. 6.32, where the (100) surface of silicon is exposed to the electrolyte. Figure 6.32a–l shows a cross section of the silicon wafer on the atomic scale, with the HF solution occupying the top portion of the figure, and the crystalline silicon in the bottom. When placed in HF solution, the silicon surface becomes passivated with hydrogen, where two hydrogen atoms, which are represented by lighter-color spheres in Fig. 6.32a, saturate the two top bonds of each silicon atom, which are shown in black. Because hydrogen is approximately as electronegative as silicon, the H—Si bonds are nonpolar, and the surface is almost impervious to F^- attack unless a hole[5] is available. If a hole, shown in Fig. 6.32b as an plus-sign (+) with a halo, reaches the surface [Fig. 6.32c], it polarizes the H—Si bond [Fig. 6.32d] and makes it susceptible to F^-. Another way to see this is to notice that the appearance of a hole at the surface is equivalent to reducing the electron density there and, in particular, the density of electrons responsible for the binding of hydrogen with silicon. As

[5]Note that in this section, we use "hole" to denote the absence of an electron in a semiconductor, following the convention of solod-state physics. This usage should not be confused with a region devoid of material.

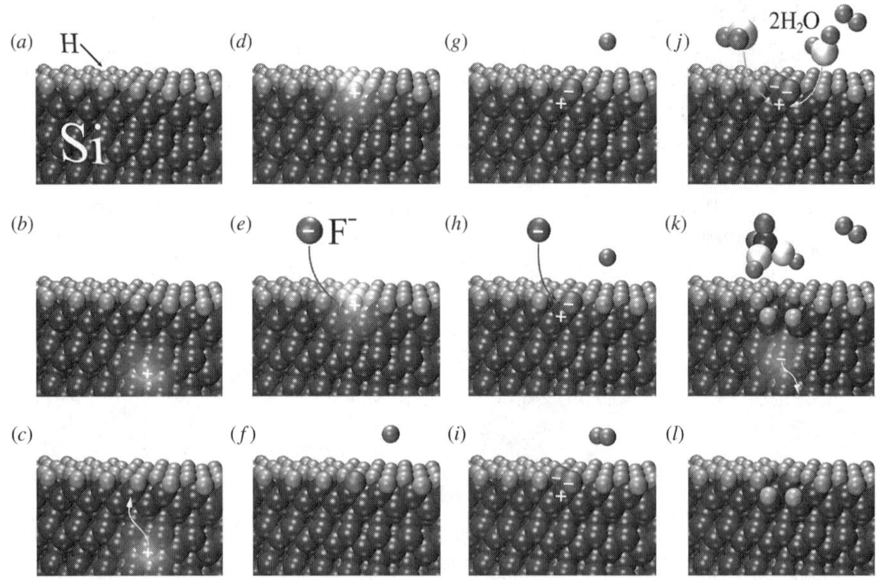

Figure 6.32 Dissolution mechanism of a silicon electrode in HF-containing electrolyte under positive bias. Silicon atoms are represented by black spheres. Gray spheres represent hydrogen, whereas darker gray and white spheres represent fluorine and oxygen atoms, respectively. A light gray halo shows positive electrical polarization caused by an electron void (hole), and a dark gray halo represents an extra electron.

the electron density is reduced, so is the binding strength of proton to the silicon atom. The positive charge attracts a negative fluoride ion present in the electrolyte [Fig. 6.32e], and the weakened bond allows for the substitution of the hydrogen with fluorine [Fig. 6.32f]. Now, the F—Si bond is highly polar [Fig. 6.32g]; the fluorine atom becomes electronegative and leaves silicon with excess positive charges. This charge redistribution reduces the density of electrons that participate in the other H—Si bond, which is weakened, and becomes susceptible to F^- attack [Fig. 6.32h]. Another fluoride ion substitutes the hydrogen atom at the silicon surface [Fig. 6.32i]. With two fluoride ions now bound to the silicon atom, the Si-Si back bonds that hold the silicon atom in the crystalline structure are weakened and can be attacked by H_2O molecules [Fig. 6.32j]. Two –OH groups of two water molecules bind to the F_2Si complex [Fig. 6.32k], leaving behind two hydrogen atoms that saturate the two dangling bonds of silicon and an excess negative charge, which is carried away from the surface. As a result, an atomic-size pit develops in the crystalline silicon surface [Fig. 6.32l]. The presence of a pit changes the electric field distribution in such a way that under positive bias, and if the reaction is limited by the supply of holes, the next hole will be attracted to the pit and will cause its deepening. As a result, an opening, or pore, develops in the silicon surface, which preferentially grows in the (100) direction perpendicular to the silicon surface. The sidewalls of the pores are depleted of holes, which are attracted to the pore tips and, as such, are protected from dissolution.

Depending on the diameter of pores thus created, the structure is classified as follows:

- *Microporous* for pore diameter less than 2 nm. In this case, the pore formation is dominated by quantum size effects, where the sidewalls that separate the pores, which essentially form quantum wires, are too small to fit the DeBroglie wavelength of a hole, thereby passivating them against dissolution [110]. Equivalently, the small lateral extent of the quantum wires causes an increase of the bandgap in silicon; this increased bandgap introduces an energy barrier for holes to enter between pores.
- *Macroporous* for pore diameter above 50 nm. Here, the pore formation is dominated by the electric field in the space-charge region in the vicinity of the silicon-electrolyte boundary.
- *Mesoporous* for pore diameter between 2 nm and 50 nm. Similar to macroporous, the pore formation is thought to be dominated by the electric field in the space-charge region.

Although the possible pore diameters span four orders of magnitude, depending on specific process parameters, in this section we will focus on macropores, as their size matches most closely the dimensions required for a photonic crystal that operates in the near-infrared region.

The details of the reaction described above may vary. For example, the fluoride attack shown in Fig. 6.32e and f may proceed through an intermediate step where Si—OH is first formed, and subsequently, F^- is substituted for the –OH group. Also, the polarized Si—Si back bonds may be attacked first, before the second hydrogen atom is displaced. Furthermore, the end product, $SiF_2(OH)_2$, shown in Fig. 6.32k may differ in composition, having an increased or reduced number of fluorine atoms at the cost of reducing or increasing of the number of $-OH$ groups; i.e., we can have $SiF_{1+x}(OH)_{3-x}$, where x can be 0, 1, 2, or 3. What does not change is that a hole is required to initiate the dissolution reaction. Consequently, the depletion of holes will passivate the respective silicon surface. For example, if n-type silicon is used in the absence of illumination or electrical bias, no reaction takes place because holes are minority charge carriers, and at room temperature, few are generated in such a substrate; effectively, n-type substrate is passivated from dissolution in hydrofluoric acid.

To provide the required holes, the backside of the substrate is illuminated, which yields electron-hole pairs. Under anodic bias, the electrons are carried away by a back-side electrode (Fig. 6.33), whereas the holes diffuse toward the front side of the substrate in contact with electrolyte. Close to the surface in contact with electrolyte, the space-charge region (SCR) forms, similar to the p-n junction under reverse bias, which extends between the pores, as illustrated in Fig. 6.33. The electric field around the pore tips focuses the holes there and promotes dissolution, thus deepening the pores. At the same time, the absence of holes between pores protects their sidewalls against dissolution.

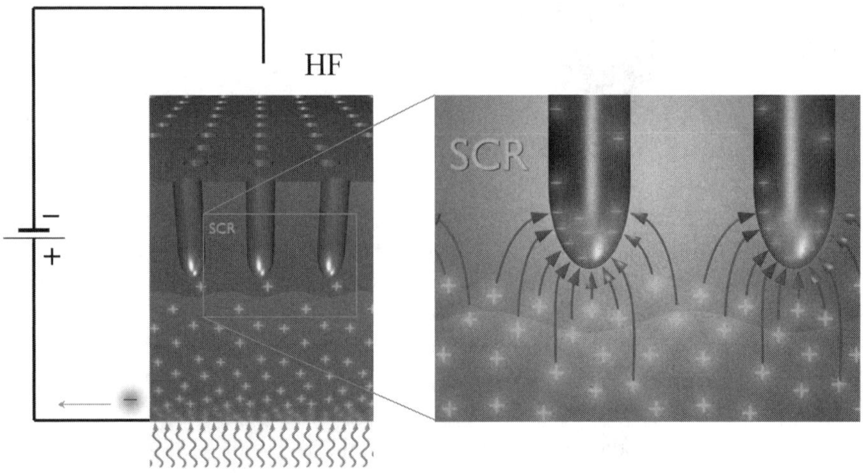

Figure 6.33 Formation of porous silicon under anodic bias and back-side illumination. Electron-hole pairs are generated in the photoabsorption process. Electrons are carried away by the back-side electrode, whereas holes diffuse toward the front side in contact with an HF-containing electrolyte. Enhanced electric field at the pore tips focuses holes there, and the SCR devoid of holes effectively passivates the pore sidewalls.

To ensure the absence of holes between pores, it is necessary for the space-charge region to extend throughout the silicon that remains between the pores. Therefore, the remaining silicon in its widest part is expected to be smaller than about twice the width of the space-charge region [110]. This general rule holds for unpatterned substrates where pores are initiated randomly on the surface due to local variability of topography or chemical composition. In this case, the density of pores in n-type silicon decreases with decreasing doping density. However, when the pores are initiated by prepatterning the surface (see the next section), a wider range of pore sizes and separations between them is possible than afforded by the simple SCR argument. Where the pore positions are defined by patterning the silicon surface lithographically, the pore diameters depend directly on the number of holes versus the number of fluoride ions available for the reaction.

Pores are produced when the reaction rate is limited by the supply of holes. On the other hand, if holes are abundant because of increased current density and the reaction is limited by the dissolution rate, i.e., by the diffusion of ions in the electrolyte, or mass transport in the Nerst diffusion layer of electrolyte near the electrode surface, electropolishing takes place. The threshold current density J_{PS} above which the silicon electrode is electropolished depends on the HF concentration c (measured in weight percent, which is denoted by w/o) and the absolute temperature T (in K) as follows [111]:

$$J_{PS} = C\, c^{3/2} \exp(-E_a/kT), \tag{6.12}$$

where $C = 3300\ \text{A/cm}^2\ (\text{w/o HF})^{-3/2}$, $E_a = 0.345$ eV, and $k = 1.38 \times 10^{-23}$ J/K is the Boltzmann constant.

For $J > J_{PS}$, the reaction is limited by mass transport in the electrolyte that is lower in surface depressions. Thus, the raised regions get dissolved first, which leads to electrode smoothing or electropolishing. In contrast, for $J < J_{PS}$ holes are depleted at the surface and fluoride ions accumulate there, which results in deepening existing depressions and, eventually, in pore formation. As pores develop, the electric field focuses holes at their tips, which thereby enhances local current density. When the local current density at the pore tips is $J = J_{PS}$, a steady state between hole supply and mass transport in electrolyte is reached, and stable pore growth ensues. In practice, the critical current density J_{PS} can be identified by the local peak in the I–V curve observed during the photoelectrochemical reaction.

The rate of pore growth v is obtained by observing that the current density at the tip J_{tip} is directly related to the dissolution rate. Thus, [111]

$$v = J_{tip}/[-enN_{Si}], \tag{6.13}$$

where $N_{Si} = 5 \times 10^{22}\,\text{cm}^{-3}$ is the atomic density of silicon; n is the dissolution valence, i.e., the number of holes consumed per one silicon atom dissolved; and $e = 1.602 \times 10^{-19}\,\text{C}$ is the elementary charge. As pointed out, for stable pore growth $J_{tip} = J_{PS}$, so for a given dissolution valence, the rate of pore growth can be readily estimated from Eq. (6.13). For 2.5 w/o hydrofluoric-acid electrolyte, the growth rate is $\sim$0.5 μm/minute over a wide range of doping, voltage, and photocurrent.

The dissolution valence n typically assumes values between 2 and 4 depending on the local current density J. The slope of the n-versus-J dependence is highest in the vicinity of the critical current density J_{PS} [111]. This behavior of the dissolution valence explains the stunning uniformity of pore depths obtained experimentally, where pores hundreds of microns deep are uniform to a fraction of a micron. Consider, for example, a pore that grew somewhat deeper than other pores in its neighborhood. Naturally, it would become more successful in attracting holes to its tip, which would produce a slightly increased local current density. However, this slightly increased current density results in a large increase of dissolution valence n for $J_{tip} \sim J_{PS}$ and, therefore, the decrease of this pore growth rate according to Eq. (6.13). Here, the increase of the denominator (n) more than offsets the increase of the numerator (J_{tip}) because of a steep dependence of n on J. This decreased growth rate brings the extended pore tip back in line with the other pores.

With the condition $J_{tip} = J_{PS}$, it is straightforward to estimate the diameter of the pores. Because all charge transfer happens at the pore tips, the total current I that flows through the silicon anode is $A_P J_{PS}$, where A_p is the total area of the sample surface occupied by pores. If we have prepatterned the surface for growth of regular square array with pitch p, the sample surface area A_t accommodates A_t/p^2 pores, which results in an area $A_P/(A_t/p^2) = I\,p^2/J_{PS}A_t$ occupied by a single pore, and its diameter

$$d = p\sqrt{I/J_{PS}A_t}\,, \tag{6.14}$$

where we approximated the circular pore with a square. The remaining pore wall thickness, $w = p - d$, is then

$$w = p\left(1 - \sqrt{I/J_{PS}A_t}\right). \quad (6.15)$$

Equation (6.15) seems to contradict the earlier remark wherein we asserted that the thickness of walls between pores has to be no larger than twice the space-charge region because the critical current density J_{PS} is independent of the latter. However, the SCR argument should only be understood as a general guideline when analyzing pore formation in prepatterned silicon. In this case, pore sizes that span an order of magnitude can be realized with commensurate variation of the silicon walls [111, 112]. Detailed analysis demonstrates that the hole concentration reaches zero at the depth of the pore tips, and therefore, no dissolution of the thus hole-depleted walls can occur, even for a space-charge region much narrower than the pore wall thickness [113]. Only the range of stable pore diameters is determined by the doping concentration: The higher the substrate resistivity, the larger the diameters of pores that can be grown. If the density of the initial pattern on the silicon surface that defines the desired pore density is too high for a given doping concentration, the pores will have the tendency to merge or die off. If the initial pattern is too sparse, the pores will branch. In either case, the porosity, i.e., the ratio of the pore area over the total area of the sample, does not change—only the size and density of pores (number of pores per unit area of the sample) changes. To select the proper doping concentration for a desired pore diameter, a good rule of thumb is to take the square of the pore diameter in μm to obtain the substrate resistivity in $\Omega \cdot$cm [114]. For example, 0.5 μm pores can be best etched using a 0.25-Ω·cm n-type substrate. This rule is consistent with the equation relating the width of the space-charge region x_{SCR} with the doping concentration N_D

$$x_{SCR} = \sqrt{\frac{2\varepsilon_{Si}\varepsilon_0(V - \phi_0)}{eN_D}}, \quad (6.16)$$

where ε_{Si} is the dielectric constant of silicon, ε_0 is the permittivity of vacuum, $e = 1.69 \times 10^{-19}$ C is the fundamental electric charge, N_D is the dopant concentration, ϕ_0 is the built-in voltage that results from the curvature of the bands at the Si/HF Schottky contact, and V is the applied voltage [115].

6.2.5.2 Patterned Growth of Pores in Silicon. In the preceding description of the electrophotochemical etch of silicon, we referred several times to prepatterning of the substrate surface as a way of controlling the placement of pores. Such control, of course, is necessary for the realization of photonic crystals, which require a periodically varying dielectric constant. Therefore here, we briefly touch on the typical method employed to initialize patterned pore growth.

The idea is to perturb the silicon surface in such a way that the electric field inside the substrate focuses holes, which migrate from the illuminated backside, on predetermined locations. Having holes where we want them promotes pore growth there and at the same time inhibits silicon dissolution in other regions depleted of holes.

To pattern the surface, we begin with standard lithography on a ⟨100⟩ silicon substrate with a thin layer of oxide [Fig. 6.34a], where either UV lithography or direct e-beam lithography can be employed to pattern an array of holes [Fig. 6.34b]. The oxide thickness should be between about 100 nm and a few microns depending on the pitch of the array. The pattern in resist is transferred to the underlying oxide by etching [Fig. 6.34c]. Dry etching is preferred as it typically produces better quality sidewalls. After stripping the resist [Fig. 6.34d], the sample is dipped into an aqueous solution of KOH, which etches crystalline silicon anisotropically. A 30% KOH solution at 85°C etches in the [100] direction as fast as 2 μm/minute. The anisotropic KOH etch leaves the {111} planes of the single-crystal silicon largely intact, which for a silicon wafer with exposed {100} patches of surface means that pyramidal pits are going to open as in Fig. 6.34e. The apexes of these pits are sufficiently sharp as to modify the electric field of the applied bias to focus incoming holes there and to promote the development of well-defined pores. The thin layer of oxide is quickly dissolved in the hydrofluoric-acid-based electrolyte used for the etching of pores [Fig. 6.34f].

Although Fig. 6.34 shows the pits arranged in a square lattice, the method is not limited to this geometry, and it can be employed just as effectively to realize a triangular lattice of openings by properly designing the lithography pattern. It also allows the placement of defects, e.g., a row of missing holes as in a waveguide, or a single

Figure 6.34 Process flow for patterning silicon surface to realize ordered pore growth. Silicon dioxide mask (a) is patterned using the standard lithography process (b)–(d). Patterned sample is immersed in an aqueous solution of KOH, which attacks silicon anisotropically, leaving planes normal to [111] crystallographic directions largely intact, thus producing an array of pyramid shaped pits. The oxide layer is stripped when the sample comes into contact with the HF-containing electrolyte.

missing hole as in a cavity, as long as the separation between adjacent holes does not exceed the stable pore growth region discussed above. If too many rows of holes are missing in the designed pattern, then the pores tend to branch as they grow in the silicon substrate. Furthermore, the pores adjacent to the missing row must pick up the excess charge carriers, which increases their diameter; this increase already becomes significant when two rows are missing [114], and therefore, it limits the available geometries of photonic-crystal patterns that can be realized this way.

6.2.5.3 Realizing 3D Photonic Crystals with Electrophotochemical Etching. Whereas the patterning of the surface for macro-pore growth in silicon provides the tool for defining the two-dimensional geometry of the photonic crystal, the dependence of the pore diameter on the applied photocurrent, Eq. (6.14), enables control in the third dimension. Combined, they allow the realization of three-dimensional photonic crystals in bulk silicon.

The modulation of the photocurrent to vary the size of pores (or material porosity) as they grow in n-type silicon under anodic bias was demonstrated by Lehman and Gruning [114] in 1997, and it was later proposed as means to realize micropumps [116]. The quality and scale of structures thus obtained make them suitable for photonic-crystal applications. However, the variety of shapes that can be achieved using this technique is limited. For example, the pore diameter can vary over its length by up to a factor of about three, and the cross sections of all pores etched at the same time are identical.[6] Furthermore, the taper of pores generated in this manner is limited to about 30° [114]. This restricts the available geometries of 3D photonic crystals obtained this way and, therefore, limits their dispersive properties; at the same time, it allows for largely independent adjustment of the dispersion relation along the pore axis and in the plane perpendicular to the pore axis [118]. To extend the range of possible geometries, extension of the technique and processing beyond macroporous silicon etching are necessary.

In addition to the photocurrent, another variable that affects the evolution of the pores in bulk n-type silicon is the width of the SCR. According to Eq. (6.16), the SCR width is a function of both the dopant concentration and the applied voltage. Although the dopant concentration is fixed the moment a particular wafer is selected for electrochemical processing (barring epitaxial growth of layers with varying dopant concentration), the applied voltage can be controlled during the electrochemical etch and is largely independent of the photocurrent, which is a function of the intensity of the backside illumination. So, high voltage can be used in conjunction with low current to initiate a narrow pore. The high voltage makes the SCR wider and focuses holes more effectively at the pore tips, which thereby allows for growth of a narrow pore whose width is determined by low photocurrent and by the current of holes produced by electrical breakdown resulting from high electric field at the pore tips. Then, the voltage is reduced to avoid breakdown-based generation of additional pores in other

[6]However, it seems that by properly offsetting the seed pattern of inverted pyramids by using different size openings, it is possible to have some vertical offset between the depths' adjacent pores and between their modulated cross sections. See Ref. 117.

⟨100⟩ crystalline directions and to allow for stable growth of larger diameter pores as determined by an increased photocurrent. This approach has been applied successfully by Matthias et al. [119, 120] to realize wider variation in pore size than previously possible as well as sharp transitions between consecutive openings; Fig. 6.35. To increase the pore sizes even more, and to obtain cubic symmetry at optimum porosity for the required dispersion properties, several cycles of oxidation at 900°C for 150 minutes and etching with hydrofluoric acid were used. This process allowed the researchers to create laterally interconnected pores.

Postprocessing with subsequent oxidation and oxide removal yields enlarged pores with smooth and curved sidewalls. Another possibility is to use a highly anisotropic KOH etch to enlarge pores and produce scaffold-like or air-octaeder structures with faceted geometries. As explained in Section 2.5.2, KOH attacks silicon anisotropically, removing the material preferentially in [100] and [110] directions, while leaving {111} crystalline planes largely intact. As a result, the enlargement of cavities produced in modulated macropore growth yields well-defined shapes of the remaining silicon. This approach was applied successfully to the realization of 3D photonic

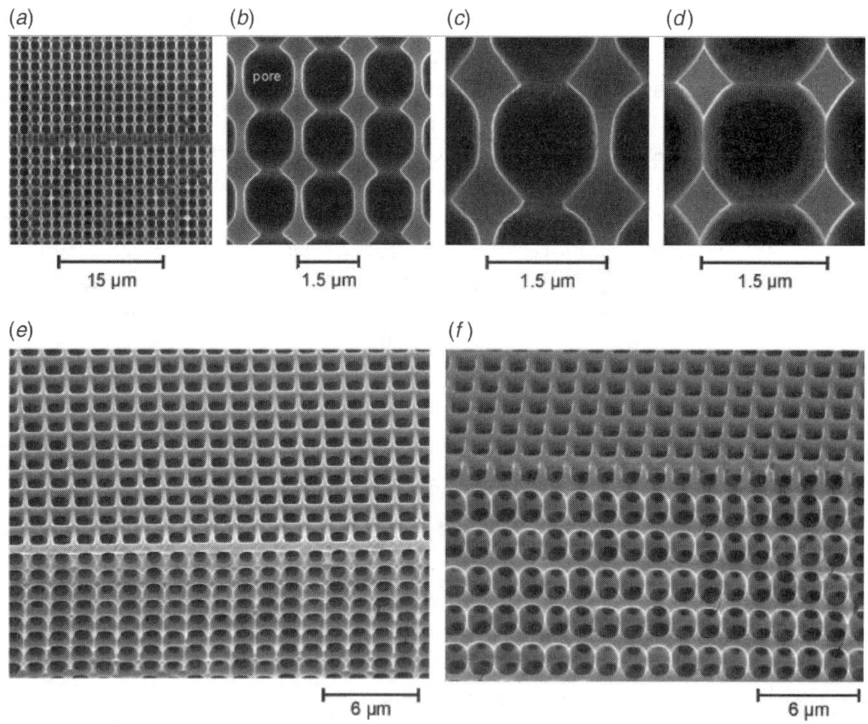

Figure 6.35 Scanning electron micrographs of 3D macroporous silicon with oxidation/etch postprocessing performed in n-type FZ-Si (100) with a resistivity of 0.5 Ω cm. (b)–(c) Result of electrophotochemical process with variable applied voltage in 5 w/o HF electrolyte at 10°. (d)–(f) The structure after pore widening with oxidation at 900°C for 150 minutes and 5 w/o HF-etch. (Reprinted with permission from Ref. 119, © 2004, Wiley-VCH Verlag GmbH & Co. KgaA.)

crystals by Matthias et al. in 2005 [121], where two distinct shapes of the unit cell have been obtained depending on the orientation of the square lattice of seed holes for pore growth with respect to the crystalline orientation of silicon.

Another way to use the high quality of macropores obtained by electrophotochemical etching of silicon is by combining this technique with focused-ion-beam (FIB) milling. The latter method relies on physical removal of material using high-energy ions, such as Ga^+, striking the surface at 5 to 50 kV. The beam of ions is focused and scanned across a sample with electromagnetic optics similar to those employed in scanning electron microscopes (SEMs). The spot focused on the sample can be as tight as a few nanometers, and the current can be as high as tens of nanoamps. As the high-energy ions strike the surface, they knock out, or sputter, the atoms of the sample in the area of impact, which thereby allows highly localized material removal. However, because no chemical reaction is involved that would make the sputtered material volatile to be carried away by the vacuum system of the instrument, the sputtered material is prone to redeposition, especially when narrow and deep openings, as required for three-dimensional photonic crystals, are drilled. The redeposition and beam self-focusing, due to beam reflection from the sidewalls, cause the narrowing of pores, which produces a noncylindrical shape [122, 123]. As a result, the pores created with FIB in bulk material are limited to an aspect ratio (the ratio of the pore depth to its diameter) of no more than about 10.

The situation changes dramatically when FIB is used on a porous sample prepared with electrophotochemical etch, where the pores run laterally with respect to the direction of the ion's impact. In this case, the sputtered material is carried away laterally and dispersed in each pore, while beam self-focusing due to reflection from the sidewalls is largely eliminated, especially when high air-filling fractions are used. This allows long, high-aspect-ratio cylindrical openings oriented at an angle with respect to the initial pores. The capabilities of FIB are complementary to macroporous etch as the latter can realize high-aspect-ratio pores that run in one direction—perpendicular to the (100) silicon substrate surface—whereas the former can drill high-aspect-ratio holes at an angle, but only when the sample is already porous. By combining the two techniques, Yablonovite-like [124] structures where the FIB pores are drilled at an oblique angle with respect to the original electrochemically etched pores [125–128], as well as structures where the FIB pores are perpendicular to the electrochemically etched pores [129–131], are possible at lattice scales compatible with near-infrared wavelengths.

In the case of Yablonovite, the electrochemically etched pores constitute one set of cylinders drilled in the silicon substrate. Because the process requires that the axes of these cylinders be perpendicular to the substrate surface, the resulting photonic crystal differs in the orientation typically shown when illustrating Yablonovite: Here, we obtain (110) orientation of the photonic-crystal top surface, as opposed to the usual (111); see Fig. 6.36. As a result, the initial pattern for electrochemical etching does not form a regular triangular lattice, but instead it requires that the elementary triangle of the grid be isosceles with vertex angles of 70.52° and 54.74°, respectively [126]. In this case, the incidence angle between the surface normal and each of the focused beams, and between the FIB beams, is 60° so that the axes of the three sets of the

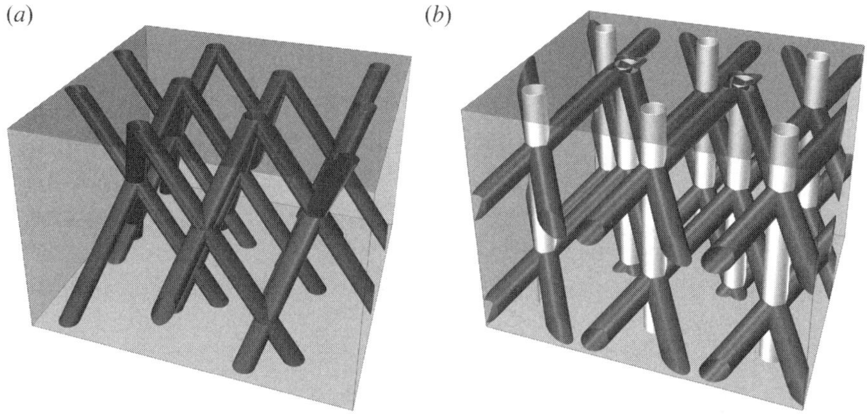

Figure 6.36 Yablonovite—the cylinders represent holes drilled in solid material. (a) Classic Yablonovite, where the holes are drilled symmetrically at an angle 35.26° with respect to the surface normal to produce a (111)-oriented photonic crystal. (b) Yablonovite obtained using a combination of electrophotochemical etch (white cylinders) and focused ion beam (black cylinders). The resulting orientation of the photonic crystal is (110) that follows the direction of white cylinders perpendicular to the substrate surface.

cylindrical openings correspond to the [110], [101], and [011] directions of the FCC lattice. The imaging capabilities of the FIB instrument allow for precise positioning of the beam used for drilling, as required to form the interconnected lattice of pores.

In their work, Lourtioz and colleagues [125–128] used a regular triangular lattice for the initial surface pattern, so the obtained structure is a slightly distorted Yablonovite. Nevertheless, the quality of the device, as shown in Fig. 6.37, demonstrates the feasibility of this approach to fabricating research-grade 3D PhCs. The challenge in using this method in high-volume production is the serial nature of the process and the relatively slow speed of FIB, which takes about 1 minute to etch a single hole for a reasonable size 3D photonic crystal [126].

6.2.6 Realizing Yablonovite for Near Infrared with Chemically Assisted Ion-Beam Etching

The photonic-crystal structure known today as Yablonovite (Fig. 6.38) was demonstrated to exhibit a full photonic bandgap in 1991. It was first realized for the microwave region [132] by mechanical drilling in low-loss dielectric material with a refractive index of $n \sim 3.6$—the Emerson & Cuming Stycast-12. In Ref. 132, the authors suggested that the same structure could be realized in semiconductor material for near-infrared wavelengths using standard planar lithography and reactive-ion etching. This approach was carried through at CalTech in 1995, and the process was improved in subsequent work [34, 133, 134] to produce a fully three-dimensional photonic crystal with six layers in the direction perpendicular to the sample surface.

The fabrication process is schematically shown in Fig. 6.39. It starts with defining a triangular array of circular holes in a planar etch mask, which is represented in Fig. 6.39 by a light-gray layer on top of a dark-gray GaAs substrate, using standard

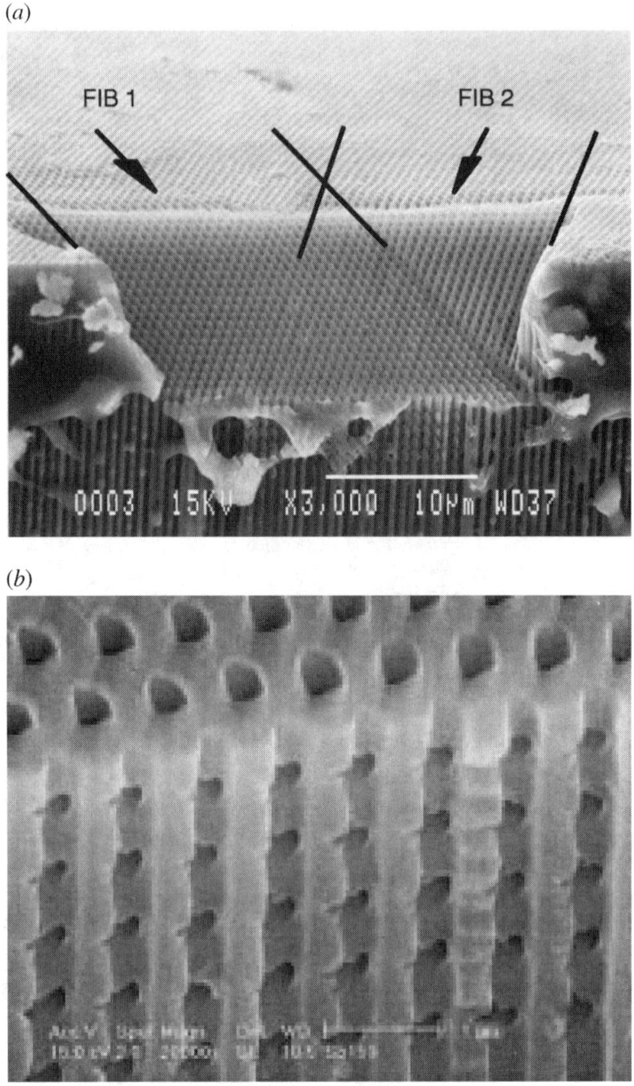

Figure 6.37 Three-dimensional Yablonovite-like photonic crystal obtained by combining patterned electrophotochemical etch of (100) n-doped silicon with subsequent FIB milling [128]. Arrows show the directions of the focused ion beams used for drilling cylinders at an angle with respect to the surface normal. The scale of structures thus obtained makes them suitable for use at near-infrared wavelengths. (Reprinted with permission from Ref. 125, © 2000, American Institute of Physics, and Ref. 127, © 2002, IEEE.)

high-resolution lithography techniques such as electron-beam lithography. This is shown as step (2) in the process. Then, dry plasma etch, using CAIBE, is performed in steps (3) through (5). For the etch, the sample is mounted at an angle with respect to the incoming flux of ions. The angle is determined by the FCC geometry of the photonic crystal and is equal to the angle between the main diagonal of a cube and its face

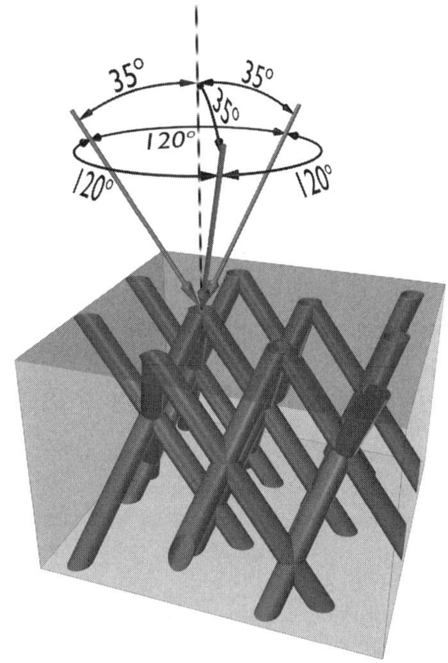

Figure 6.38 Geometry of Yablonovite photonic crystal. The cylinders represent holes drilled in solid material shown here as translucent block.

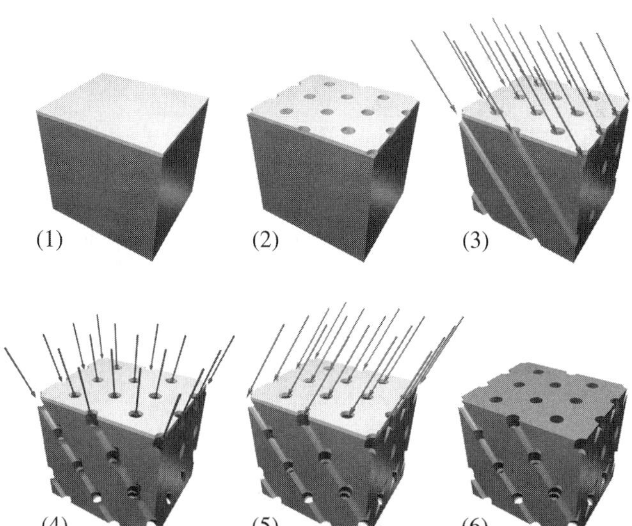

Figure 6.39 Fabrication process using a planar etch mask and three highly anisotropic reactive-ion etches. Arrows in steps (3) through (5) indicate the direction of the flux of incoming ions. The final structure, step (6) shows 1.33 periods in the vertical [111] direction.

diagonal, or approximately 35°. Because the removal of substrate material is highly anisotropic; i.e., the reaction takes place mostly at the bottom of the trenches rather than at the sidewalls, this results in opening deep, almost cylindrically shaped holes in the GaAs substrate where the cylinders follow the direction of the ion flux and are positioned where the etch mask material has openings. Then, the substrate is rotated by 120° around surface normal, and the etch step is repeated, as in step (4) of Fig. 6.39. After another rotation by 120° and CAIBE etch, shown in step (5), three sets of deep cylindrical trenches are opened in the GaAs substrate. Removing the etch mask, Fig. 6.39 step (6), reveals a three-dimensional FCC photonic crystal oriented such that the main diagonal of the cube is perpendicular to the substrate surface, i.e., $\langle 111 \rangle$ orientation. Although the crystalline lattice has FCC symmetry, the symmetry of the photonic crystal is broken by the asymmetric "atoms." The Yablonovite is realized by opening only three sets of holes, whereas a fully symmetric FCC structure requires six—the remaining three are parallel to the substrate surface, which makes them inaccessible in this fabrication approach.

The process uses standard techniques from the semiconductor industry, including high-resolution lithography and plasma etching, which makes it suitable for batch fabrication and thus viable for industrial-scale realization of photonic crystals. However, it is not entirely free of difficulties and drawbacks. First, it does not easily allow for placing arbitrary defects inside the three-dimensional photonic crystal. This issue can be overcome by using wafer bonding, which was discussed previously in this chapter, to sandwich the defect layer in between two 3D photonic crystals. This way, arbitrary defects can be inserted inside the volume of a three-dimensional GaAs photonic crystal potentially allowing for realizing one of the primary goals of the early research in this field—a thresholdless laser. Another difficulty has to do with the details of the etch process itself. To etch holes at an angle, the etch mask must be thin so as to allow line-of-sight access to the substrate for the incoming ion flux. The thicker the etch-mask layer, the smaller the resulting cylindrical holes in GaAs for the same size openings in the mask. However, to etch deep into the material, the mask must withstand the bombardment of ions for three consecutive etches. This means that very high etch selectivity between the mask and the substrate material is required to achieve even modest-size 3D photonic crystals. Although tremendous advances have been made in this regard [34, 133, 134], the etch selectivity over the mask remains the main limitation of this process, and the size of obtained 3D photonic crystals is restricted to only a couple of periods.

The high etch selectivity has been obtained by transferring the pattern to AlAs, which is subsequently oxidized to form a robust Al_2O_3 layer. The process, as described in Ref. 134, is depicted in Fig. 6.40. First, the substrate is prepared by epitaxially growing a 200-nm layer of AlAs followed by a thin GaAs cap layer to protect AlAs from oxidation in the ambient atmosphere [Fig. 6.40(1)]. Then, 50 nm of silicon dioxide, followed by a Cr/Au layer, are deposited, and PMMA is spun on the substrate—chromium is used for better adhesion of the metal to SiO_2. The triangular array of circular holes, which correspond to the (111) plane of the photonic crystal, is defined in PMMA using electron-beam lithography and transferred to the underlying metal layer by ion milling [Fig. 6.40(4)]. The pattern is then transferred to SiO_2 by reactive

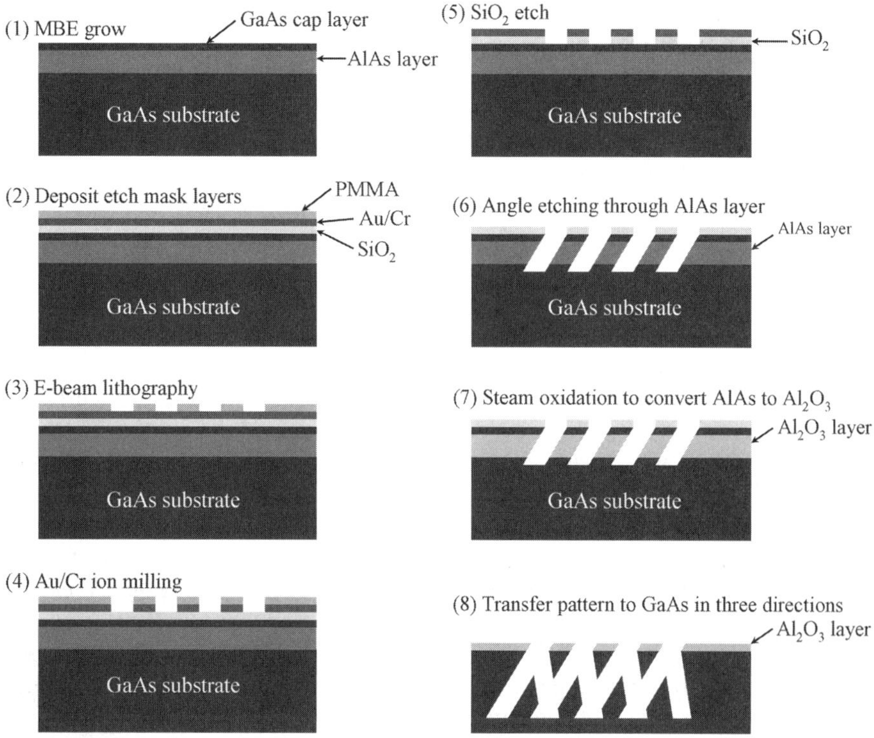

Figure 6.40 Preparation of an etch mask for the realization of Yablonovite in GaAs semiconductor using three directional anisotropic etches. High etch selectivity is obtained by transferring the pattern to AlAs layer, which is oxidized using water vapor at an elevated temperature to produce a robust Al_2O_3 etch mask.

ion etching with C_2F_6 plasma [Fig. 6.40(5)]. After stripping the metal layer, SiO_2 serves as an etch mask for CAIBE pattern transfer to the underlying GaAs and AlAs layers [Fig. 6.40(6)]. AlAs is then oxidized using water vapor to convert it to Al_2O_3, which is a hard, etch-resistant material. Having the etch mask now defined in alumina allows deep anisotropic etch in GaAs with little loss of the mask material. Indeed, Cheng et al. [134] report that the etch selectivity of GaAs over Al_2O_3 is better than 30, which makes it well suited for the job.

6.2.7 Sculpting Bulk Silicon with Reactive Plasma

The electrophotochemical etch of silicon described in Section 6.2.5 takes place in a wet environment where fluorine ions are dispersed in aqueous electrolyte. However, in the industrial setting, it is preferred to carry out the processing under dry conditions using reactive ion, or plasma, etching. To address the industry needs, we have developed a process that can realize three-dimensional photonic crystals in bulk silicon material using a conventional planar etch mask and dry etching with an ICP reactor.

The ICP, where plasma is ignited with magnetic field produced by RF coils, is superior to an RIE, where plasma is coupled capacitively to the RF source, as it allows independent control of plasma density and the energy at which ions strike the substrate. This makes an ICP a more flexible instrument that can provide a wide range of process conditions for a variety of applications. What proved useful in the application of an ICP to the fabrication of three-dimensional photonic crystals was its ability to etch bulk silicon both isotropically and anisotropically. In the isotropic etch, a small opening in the planar etch mask produces a spherical void in the bulk material, whereas in anisotropic etch, vertical sidewalls that follow the opening in the mask ensue; see Fig. 6.41. The two etch regimes are reached by properly setting the plasma chemistry and balancing the ICP and bias powers. However, to achieve a highly anisotropic etch with vertical sidewalls, it is necessary to passivate the sidewalls against reaction with the plasma. In contrast to electrophotochemical etch, where the sidewall passivation is achieved by depleting the silicon region of holes, in plasma etching, we deposit an inert layer on surfaces to protect them. In a continuous etch, this is achieved by having polymer precursors and reactive species at the same time in the chamber. For protection against etching, polymer gets deposited on all surfaces, but it is continuously removed from the bottoms of trenches by bombardment of ions accelerated by the bias power, which thereby makes them susceptible to attack by reactive species. Because the ion bombardment is directional—perpendicular to the surface—the sidewalls stay protected and the process can produce highly anisotropic etches.

The difficulty with this approach is that a delicate balance has to be maintained between the deposition of the passivating polymer and its removal from the trench bottom by the ion bombardment. If the removal rate is too high, we get undercut

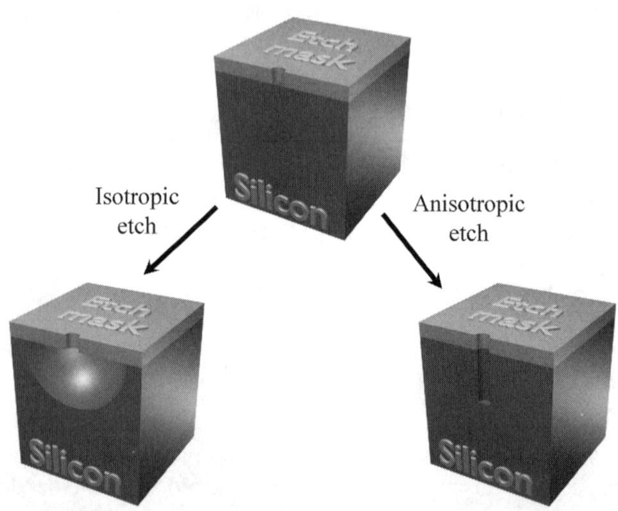

Figure 6.41 The same pattern in the etch mask can lead to different shapes of voids produced in silicon depending on the plasma conditions in the ICP reactor. Assuming the opening in the etch mask is a small circle, for perfectly isotropic etch, the etch front is a sphere (left image), whereas a cylinder is produced for anisotropc etch (right image), where the sidewalls follow the contour of the mask opening.

sidewalls; if it is too low, we get V-shaped openings. The process parameter window to maintain such a balance is narrow and may actuallly shift between runs as the condition of the reactor chamber undergoes variations as a result of environmental variability and the buildup of deposited material. Furthermore, the obtained etch rate is relatively low as the competing etch and passivation deposition take place simultaneously.

To address these issues, a more robust and efficient process has been developed where the deposition and etching do not take place simultaneously, but rather they are alternated. This method gives rise to the time-multiplexed reactive ion etch (TM-RIE), which is often referred to as the Bosch process [Fig. 6.42]. In the first step, Fig. 6.42b, the etch is allowed to proceed (almost) isotropically under relatively low bias, where for silicon we would use for example SF_6 as the reactive gas. The etch step is followed by conformal deposition of Teflon-like material generated in the reactor by flowing appropriate precursor gas, such as C_4F_8, under very low bias [Fig. 6.42c]. Then, high-energy ion bombardment is used to remove the passivation from the bottom of the trenches [Fig. 6.42d], which exposes silicon to allow its removal in the low-energy etch that follows [Fig. 6.42e]. The sequence is then repeated as many times as necessary to achieve the required depth.

By using low bias in the etch steps [Fig. 6.42b and e], and properly timing the high-energy directional etch [Fig. 6.42d], high selectivity over even soft masks, such as a photoresist, can be achieved. Furthermore, considerably higher etch rates than those available using direct etch have been achieved. The disadvantage of this process is that it leaves the telltale scalloping, or ripples, on the sidewalls that may affect the performance of the finished device. This issue can be alleviated by using short etch steps, thus reducing the scalloping period to a deep subwavelength regime where it is seen by the electromagnetic wave as an effective medium blurring the boundary between the high-refractive-index silicon and the low-refractive-index air. Another way to minimize the amplitude of the ripples is by increasing the bias power of the

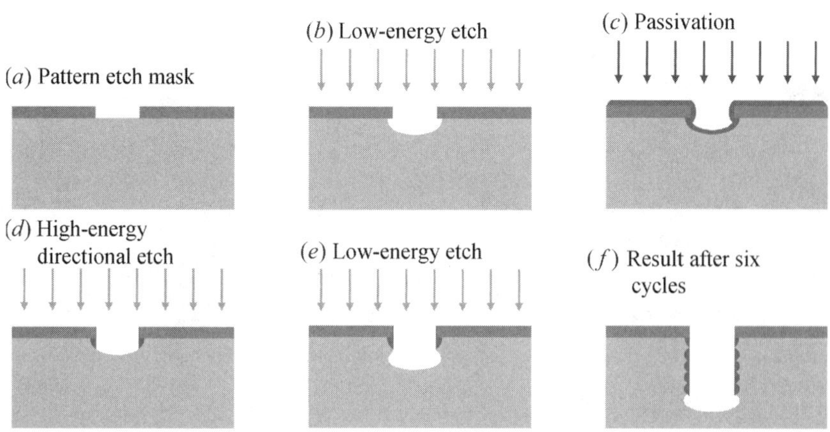

Figure 6.42 Highly anisotropic trenches with vertical sidewalls obtained by time-multiplexed reactive-ion etching.

low-energy etch in steps (b) and (e). This makes the etch more directional, which minimizes the lateral extent of the ripples at the cost of reducing the selectivity over the etch mask that is bound to erode faster under the increased bias. However, the reduced selectivity is typically not a problem when a hard mask, such as silicon dioxide or metal, is used, as is the case in the fabrication of photonic crystals outlined below.

Combining the isotropic etch obtained at low bias when the energy of ions striking the surface is low with the highly directional anisotropic etch afforded by the Bosch process allows the realization of a variety of shapes of voids in bulk silicon material. Some examples are shown in Fig. 6.43, where (c), (d), and (e) illustrate the combination of isotropic with anisotropic etches in different orders. The combination of the two etching regimes and an additional sidewall passivation step allow the realization of a three-dimensional array of spherical voids buried inside silicon to form a fully 3D photonic crystal. Conceptually, the process is presented in Fig. 6.44.

The starting point is a planar etch mask patterned on the silicon surface, which in Fig. 6.44a is illustrated as an array of holes. When a sample thus prepared is subjected to anisotropic etch, cylindrical holes are produced in the underlying substrate [Fig. 6.44b]. The sidewalls of the cylindrical openings are passivated as part of the process to inhibit lateral etching of silicon and to ensure that the sidewalls are vertical. The passivation is effected by introducing a polymer precursor, such as C_4F_8, to the mixture of gases in the plasma reactor and using the time-multiplexed process described above. The bottoms of the cylindrical openings are free of passivation and are the only exposed regions of silicon for the next step, which is an isotropic etch. By reducing the bias power, which controls how fast the ions are accelerated toward the substrate, the etch, which is carried out mostly by neutral reactive species,

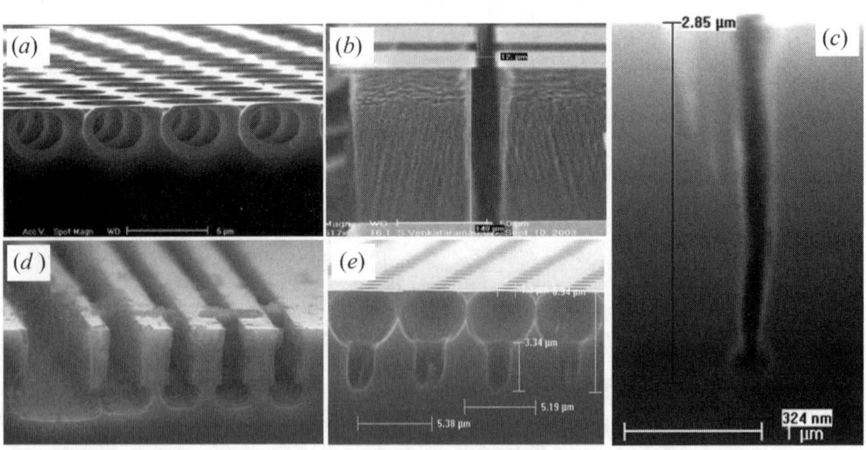

Figure 6.43 Variety of shapes obtained by combining isotropic and anisotropic etching of bulk silicon with a planar etch mask. (a) Two-dimensional array of spherical voids results from isotropic etch at low bias. (b) Highly anisotropic etch with straight vertical sidewalls. (c) Long anisotropic etch followed by isotropic that opens a spherical void at the bottom of a deep pore. (d) Anisotropic etch followed by isotropic etch. (e) Isotropic etch that opens a two-dimensional array of spherical voids beneath the surface is followed by an anisotropic etch to produce lollipop-shaped voids.

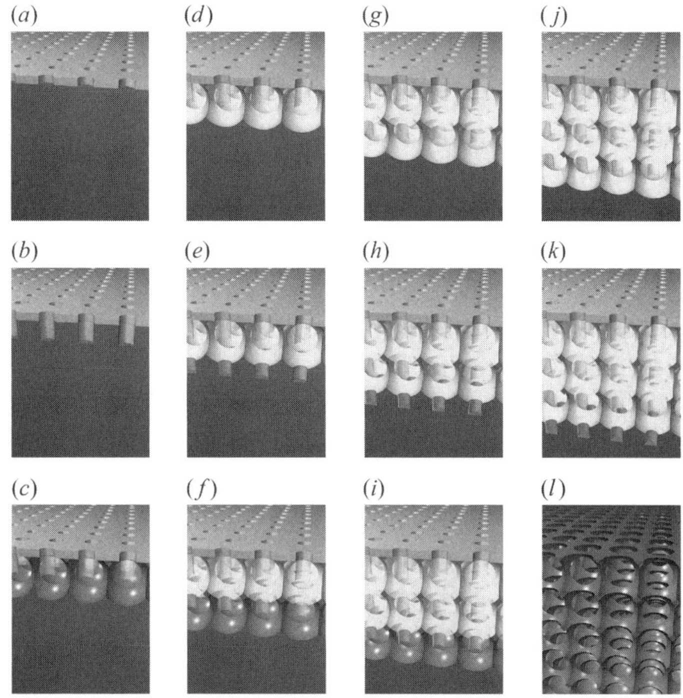

Figure 6.44 Dry fabrication of a three-dimensional photonic crystal in bulk silicon. A sequence of anisotropc etch, isotropic etch, and passivation produces an array of spherical voids in bulk silicon.

proceeds at equal rates in all directions and produces a spherical void as shown in Fig. 6.44c. Note that the passivation lining the original cylinder remains intact, as the reaction is highly selective in removing silicon. The isotropic etch is followed by the passivation of the entire silicon surface of the voids [Fig. 6.44d]. Although in principle it can be achieved by using a polymer precursor in the ICP reactor, we found that oxidation results in a more uniform and more robust passivation layer, which allows for wider process-parameter latitude in the subsequent processing. In this case, as little as 5 nm of oxide is sufficient to protect the silicon surface from etching.

After passivation, the ICP is again used in anisotropic-etch mode where first a highly directional flux of energetic ions is used to open a window in the oxide layer at the bottom of the spherical void, and then underlying silicon is etched vertically [Fig. 6.44e]. The shape of the opening in the passivation layer follows the shape of the planar mask on top of the silicon substrate. Additional guiding for stray ions is provided by the now suspended passivation layer of the first anisotropic etch. As in step (b), the bottoms of the cylindrical openings in step (e) are exposed and are susceptible to low-energy anisotropic etch that follows [Fig. 6.44f]. While the second layer of spherical voids develops, the first layer is protected by the oxide grown in step (d). Then, the inner surface of the voids in the second layer is passivated as in Fig. 6.44g, and the entire cycle is repeated to produce as many layers of the photonic crystal as required for a given application. Finally, the oxide is removed with hydrofluoric acid solution to reveal a regular, three-dimensional array of spherical voids in bulk silicon [Fig. 6.44l].

The result of the process with six periods of thus fabricated photonic crystal is shown in Fig. 6.45—see also Ref. 135. Although the photonic crystal is fairly regular, it can be noticed that the spherical openings of the layers closer to the bottom are somewhat smaller than the top-layer voids. This phenomenon is the result of a reduced etch rate as the trenches grow deeper, which can be compensated for by increasing the etch time or plasma density during the isotropic etch step for each subsequent layer. Indeed, the result of Fig. 6.45 already incorporates such compensation, although clearly not enough, as the etch time of the bottom layer was more than twice as long as the first layer.

The advantage of this approach to fabricating 3D photonic crystals lies in its compatibility with standard complementary metal-oxide-semiconductor (CMOS) processes—the etch mask is an ordinary planar silicon dioxide mask that can be patterned using high-resolution UV lithography, and the process involves dry etching and oxidation, which are commonly used in semiconductor foundries. The disadvantage is the difficulty in introducing arbitrary defects inside the volume of the photonic crystal, which can be realized by extending the method to include wafer bonding where the defect layer is sandwiched between two 3D photonic crystals. However, it is possible to introduce defects by properly patterning the 2D etch mask by, for example, leaving out a single hole or a row of holes or by varying the diameter of holes in the mask. The defect created this way will propagate throughout all layers of the photonic crystal. Alternatively, an entire layer can be modified by, for example, increasing or decreasing the etch time opening the spherical voids, or by combining a spherical-void etch with a cylindrical-void etch (Fig. 6.46). In extreme

Figure 6.45 Six layers of voids in silicon forming a three-dimensional photonic crystal obtained by sequencing isotropic etch, passivation, and anisotropic etch in a process shown in Fig. 6.44.

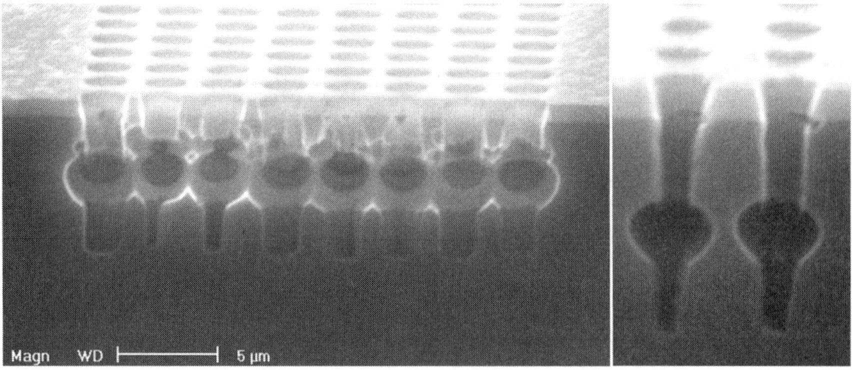

Figure 6.46 Spherical voids placed in between cylindrical voids obtained with a single planar mask and time-multiplexed plasma etching.

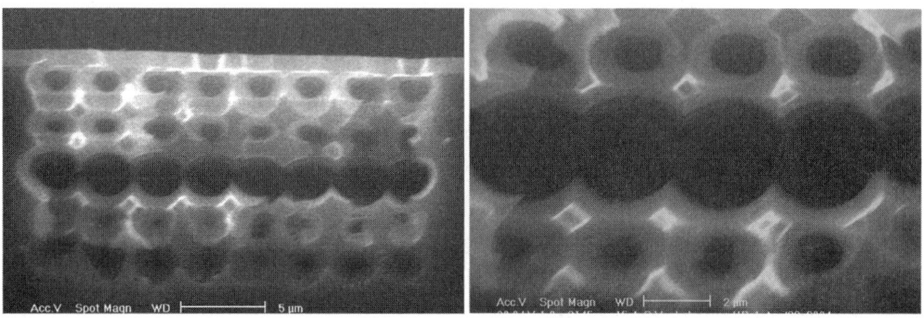

Figure 6.47 An entire layer of spherical voids can be made into a single air cavity by extending the etch time. Beneath, the individual spherical voids can still be realized.

cases, this can lead to opening a large air cavity buried underneath several layers of photonic crystal as in Fig. 6.47. This way, we can have a layer of photonic crystal that exhibits self-collimation property sandwiched between several layers of 3D photonic crystal with photonic bandgap. As a result, light is guided by the combination of self-collimation and photonic bandgap in the absence of linear waveguides. Of course, the patterning of the etch mask and modifying individual layers can be done when processing the same sample to realize a wider variety of defects.

6.3 THREE-DIMENSIONAL PHOTONIC CRYSTALS: HOLOGRAPHIC LITHOGRAPHY

Berger et al. [136, 137] first proposed and demonstrated that a standing wave created by interfering several coherent optical waves within a photosensitive material might be used to fabricate photonic crystals; this method is referred to as *holographic lithography*. The concept is appealing because several of the most important criteria for realizing a photonic bandgap, such as long-range periodicity and features comparable with optical wavelengths, are satisfied automatically, with no need for complicated alignment and virtually none of the structural defects that plague other fabrication methods, such as multiphoton polymerization or self-assembly described in Sections 6.4 and 6.5, respectively. It should be noted that although Berger et al. originally proposed the idea for PhCs of higher dimensionality than one, holographic methods had for many years been, and continue to be, used for fabricating structures that are in fact PhCs (in the most general sense of a periodic arrangement of dielectric material), namely gratings. This includes diffraction gratings, fiber Bragg gratings, distributed Bragg reflectors, as well as 2D crossed-grating structures, which are usually fabricated by multiple two-beam exposures separated by a rotation of the sample.

6.3.1 Interference of Coherent Waves

Before discussing holographic methods for PhC fabrication in detail, it is worthwhile to discuss briefly the nature of electromagnetic interference. Electromagnetic waves obey the *principle of linear superposition*, which means (among other things) that the total field at any point in space and time can be obtained by simply (vector) summing the contributions from all source fields. An interesting phenomenon that arises as a consequence of linear superposition is the *standing wave*. A standing wave forms when several *coherent* plane waves of equal frequency are combined in a common region of space. Coherence is defined as the condition wherein the several waves all have a fixed phase difference between them. Lasers provide a readily available source of coherent optical waves.

In a standing-wave condition, the time-averaged field intensity is a constant for any point $\mathbf{r}$ in space; i.e.,

$$\frac{d}{dt}\left(\frac{1}{T}\int_{t_0}^{t_0+T} \mathbf{E}(\mathbf{r}, t)^{*}\mathbf{E}(\mathbf{r}, t)\, dt\right) = 0, \tag{6.17}$$

where $T = 2\pi/\omega$ is the optical period of the waves, $\mathbf{E}$ is the electric field amplitude, "*" denotes the complex conjugate, and t is time. The standing-wave pattern consists of an array of points of maximum and minimum intensity. From the properties of Fourier series, we know that the standing-wave pattern will be periodic, so long as it arises from the superposition of a finite number of waves, with a period comparable with the wavelength of the waves.

We first consider how to define the symmetry of the standing-wave pattern that will result from a set of plane waves. A plane electromagnetic wave has an electric field that can be described by $\mathbf{E}(\mathbf{r}, t) = \hat{\mathbf{e}}\varphi \exp[j(\mathbf{k}\cdot\mathbf{r} - \omega t)]$, where $\hat{\mathbf{e}}$ is a unit vector in the direction of polarization, φ is the complex amplitude, $\mathbf{k}$ is the wave vector, and $j = (-1)^{1/2}$ is the imaginary unit. The total time-averaged intensity from the interference of N plane waves can be expressed, by superposition, as

$$\Psi(\mathbf{r}) = \left|\sum_{m=1}^{N} \hat{\mathbf{e}}_m \varphi_m \exp(j\mathbf{K}_m\cdot\mathbf{r})\right|^2. \tag{6.18}$$

given by

$$\mathbf{R} = m\mathbf{a} + n\mathbf{b} + p\mathbf{c}, \tag{6.26}$$

reciprocal lattice vectors are given by

$$\mathbf{G}' = h'\mathbf{A} + k'\mathbf{B} + \ell'\mathbf{C}. \tag{6.27}$$

on that relates a real-space lattice to its reciprocal lattice is

$$\mathbf{G}' \cdot \mathbf{R} = 2\pi L', \tag{6.28}$$

n integer, which is satisfied according to Eqs. (6.25)–(6.27) with L' given $' + p\ell$.

erning PhCs with Interference Lithography

e discussion in the previous section showed how a 3D lattice of maximum-nts can be formed by four coherent electromagnetic plane waves, the first l demonstrations of the method used the interference of three laser beams D lattice of holes in a plane, which could subsequently be used for the of planar-slab PhCs.

rimental demonstration of the method was reported in the previously apers by Berger et al. [136, 137]. Their experiment used a set of oriented ratings, as pictured schematically in Fig. 6.48a. A laser beam incident on a of the gratings generates six sets of first-order diffracted beams (+1, −1 ing), three of which intersect on the far side of the gratings because of the ative orientations. Within the intersection, the beams interfere, forming a intensity minima. The symmetry of the interference pattern is triangular, ne specific orientation of the gratings. (A similar scheme can be used with entations to produce different symmetries, as will be shown later in this e interference pattern was used to expose a layer of photoresist on a ate, which was subsequently etched to make a planar PhC. The result is g. 6.48b.

experimental result had triangular symmetry because of the specific f the diffraction gratings used to generate the interfering beams. A similar sed on oriented diffraction gratings can be used to fabricate a variety of PhC lattices. For example, the arrangement of gratings pictured in roduces a set of beams whose k-vector differences are perpendicular Fig. 6.49b), yielding a square lattice.

neering the Interference Pattern

t, we have not considered the relative polarizations of the interfering it is clear from Eqs. (6.18) and (6.19) that they play a key role in

The cross-terms in the squared sum are responsil
can be seen by rewriting Eq. (6.18) as

$$\Psi(\mathbf{r}) = \frac{1}{2}\sum_{m=1}^{N}\sum_{n=1}^{N} \hat{\mathbf{e}}_m \cdot \hat{\mathbf{e}}_n |\varphi_m||\varphi_n| \cos\{(\mathbf{k}_m$$

where $\Delta\phi_{mn} = \arg(\varphi_m) - \arg(\varphi_n)$. This expressi
basis of the periodic intensity distribution is dete
the wave vectors of the incident plane waves. We
of this important result for the case of four inciden
can be found in Ref. 138.

The condition for maximum intensity for the in
with wave vectors $\mathbf{k}_0$ and $\mathbf{k}_i$ is

$$(\mathbf{k}_i - \mathbf{k}_0) \cdot \mathbf{r} = \Delta\mathbf{k}_{i0} \cdot \mathbf{r}$$

where i is an index over the three beams other
Equation (6.20) represents three equations that ca

$$(\Delta\mathbf{k}_{10} + \Delta\mathbf{k}_{20} + \Delta\mathbf{k}_{30}) \cdot$$

where L is another integer.

Vectors in three-dimensional k-space, such as
expressed in terms of a set of basis vectors $\{\mathbf{A},$

$$\Delta\mathbf{k}_{i0} = h_i\mathbf{A} + k_i\mathbf{B}$$

where again i is the index from 1 to 3, and h_i
Eq. (6.22) and Eq. (6.20) yields

$$(h_i\mathbf{A} + k_i\mathbf{B} + \ell_i\mathbf{C}) \cdot \mathbf{r}$$

and finally, with Eq. (6.21), this becomes

$$(h'\mathbf{A} + k'\mathbf{B} + \ell'\mathbf{C}) \cdot$$

where primes indicate sums over $i = 1$ to 3; i
definition of the reciprocal basis vectors from C

$$\mathbf{A} = 2\pi\frac{\mathbf{b}\times\mathbf{c}}{\mathbf{a}\cdot\mathbf{b}\times\mathbf{c}}, \quad \mathbf{B} = 2\pi\frac{\mathbf{c}\times}{\mathbf{a}\cdot\mathbf{b}\times}$$

we can identify **a**, **b**, and **c** as real-space basis
lattice defined by **A**, **B**, and **C**.

We are now left to show that the intensity ma
at lattice sites associated with a reciprocal lattice

vectors are

whereas th

The condit

where L' is
by $mh' + n$

6.3.2 Pat

Although t
intensity p
experiment
to create a
fabrication

An exp
mentioned
diffraction
the entire a
for each gra
gratings' re
2D array of
because of
different or
section.) T
GaAs subst
shown in F

Berger's
orientation
technique b
2D and 3D
Fig. 6.49a
(indicated i

6.3.3 Eng

To this poi
beams; yet

(a)

(b)

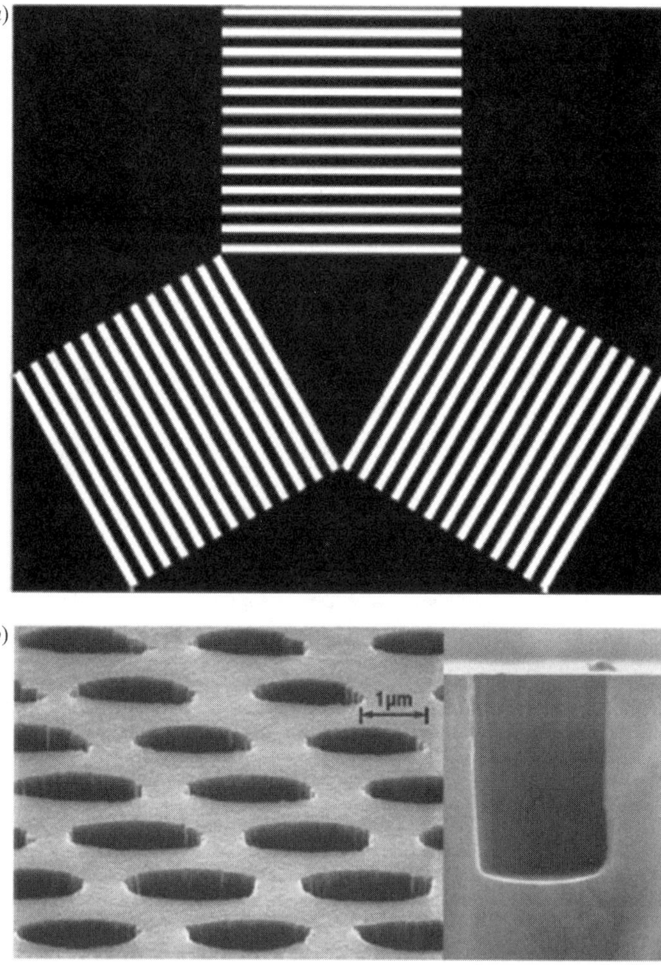

Figure 6.48 Photonic crystal etched in GaAs through a photoresist mask patterened by interference lithography. (Reprinted with permission from Ref. 136, © 1997, American Institute of Physics.)

determining the interference pattern. We have been able to ignore them safely thus far because, for 2D interference patterns, determined by three or more interfering beams whose *k*-differences are coplanar, the polarizations can always be chosen so that their projections on to the plane of the pattern are equal and parallel. However, this may not always be the optimum case. For example, if the desired structure is a square lattice of circles, the arrangement of beams described by Fig. 6.49 yields less-than-ideal results. If all three beams are derived from a single parent beam illuminating the gratings shown in Fig. 6.49a, and are of equal intensity, then the desired square lattice obtained will have "atoms" that lack square symmetry, as seen in Fig. 6.50a, owing to the contribution of the difference between beams labeled 2 and 3 in Fig. 6.49b. This asymmetry can be alleviated by adjusting the beams' relative

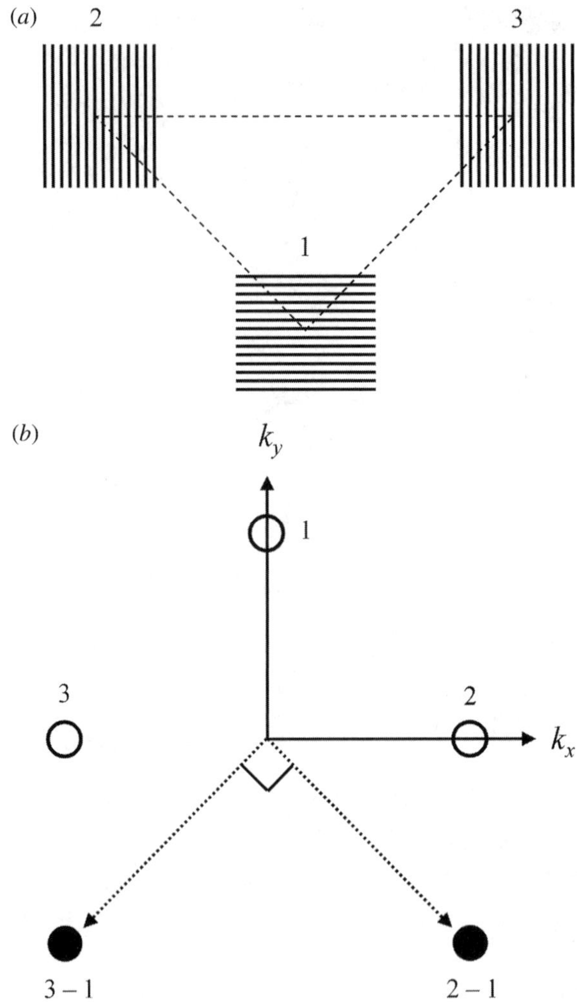

Figure 6.49 (a) Set of diffraction gratings that produce a square-lattice interference pattern. The gratings are oriented at right angles and positioned at the corners of a square. (b) *k*-space representation of the diffracted beams (open circles) and the differences (filled circles) that determine the reciprocal lattice.

intensities, as depicted in Fig. 6.50b. As is indicated clearly by Fig. 6.50b, this symmetry comes at the expense of severely reducing the contrast. Whereas in (a) all beams have equal intensities, in (b) the intensity of beam 1 is 225 times that of beams 2 and 3. This diminishes the relative contribution of the 2–3 *k*-difference, which results in greater symmetry because the two perpendicular *k*-differences dominate. The visibility (defined as $I_{\max} - I_{\min}/I_{\max} + I_{\min}$) in (a) is 0.97, while in (b), it is 0.36.

Although any pattern's symmetry is defined solely by the *k*-vectors, the contrast of the pattern and the shape of the atoms" are strongly dependent on the polarizations. Each term in the sum in Eq. (6.19) arises from the superposition of one pair of the

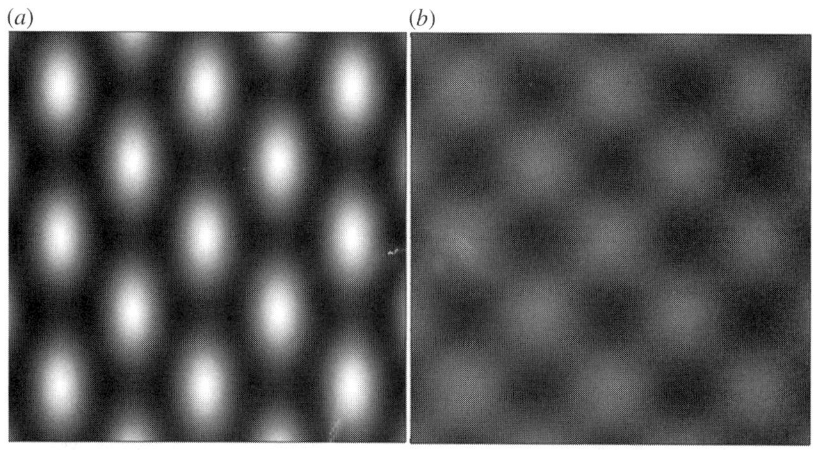

Figure 6.50 Two gray-level plots of the intensity of the interference pattern generated by the beam configuration of Fig. 6.49. In (a), all beams have equal intensity and in-plane polarization along y. Note the "atoms" are distorted vertically. The pattern visibility is 0.97. In (b), the intensity of beam 1 is 225 times that of beams 2 and 3. The atoms share the square symmetry of the lattice, but the visibility is only 0.36. Note that the two plots share the same gray levels to indicate the intensity, and they are normalized to equal mean intensity.

contributing beams and has an amplitude equal to the product of the beam amplitudes multiplied by the scalar product of the polarization unit vectors: $(\hat{\mathbf{e}}_m \cdot \hat{\mathbf{e}}_n)|\varphi_m||\varphi_n|$. Hence, the contribution to the pattern of each k-difference is weighted by the degree of overlap between the pair of beams associated with that k-difference. This is why the symmetry improved in Fig. 6.50b when the intensity of beam 1 increased: The k-differences that involve beam 1 dominate the diagonal contribution of beams 2–3. Of course, when interfering waves are unequal in amplitude, the contrast is diminished.

A better trade off to realize a square lattice can be obtained by adding an additional beam. In general, adding a fourth beam is useful to generate 3D interference patterns, as will be shown later. However, if all the possible k-differences are coplanar, the pattern is 2D regardless of the number of beams. To illustrate this situation, consider the calculated interference patterns shown in Fig. 6.51. In this figure, the upper panels indicate the relative orientations of the beams' k-vectors by solid lines, and their polarizations are indicated by dashed lines. The amplitudes of the beams are indicated by the lengths of the dashed lines. The lower panels contain intensity colormaps calculated for each beam configuration (each map is normalized to the mean intensity for that beam configuration). The figure illustrates how the pattern can be engineered to be more symmetrical by changing the relative beam intensities from $1:1:1:1$ to $13^2:5^2:5^2:13^2$, with a subsequent mild trade off in the contrast of the pattern: The visibility of the pattern in (a) is 1.0, whereas in (b), it is 0.9.

We can see the effect of the additional k-differences in the comparison between Fig. 6.51a and 6.51b. In the beam orientation diagrams (top), we see that the polarizations of the beams whose k-differences are along the x direction are almost mutually

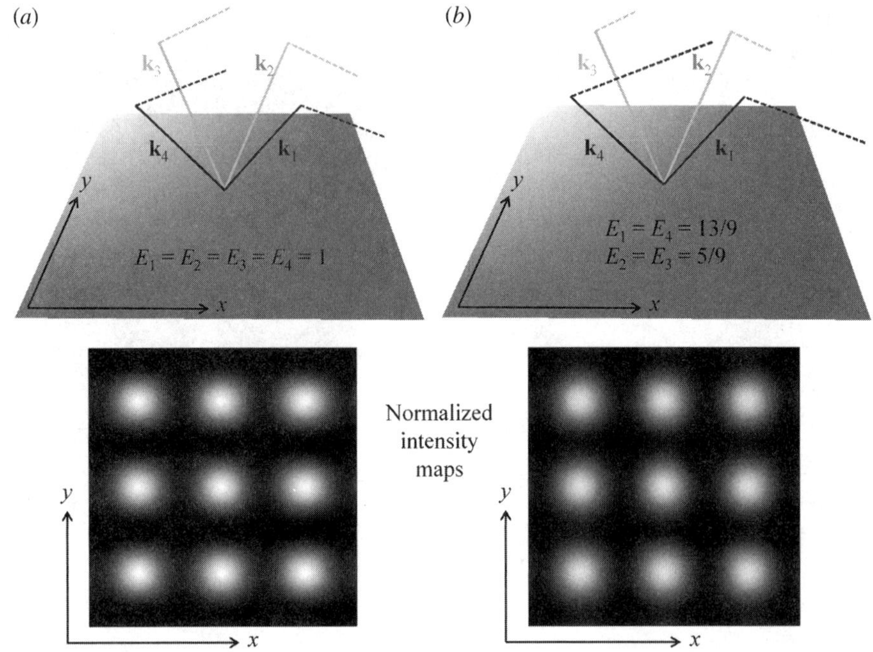

Figure 6.51 Engineering the symmetry and contrast of a 2D square-lattice interference pattern. (a) Calculated pattern that results from four interfering beams of equal intensity. The top pane contains a diagram of the relative orientation of the beams' k-vectors (solid lines) and polarizations (dashed lines) that result from the diffraction of a single beam by four gratings located at the corners of a square. The visibility of this pattern is very close to unity. Note the elliptical shape of the antinodes that develops from the stronger overlap between pairs of beams whose differences are along the y direction compared with those whose differences are along the x direction. (b) A more symmetric pattern that results from the same set of incident beam orientations, but unequal intensities; the relative electric field amplitudes are indicated by the lengths of the dashed lines. The amplitude ratio for this set of beams is 13 : 5 : 5 : 13. Modifying the intensities in this way has the effect of reducing the strength of the interference between the pairs of beams whose polarizations are nearly parallel. The contours are nearly circular, which may be more desirable for photonic-crystal applications. The cost of the improved symmetry is modestly reduced contrast: The visibility of this pattern is 0.9.

perpendicular, whereas the pairs whose differences are along the y direction are nearly parallel. Hence in (a) where all amplitudes are equal, the interference is dominated by the nearly parallel pairs, and the resulting irradiance pattern shows a pattern of horizontal lines superimposed on the square array of antinodes (i.e., the intensity modulation is greater along y). In (b), however, the influence of the parallel-polarization beam pairs is diminished by the changed relative amplitudes of the beams. By increasing the amplitudes of one pair of beams, whose k-difference is in the x direction, square symmetry has been improved. The trade off in pattern contrast mentioned above can be understood as arising from the fact that beam pairs with unequal amplitudes cannot destructively interfere perfectly, even when the polarizations are parallel. This leads to a bias level of exposure on top of which the variations are present. As a result, when using such a pattern to expose a resist, the results will generally be more sensitive to variations in the dose.

Engineering the interference pattern used to pattern PhCs can be a subtle and complicated task, with many coupled parameters that can influence the outcome. The examples just discussed were selected to introduce just some issues that arise, but many more have not been illustrated fully for brevity's sake. For example, in the case of Fig. 6.50, it was shown that the symmetry can be broken by the unavoidable additional k-differences, apart from those that define the PhC lattice. Symmetry was restored by sacrificing contrast, making one beam much stronger than the others. However, the symmetry can also be recovered another way: If the three beams are all incident at 45°, the contribution of the unwanted k-difference will vanish because the polarization vectors of those beams will be perpendicular. This restriction sacrifices the ability to adjust the lattice spacing independently of the exposure wavelength.

6.3.4 Holographic Fabrication Methods for 3D PhCs

It has been shown by Cai et al. [139] that all the fundamental 3D lattices (the 14 Bravais lattices) can be realized by the combination of no more than four beams. Thus, the challenge is to come up with an experimental scheme to realize such a beam configuration. Numerous researchers have done so, with different trade offs made between versatility, simplicity, repeatability, and other various criteria.

The first experimental demonstration of 3D PhC fabrication by interference lithography was from Campbell et al. [140]. In their experimental configuration, the four beams were derived from a common laser source and directed into the volume of photoresist by a combination of beamsplitters, mirrors, and wave plates to provide independent control of all beams' incident k-vector, intensity, phase, and polarization. They used a frequency-tripled, Q-switched Nd:YAG laser and SU8 photoresist. They also demonstrated that 3D PhCs patterned in photoresist using interference lithography could be used as templates for subsequent backfilling with high-index material [141]. In their case, they backfilled their PhC with titania via a sol-gel process. An SEM micrograph of a sample they fabricated is shown in Fig. 6.52. The beams were arranged such that the angle between the three outer beams and the central beam

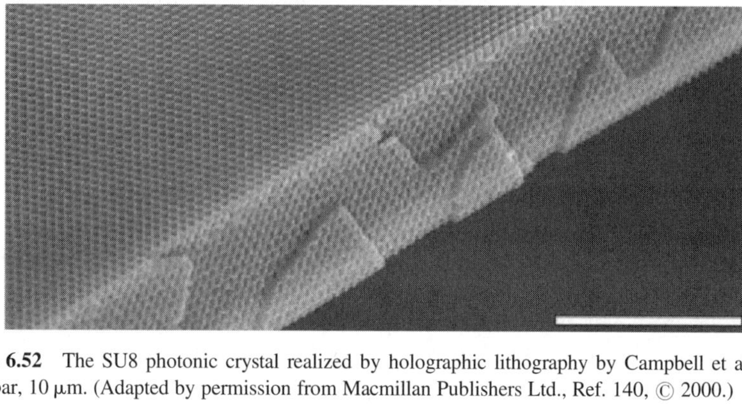

Figure 6.52 The SU8 photonic crystal realized by holographic lithography by Campbell et al. [140]. Scale bar, 10 μm. (Adapted by permission from Macmillan Publishers Ltd., Ref. 140, © 2000.)

was 39.8°, and the outer beams were distributed azimuthally about the central beam at 120° intervals. This arrangement resulted in an FCC lattice with diamond-like symmetry, a structure that is favorable for realizing photonic bandgaps.

Although the use of beamsplitters, mirrors, and wave plates allows for maximum flexibility of engineering the interference pattern, many research groups have sought methods that sacrifice some flexibility for increased simplicity, lattice tunability, or exposure contrast. For example, a simpler and more compact method of generating several beams is to use one or more diffractive elements, such as a holographic beamsplitter or an array of diffraction gratings; these methods usually sacrifice control over the beams' polarizations. Furthermore, several groups have shown that multiple exposures using fewer than four beams per exposure, while reorienting either the sample or the beam between exposures, can produce arbitrary lattices with maximum exposure contrast (i.e., pattern visibility) [142]. The following sections describe several such results.

6.3.4.1 Diffractive Element Methods. At the University of Delaware, a system was designed for generating a set of four beams, automatically aligned to produce a trigonal (rhombohedral) lattice [143, 144]. This is accomplished by patterning three gratings surrounding a window on a mask plate and illuminating the entire mask with a single wide-area laser beam, as depicted in Fig. 6.53a. The gratings and window were patterned on an Al-coated mask by e-beam lithography and then the Al coating was wet etched; one such mask is shown in Fig. 6.53b. The gratings were positioned such that the first diffracted orders from each of the three gratings intersect at a common location behind the window on the back side of the mask plate. To expose a sample, a substrate was coated with resist (Microchem SU-8) and positioned in flush contact with the back side of the mask. No alignment is necessary to generate a 3D photonic crystal by illuminating the mask with UV laser radiation. The laser was a frequency-tripled, Q-switched Nd:YAG that delivers up to 250 mJ per 5-ns pulse. Aluminum was chosen for the mask material because of its high reflectance at ultraviolet wavelengths; a chromium mask was used initially, but because of the absorption and subsequent rapid heating of the chrome layer, the mask was damaged by the high peak power of the laser pulses.

One drawback to this approach is that one cannot control the orientation of the beams' polarizations independently, and hence, the visibility contrast of the interferometric exposure cannot be optimized fully. Several degrees of freedom are available with which to compensate for this problem: One can vary the duty cycles of the three gratings to modify their diffraction efficiencies, and one can rotate the entire mask relative to the polarization of the illuminating laser beam. Calculations indicate that very good visibility contrast and a symmetric primitive structure can be realized with this approach when the polarization of the laser source is oriented vertically with respect to the grating pattern as depicted in Fig. 6.53b. Another drawback is the relatively small size of the gratings, whose size is limited to a few hundred microns across by the e-beam lithography write-field size. The consequences of the small gratings are as follows: First, the size of the PhC is limited to the size of the gratings. Second, the outer boundary of the grating produces significant diffraction, which

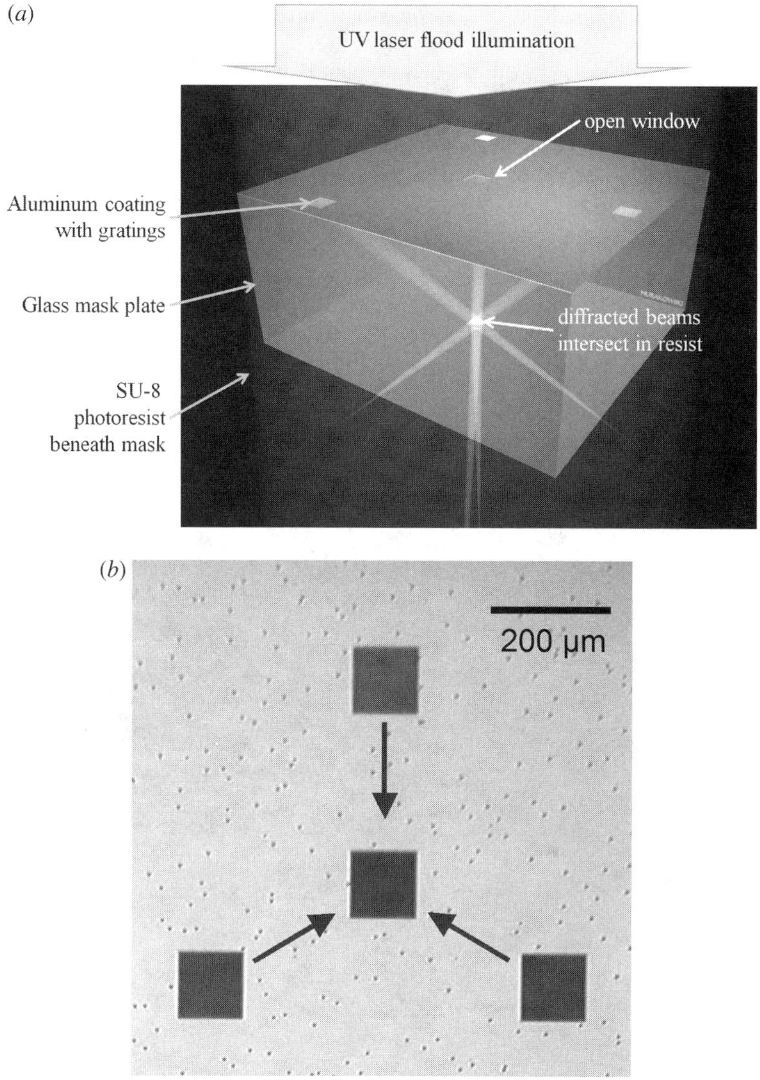

Figure 6.53 (a) Diagram of the system used to pattern 3D photonic crystals by interference lithography. (b) Optical micrograph of an Al mask plate, which contains 100-μm square gratings surrounding an open window.

causes the exposed area to be patterned with uneven dose and thus leads to nonuniformity of the final PhC.

The mask shown in Fig. 6.53b contains gratings with a period of 0.6 μm, which corresponds to a diffracted angle in the resist of 24.2°, taking into account the refraction at the interface between the resist and the mask plate. This angle is less than the 38.9° angle that would produce an FCC lattice. Hence, the structures created with it are not cubic but rhombohedral: FCC lattices that have been stretched along the [111] direction.

An example of a 3D photonic crystal fabricated by the method described above is shown in Fig. 6.54a. This sample was exposed using gratings with a 0.4-μm period and a 10-mJ laser pulse. We have calculated the 3D intensity profile resulting from the exposure that was used to fabricate this structure; a surface of constant intensity is plotted in Fig. 6.54b. The SEM image in (a) corresponds to a top-down view along the z direction of the structure as depicted in (b). The length scales of the features in the fabricated structure agree well with the calculation.

The method just described is very similar to the approach independently proposed and demonstrated by Divliansky et al. [145]. In this work, a diffraction grating array was fabricated with much larger gratings, which were themselves patterned

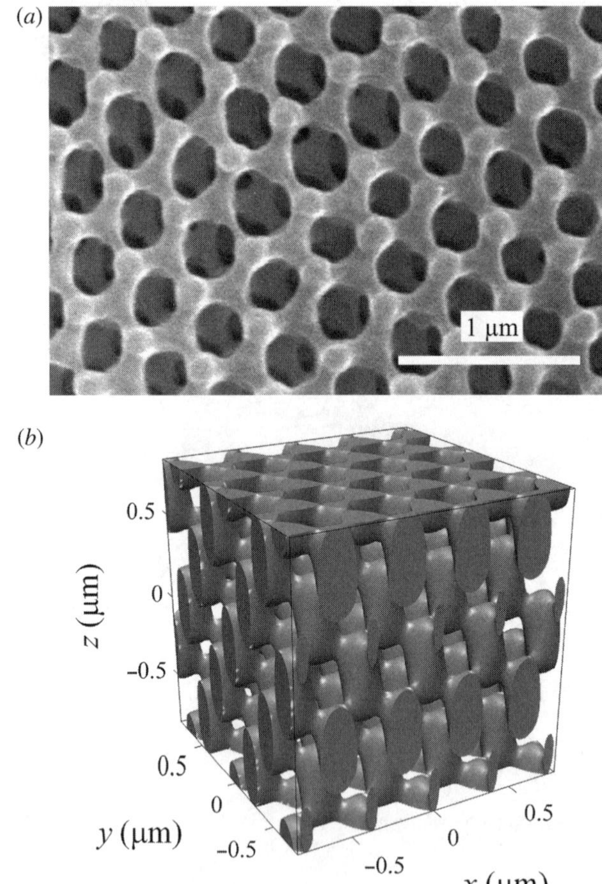

Figure 6.54 (a) SEM micrograph of a photonic crystal fabricated in SU-8 photoresist by interference lithography. (b) Calculated surface of constant intensity for the interference pattern used to expose the sample shown in (a).

with two-beam holographic lithography, as depicted in the diagram of Fig. 6.55a. These gratings were ~8 mm across, and therefore, diffraction by the aperture of the grating boundaries can be neglected over a very large area of the final sample. However, with gratings this large, a comparable distance must exist between the gratings and the sample; hence, the mask and the sample were not in contact. This makes the exposure vulnerable to vibration in principle; however, in practice the researchers used a nanosecond-pulsed laser to eliminate this concern. The SEM image in Fig. 6.55b shows the results that were obtained using this approach.

One of the earliest reported uses of a diffractive element for simple generation of multiple coherent beams for interference-lithographic PhC fabrication was the work of Kondo et al. [146]. They used a commercially supplied diffractive beam splitter

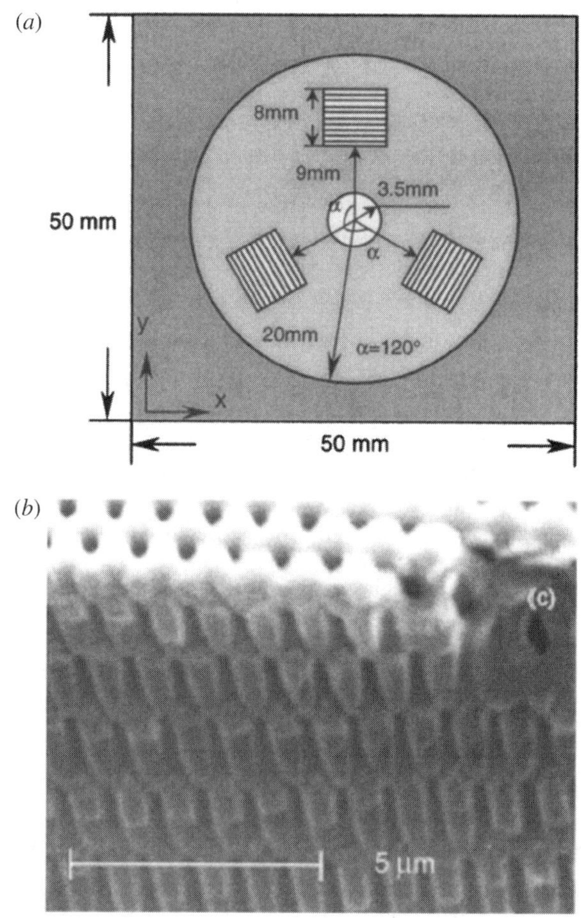

Figure 6.55 (a) Schematic of the diffraction grating array used by Divliansky et al. [145] to fabricate 3D photonic crystal structure by interference lithography. (b) SEM image of the sample obtained, with a cross section showing the depth of the structure. (Reprinted with permission from Ref. 145, © 2003, American Institute of Physics.)

that produced a set of eight beams arranged in a circle around the central beam. To select the five beams they used for patterning, they used an aperture array to pass only four of the outer beams along with the central beam. After the apertures, an objective lens was used to direct the beams back toward a region of intersection within a photocurable sample, namely SU8 photoresist. A small concentration of fluorescent dye was added to the resist for subsequent characterization by scanning confocal microscopy. This experimental arrangement is shown schematically in Fig. 6.56a. The exposure source was a frequency-doubled femtosecond laser with a wavelength of 380 nm; an average laser power of ~100 μW was used with an exposure time of ~20 s. Under these conditions, care must be exercised to avoid the interference being disturbed by vibrations and air currents.

As a result of this beam configuration, the samples fabricated by this method possess square symmetry in the plane transverse to the illumination direction. Furthermore, the angle between the central and outer beams was only ~14°, leading to a large discrepancy between the periodicity in the transverse plane and that in the longitudinal direction, which is undesirable for the formation of a bandgap. This discrepancy between periods is clearly seen in the scanning confocal microscope images shown in Fig. 6.56b, c.

6.3.4.2 Multiple Exposure Methods. Shoji et al. [147, 148] were the first to demonstrate that exposures using fewer than four beams, combined over multiple exposures, also can realize 3D PhCs, often more conveniently than with a single

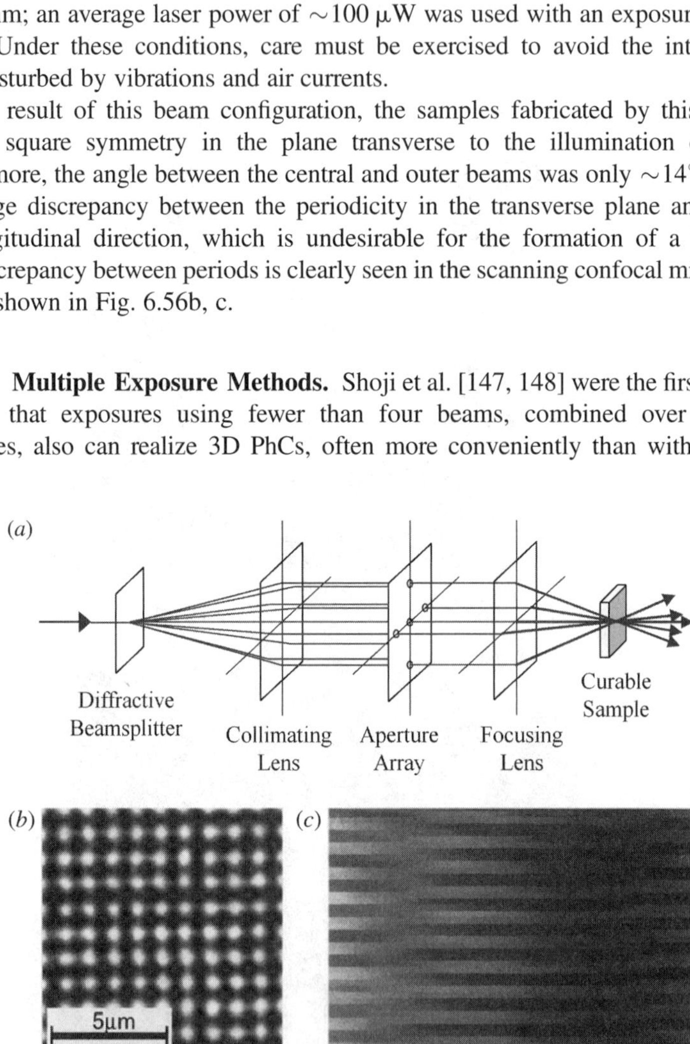

Figure 6.56 (a) Schematic of the diffractive-beamsplitter/aperture-array based method for interferometric fabrication of photonic crystals, as implemented by Kondo et al. [146]. (b) Top-down and (c) side views of a sample fabricated by this method, obtained by scanning confocal microscopy. (Reprinted with permission from Ref. 146, © 2001, American Institute of Physics.)

exposure. A significant advantage of using multiple exposures, each producing a 1D or 2D interference pattern (array of planes or rods), is that the lattice period can be varied without deforming the lattice symmetry. As a result, UV or visible lasers can be used to expose 3D PhCs with lattice constants of several micrometers (Fig. 6.57). A drawback to this approach comes from its sensitivity to the relative phases of the interference patterns formed during the several exposures, which is a symptom of the need to align the patterns from each of the multiple exposures. For example, a phase shift of π in one beam, as caused by a $\lambda/2$ displacement of a beam-steering mirror, will displace the antinodes of the interference pattern by one half the lattice period, compromising the alignment of the several exposures. Alignment and stability problems also arise if the sample substrate/container must be repositioned or reoriented between exposures.

The material system that was used in their work was a photopolymerizable resin that consisted of urethane acrylate monomers, urethane acrylate oligomers, and photoinitiators for the photopolymerization reaction. The resin absorbs ultraviolet or blue light, and solidifies by a photochemical reaction. In their experiments, a He-Cd laser operating at 442 nm was used for solidifying the resin.

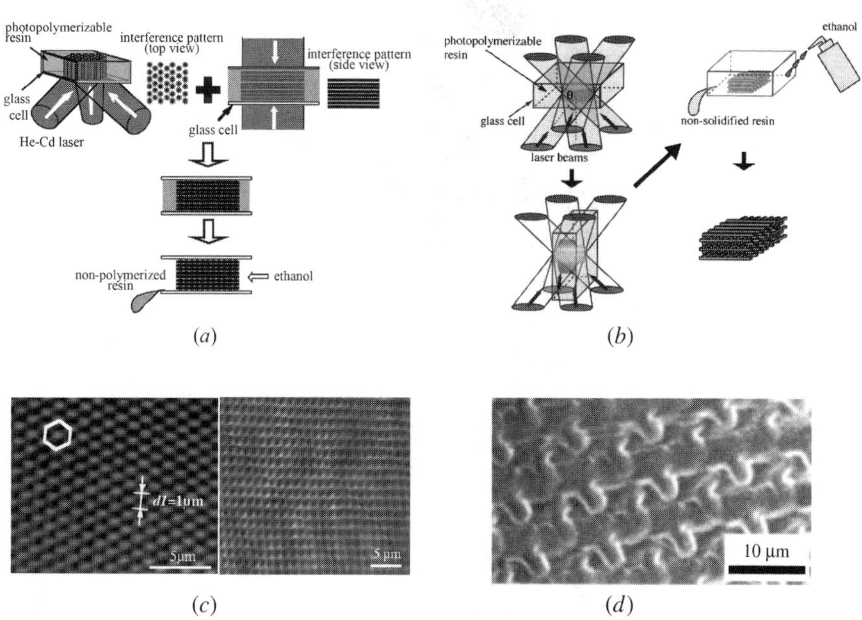

Figure 6.57 The interferometric fabrication schemes of Shoji et al. [147] and Shoji and Kawata [148], using multiple exposures to combine overlapping interference patterns of dimensionality lower than 3 to realize 3D structures. (a) Scheme to realize a simple hexagonal lattice with one 3-beam exposure and one 2-beam exposure. (b) Scheme to realize a woodpile (tetragonal) lattice with two identical four-beam exposures and sample rotation between. (c), (d) SEM images of fabricated samples using schemes in (a) and (b), respectively. [(a), (c): Reprinted with permission from Ref. 148, © 2000, American Institute of Physics. (b), (d) Reprinted with permission from Ref. 147, © 2003, American Institute of Physics.]

An alternative method to the one developed by Shoji et al. has been demonstrated by Schneider et al. [149] at the University of Delaware. This approach is immune to the problems described above. Using multiple three-beam exposures and a monolithic diffraction mask to generate all beams (see Section 6.3.4.1), an interference analog of the drilling strategy originally proposed by Yablonovitch et al. [124] (see also Section 6.2.6) was developed. Each three-beam exposure is generated by a set of three diffraction gratings, oriented to produce an interference pattern that consists of a hexagonal array of rods, which are oriented 35.26° from the normal to the gratings (Fig. 6.58). Three such grating triplets are positioned on the same grating mask, with

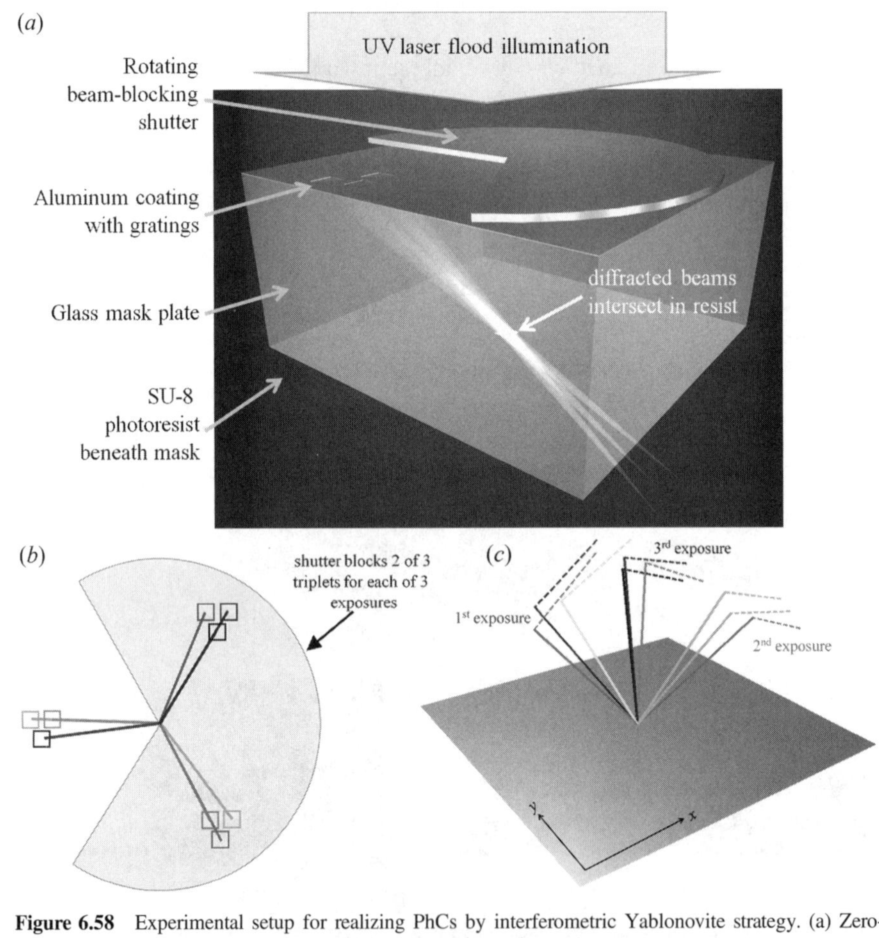

Figure 6.58 Experimental setup for realizing PhCs by interferometric Yablonovite strategy. (a) Zero-alignment setup using monolithic grating array plus rotating shutter/beam blocker. (b) Layout of gratings and rotating blocker; squares indicate grating location and lines indicate the direction along which the first-order diffracted beam will travel (projected onto the plane of the mask) until they intersect in the target area of the photoresist at the center. (c) 3D depiction of the diffracted beam geometry; solid lines indicate *k*-vectors, dotted lines indicate polarization vectors, assuming that the beam illuminating the mask is polarized along the *x* direction. (The lines are colored individually only so that they may be more easily distinguished.)

each triplet positioned at 120° azimuthal intervals, and they are illuminated one at a time with a parent laser beam. A rotating shutter blocks two triplets for each exposure. The sum of the exposure doses, each generating a set of rods, yields an FCC PhC lattice with diamond-like symmetry. The lattice period is determined by the period of the rod arrays, which can be adjusted by changing the spacings of the gratings in each triplet. Because all the beams are generated by gratings on a single, monolithic mask plate, there is no possibility for phase drift to compromise the pattern as described earlier. Also, no alignment is required, as the substrate is placed in direct contact with the back side of the mask plate. This method has been used to fabricate FCC PhCs with lattice parameters of ~2–3 μm.

6.3.5 Summary

Interference lithography has the potential to fabricate nearly perfect crystal lattices over areas limited only by the size of the optics used to condition the laser beam. Moreover, with powerful lasers, samples can often be patterned by an exposure that consists of a single nanosecond laser pulse.

The interference lithography approach offers considerable advantages over other methods that have been attempted to fabricate 3D photonic crystals. The most common alternative methods are layer-by-layer micromachining (discussed in Section 6.2) and self-assembly of monodisperse colloidal particles (see Section 6.5). Compared with the layer-by-layer approaches, interference lithography has considerable advantages in cost and the size of the samples that can easily be fabricated. Micromachining and deposition usually require expensive equipment, and structures must be processed one layer at a time, which makes the fabrication of many layers very time consuming. Self-assembly, on the other hand, is capable of creating large volumes in a single process. It is limited, however, to a very small range of crystal lattices (primarily FCC and hexagonal close-packed (HCP)), and samples typically contain a large number of defects, which have proven very difficult to control. More defect-free self-assembled photonic crystals can be fabricated at the expense of increasing the process complexity and time.

There are two primary shortcomings of the interference lithography method. First, because of the large-scale, parallel nature of the method, it is difficult to incorporate engineered defects, such as waveguide channels and cavities. Second, because it is based on photopolymers, the resulting air-polymer structures have low dielectric contrast. In both of these respects, layer-by-layer semiconductor micromachining has a definite advantage over both interference lithography and self-assembly. Lee et al. [150] have proposed a method to incorporate defects into self-assembled lattices by filling the lattice with an index-matched monomer and using multiphoton absorption to polymerize defects, whereas a combination e-beam/UV-interference processes to introduce designed defects into interferometrically patterned structures has been demonstrated for 2D [143] and 3D [144]. Both self-assembly and interference lithography can be combined with one of many methods developed to infiltrate templates with high-index materials like ceramics, semiconductors, and metals, thereby improving the dielectric contrast. These techniques add cost and complexity to the processes.

6.4 THREE-DIMENSIONAL PHOTONIC CRYSTALS: MULTIPHOTON POLYMERIZATION

6.4.1 Stereolithography/Laser Rapid Prototyping to Fabricate Arbitrary 3D Structures

Stereolithography is a 3D fabrication process that builds a structure layer by layer in a photocurable resin system. It is most commonly implemented using a computer-controlled movement stage and an ultraviolet laser whose focused beam can be scanned in a horizontal plane. The stage is lowered into a tank that contains an uncured liquid photopolymer precursor until it is covered by a thin layer of the liquid. Then the laser scans across the surface, curing the liquid in those regions that correspond to the bottom "slice" of the desired structure. The stage is lowered by a small distance, and the process repeats until the entire structure has been realized. The movement of the stage and laser scanning are controlled by a computer running software that can read the structure from a standard computer-assisted design (CAD) file, and calculates the shape of the slices. The process is entirely automated and can realize very large ($\sim$25 cm) structures in periods of several hours.

Such a system could fabricate a chip-scale photonic-crystal structure quickly. Unfortunately, the resolution is not sufficient to pattern the submicron features needed for photonic crystals that operate in the visible/near-IR range. In a typical commercial stereolithography system, the resolution is limited to a few hundred micrometers by the precision of the moving stage and beam steering hardware; these systems are targeted for rapid tool-free prototyping of custom parts to reduce product development time and cost. However, even if such a system were optimized with submicron precision stages, vertical resolution would still be limited by the viscosity and surface tension of the liquid polymer precursor.

A superior approach, similar in concept, would be simply to focus the laser beam within the volume of the photopolymer. However, this approach is ill-suited for 3D fabrication because, as will be shown later, the dose cannot be confined to the focus region when the beam is scanned horizontally, which severely limits the type of structures that can be obtained. A solution to this limitation was devised by Maruo and Kawata [151], who demonstrated a 3D micropatterning process based on two-photon absorption, a special case of multiphoton absorption. The following section will describe the general process of multiphoton absorption and explain why it works for 3D fabrication.

6.4.2 Multiphoton Absorption

Multiphoton absorption (MPA) is a *nonlinear* optical effect: Nonlinear optics is the regime of optical phenomena that occur when optical fields applied to matter are sufficiently large so that the material's response, i.e., its *polarization*, or induced dipole moment per unit volume (not to be confused with the orientation of the field in an electromagnetic wave, which goes by the same name!) is not linearly proportional to the applied field amplitude. Such phenomena are made accessible for study and applications only by the intense optical fields generated by lasers. Applications of nonlinear optics include harmonic generation (frequency doubling, tripling, etc.), parametric fluorescence for continuum generation, parametric amplification for tunable sources of coherent radiation, phase-conjugate reflection for wavefront correction, absorption saturation for passive mode-locking, optical bistability (i.e., optical switching) and optical limiting by intensity-dependent absorption or refractive index, electro-optic modulation, and dispersion compensation for optical telecommunications.

Nonlinear optical phenomena are described mathematically by expanding the induced dipole moment in powers of the incident electric field(s). But, first, let us review the linear case: In linear optics, the polarization P is given by

$$P = \varepsilon_0 \chi^{(1)} E, \tag{6.29}$$

where $\chi^{(1)}$ is the linear susceptibility. The superscript in parenthesis, as we shall see below, denotes the order of the interaction in the expansion. From this relation, we can derive the dependence on $\chi^{(1)}$ of other familiar optical material parameters. For example, recalling the displacement field $D = \varepsilon E = \varepsilon_0 E + P$, we deduce that the dielectric permittivity ε and linear refractive index n are related to $\chi^{(1)}$ by

$$\varepsilon = \varepsilon_0 (1 + \chi^{(1)}) \tag{6.30}$$

and

$$n = (\varepsilon/\varepsilon_0)^{1/2} = \sqrt{1 + \chi^{(1)}}, \tag{6.31}$$

respectively. In general, these quantities are complex; the real parts describe refraction whereas the imaginary parts are used to describe absorption and/or gain. From Eqs. (6.30) and (6.31) it is clear that the imaginary parts of n and ε, which are denoted by $\Im(n)$ and $\Im(\varepsilon)$, come from $\Im(\chi^{(1)})$. To compute the absorption coefficient in terms of $\chi^{(1)}$, consider a plane wave of the form $\sim e^{-j(kz-\omega t)}$. Here, ω is the angular frequency, and k is the complex wavenumber, equal to $nk_0 = n\omega/c = 2\pi n/\lambda_0$, with λ_0 the vacuum wavelength and c the speed of light in vacuum (note that χ, ε, n are in general functions of ω). If we make n complex: $n = \Re(n) + j\Im(n)$, where $\Re$ denotes the real part, we find that the wave is attenuated as $\sim e^{-\Im(n)k_0 z}\, e^{-j(\Re(n)k_0 z - \omega t)}$. Hence, the linear absorption coefficient is

$$\alpha_0 = (2k_0)\Im(n) = (2\omega/c)\Im\left(\sqrt{1 + \chi^{(1)}}\right) \approx (\omega/c)\Im(\chi^{(1)}), \tag{6.32}$$

where the factor of 2 arises because absorption is defined in terms of intensity, which is proportional to the square of the wave amplitude.

In the case of nonlinear interactions, we expand the polarization as

$$\begin{aligned} P &= \varepsilon_0 \chi^{(1)} E + \varepsilon_0 \chi^2 E^{(2)} + \varepsilon_0 \chi^{(3)} E^3 + \cdots \\ &\equiv P^{(1)} + P^{(2)} + P^{(3)} + \cdots, \end{aligned} \tag{6.33}$$

where $\chi^{(2)}$ and $\chi^{(3)}$ are the second- and third-order nonlinear susceptibilities, respectively, and likewise $P^{(m)}$ is called the mth order nonlinear polarization (when $m > 1$; $m = 1$ corresponds to the linear polarization). It should be noted that in general, the susceptibilities are complex tensors that relate the Cartesian components of the vector polarization to those of the electric field, and that the applied field is frequently the sum of several components with different frequencies; hence, the powers of the fields contain numerous cross terms that oscillate at many different frequencies. For simplicity's sake, we disregard these complications in this discussion.

Not surprisingly, just as linear absorption can be described in terms of the imaginary part of $\chi^{(1)}$, nonlinear absorption can arise from the imaginary parts of the nonlinear susceptibilities. Specifically, two-photon absorption is a third-order nonlinear effect that arises from the imaginary part of $\chi^{(3)}$. Third-order contributions to the nonlinear polarization are proportional to the cube of the incident field amplitude, which results in a nonlinear contribution to n that is proportional to the intensity. We can show this by defining a total refractive index $n = n_0 + n^{\mathrm{NL}}$, where we shall call $n_0 = (1 + \chi^{(1)})^{1/2}$ the linear refractive index and n^{NL} the nonlinear index. We determine n by replacing $\chi^{(1)}$ in Eq. (6.31), with an effective total susceptibility χ_{eff}, defined simply as $P/\varepsilon_0 E$. Then with Eq. (6.33), we have

$$\begin{aligned} n &= \sqrt{1 + \chi_{\mathrm{eff}}} = \sqrt{1 + P/\varepsilon_0 E} \\ &= \sqrt{1 + \chi^{(1)} + \chi^{(2)} E + \chi^{(3)} E^2} \\ &= \sqrt{n_0^2 + \chi^{(2)} E + \chi^{(3)} E^2} \\ &= n_0 \sqrt{1 + \frac{1}{n_0^2} (\chi^{(2)} E + \chi^{(3)} E^2)}. \end{aligned} \tag{6.34}$$

Let us assume that we can neglect the second-order susceptibility, since it plays no role in two-photon absorption. Then, after expanding the root, safely assuming the nonlinear terms are small compared with n_0, we obtain

$$n \simeq n_0 + \frac{\chi^{(3)}}{2n_0} E^2 = n_0 + \frac{\chi^{(3)}}{n_0 \varepsilon_0 c} I, \tag{6.35}$$

where we have identified the intensity as $I = (\varepsilon_0 c/2) E^2$. If we isolate the imaginary part of this result, as we did to obtain Eq. (6.32), we obtain the total absorption coefficient

$$\alpha = \alpha_0 + \alpha^{\mathrm{NL}} = \alpha_0 + \beta I, \tag{6.36}$$

where β is the two-photon absorption coefficient, related to $\chi^{(3)}$ by $\beta = \omega\Im(\chi^{(3)})/n_0\varepsilon_0 c^2$. Thus, we see that the total absorption is intensity dependent. (Note also that $\Re(\chi^{(3)})$ leads to intensity-dependent refraction.)

The total absorption appears in the following differential equation for the intensity as a function of propagation distance (for one-dimensional propagation along the z direction):

$$\frac{dI}{dz} = -\alpha(I)I, \tag{6.37}$$

where we note explicitly that the absorption can be a function of intensity. The solution of this equation in the case of purely linear absorption is simple exponential decay according to Beer's Law: $I(z) = I_0\exp(-\alpha z)$.

The property of MPA that makes it so extraordinarily useful for fabrication of 3D microstructures is the intensity dependence of the absorption. This is because of the behavior of photosensitized polymers like photoresists and other photocurable resins, liquid or solid. These materials require a threshold dose of energy to be patterned: For surface exposures in lithography, doses are expressed in units of energy per unit area; thus for optical photolithography, the dose is proportional to the source irradiance (intensity) and the exposure time. An important distinction that is sometimes overlooked in planar photolithography, where many resist formulations are standardized by their commercial suppliers, is the dependence on the material absorption. The *true* dose that determines whether a certain region of the sample is patterned is not the energy to which that region is *exposed*, but rather the energy that is *absorbed* by the photoactive component of the material (e.g., photoinitiator for free-radical polymerization and photoacid generator for chemically amplified resists). So, although one may commonly assert that increasing the initiator concentration in a certain photocurable material system will reduce the "dose" required to pattern it, this language refers to the energy applied to the sample, not the energy actually absorbed by the sample. For a given material system, the required "true dose" is the same, regardless of initiator concentration. For example, if the initiator concentration, and thus α, is doubled, then the threshold exposure time t for a source of fixed intensity I will be halved, whereas the energy deposited in the sample, proportional to αIt (and to the final concentration of activated initiator), will be the same.

Consider a Gaussian beam that propagates along the z direction. The intensity of this beam as a function of position is

$$I(\rho, z) = I_0\left(\frac{w_0}{w(z)}\right)^2 \exp\left(-\frac{2\rho^2}{w(z)^2}\right), \tag{6.38}$$

where $\rho = (x^2 + y^2)^{1/2}$ is the radial position in cylindrical coordinates (with the axial direction along z), $w(z)$ is the Gaussian beam width which varies along z according to

$$w(z) = w_0\sqrt{1 + (z/z_0)^2}, \tag{6.39}$$

and w_0 is the minimum beam width, or beam waist, at $z = 0$. z_0 is a characteristic length called the Rayleigh range, given by $z_0 = (n/\lambda_0)\pi w_0^2$.

Let us now compute the energy dose that this beam will deposit in a volume of material dV, based on Eq. (6.36) in a fixed exposure time t' (i.e., we assume the intensity is given by Eq. (6.38) from $0 < t < t'$, and is zero otherwise). We further assume that absorption, whether linear or nonlinear, is sufficiently small that Eq. (6.38) is accurate everywhere, which means that the total beam power is unchanged as it propagates through the material. Also, assume for the moment that our volume element is near the beam axis, so that we can neglect the radial dependence of I. Under these conditions, the energy deposited in $dV = dx\,dy\,dz$ is

$$\begin{aligned} dE &= -[I(z + dz) - I(z)]t'\,dx\,dy \\ &= [-dI/I]It'\,dx\,dy \\ &= [\alpha\,dz]It'\,dx\,dy \\ &= \alpha\,It'\,dV. \end{aligned} \tag{6.40}$$

To obtain this result, we have rearranged Eq. (6.37) to identify the total absorption as $-I^{-1}(dI/dz)$. Then the dose $\mathfrak{D}$ is just

$$\mathfrak{D} = dE/dV = \alpha It'. \tag{6.41}$$

In the case of linear absorption, the dose is simply proportional to I. A focused laser beam's intensity profile, depicted in Fig. 6.59a, can be used to pattern a small volume of photosensitive material within the bulk. By scaling the exposure time, the region receiving at least the threshold dose of energy can be adjusted to correspond to any of the contours in the figure.

For two-photon absorption, it is generally the case that the incident frequency, being half that corresponding to a resonant electronic transition, is far from a single-photon resonance, and hence α_0 is negligible. Therefore let us consider only the second term in Eq. (6.36) so that $\alpha = \beta I$, in which case we see that the dose becomes $\beta I^2 t'$. Examining this dose profile, which is depicted in Fig. 6.59b, it is clear that the dose is more strongly localized in the two-photon case. Note, however, that for a given material system, the two-photon photopatterning is performed using a source with double the wavelength, which means that the focused spot size is proportionally greater, as shown in Fig. 6.59c,d.

But even if we compare the two at the same wavelength, the advantage of two-photon patterning is more significant than merely somewhat-better feature resolution. To see why, consider that, in general, one does not wish merely to pattern isolated points within a volume; rather, the goal is to scan the laser beam to pattern complex interconnected structures, such as 3D photonic crystals. So, what happens to the volume of material being patterned when we scan the laser beam in, e.g., the transverse (xy) plane? It turns out that, for linear absorption, the result of scanning the beam is to deliver a constant dose to all levels, independent of z. This means that a pattern cannot be defined within the bulk (unless the features are simply isolated points separated by distances comparable to the beam width $w(z)$ at the limits of the

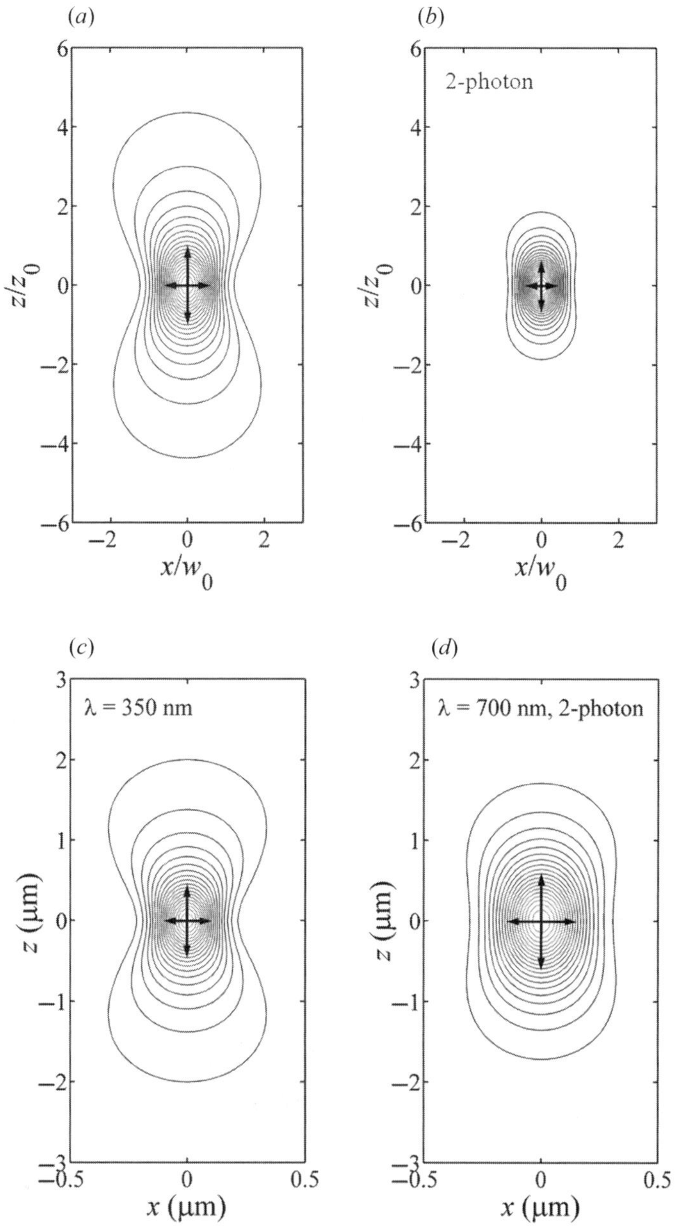

Figure 6.59 Comparison of intesnsity/dose profiles for a stationary focused Gaussian beams. (a) Intensity profile, corresponding to linear absorption (one-photon). (b) Squared intensity, corresponding to nonlinear absorption (two-photon). In (a) and (b), the x and z axes are scaled to the beam waist and Rayleigh range, respectively, and the wavelength is the same for both. The two-photon dose is seen to be confined to a smaller volume. (c), (d) Same as (a) and (b), respectively, except that axes are scaled in microns, based on a real refractive index of $n_0 = 1.67$ and wavelengths of 350 nm in (c), 700 nm in (d). Note that in this, more realistic case, the two-photon dose is not better confined, because of the larger focused spot size when the wavelength is doubled. In all plots, the contours correspond to intervals of 5% of the maximum intensity at focus. Crossed arrows indicate the horizontal and vertical extent of the 50% contour.

sample thickness, i.e., $w(\pm L/2)$ for a sample thickness of L; note $w(z)$ scales linearly with z when $z \gg z_0$). This can be shown by integrating the dose over the transverse plane. For the linear absorption case we obtain the following:

$$\int \mathfrak{D}(x, y, z)dx\,dy = 2\pi \int \mathfrak{D}(p, z)\rho\,d\rho$$
$$= 2\pi\alpha_0 t' \int I(\rho, z)\rho\,d\rho = 2\pi\alpha_0 I_0 t' \left(\frac{w_0^2}{[w(z)]^2}\right) \left[\int_0^\infty \exp\left(-\frac{2\rho^2}{[w(z)]^2}\right)\rho\,d\rho\right]. \tag{6.42}$$

It is simple to show that the integral in square brackets yields $[w(z)]^2/4$, and thus the z dependence cancels out of the integrated dose, leaving a constant. Contrast this with the same calculation in the two-photon absorption case, which yields the following:

$$\int \mathfrak{D}^{\mathrm{NL}}(x, y, z)\,dx\,dy = 2\pi t' \int \alpha(I)I(\rho, z)\rho d\rho$$
$$= 2\pi\beta I_0^2 t' \left(\frac{w_0^4}{[w(z)]^4}\right) \left[\int_0^\infty \exp\left(-\frac{4\rho^2}{[w(z)]^2}\right)\rho d\rho\right]. \tag{6.43}$$

The integral is again proportional to $[w(z)]^2$, but there is no cancellation because of the factor of $[w(z)]^{-4}$ outside the integral. Hence, even when integrating over the entire xy plane, the dose exceeding threshold can be confined to a narrow range of z within the bulk.

The preceding analysis demonstrates that two-photon absorption can be used to apply above-threshold dose to confined regions of photocurable materials within the bulk, while scanning laterally and vertically. Therefore it can be used in a manner similar to stereolithography, in that a 3D object can be rasterized and created point by point, layer by layer. But significantly, it does not require the structure to be immersed in liquid precursor as each layer is formed. The patterning can be done at any depth that is practical given the working distance of the focusing optics, and the precursor can be either a volume of liquid or even a solid film of photoresist. In fact, transparent chemically amplified photoresists are well suited to the process, as the linear absorption is low, and the threshold dose is small because of chemical amplification, hence patterning can be rapid.

6.4.3 PhC Fabrication Using Multiphoton Absorption

In this section, we review the major literature results using MPA for PhC fabrication. The first reported use of the two-photon direct-writing technique for PhC fabrication was the work of Cumpston et al. [152]. This work consisted of a small volume patterned in the standard PhC woodpile, as depicted in Fig. 6.60. Some polymer shrinkage is visible, manifested as the tapering in the vertical direction as the layers go up

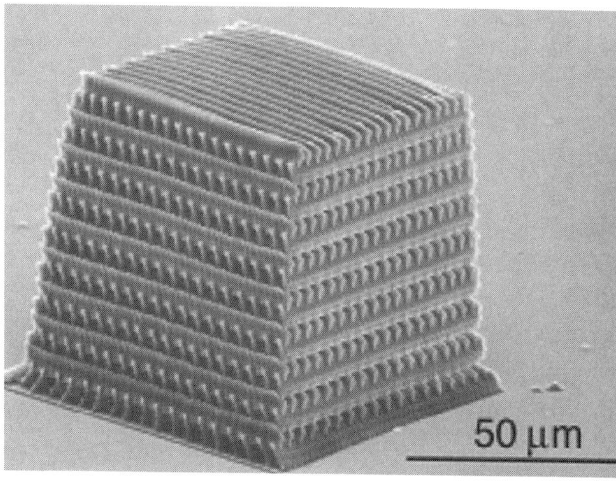

Figure 6.60 Photonic-crystal woodpile fabricated by Cumpston et al. [152], using a two-photon-absorption-based direct-write method. (Adapted by permission from Macmillan Publishers Ltd., Ref. 152, © 1999.)

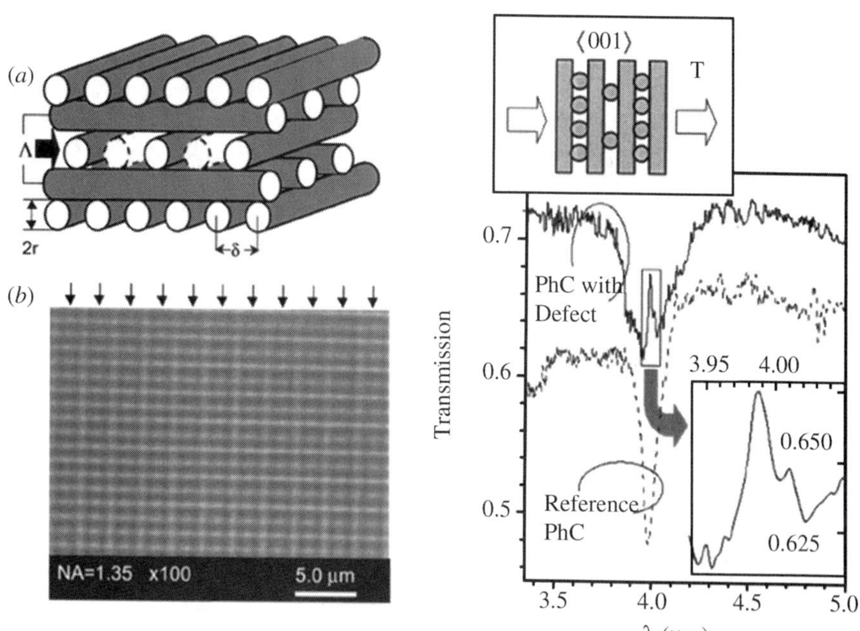

Figure 6.61 PhC fabricated by Sun et al. [154] incorporating intentional defects: Structure and SEM images appear on the left, and transmission spectra for samples with and without defects appear on the right. (Reprinted with permission from Ref. 154, © 2001, American Institute of Physics.)

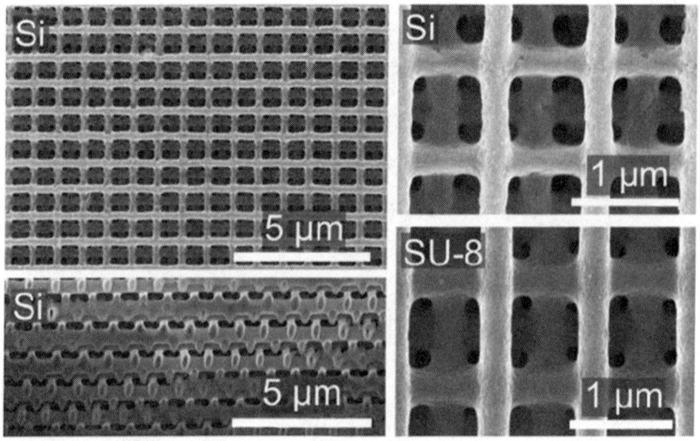

Figure 6.62 Silicon woodpile PhC fabricated by two-photon direct-writing and double-inversion CVD technique [157]. The SEM image at the lower right shows the original SU-8 polymer template. (Reproduced with permission from Ref. 157, © 2006, Wiley-VCH Verlag GmbH & Co.)

away from the substrate that anchors the bottom layer. Unlike in later work, where researchers used commercial photopolymers and/or photoresists, this work used custom-developed photocurable polymers, to which the undesirable shrinkage can be attributed.

Similar work from Sun et al. [153 and 154] followed, in which smaller features and designed defects were realized. In their work, Sun et al. used Nopcocure 800 resin, which consisted of radical photoinitiator and acrylic acid ester, as the photocurable material. The fabrication wavelength was 400 nm, which is beyond the absorption range. A Ti : sapphire femtosecond laser pulse generation and chirp pulse amplification system was used as their irradiation source, with a repetition rate of 5 kHz, and pulse width of approximately 130 fs. The output wavelength was 800 nm, which corresponds to a curing wavelength of 400 nm due to the two-photon absorption. Their results are summarized in Fig. 6.61.

More recently, the direct write method has been refined to yield impressive results by Deubel et al. and subsequently by Tetreault et al. [155–157]. In this work, the material system was commercial SU8 photoresist, exposed using a regeneratively amplified Ti : sapphire laser system with a pulse duration of 120 fs. The repetition rate can be computer controlled from 1 kHz to single shot mode. The wavelength was tuned to 800 nm. Features of less than a 150-nm lateral dimension can be created. Figure 6.62 depicts the remarkable quality of the structures that can be realized with this method.

6.5 THREE-DIMENSIONAL PHOTONIC CRYSTALS: SELF-ASSEMBLY

Self-assembly refers to a category of processes that lead to the spontaneous formation of order in randomly arranged systems. In the most general sense, a diverse range of systems, from atomic crystals to living organisms, are self-assembling. Order arises when a system settles into a state of minimum of energy, often driven by the application of annealing energy to assist the system in transitioning from local energy minima to a more global minimum, such as when polycrystalline metals are thermally annealed into single crystals.

In the case of photonic crystals, relevant length scales are determined by the wavelength of light and are over three orders of magnitude greater than atomic/molecular sizes. Therefore, self-assembly processes for fabrication of photonic crystals must be driven by different annealing mechanisms. As we will learn throughout this section, a wide variety of methods have been developed to drive the ordering of systems that consist of many nearly identical, wavelength-sized objects.

6.5.1 Monodisperse Colloidal Suspensions

Naturally occurring monodisperse silica particles can sediment into periodic close-packed structures that strongly diffract light because the particles' size is comparable with the wavelength of visible light. These natural formations are commonly known as opals. The strong interaction of opals with light reveals that they are in fact photonic crystals, and the light they diffract is an indication of stop bands, which are precursors to full photonic bandgaps. A thorough overview of the subject can be found in Ref. 158; for more details the literature is abundant, for example see Refs. 159–179.

6.5.1.1 Synthesis of Monodisperse Particle Suspensions

Silica. Silica is one of the most common materials used by researchers for self-assembly of photonic crystals. Besides being found in natural opals, monodisperse silica particles can be synthesized by a variety of methods to match specifications for photonic-crystal researchers. Herein is described one common fabrication method, known as the Stöber method [180]. It is based on the formation of silica by hydrolysis of alkyl silicates (such as tetraethylorthosilicate, or TEOS) and subsequent condensation of silicic acid in alcohol solutions.

Figure 6.63a depicts the reaction at the heart of the silica synthesis. Hydrolysis of the silicates occurs in a solution of ethanol, ammonia, and water. When the process is completed, each tetravalent Si atom is bound to four bivalent oxygen atoms, hence the final macromolecule is $(SiO_2)_n$.

An imbalance in the number of silicate molecules can lead to the incomplete termination of the macromolecule formation, resulting in a surface composed of —Si—OH groups, which can ionize in acid solution yielding a net surface charge density. These surface groups can also be exploited for the synthesis of self-assembled monolayers (SAMs), which enhance the particles with useful functionality.

Polystyrene. Monodisperse polymeric colloidal particles are typically fabricated by *emulsion polymerization*. An overview of the process is presented here; more details can be found in numerous sources in the literature. In the emulsion polymerization process, a solution of hydrophobic monomers is suspended in water to form an emulsion.

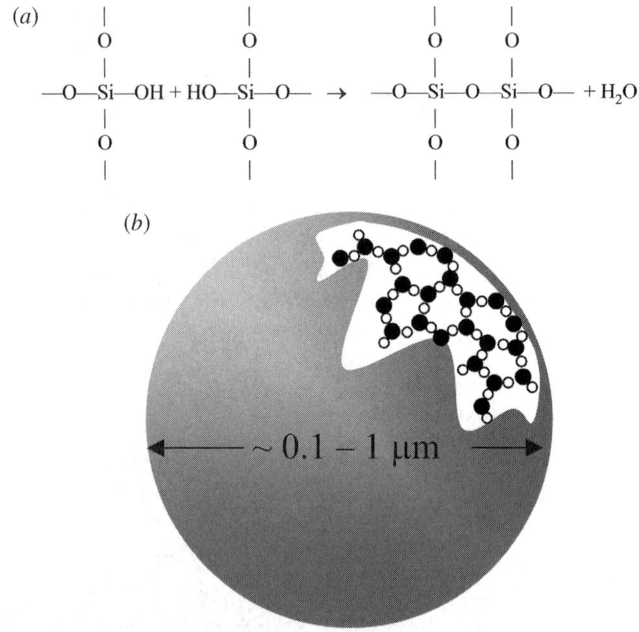

Figure 6.63 (a) Formation of silica by condensation of silicic acid. (b) Structure of the final silica particle. Open circles represent silicon; filled circles represent oxygen. The tetravalent silicon atoms are connected to each other by oxygen atoms, forming a single silica *macromolecule* [158].

The dispersal of monomers is often assisted by another species in the solution called an emulsifier. Also present in the solution are initiators—chemical species that break into two radicals and trigger the polymerization process, called free-radical polymerization. The reaction proceeds is depicted in Fig. 6.64(a), and described below:

$$\begin{array}{rcll} X - X & \longrightarrow & 2X^* & \text{free radical formation} \\ M + X^* & \longrightarrow & XM^* & \text{initiation} \\ XM_n^* + M & \longrightarrow & XM_{n+1}^* & \text{propagation} \\ XM_n^* + M_m^* X & \longrightarrow & XM_{n+m}X & \text{termination} \end{array}$$

Monomers (M) react with free radicals (X^*) derived from the initiator (X–X) yielding a reactive radical-monomer, (XM^*). The product can then combine with another monomer, a reaction that can then repeat itself many times, comprising the chain-growth or propagation phase of the reaction. Finally, the growth is terminated when two chains combine to yield a polymer chain.

The mechanism that leads to particle formation is based on the relative solubilities of the reacting species: The initiator is very strongly soluble, dissolving to form radicals in the solution. The monomers and polymer chains are weakly

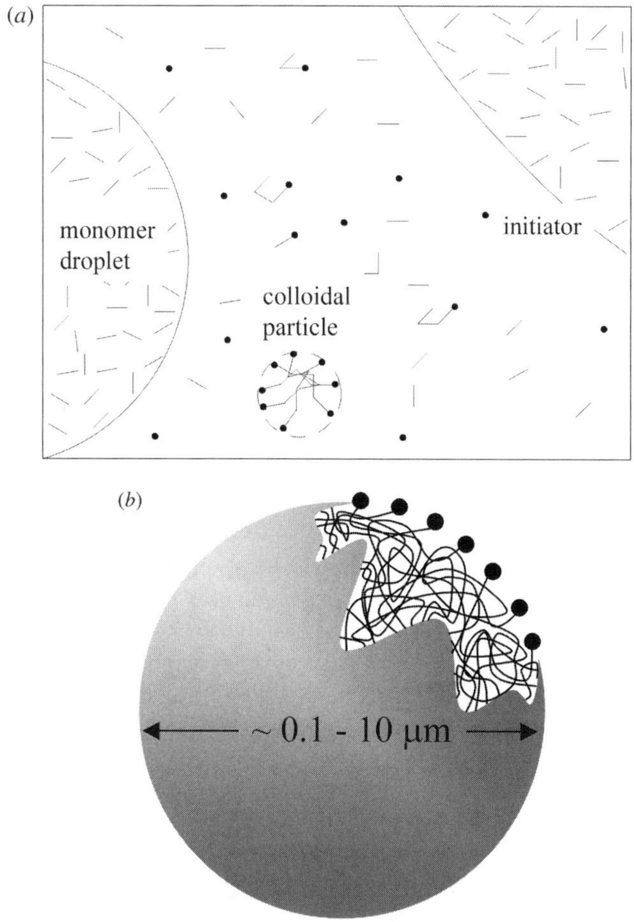

Figure 6.64 (a) Schematic picture of the emulsion polymerization reaction. Low-solubility hydrophobic monomers are largely confined to droplets, although a small amount is present in the water, assisted by a surfactant (not shown). Highly soluble initiators react with monomers to form head-tail structures, which aggregate into spherical colloidal particle precursors called micelles. Polymer chains grow within the micelle, swelling the particle until the reaction terminates with the joining of two chains. (b) Final structure of the polymeric colloidal particle. Highly entangled chains comprise the interior, while initiator-derived end groups reside on the surface. After Ref. 158.

soluble, with solubility decreasing with polymer chain length. This means that when the radicals combine with the monomers, the result is a structure with a hydrophilic head (the radical) and a hydrophobic tail (the monomer). Such head-tail structures naturally self-organize into spherical *micelles* possessing charged end groups on their surface and monomers inside. Monomers diffuse into the micelles during the polymer growth phase, and the micelles swell while the polymer chains entangle inside.

By carefully controlling the experimental parameters, such as the relative concentrations of the reacting species and temperature, the growth of polystyrene spherical particles can be tailored to produce particles with diameters that range from less than 100 nm to several tens of microns. The monomer is typically styrene, which of course yields polystyrene particles, but many other possibilities are available, such as methyl methacrylate, which yields PMMA (poly-methyl methacrylate, a.k.a. plexiglass). A common choice for the initiator is $K_2S_2O_8$, which leads to sulfate ($-SO_4$) surface groups. Other choices can produce particles with different surface groups and charge densities. Depending on the self-assembly method being used, different charge densities can be useful to tailor the interparticle interactions and optimize photonic-crystal fabrication.

6.5.1.2 Considerations for PhC Fabrication. Owing to their usefulness in biological and chemical applications, which arises in large part from their uniform sizes and flexible surface functionalization, colloidal suspensions of both polystyrene and silica particles are available from several commercial suppliers. This situation has been of great benefit to researchers in the area of photonic crystals, many of whom come from backgrounds in physics and electrical engineering and thus have little expertise in the realm of chemical synthesis of monodisperse microparticles.

Silica particles are more dense, more rigid, and more chemically inert, and they can withstand higher temperatures than those formed by entangled polymer chains. These differences have an impact on the kinds of crystallization and postcrystallization processes that have been implemented to fabricate photonic crystals with optimal optical properties.

6.5.2 Colloidal Crystallization

As we have just discussed, a variety of chemical surface functionalities can be present after the completion of the colloidal particle synthesis process. Likewise, there exists a variety of dynamic interactions that can lead to crystallization. We briefly consider a few basic interactions below, following the notation and discussion of Pieranski [158].

The simplest model that can be considered is a macroscopic hard-sphere potential, in which neutral particles interact only when they are in direct mechanical contact, i.e., when the center-to-center distance between particles is equal to twice their radius. Such a system can exist in either a solid phase with long-range order, or a disordered liquid phase. The system can be completely characterized in terms of only two dimensionless parameters: (1) the volume fraction, $\Phi = nv$, where n is the concentration of particles in the suspension, and $v = (4/3)\pi a^3$ is the volume of a single particle with radius a; and (2) the reduced pressure $p = P/n_{cp}kT$, where P is the absolute pressure, $n_{cp} = \Phi_{cp}/v$ is the close-packed concentration (the close-packed volume fraction is $\Phi_{cp} = \sqrt{2}(\pi/6) \sim 0.74$), k is the Boltzmann constant, and T is the absolute temperature. In other words, p is the pressure in units of a characteristic scale pressure P_0 obtained from an ideal-gas-law approximation, $P_{ideal}(n) = nkT$, when $n = n_{cp}$: $p = P/P_0 = P/P_{ideal}(n_{cp})$.

Numerical simulations show that this system undergoes a phase transition in the range $0.667\Phi_{\rm cp} < \Phi < 0.736\Phi_{\rm cp}$. Below this range is the liquid phase, above it is the solid, and within it both phases coexist [181]. The phase transition occurs at $p = p_{\rm c} \approx 8.27$. The equation of state is therefore

$$\Phi = \begin{cases} \Phi^{\rm L}(p), & 0 < p < p_c \\ \Phi^{\rm S}(p), & p > p_c \end{cases}, \tag{6.44}$$

where a jump discontinuity exists between the liquid (L) and solid (S) curves at $p_{\rm c}$. (Note that "liquid" here refers to the liquid phase of the colloid, not the host liquid component of the colloidal suspension.) These curves are determined numerically by Monte-Carlo simulations.

Based on the results of this hard-sphere theory, the simplest apparent method to obtain crystals of colloidal particles is to increase the particle concentration until $\Phi > 0.5$. However, the reaction conditions required for monodisperse synthesis yield volume fractions that are typically below 0.1; hence, crystallization methods amount to postsynthesis densification of the particle suspensions. Several such methods are discussed in later sections, with emphasis on their use for fabricating photonic crystals, including: sedimentation under gravity or centrifuge, electrophoresis, evaporation of the host fluid, and filtration.

Sedimentation Dynamics. A monodisperse colloidal suspension of neutral hard sphere particles will condense until sedimentation equilibrium is reached, provided that the particles' density is different from that of the host liquid. To determine the conditions of this equilibrium, consider a column of the suspension, with constant cross-sectional area A. A volume element of the column $dV = Adz$ is suspended against gravity by a pressure differential

$$\begin{aligned} dP &= P(z + dz) - P(z) \\ &= A^{-1}(\bar{\rho} n v \, dV) g \\ &= -\bar{\rho} g \Phi \, dz, \end{aligned} \tag{6.45}$$

where $\bar{\rho}$ is the relative mass density of the particles: $\bar{\rho} = \rho_{\rm part} - \rho_{\rm liq}$, and g is the acceleration due to gravity, 9.8 m/s^2. This leads to the following differential equation for the pressure:

$$\frac{dP}{dz} = -\bar{\rho} g \Phi(P), \tag{6.46}$$

involving the equation of state $\Phi(P)$. This can be rewritten in terms of dimensionless quantities as

$$\frac{dp}{d\zeta} = -\tilde{\Phi}(P), \tag{6.47}$$

where $\tilde{\Phi} = \Phi/\Phi_{cp}$ is the volume fraction in units of close packing, and $\zeta = z/h_0$, with $h_0 = kT/\mu g$ the scale height at temperature T, and $\mu = \bar{\rho}v$ the relative mass of the particles with volume v. This differential equation can be integrated numerically from the equation of state, Eq. (6.44).

Electrostatic Interaction Dynamics. The previous discussions of the methods of synthesizing monodisperse particle suspensions indicated that these methods may produce a surface covered with functional groups, such as sulfate or hydroxide groups. These groups can dissociate in polar solvents such as water, which leads to a net charge on the particles and adds an equivalent but opposite total charge to the solution in the form of dissolved counterions. This situation is in fact the rule rather than the exception, so the neutral hard-sphere model presented above must be regarded as a useful but imprecise approximation to all but the most carefully designed suspensions.

In any solvent, the number of surface groups that dissociate, and hence the net charge on a particle, will be dependent on many parameters, which include the strength and purity of the solvent, the number of surface groups per particle, the temperature, and the type and concentration of ions from other sources, whether intentional or the result of contamination. All these various charges in the solution, from whatever source, interact via long-range Coulomb forces.

6.5.3 Self-Assembly Methods

Although the dynamics of crystallization of colloidal particle suspensions have been studied for nearly a century, only somewhat recently has the application of such crystalline arrays to fabrication of 3D photonic crystals been considered, owing to the relative youth of the field of photonic crystals itself. Hence, the following discussion is limited to select recent works and those performed specifically (but not exclusively) toward the goal of PhC fabrication.

As described earlier in the discussion of the simple theory of colloidal crystal sedimentation dynamics (Section 6.5.2), the simplest way to obtain a self-assembled PhC lattice from a monodisperse particle suspension is simply to wait while gravity, owing to the slight difference in density between the particles and the surrounding fluid, drives the sedimentation. In practice, this method is extremely time consuming and provides virtually no control over the crystalline structure or orientation. Therefore, numerous experimental methods have been devised to speed things up and/or afford varying degrees of control over the process. Several examples of such methods are discussed next.

6.5.3.1 Electrophoretic Assembly. Holgado et al. [182] developed a method called electrophoretic assembly, in which they arranged their cylinder of suspended particles between two conuducting plates and applied an electric field across the column. Silica particles larger than ~500 nm in diameter tend to sediment too quickly, solidifying in a disordered array, whereas particles smaller than 300 nm sediment too slowly to be practical for many applications. Because the silica particles possess a

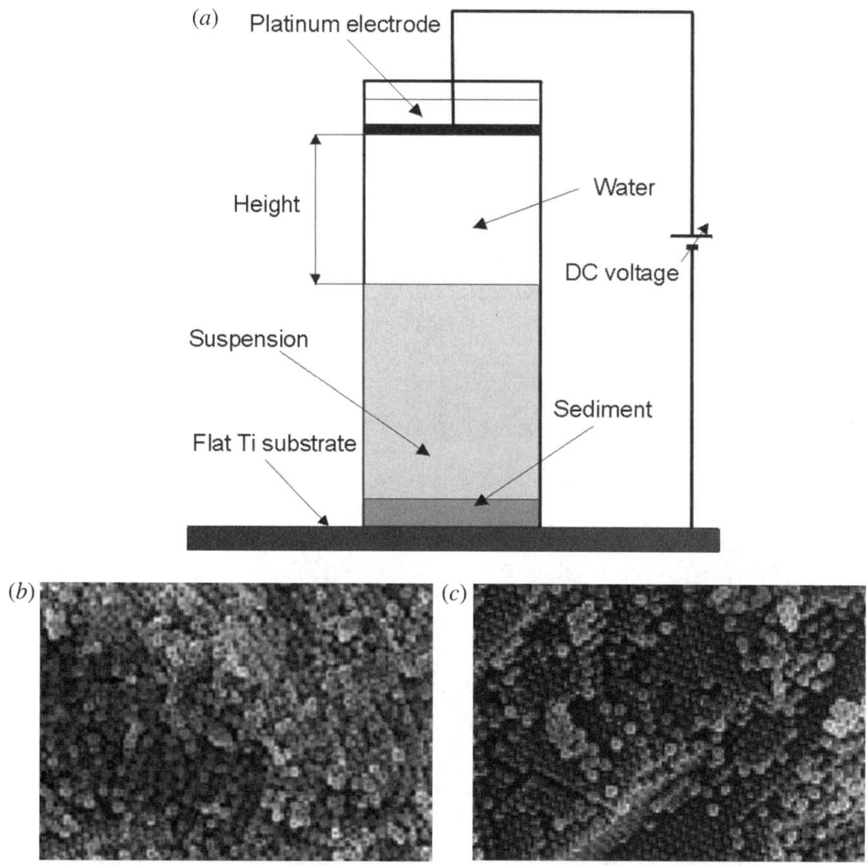

Figure 6.65 Electrophoretic assembly for improving sedimentation of silica particles. (a) Experimental setup. (b), (c) SEM images of samples fabricated with and without applied electric field, respectively. In both samples, the particle diameter is 870 nm. (Reprinted with permission from Ref. 182, ©, 1999, American Chemical Society.)

weak surface-charge density, the sedimentation rate can be increased or reduced by modulating the applied electric field. Figure 6.65b, c shows an example of the improved results they obtained with this method.

6.5.3.2 Vertical Meniscus Deposition/Convective Assembly. Convective assembly turns sedimentation upside down, so to speak [183]. Rather than use gravity to assemble the array at the bottom of a suspension, convective assembly takes advantage of the increased rate of evaporation of the suspending fluid that occurs in a meniscus formed at the interface of a partially submerged host-fluid-philic substrate. The evaporation rate can be enhanced by using a high vapor-pressure fluid such as ethanol instead of water, and by heating the suspension. Furthermore, heating can be used to generate and maintain a thermal gradient, which leads to large-scale convective flow

(a)

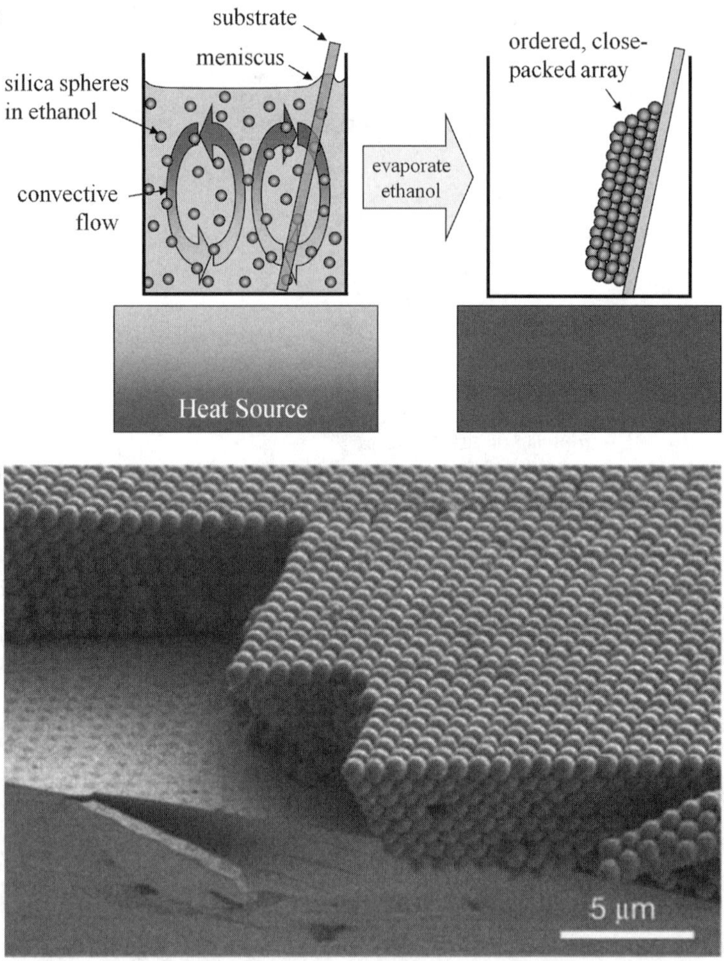

(b)

Figure 6.66 (a) Schematic of convective assembly for self-assembled colloidal crystals. (b) SEM image of sample fabricated by Vlasov et al. [184]. (Adapted by permission from Macmillan Publishers Ltd., Ref. 184, © 2001.)

that can be used to enhance the rate of particle deposition onto the substrate. The effects of evaporation and convection, independently or in tandem, lead to a net flow of particles, carried by the fluid flow, toward the meniscus. There, the particles are deposited in ordered layers and the fluid evaporates. This process is depicted schematically in Fig. 6.66a. By controlling the rate of evaporation and/or gradually withdrawing the substrate from the container, a precisely controllable number of layers can be grown with crystalline ordering superior to many other methods.

Impressive results were obtained using convective assembly by Vlasov et al. [184] in work reported in 2001. These results are depicted in Fig. 6.66b. A similar method was demonstrated in Ref. 185. The dynamics of convective assembly for particle monolayers was studied in Ref. 186.

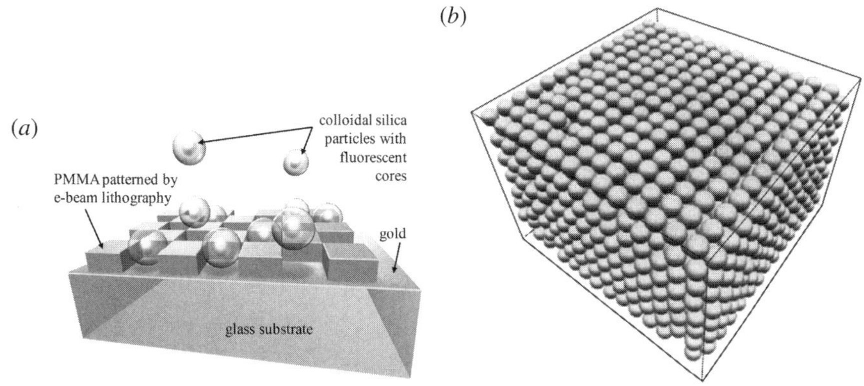

Figure 6.67 (a) Colloidal epitaxy illustrated schematically. The glass substrate was patterned by e-beam lithography to seed the growth of the first layer. Fluorescent spheres were used so that confocal fluorescence microscopy could be used to image the completed sample layer-by-layer through the depth, verifying the crystalline ordering. A computer was used to reconstruct the 3D structure, depicted in (b). (Adapted by permission from, Macmillan Publisher Ltd., Ref. 187, © 1997.)

6.5.3.3 Template-Directed Assembly—Colloidal Epitaxy. As a means to obtain some control over the orientation and crystalline perfection of sedimented colloidal photonic crystals, vanBlaaderen et al. [187] demonstrated that a substrate prepatterened with a template of preferred binding locations can be used to seed the epitaxial growth of colloidal crystals. They patterned a substrate with an array of holes in a PMMA film by electron-beam lithography, which allowed precise control of the pattern dimensions to match the particle size (Fig. 6.67). By allowing the bottom layer of 1.0-μm particles the opportunity to reduce their gravitational energy by falling into the 0.5-μm-deep holes in the PMMA layer, the patterned substrate was shown to induce the ordering of the first layer of the structure, which thereby caused growth along a known crystalline axis, and with a known orientation over a controlled domain area. Thus, not only is the orientation of the crystal structure controlled, but the likelihood of structural defects is reduced as well. Specifically, epitaxial crystal growth was demonstrated along the FCC [100] direction, when growth on unpatterned substrates almost always occurs along the close-packing [111] direction. This distinction is significant for PhCs because unpatterned growth of close-packed planes suffers from the potential for stacking faults, since neither of the two close-packed structures (FCC or HCP) is strongly thermodynamically preferred [188–191]. Only the FCC structure is known to possess the potential for a photonic bandgap; templating for directed growth along [100] ensures that the FCC structure will result.

6.5.3.4 Packing Cells. Another means of acceleration the condensation of a monodisperse particle suspension was demonstrated by Xia et al. [192]. They constructed a simple flow cell, into which a suspension of particles was flowed under pressure. Their flow cell is depicted schematically in Fig. 6.68a. The sidewalls are composed of

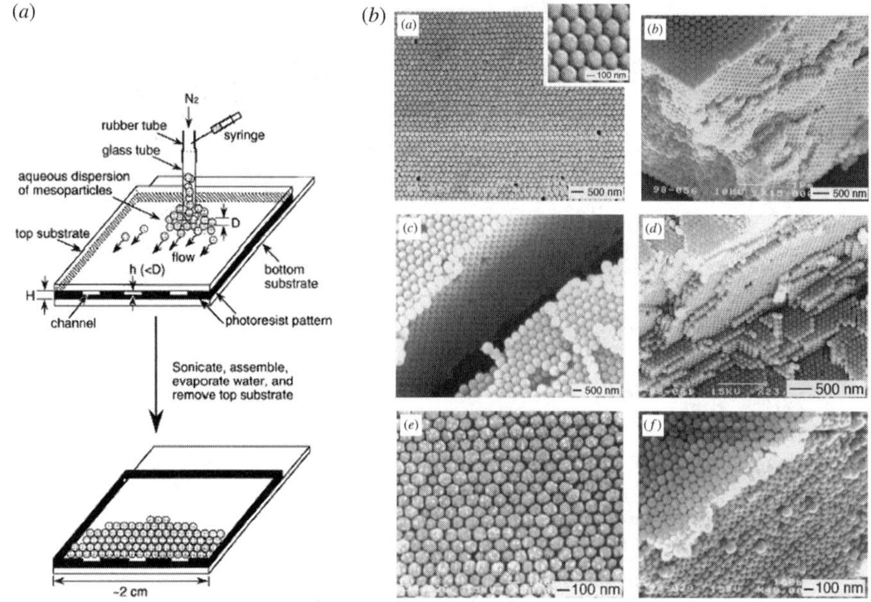

Figure 6.68 (a) Schematic illustration of the fluid packing cell designed and implemented by Xia et al. (b) Examples of colloidal PhC structures fabricated using the packing cell. Note the large volume of the ordered arrays, which were obtained rapidly using this method. (Reprinted with permission from Ref. 192, © 1999, IEEE.)

lithographically patterned photoresist; drain channels were present in the sidewalls on one side that allowed the host fluid to pass through, but retained the colloidal particles. The directional flow of host fluid provides simultaneous condensation and annealing energy, and it results in well-ordered single close-packed domains over very large areas ($\sim 1\ \text{cm}^2$), using a range of particle sizes and compositions, as shown in Fig. 6.68b.

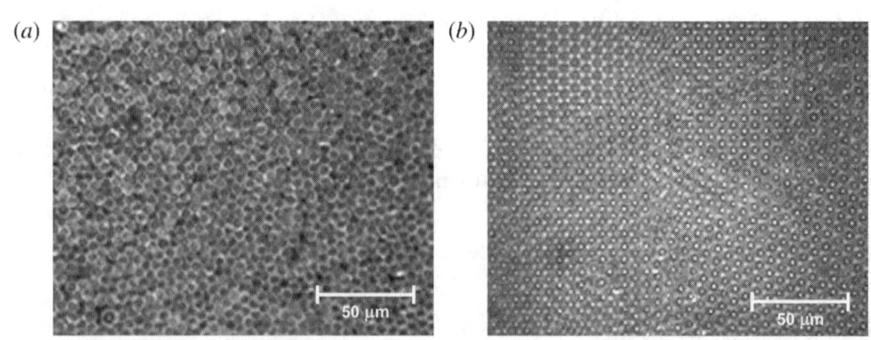

Figure 6.69 (a) Optical micrograph of an opal sample prepared using the centrifuge step only, omitting the spin-annealing step, where the microspheres show a high level of disorder, and (b) optical micrograph of opal sample after spin annealing. The spinning results in the formation of large, ordered domains.

Table 6.1 Sample Spinning Parameters Used for Spin-Annealing Colloidal Photonic Crystal Templates

Stage	Spin Speed (rpm)	Spin Time (s)
1	150	30
2	250	30
3	300	30
4	400	30

Note: This recipe is representative only; precise adherence to the stated speeds and times is not critical.

6.5.3.5 Spin-Annealing. Sedimentation can be accelerated significantly (i.e., from months to minutes) by centrifugation. However, such rapid sedimentation generally yields poorly ordered polycrystalline or amorphous lattices, because the effective g-force in a centrifuge ($\sim 10^3 - 10^4 g$) means that thermal energy $\sim kT$ is not sufficient to anneal the sedimented particles once they arrive on the solid-phase interface. One method to anneal them after condensation is by spinning the substrate [193].

The process begins with an aqueous solution of monodisperse polystyrene (ps) microspheres. The solution is spun in a centrifuge to create sedimentation at the bottom of the tube. Once extracted from the tube, the concentrated sample is placed on a spin-coating machine, whereupon the actual crystal lattice, or opal structure, is formed. The ordering arises from shear annealing [194] due to the flow of the sample during the spinning process rather than from the centrifuge step, which only serves to concentrate the solution, see Fig. 6.69. After the spinning step, the opal structure is dried and (optionally) sintered in an oven, and finally back-filled with UV-curable polymer precursors. Once cured, the opal structure is immersed in a solution that selectively dissolves the PS microspheres, thereby producing an inverse-opal structure.

By controlling the spin speed and time (see Table 6.1), we obtained a very uniform distribution of PS beads on the glass substrate; a typical sample size after spinning is $\sim 1\ cm^2$, with a sample thickness of a few hundred micrometers near the center, which tapers to less than 100 μm at the outer edges. The final thickness is determined by the spin speed and the viscosity of the centrifuged solution, and it is difficult to control precisely.

REFERENCES

1. F. R. Flory, "Thin Films for Optical Systems," *Optical Engineering*, vol. 49, 1995, New York: Marcel Dekker, Inc.
2. F. Zolla, G. Renversez, A. Nicolet, B. Kuhlmey, S. Guenneau, and D. Felbacq, *Foundations of Photonic Crystal Fibres*. London: Imperial College Press, 2005.
3. A. Bjarklev, J. Broeng, and A. S. Bjarklev, *Photonic Crystal Fibers*. New York: Springer, 2003.

4. A. Chutinan, S. John, and O. Toader, "Diffractionless Flow of Light in All-Optical Microchips," *Phys. Rev. Lett.*, vol. 90, pp. 123901, 2003.
5. M. J. Madou, *Fundamentals of Microfabrication: The Science of Miniaturization*, 2nd ed. Boca Raton, FL: CRC Press, 2002.
6. K. A. Valiev, *The Physics of Submicron Lithography*. New York. Plenum Press, 1992.
7. M. Kohler, *Etching in Microsystem Technology*. Weinheim: Wiley-VCH Verlag, 1999.
8. R. Ferrini, B. Lombardet, B. Wild, R. Houdre, and G. H. Duan, "Hole Depth- and Shape-Induced Radiation Losses in Two-Dimensional Photonic Crystals," *Appl. Phys. Lett.*, vol. 82, pp. 1009–1011, 2003.
9. U. Grüning, V. Lehmann, and C. M. Engelhardt, "2-Dimensional Infrared Photonic Band-Gap Structure-Based on Porous Silicon," *Appl. Phys. Lett.*, vol. 66, pp. 3254–3256, 1995.
10. U. Grüning, V. Lehmann, S. Ottow, and K. Busch, "Macroporous Silicon with a Complete Two-Dimensional Photonic Band Gap Centered at 5 μm," *Appl. Phys. Lett.*, vol. 68, pp. 747–749, 1996.
11. U. Grüning and V. Lehmann, "Two-Dimensional Infrared Photonic Crystal Based on Macroporous Silicon," *Thin Solid Films*, vol. 276, pp. 151–154, 1996.
12. A. Birner, U. Grüning, S. Ottow, A. Schneider, F. Müller, V. Lehmann, H. Foll, and U. Gösele, "Macroporous Silicon: A Two-Dimensional Photonic Bandgap Material Suitable for the Near-Infrared Spectral Range," *Phys. Status Solidi A*, vol. 165, pp. 111–117, 1998.
13. S. W. Leonard, H. M. van Driel, K. Busch, S. John, A. Birner, A. P. Li, F. Müller, U. Gösele, and V. Lehmann, "Attenuation of Optical Transmission within the Band Gap of Thin Two-Dimensional Macroporous Silicon Photonic Crystals," *Appl. Phys. Lett.*, vol. 75, pp. 3063–3065, 1999.
14. S. Rowson, A. Chelnokov, and J. M. Lourtioz, "Two-Dimensional Photonic Crystals in Macroporous Silicon: From Mid-Infrared (10 μm) to Telecommunication Wavelengths (1.3–1.5 μm)," *J. Lightwave Technol.*, vol. 17, pp. 1989–1995, 1999.
15. F. Müller, A. Birner, U. Gösele, V. Lehmann, S. Ottow, and H. Foll, "Structuring of Macroporous Silicon for Applications as Photonic Crystals," *J. Porous Mater.*, vol. 7, pp. 201–204, 2000.
16. A. Birner, R. B. Wehrspohn, U. M. Gösele, and K. Busch, "Silicon-Based Photonic Crystals," *Adv. Mat.*, vol. 13, pp. 377–388, 2001.
17. P. Bettotti, L. Dal Negro, Z. Gaburro, L. Pavesi, A. Lui, M. Galli, M. Patrini, and F. Marabelli, "P-Type Macroporous Silicon for Two-Dimensional Photonic Crystals," *J. Appl. Phys.*, vol. 92, pp. 6966–6972, 2002.
18. E. V. Astrova, T. N. Borovinskaya, V. A. Tolmachev, and T. S. Perova, "Technique for Patterning Macroporous Silicon and the Fabrication of Bars of 2D Photonic Crystals with Vertical Walls," *Semiconductors*, vol. 38, pp. 1088–1091, 2004.
19. M. Haurylau, A. R. Shroff, and P. M. Fauchet, "Optical Properties and Tunability of Macroporous Silicon 2-D Photonic Bandgap Structures," *Phys. Status Solidi A*, vol. 202, pp. 1477–1481, 2005.
20. T. Baba and T. Matsuzaki, "Polarization Changes in Spontaneous Emission from GaInAsP/InP 2-Dimensional Photonic Crystals," *Electron. Lett.*, vol. 31, pp. 1776–1778, 1995.

21. T. Krauss, Y. P. Song, S. Thoms, C. D. W. Wilkinson, and R. M. Delarue, "Fabrication of 2-D Photonic Bandgap Structures in GaAs/AlGaAs," *Electron. Lett.*, vol. 30, pp. 1444–1446, 1994.

22. T. F. Krauss, R. M. DeLaRue, and S. Brand, "Two-Dimensional Photonic-Bandgap Structures Operating at Near Infrared Wavelengths," *Nature*, vol. 383, pp. 699–702, 1996.

23. J. O'Brien, O. Painter, R. Lee, C. C. Cheng, A. Yariv, and A. Scherer, "Lasers Incorporating 2D Photonic Bandgap Mirrors," *Electron. Lett.*, vol. 32, pp. 2243–2244, 1996.

24. D. Labilloy, H. Benisty, C. Weisbuch, T. F. Krauss, D. Cassagne, C. Jouanin, R. Houdre, U. Oesterle, and V. Bardinal, "Diffraction Efficiency and Guided Light Control by Two-Dimensional Photonic-Bandgap Lattices," *IEEE J. Quantum Electron.*, vol. 35, pp. 1045–1052, 1999.

25. D. Labilloy, H. Benisty, C. Weisbuch, C. J. M. Smith, T. F. Krauss, R. Houdre, and U. Oesterle, "Finely Resolved Transmission Spectra and Band Structure of Two-Dimensional Photonic Crystals Using Emission from InAs Quantum Dots," *Phys. Rev. B*, vol. 59, pp. 1649–1652, 1999.

26. C. J. M. Smith, H. Benisty, D. Labilloy, U. Oesterle, R. Houdre, T. F. Krauss, R. M. De la Rue, and C. Weisbuch, "Near-Infrared Microcavities Confined by Two-Dimensional Photonic Bandgap Crystals," *Electron. Lett.*, vol. 35, pp. 228–230, 1999.

27. R. Ferrini, D. Leuenberger, M. Mulot, M. Qiu, J. Moosburger, M. Kamp, A. Forchel, S. Anand, and R. Houdre, "Optical Study of Two-Dimensional InP-Based Photonic Crystals by Internal Light Source Technique," *IEEE J. Quantum Electron.*, vol. 38, pp. 786–799, 2002.

28. F. Pommereau, L. Legouezigou, S. Hubert, S. Sainson, J. P. Chandouineau, S. Fabre, G. H. Duan, B. Lombardet, R. Ferrini, and R. Houdre, "Fabrication of Low Loss Two-Dimensional InP Photonic Crystals by Inductively Coupled Plasma Etching," *J. Appl. Phys.*, vol. 95, pp. 2242–2245, 2004.

29. M. Mulot, S. Anand, R. Ferrini, B. Wild, R. Houdre, J. Moosburger, and A. Forchel, "Fabrication of Two-Dimensional InP-Based Photonic Crystals by Chlorine Based Chemically Assisted Ion Beam Etching," *J. Vac. Sci. Technol. B*, vol. 22, pp. 707–709, 2004.

30. M. Mulot, S. Anand, C. F. Carlstrom, M. Swillo, and A. Taineau, "Dry Etching of Photonic Crystals in InP Based Materials," *Physica Scripta*, vol. T101, pp. 106–109, 2002.

31. M. V. Kotlyar, L. O'Faolain, R. Wilson, and T. F. Krauss, "High-Aspect-Ratio Chemically Assisted Ion-Beam Etching for Photonic Crystals Using a High Beam Voltage-Current Ratio," *J. Vac. Sci. Technol. B*, vol. 22, pp. 1788–1791, 2004.

32. M. V. Kotlyar, T. Karle, M. D. Settle, L. O'Faolain, and T. F. Krauss, "Low-Loss Photonic Crystal Defect Waveguides in InP," *Appl. Phys. Lett.*, vol. 84, pp. 3588–3590, 2004.

33. S. E. Hicks, S. K. Murad, I. Sturrock, and C. D. W. Wilkinson, "Improving the Resistance of PECVD Silicon Nitride to Dry Etching Using an Oxygen Plasma," *Microelectron. Eng.*, vol. 35, pp. 41–44, 1997.

34. C. C. Cheng, A. Scherer, V. Arbet-Engels, and E. Yablonovitch, "Lithographic Band Gap Tuning in Photonic Band Gap Crystals," *J. Vac. Sci. Technol. B*, vol. 14, pp. 4110–4114, 1996.

35. M. Mulot, S. Anand, M. Swillo, M. Qiu, B. Jaskorzynska, and A. Talneau, "Low-Loss InP-Based Photonic-Crystal Waveguides Etched with Ar/Cl_2 Chemically Assisted Ion Beam Etching," *J. Vac. Sci. Technol. B*, vol. 21, pp. 900–903, 2003.

36. T. D. Happ, A. Markard, M. Kamp, J. L. Gentner, and A. Forchel, "InP-Based Short Cavity Lasers with 2D Photonic Crystal Mirror," *Electron. Lett.*, vol. 37, pp. 428–429, 2001.

37. J. Zimmermann, H. Scherer, M. Kamp, S. Deubert, J. P. Reithmaier, A. Forchel, R. Marz, and S. Anand, "Photonic Crystal Waveguides with Propagation Losses in the 1 dB/mm Range," *J. Vac. Sci. Technol. B*, vol. 22, pp. 3356–3358, 2004.

38. T. D. Happ, M. Kamp, F. Klopf, J. P. Reithmaier, and A. Forchel, "Two-Dimensional Photonic Crystal Laser Mirrors," *Semiconductor Sci. Technol.*, vol. 16, pp. 227–232, 2001.

39. T. D. Happ, A. Markard, M. Kamp, A. Forchel, S. Anand, J. L. Gentner, and N. Bouadma, "Nanofabrication of Two-Dimensional Photonic Crystal Mirrors for 1.5 μm Short Cavity Lasers," *J. Vac. Sci. Technol. B*, vol. 19, pp. 2775–2778, 2001.

40. L. Raffaele, R. M. De La Rue, and T. F. Krauss, "Ultrashort In-Plane Semiconductor Microlasers with High-Reflectivity Microstructured Mirrors," *Opt. Quantum Electron.*, vol. 34, pp. 101–111, 2002.

41. L. Raffaele, R. M. De La Rue, J. S. Roberts, and T. E. Krauss, "Edge-Emitting Semiconductor Microlasers with Ultrashort-Cavity and Dry-Etched High-Reflectivity Photonic Microstructure Mirrors," *IEEE Photonics Technol. Lett.*, vol. 13, pp. 176–178, 2001.

42. C. F. Carlstrom, R. van der Heijden, F. Karouta, R. W. van der Heijden, H. W. M. Salemink, and E. van der Drift, "Cl_2/O_2-Inductively Coupled Plasma Etching of Deep Hole-Type Photonic Crystals in InP," *J. Vac. Sci. Technol. B*, vol. 24, pp. L6–L9, 2006.

43. T. F. Krauss, C. J. M. Smith, B. Vogele, S. K. Murad, C. D. W. Wilkinson, R. S. Grant, M. G. Burt, and R. M. DeLaRue, "Two-Dimensional Waveguide Based Photonic Microstructures in GaAs and InP," *Microelectron. Eng.*, vol. 35, pp. 29–32, 1997.

44. R. Wuest, P. Strasser, F. Robin, D. Erni, and H. Jackel, "Fabrication of a Hard Mask for InP Based Photonic Crystals: Increasing the Plasma-Etch Selectivity of Poly(Methyl Methacrylate) Versus SiO_2 and SiNx," *J. Vac. Sci. Technol. B*, vol. 23, pp. 3197–3201, 2005.

45. R. Ferrini, B. Lombardet, B. Wild, R. Houdre, S. Olivier, H. Benisty, A. Djoudi, L. Legouezigou, S. Hubert, S. Sainson, J. P. Chandouineau, S. Fabre, F. Pommereau, and G. H. Duan, "Optical Characterisation of 2D InP-Based Photonic Crystals Fabricated by Inductively Coupled Plasma Etching," *Electron. Lett.*, vol. 38, pp. 962–964, 2002.

46. S. K. Murad, N. I. Cameron, S. P. Beaumont, and C. D. W. Wilkinson, "Effects of O_2 Addition to $SiCl_4/SiF_4$ and the Thickness of the Capping Layer on Gate Recess Etching of GaAs-Pseudomorphic High Electron Mobility Transistors," *J. Vac. Sci. Technol. B*, vol. 14, pp. 3668–3673, 1996.

47. S. G. Johnson and J. D. Joannopoulos, "Three-Dimensionally Periodic Dielectric Layered Structure with Omnidirectional Photonic Band Gap," *Appl. Phys. Lett.*, vol. 77, pp. 3490–3492, 2000.

48. M. L. Povinelli, S. G. Johnson, S. H. Fan, and J. D. Joannopoulos, "Emulation of Two-Dimensional Photonic Crystal Defect Modes in a Photonic Crystal with a Three-Dimensional Photonic Band Gap," *Phys. Rev. B*, vol. 6407, art. no. 075313, 2001.

49. T. Mogami, M. Morimoto, H. Okabayashi, and E. Nagasawa, "SiO_2 Planarization by 2-Step Rf Bias-Sputtering," *J. Vac. Sci. Technol. B*, vol. 3, pp. 857–861, 1985.
50. M. H. Qi and H. I. Smith, "Achieving Nanometer-Scale, Controllable Pattern Shifts in X-Ray Lithography Using an Assembly-Tilting Technique," *J. Vac. Sci. Technol. B*, vol. 20, pp. 2991–2994, 2002.
51. E. E. Moon, J. Lee, P. N. Everett, and H. I. Smith, "Application of Interferometric Broadband Imaging Alignment on an Experimental X-Ray Stepper," *J. Vac. Sci. Technol. B*, vol. 16, pp. 3631–3636, 1998.
52. D. Roundy and J. Joannopoulos, "Photonic Crystal Structure with Square Symmetry Within Each Layer and A Three-Dimensional Band Gap," *Appl. Phys. Lett.*, vol. 82, pp. 3835–3837, 2003.
53. S. G. Johnson, C. Manolatou, S. H. Fan, P. R. Villeneuve, J. D. Joannopoulos, and H. A. Haus, "Elimination of Cross Talk in Waveguide Intersections," *Opt. Lett.*, vol. 23, pp. 1855–1857, 1998.
54. E. Özbay, A. Abeyta, G. Tuttle, M. Tringides, R. Biswas, C. T. Chan, C. M. Soukoulis, and K. M. Ho, "Measurement of a 3-Dimensional Photonic Band-Gap in a Crystal-Structure Made of Dielectric Rods," *Phys. Rev. B*, vol. 50, pp. 1945–1948, 1994.
55. E. Özbay, E. Michel, G. Tuttle, R. Biswas, M. M. Sigalas, and K. M. Ho, "Micromachined Millimeter-Wave Photonic Band-Gap Crystals," *Appl. Phys. Lett.*, vol. 64, pp. 2059–2061, 1994.
56. E. Özbay, E. Michel, G. Tuttle, R. Biswas, K. M. Ho, J. Bostak, and D. M. Bloom, "Teraherz Spectroscopy of 3-Dimensional Photonic Band-Gap Crystals," *Opt. Lett.*, vol. 19, pp. 1155–1157, 1994.
57. K. Aoki, H. T. Miyazaki, H. Hirayama, K. Inoshita, T. Baba, N. Shinya, and Y. Aoyagi, "Three-Dimensional Photonic Crystals for Optical Wavelengths Assembled by Micromanipulation," *Appl. Phys. Lett.*, vol. 81, pp. 3122–3124, 2002.
58. H. T. Miyazaki, H. Miyazaki, K. Ohtaka, and T. Sato, "Photonic Band in Two-Dimensional Lattices Micrometer-Sized Spheres Mechanically Arranged Under a Scanning Electron Microscope," *J. Appl. Phys.*, vol. 87, pp. 7152–7158, 2000.
59. S. Y. Lin, J. G. Fleming, D. L. Hetherington, B. K. Smith, R. Biswas, K. M. Ho, M. M. Sigalas, W. Zubrzycki, S. R. Kurtz, and J. Bur, "A Three-Dimensional Photonic Crystal Operating at Infrared Wavelengths," *Nature*, vol. 394, pp. 251–253, 1998.
60. J. G. Fleming and S. Y. Lin, "Three-Dimensional Photonic Crystal with a Stop Band from 1.35 to 1.95 mu m," *Opt. Lett.*, vol. 24, pp. 49–51, 1999.
61. S. Y. Lin and J. G. Fleming, "A Three-Dimensional Optical Photonic Crystal," *J. Lightwave Technol.*, vol. 17, pp. 1944–1947, 1999.
62. S. Y. Lin, J. G. Fleming, M. M. Sigalas, R. Biswas, and K. M. Ho, "Photonic Band-Gap Microcavities in Three Dimensions," *Phys. Rev. B*, vol. 59, pp. R15579–R15582, 1999.
63. S. Y. Lin, J. G. Fleming, R. Lin, M. M. Sigalas, R. Biswas, and K. M. Ho, "Complete Three-Dimensional Photonic Bandgap in a Simple Cubic Structure," *J. Opt. Soc. Am. B*, vol. 18, pp. 32–35, 2001.
64. S. Y. Lin, J. G. Fleming, and E. Chow, "Two- and Three-Dimensional Photonic Crystals Built with VLSI Tools," *Mrs Bull.*, vol. 26, pp. 627–631, 2001.

65. W. J. Patrick, W. L. Guthrie, C. L. Standley, and P. M. Schiable, "Application of Chemical Mechanical Polishing to the Fabrication of VLSI Circuit Interconnections," *J. Electrochem. Soc.*, vol. 138, pp. 1778–1784, 1991.
66. S. Noda, N. Yamamoto, and A. Sasaki, "New Realization Method for Three-Dimensional Photonic Crystal in Optical Wavelength Region," *Jpn. J. Appl. Phys. Part 2*, vol. 35, pp. L909–L912, 1996.
67. N. Yamamoto, S. Noda, and A. Chutinan, "Development of One Period of a Three-Dimensional Photonic Crystal in the 5–10 μm Wavelength Region by Wafer Fusion and Laser Beam Diffraction Pattern Observation Techniques," *Jpn. J. Appl. Phys. Part 2*, vol. 37, pp. L1052–L1054, 1998.
68. N. Yamamoto, S. Noda, and A. Sasaki, "New Realization Method for Three-Dimensional Photonic Crystal in the Optical Wavelength Region: Experimental Consideration," *Jpn. J. Appl. Phys. Part 1*, vol. 36, pp. 1907–1911, 1997.
69. N. Yamamoto and S. Noda, "100-nm-Scale Alignment Using Laser Beam Diffraction Pattern Observation Techniques and Wafer Fusion for Realizing Three-Dimensional Photonic Crystal Structure," *Jpn. J. Appl. Phys. Part 1*, vol. 37, pp. 3334–3338, 1998.
70. N. Yamamoto and S. Noda, "Fabrication and Optical Properties of One Period of a Three-Dimensional Photonic Crystal Operating in the 5–10 μm Wavelength Region," *Jpn. J. Appl. Phys. Part 1*, vol. 38, pp. 1282–1285, 1999.
71. S. Noda, N. Yamamoto, H. Kobayashi, M. Okano, and K. Tomoda, "Optical Properties of Three-Dimensional Photonic Crystals Based on III–V Semiconductors at Infrared to Near-Infrared Wavelengths," *Appl. Phys. Lett.*, vol. 75, pp. 905–907, 1999.
72. S. Noda, N. Yamamoto, M. Imada, H. Kobayashi, and M. Okano, "Alignment and Stacking of Semiconductor Photonic Bandgaps by Wafer-Fusion," *J. Lightwave Technol.*, vol. 17, pp. 1948–1955, 1999.
73. M. Imada, S. Noda, A. Chutinan, T. Tokuda, M. Murata, and G. Sasaki, "Coherent Two-Dimensional Lasing Action in Surface-Emitting Laser with Triangular-Lattice Photonic Crystal Structure," *Appl. Phys. Lett.*, vol. 75, pp. 316–318, 1999.
74. S. Noda, K. Tomoda, N. Yamamoto, and A. Chutinan, "Full Three-Dimensional Photonic Bandgap Crystals at Near-Infrared Wavelengths," *Science*, vol. 289, pp. 604–606, 2000.
75. S. Noda, M. Imada, A. Chutinan, and N. Yamamoto, "III–V Based-Semiconductor Photonic Crystals," *Opt. Quantum Electron.*, vol. 34, pp. 723–736, 2002.
76. S. Noda, M. Imada, M. Okano, S. Ogawa, M. Mochizuki, and A. Chutinan, "Semiconductor Three-Dimensional and Two-Dimensional Photonic Crystals and Devices," *IEEE J. Quantum Electron.*, vol. 38, pp. 726–735, 2002.
77. S. Ogawa, M. Imada, and S. Noda, "Analysis of Thermal Stress in Wafer Bonding of Dissimilar Materials for the Introduction of an InP-Based Light Emitter into a GaAs-Based Three-Dimensional Photonic Crystal," *Appl. Phys. Lett.*, vol. 82, pp. 3406–3408, 2003.
78. S. Kawakami, "Fabrication of Submicrometre 3D Periodic Structures Composed of Si/SiO_2," *Electron. Lett.*, vol. 33, pp. 1260–1261, 1997.
79. S. Kawakami, T. Kawashima, and T. Sato, "Mechanism of Shape Formation of Three-Dimensional Periodic Nanostructures by Bias Sputtering," *Appl. Phys. Lett.*, vol. 74, pp. 463–465, 1999.

80. S. Kawakami, O. Hanaizumi, T. Sato, Y. Ohtera, T. Kawashima, N. Yasuda, Y. Takei, and K. Miura, "Fabrication of 3D Photonic Crystals by Autocloning and its Applications," *Electron. Commun. Jap. Part II-Electron.*, vol. 82, pp. 43–52, 1999.

81. S. Tazawa, S. Matsuo, and K. Saito, "A General Characterization and Simulation Method for Deposition and Etching Technology," *IEEE Trans. Semiconductor Manufac.*, vol. 5, pp. 27–33, 1992.

82. O. Hanaizumi, Y. Ohtera, T. Sato, and S. Kawakami, "Propagation of Light Beams Along Line Defects Formed in a- Si/SiO_2 Three-Dimensional Photonic Crystals: Fabrication and Observation," *Appl. Phys. Lett.*, vol. 74, pp. 777–779, 1999.

83. C. T. Chan, S. Datta, K. M. Ho, and C. M. Soukoulis, "A7 Structure—a Family of Photonic Crystals," *Phys. Rev. B*, vol. 50, pp. 1988–1991, 1994.

84. M. Notomi, T. Tamamura, T. Kawashima, and S. Kawakami, "Drilled Alternating-Layer Three-Dimensional Photonic Crystals Having a Full Photonic Band Gap," *Appl. Phys. Lett.*, vol. 77, pp. 4256–4258, 2000.

85. E. Kuramochi, M. Notomi, T. Tamamura, T. Kawashima, S. Kawakami, J. Takahashi, and C. Takahashi, "Drilled Alternating-Layer Structure for Three-Dimensional Photonic Crystals with a Full Band Gap," *J. Vac. Sci. Technol. B*, vol. 18, pp. 3510–3513, 2000.

86. E. Kuramochi, M. Notomi, T. Kawashima, J. Takahashi, C. Takahashi, T. Tamamura, and S. Kawakami, "A New Fabrication Technique for Photonic Crystals: Nanolithography Combined with Alternating-Layer Deposition," *Opt. Quantum Electron.*, vol. 34, pp. 53–61, 2002.

87. E. Kuramochi, M. Notomi, I. Yokohama, J. Takahashi, C. Takahashi, T. Kawashima, and S. Kawakami, "Transmission Characterization of Drilled Alternating-Layer Three-Dimensional Photonic Crystals," *J. Appl. Phys.*, vol. 93, pp. 8848–8851, 2003.

88. T. Sato, Y. Ohtera, N. Ishino, K. Miura, and S. Kawakami, "In-Plane Light Propagation in Ta_2O_5/SiO_2 Autocloned Photonic Crystals," *IEEE J. Quantum Electron.*, vol. 38, pp. 904–908, 2002.

89. M. Shirane, A. Gomyo, K. Miura, H. Yamada, and S. Kawakami, "Optical Directional Couplers Based on Autocloned Photonic Crystals," *Electron. Lett.*, vol. 39, pp. 53–54, 2003.

90. K. Miura and Y. Ohtera, "Reduction of Propagation and Bending Losses of Heterostructured Photonic Crystal Waveguides by Use of a High-Delta Structure," *Opt. Lett.*, vol. 28, pp. 734–736, 2003.

91. S. Kawakami, T. Sato, K. Miura, Y. Ohtera, T. Kawashima, and H. Ohkubo, "3-D Photonic-Crystal Hetero Structures: Fabrication and In-Line Resonator," *IEEE Photon. Technol. Lett.*, vol. 15, pp. 816–818, 2003.

92. Jm. Nieuwenh and H. B. Haanstra, "Microfractography of Thin Films," *Philips Tech. Rev.*, vol. 27, pp. 87–&, 1966.

93. A. G. Dirks and H. J. Leamy, "Columnar Microstructure in Vapor-Deposited Thin-Films," *Thin Solid Films*, vol. 47, pp. 219–233, 1977.

94. R. N. Tait, T. Smy, and M. J. Brett, "Modeling and Characterization of Columnar Growth in Evaporated-Films," *Thin Solid Films*, vol. 226, pp. 196–201, 1993.

95. L. Abelmann and C. Lodder, "Oblique Evaporation and Surface Diffusion," *Thin Solid Films*, vol. 305, pp. 1–21, 1997.

96. S. R. Kennedy, M. J. Brett, O. Toader, and S. John, "Fabrication of Tetragonal Square Spiral Photonic Crystals," *Nano Lett.*, vol. 2, pp. 59–62, 2002.

97. O. Toader and S. John, "Square Spiral Photonic Crystals: Robust Architecture for Microfabrication of Materials with Large Three-Dimensional Photonic Band Gaps," *Phys. Rev. E*, vol. 66, art. no. 016610, 2002.
98. M. Malac, R. F. Egerton, M. J. Brett, and B. Dick, "Fabrication of Submicrometer Regular Arrays of Pillars and Helices," *J. Vac. Sci. Technol. B*, vol. 17, pp. 2671–2674, 1999.
99. R. M. A. Azzam, "Chiral Thin Solid Films—Method of Deposition and Applications," *Appl. Phys. Lett.*, vol. 61, pp. 3118–3120, 1992.
100. K. Robbie, M. J. Brett, and A. Lakhtakia, "First Thin-Film Realization of a Helicoidal Bianisotropic Medium," *J. Vac. Sci. Technol. A*, vol. 13, pp. 2991–2993, 1995.
101. K. Robbie and M. J. Brett, "Sculptured Thin Films and Glancing Angle Deposition: Growth Mechanics and Applications," *J. Vac. Sci. Technol. A*, vol. 15, pp. 1460–1465, 1997.
102. K. Robbie, J. C. Sit, and M. J. Brett, "Advanced Techniques for Glancing Angle Deposition," *J. Vac. Sci. Technol. B*, vol. 16, pp. 1115–1122, 1998.
103. O. Toader and S. John, "Proposed Square Spiral Microfabrication Architecture for Large Three-Dimensional Photonic Band Gap Crystals," *Science*, vol. 292, pp. 1133–1135, 2001.
104. N. Tétreault, G. von Freymann, M. Deubel, M. Hermatschweiler, F. Pérez-Willard, S. John, M. Wegener, and G. A. Ozin, "New Route to Three-Dimensional Photonic Bandgap Materials: Silicon Double Inversion of Polymer Templates," *Adv. Mat.*, vol. 18, pp. 457, 2006.
105. K. D. Harris, J. C. Sit, and M. J. Brett, "Fabrication and Optical Characterization of Template-Constructed Thin Film with Chiral Nanostructures," *IEEE Trans. Nanotechnol.*, vol. 1, pp. 122, 2002.
106. A. Uhlir, "Electrolytic Shaping of Germanium and Silicon," *Bell Sys. Tech. J.*, vol. 35, pp. 333–347, 1956.
107. D. R. Turner, "Electropolishing Silicon in Hydrofluoric Acid Solutions," *J. Electrochem. Soc.*, vol. 105, pp. 402–408, 1958.
108. H. Foll, M. Christophersen, J. Carstensen, and G. Hasse, "Formation and Application of Porous Silicon," *Materials Sci. Eng. R-Reports*, vol. 39, pp. 93–141, 2002.
109. X. G. Zhang, "Morphology and Formation Mechanisms of Porous Silicon," *J. Electrochem. Soc.*, vol. 151, pp. C69–C80, 2004.
110. V. Lehmann and U. Gösele, "Porous Silicon Formation—a Quantum Wire Effect," *Appl. Phys. Lett.*, vol. 58, pp. 856–858, 1991.
111. V. Lehmann, "The Physics of Macropore Formation in Low Doped N-Type Silicon," *J. Electrochem. Soc.*, vol. 140, pp. 2836–2843, 1993.
112. V. Lehmann and H. Foll, "Formation Mechanism and Properties of Electrochemically Etched Trenches in N-Type Silicon," *J. Electrochem. Soc.*, vol. 137, pp. 653–659, 1990.
113. V. Lehmann and H. Foll, "Minority-Carrier Diffusion Length Mapping in Silicon-Wafers Using a Si-Electrolyte-Contact," *J. Electrochem. Soc.*, vol. 135, pp. 2831–2835, 1988.
114. V. Lehmann and U. Grüning, "The Limits of Macropore Array Fabrication," *Thin Solid Films*, vol. 297, pp. 13–17, 1997.
115. T. Geppert, J. Schilling, R. Wehrspohn, and U. Gösele, "Silicon-Based Photonic Crystals," *Silicon Photon.*, vol. 94, pp. 295–322, 2004.

116. M. Müller, R. Zentel, T. Maka, S. G. Romanov, and C. M. S. Torres, "Photonic Crystal Films with High Refractive Index Contrast," *Adv. Mat.*, vol. 12, pp. 1499–1503, 2000.

117. S. Matthias, F. Müller, and U. Gösele, "Controlled Nonuniformity in Macroporous Silicon Pore Growth," *Appl. Phys. Lett.*, vol. 87, pp. 224106, 2005.

118. J. Schilling, F. Müller, S. Matthias, R. B. Wehrspohn, U. Gösele, and K. Busch, "Three-Dimensional Photonic Crystals Based on Macroporous Silicon with Modulated Pore Diameter," *Appl. Phys. Lett.*, vol. 78, pp. 1180–1182, 2001.

119. S. Matthias, F. Müller, C. Jamois, R. B. Wehrspohn, and U. Gösele, "Large-Area Three-Dimensional Structuring by Electrochemical Etching and Lithography," *Adv. Mater.*, vol. 16, pp. 2166–2170, 2004.

120. S. Matthias, F. Müller, J. Schilling, and U. Gösele, "Pushing the Limits of Macroporous Silicon Etching," *Appl. Phys. A*, vol. 80, pp. 1391–1396, 2005.

121. S. Matthias, F. Müller, and U. Gösele, "Simple Cubic Three-Dimensional Photonic Crystals Based on Macroporous Silicon and Anisotropic Posttreatment," *J. Appl. Phys.*, vol. 98, pp. 023524, 2005.

122. R. L. Kubena, R. L. Seliger, and E. H. Stevens, "High-Resolution Sputtering Using a Focused Ion-Beam," *Thin Solid Films*, vol. 92, pp. 165–169, 1982.

123. H. Yamaguchi, A. Shimase, S. Haraichi, and T. Miyauchi, "Characteristics of Silicon Removal by Fine Focused Gallium Ion-Beam," *J. Vac. Sci. Technol. B*, vol. 3, pp. 71–74, 1985.

124. E. Yablonovitch, T. J. Gmitter, and K. M. Leung, "Photonic Band-Structure—the Face-Centered-Cubic Case Employing Nonspherical Atoms," *Phys. Rev. Lett.*, vol. 67, pp. 2295–2298, 1991.

125. A. Chelnokov, K. Wang, S. Rowson, P. Garoche, and J. M. Lourtioz, "Near-Infrared Yablonovite-Like Photonic Crystals by Focused-Ion-Beam Etching of Macroporous Silicon," *Appl. Phys. Lett.*, vol. 77, pp. 2943–2945, 2000.

126. K. Wang, A. Chelnokov, S. Rowson, P. Garoche, and J. M. Lourtioz, "Focused-Ion-Beam Etching in Macroporous Silicon to Realize Three-Dimensional Photonic Crystals," *J. Phys. D*, vol. 33, pp. L119–L123, 2000.

127. A. Chelnokov, S. David, K. Wang, F. Marty, and J. M. Lourtioz, "Fabrication of 2-D and 3-D Silicon Photonic Crystals by Deep Etching," *IEEE J. Sel. Topics Quantum Electron.*, vol. 8, pp. 919–927, 2002.

128. K. Wang, A. Chelnokov, S. Rowson, and J. M. Lourtioz, "Extremely High-Aspect-Ratio Patterns in Macroporous Substrate by Focused-Ion-Beam Etching: The Realization of Three-Dimensional Lattices," *Appl. Phys. A*, vol. 76, pp. 1013–1016, 2003.

129. R. Hillebrand, S. Senz, W. Hergert, and U. Gösele, "Macroporous-Silicon-Based Three-Dimensional Photonic Crystal with a Large Complete Band Gap," *J. Appl. Phys.*, vol. 94, pp. 2758–2760, 2003.

130. J. Schilling and A. Scherer, "3D Photonic Crystals Based on Macroporous Silicon: Towards a Large Complete Photonic Bandgap," *Photon. Nanostruct.-Fundamentals Appl.*, vol. 3, pp. 90–95, 2005.

131. J. Schilling, J. White, A. Scherer, G. Stupian, R. Hillebrand, and U. Gösele, "Three-Dimensional Macroporous Silicon Photonic Crystal with Large Photonic Band Gap," *Appl. Phys. Lett.*, vol. 86, pp. 011101, 2005.

132. E. Yablonovitch, T. J. Gmitter, and K. M. Leung, "Photonic Band-Structure—the Face-Centered-Cubic Case Employing Nonspherical Atoms," *Phys. Rev. Lett.*, vol. 67, pp. 2295–2298, 1991.

133. C. C. Cheng and A. Scherer, "Fabrication of Photonic Band-Gap Crystals," *J. Vac. Sci. Technol. B*, vol. 13, pp. 2696–2700, 1995.

134. C. C. Cheng, A. Scherer, R.-C. Tyan, Y. Fainman, G. Witzgall, and E. Yablonovitch, "New Fabrication Techniques for High Quality Photonic Crystals," *J. Vac. Sci. Technol. B*, vol. 15, pp. 2764–2767, 1997.

135. S. Venkataraman, "Fabrication of Two-Dimensional and Three-Dimensional Photonic Crystal Devices for Applications in Chip-Scale Optical Interconnects," PhD thesis, Department of Electrical and Computer Engineering, Newark, University of Delaware, 2005.

136. V. Berger, O. Gauthier-Lafaye, and E. Costard, "Photonic Band Gaps and Holography," *J. Appl. Phys.*, vol. 82, pp. 60–64, 1997.

137. V. Berger, O. Gauthier-Lafaye, and E. Costard, "Fabrication of a 2D Photonic Bandgap by a Holographic Method," *Electron. Lett.*, vol. 33, pp. 425–426, 1997.

138. Y. Ono and K. Ikemoto, "Fabrication of Arbitrary Three-Dimensional Photonic Crystals by Four Plane-Waves Interference," *Proc. SPIE*, vol. 4984, pp. 70–78, 2003.

139. L. Z. Cai, X. L. Yang, and Y. R. Wang, "All Fourteen Bravais Lattices can be Formed by Interference of Four Noncoplanar Beams," *Opt. Lett.*, vol. 27, pp. 900–902, 2002.

140. M. Campbell, D. N. Sharp, M. T. Harrison, R. G. Denning, and A. J. Turberfield, "Fabrication of Photonic Crystals for the Visible Spectrum by Holographic Lithography," *Nature*, vol. 404, pp. 53–56, 2000.

141. D. N. Sharp, M. Campbell, E. R. Dedman, M. T. Harrison, R. G. Denning, and A. J. Turberfield, "Photonic Crystals for the Visible Spectrum by Holographic Lithography," *Opt. Quantum Electron.*, vol. 34, pp. 3–12, 2002.

142. Y. Ono and T. Ochi, "All Fourteen Bravais Lattices Can Be Fabricated by Triple Exposure of Two-Beam Interference Fringes," *Proc. SPIE*, vol. 6327, pp. 632709, 2006.

143. G. J. Schneider, J. Murakowski, S. Venkataraman, and D. W. Prather, "Combination Lithography for Photonic-Crystal Circuits," *J. Vac. Sci. Technol. B*, vol. 22, pp. 146–151, 2004.

144. J. Murakowski, G. J. Schneider, and D. W. Prather, "Fabrication of 3D Photonic Crystals with Embedded Defects," *Proc. SPIE*, vol. 5347, pp. 181–198, 2003.

145. I. Divliansky, T. S. Mayer, K. S. Holliday, and V. H. Crespi, "Fabrication of Three-Dimensional Polymer Photonic Crystal Structures Using Single Diffraction Element Interference Lithography," *Appl. Phys. Lett.*, vol. 82, pp. 1667–1669, 2003.

146. T. Kondo, S. Matsuo, S. Juodkazis, and H. Misawa, "Femtosecond Laser Interference Technique with Diffractive Beam Splitter for Fabrication of Three-Dimensional Photonic Crystals," *Appl. Phys. Lett.*, vol. 79, pp. 725–727, 2001.

147. S. Shoji, H. B. Sun, and S. Kawata, "Photofabrication of Wood-Pile Three-Dimensional Photonic Crystals Using four-Beam Laser Interference," *Appl. Phys. Lett.*, vol. 83, pp. 608–610, 2003.

148. S. Shoji and S. Kawata, "Photofabrication of Three-Dimensional Photonic Crystals by Multibeam Laser Interference into a Photopolymerizable Resin," *Appl. Phys. Lett.*, vol. 76, pp. 2668–2670, 2000.

149. G. J. Schneider, E. D. Wetzel, J. A. Murakowski, and D. W. Prather, "Fabrication of Three-Dimensional Yablonovite Photonic Crystals by Multiple-Exposure UV Interference Lithography," *Proc. SPIE*, vol. 5720, pp. 9–17, 2005.

150. W. M. Lee, S. A. Pruzinsky, and P. V. Braun, "Multi-Photon Polymerization of Waveguide Structures Within Three-Dimensional Photonic Crystals," *Adv. Mater.*, vol. 14, pp. 271–274, 2002.

151. S. Maruo and S. Kawata, "Two-Photon-Absorbed Near-Infrared Photopolymerization for Three-Dimensional Microfabrication," *J. Microelectromechan. Syst.*, vol. 7, pp. 411–415, 1998.

152. B. H. Cumpston, S. P. Ananthavel, S. Barlow, D. L. Dyer, J. E. Ehrlich, L. L. Erskine, A. A. Heikal, S. M. Kuebler, I. Y. S. Lee, D. McCord-Maughon, J. Q. Qin, H. Rockel, M. Rumi, X. L. Wu, S. R. Marder, and J. W. Perry, "Two-Photon Polymerization Initiators for Three-Dimensional Optical Data Storage and Microfabrication," *Nature*, vol. 398, pp. 51–54, 1999.

153. H. B. Sun, S. Matsuo, and H. Misawa, "Three-Dimensional Photonic Crystal Structures Achieved with Two-Photon-Absorption Photopolymerization of Resin," *Appl. Phys. Lett.*, vol. 74, pp. 786–788, 1999.

154. H. B. Sun, V. Mizeikis, Y. Xu, S. Juodkazis, J. Y. Ye, S. Matsuo, and H. Misawa, "Microcavities in Polymeric Photonic Crystals," *Appl. Phys. Lett.*, vol. 79, pp. 1–3, 2001.

155. M. Deubel, G. Von Freymann, M. Wegener, S. Pereira, K. Busch, and C. M. Soukoulis, "Direct Laser Writing of Three-Dimensional Photonic-Crystal Templates for Telecommunications," *Nature Mater.*, vol. 3, pp. 444–447, 2004.

156. M. Deubel, M. Wegener, S. Linden, G. von Freymann, and S. John, "3D-2D-3D Photonic Crystal Heterostructures Fabricated by Direct Laser Writing," *Opt. Lett.*, vol. 31, pp. 805–807, 2006.

157. N. Tétreault, G. von Freymann, M. Deubel, M. Hermatschweiler, F. Perez-Willard, S. John, M. Wegener, and G. A. Ozin, "New Route to Three-Dimensional Photonic Bandgap Materials: Silicon Double Inversion of Polymer Templates," *Adv. Mater.*, vol. 18, pp. 457, 2006.

158. P. Pieranski, "Colloidal Crystals," *Contemp. Phys.*, vol. 24, pp. 25–73, 1983.

159. R. S. Crandall and R. Williams, "Gravitational Compression of Crystallized Suspensions of Polystyrene Spheres," *Science*, vol. 198, pp. 293–295, 1977.

160. M. K. Udo and M. F. Desouza, "The Ordering of Aqueous Suspensions of Polystyrene Spheres," *Solid State Commun.*, vol. 35, pp. 907–910, 1980.

161. R. Kesavamoorthy and A. K. Arora, "Gravitational Compression in Colloidal Suspension," *J. Phys. A*, vol. 18, pp. 3389–3398, 1985.

162. S. Melone and F. Rustichelli, "Dynamical Light-Diffraction by Colloidal Crystals," *Solid State Commun.*, vol. 55, pp. 97–100, 1985.

163. B. U. Felderhof and R. B. Jones, "Dynamics of Colloidal Crystals," *Zeitschrift Fur Physik B*, vol. 64, pp. 393–400, 1986.

164. A. K. Sood, "Light-Scattering from Colloids," *Hyperfine Interact.*, vol. 37, pp. 365–384, 1987.

165. R. O. Rosenberg and D. Thirumalai, "Order-Disorder Transition in Colloidal Suspensions," *Phys. Rev. A*, vol. 36, pp. 5690–5700, 1987.

166. W. Y. Shih, I. A. Aksay, and R. Kikuchi, "Phase-Diagrams of Charged Colloidal Particles," *J. Chem. Phys.*, vol. 86, pp. 5127–5132, 1987.

167. T. Yoshiyama and I. S. Sogami, “Kossel Images as Direct Manifestations of the Gap Structure of the Dispersion Surface for Colloidal Crystals,” *Phys. Rev. Lett.*, vol. 56, pp. 1609–1612, 1986.

168. T. Yoshiyama, “Ordering Process of Colloidal Crystal in Semidilute Aqueous Suspensions,” *Polymer*, vol. 27, pp. 827–833, 1986.

169. A. K. Sood, “Structural Ordering in Colloidal Suspensions,” in *Solid State Physics*, vol. 45, Eds. H. Ehrenreich and D. Turnbull. Academic Press, New York, pp. 1–73, 1991.

170. T. Okubo, “Centrifugal Compression of Crystal-Like Structures of Deionized Colloidal Spheres,” *J. Am. Chem. Soc.*, vol. 112, pp. 5420–5424, 1990.

171. W. G. Xue and G. S. Grest, “Shear-Induced Alignment of Colloidal Particles in the Presence of a Shear-Flow,” *Phys. Rev. Lett.*, vol. 64, pp. 419–422, 1990.

172. C. A. Murray, “Colloids—When Like Charges Attract,” *Nature*, vol. 385, pp. 203–204, 1997.

173. T. Okubo, “Giant Single-Crystals of Colloidal Spheres in Deionized and Diluted Suspension,” *Naturwissenschaften*, vol. 79, pp. 317–320, 1992.

174. T. Okubo, “Melting Temperature of Colloidal Crystals of Monodisperse Silica Spheres,” *J. Chem. Phys.*, vol. 96, pp. 2261–2268, 1992.

175. K. Hoshino and W. H. Young, “Structure of Polystyrene Colloidal Solutions,” *Molec. Phys.*, vol. 62, pp. 1119–1125, 1987.

176. T. Okubo, “The Viscosity of Colloidal Spheres in Deionized Suspensions,” *J. Chem. Phys.*, vol. 87, pp. 6733–6739, 1987.

177. W. G. Xue and G. S. Grest, “Brownian Dynamics Simulations for Interacting Colloids in the Presence of a Shear-Flow,” *Phys. Rev. A*, vol. 40, pp. 1709–1712, 1989.

178. B. J. Ackerson, “Shear Induced Order and Shear Processing of Model Hard-Sphere Suspensions,” *J. Rhe.*, vol. 34, pp. 553–590, 1990.

179. R. Kesavamoorthy, “Evolution of the Sedimentation Equilibrium in an Interacting Colloidal Suspension,” *J. Phys.: Condens. Matter*, vol. 2, pp. 4273–4280, 1990.

180. W. Stöber, A. Fink, and E. Bohn, “Controlled Growth of Monodisperse Silica Spheres in Micron Size Range,” *J. Colloid Interface Sci.*, vol. 26, pp. 62–69, 1968.

181. W. G. Hoover and F. H. Ree, “Melting Transition and Communal Entropy for Hard Spheres,” *J. Chem. Phys.*, vol. 49, pp. 3609, 1968.

182. M. Holgado, F. Garcia-Santamaria, A. Blanco, M. Ibisate, A. Cintas, H. Miguez, C. J. Serna, C. Molpeceres, J. Requena, A. Mifsud, F. Meseguer, and C. Lopez, “Electrophoretic Deposition to Control Artificial Opal Growth,” *Langmuir*, vol. 15, pp. 4701–4704, 1999.

183. A. F. Routh and W. B. Zimmerman, “Distribution of Particles During Solvent Evaporation from Films,” *Chem. Eng. Sci.*, vol. 59, pp. 2961–2968, 2004.

184. Y. A. Vlasov, X. Z. Bo, J. C. Sturm, and D. J. Norris, “On-Chip Natural Assembly of Silicon Photonic Bandgap Crystals,” *Nature*, vol. 414, pp. 289–293, 2001.

185. P. Jiang, J. F. Bertone, K. S. Hwang, and V. L. Colvin, “Single-Crystal Colloidal Multilayers of Controlled Thickness,” *Chem. Mater.*, vol. 11, pp. 2132–2140, 1999.

186. N. D. Denkov, O. D. Velev, P. A. Kralchevsky, I. B. Ivanov, H. Yoshimura, and K. Nagayama, “Mechanism of Formation of 2-Dimensional Crystals from Latex-Particles on Substrates,” *Langmuir*, vol. 8, pp. 3183–3190, 1992.

187. A. van Blaaderen, R. Ruel, and P. Wiltzius, "Template-Directed Colloidal Crystallization," *Nature*, vol. 385, pp. 321–324, 1997.

188. H. Koch, C. Radin, and L. Sadun, "Most Stable Structure for Hard Spheres," *Phys. Rev. E*, vol. 72, 2005.

189. A. D. Bruce, N. B. Wilding, and G. J. Ackland, "Free Energy of Crystalline Solids: A Lattice-Switch Monte Carlo Method," *Phys. Rev. Lett.*, vol. 79, pp. 3002–3005, 1997.

190. L. V. Woodcock, "Entropy Difference Between Crystal Phases—Reply," *Nature*, vol. 388, pp. 236–237, 1997.

191. L. V. Woodcock, "Entropy Difference Between the Face-Centred Cubic and Hexagonal Close-Packed Crystal Structures," *Nature*, vol. 385, pp. 141–143, 1997.

192. Y. N. Xia, B. Gates, and S. H. Park, "Fabrication of Three-Dimensional Photonic Crystals for Use in the Spectral Region from Ultraviolet to Near-Infrared," *J. Lightwave Technol.*, vol. 17, pp. 1956–1962, 1999.

193. Y. Xu, G. J. Schneider, E. D. Wetzel, and D. W. Prather, "Centrifugation and Spin-Coating Method for Fabrication of Three-Dimensional Opal and Inverse-Opal Structures as Photonic Crystal Devices," *J. Microlith. Microfabr. Microsys.*, vol. 3, pp. 168–173, 2004.

194. B. J. Ackerson and N. A. Clark, "Shear-Induced Partial Translational Ordering of a Colloidal Solid," *Phys. Rev. A*, vol. 30, pp. 906–918, 1984.

Index

Photonic Crystals: Theory, Applications, and Fabrication. By D. W. Prather, S. Shi, A. Sharkawy, J. Murakowski, and G. J. Schneider